Advanced Structured Materials

Volume 12

Series Editors

Andreas Öchsner
Lucas F. M. da Silva
Holm Altenbach

For further volumes:
http://www.springer.com/series/8611

P. M. Visakh · Sabu Thomas
Arup K. Chandra · Aji. P. Mathew
Editors

Advances in Elastomers II

Composites and Nanocomposites

Springer

Editors
P. M. Visakh
Centre for Nanoscience
and Nanotechnology
Mahatma Gandhi University
Kottayam, Kerala
India

Sabu Thomas
Kottayam, Kerala
India

Arup K. Chandra
Apollo Tyres Ltd
Gujarat
India

Aji. P. Mathew
Department of Applied Physics
and Mechanical Engineering
Lulea University of Technology
Luleå
Sweden

ISSN 1869-8433 ISSN 1869-8441 (electronic)
ISBN 978-3-642-20927-7 ISBN 978-3-642-20928-4 (eBook)
DOI 10.1007/978-3-642-20928-4
Springer Heidelberg New York Dordrecht London

Library of Congress Control Number: 2013934535

Printed on acid-free paper

Springer is part of Springer Science+Business Media (www.springer.com)

Preface

This book "Advances in Elastomers-II: Their Composites and Nanocomposites" summarizes many of the recent technical research accomplishments in the area of elastomer-based composites and nanocomposites. As the title indicates, the book emphasizes the various aspects of preparation, structure, processing, morphology, properties, and applications of elastomer-based composites and nanocomposites, in a systematic and comprehensive manner. Recent advances in the development and characterization of multicomponent polymer composites and nanocomposites based on elastomers are discussed in detail. It is important to mention that till date, there are not many books published on the recent advances in the synthesis, morphology, structure, properties, and applications of elastomer-based composites and nanocomposites. The book discussed in various chapters topics such as elastomer macrocomposites, elastomer-based nanocomposites, interphase modification and compatiblization of rubber-based nanocomposites, fully green elastomer nanocomposites, elastomeric micro and nanocomposites for tire applications: past, present, and future elastomer-based bionanocomposites, and physical phenomena related to free volumes in rubbers and blends.

It covers an up-to-date record on the major findings and observations in the field of elastomer-based composites and nanocomposites. The book is intended to serve as a "one stop" reference resource for important research accomplishments in this area. The various chapters in the book have been contributed by prominent researchers from the industry, academia, and government/private research laboratories across the globe. This book will be a valuable reference source for university and college faculties, professionals, post-doctoral research fellows, senior graduate students, polymer technologists, and researchers from R&D laboratories working in the area of elastomer-based composites and nanocomposites.

The first chapter on "Advances in Elastomers-II: Their Composites and Nanocomposites", gives an overview of the advances in composites and nanocomposites on state-of-art new challenges and opportunities. This chapter is essential for beginners in these fields as it provides a thorough basic understanding of the elastomer-based composites and nanocomposites. The following chapter on "Elastomer Macrocomposites" gives an introduction to elastomer macrocomposites, elastomer-based macrocomposites, recent studies on elastomer-based macrocomposites, different manufacturing methods of elastomer-based macrocomposites, and characterization methods of elastomer-based macrocomposites. Finally, applications of elastomer macrocomposites are discussed in detail. The third chapter analyzes "Rubber Nanocomposites"; the first part of this chapter gives an introduction and then discusses different types of nanocomposites such as elastomer-based thermoplastic nanocomposites and elastomer-based thermoset nanocomposites.

The fourth chapter, on "Reinforced Elastomers: Interphase Modification and Compatibilization in Rubber-Based Nanocomposites", is well explained. The chapter opens with a good introduction and later, the authors present several topics such as current trends in interphase modification and compatibilization, development of grafts/block copolymers, and crosslinking. Different compatibilizing agents, surface modifiers, plasticizer properties of interphase modified/compatibilized elastomeric composites, and future prospects are also discussed in detail.

A review of "Fully Green Elastomer Composites" is done in the fifth chapter. The authors deal with the recent development in the area of green elastomers. The chapter comprises an introduction, green elastomers, green elastomeric blends, green elastomer-based composites, processing of green elastomer-based composites and blends, characterization methods, properties and applications, and industrial products based on green elastomers. Finally, the current trends and the path forward is surveyed.

The next chapter is on "Nanocomposites for Tyre Applications": past, present, and future. The first section contains an introduction on tire manufacture and processing and recent studies of nanocomposites in the tire industry. The last section of this chapter discusses the different techniques used in the tire industry. The following chapter on "Elastomer-Based Bio-Nanocomposites" focuses on different studies on elastomer-based bionanocomposites. This chapter details various topics such as recent studies on cellulose, starch, and chitin/chitosan-based elastomers nanocomposites. Finally, the biomedical application of elastomer-based bionanocomposites is discussed. The chapter on "Bio-Medical Applications of Elastomeric Blends, Composites", discusses recent developments on elastomeric composite and blends for medical applications and an overview of elastomer-based products for biomedical application. Explanations are given with method and fabrication of elastomeric composite and blends for medical applications. Finally, the current trends and future prospects are discussed. The chapter "Other Applications: Engineering" discusses some of the applications of elastomer composites and nanocomposites, the author-coined title of the chapter, as well as other applications such as Engineering, foot ear, belting, sports, etc. In this chapter an introduction with some theoretical models of elastomeric materials is given. Finally, some industrial applications such as window seal, impact absorbing devices, silent block, hydraulic seals, elastic wheel of railway coach are discussed. The last chapter discusses the "Physical Phenomena Related to Free Volumes in Rubber and Blends".

Finally, the editors wish to express their sincere gratitude to all the contributors of this book, who provided excellent support for the successful completion of this venture. We are grateful to them for the commitment and sincerity they showed in contributing to this book. Without their enthusiasm and support the compilation of this book series could not have been possible. We thank all the reviewers who spent valuable time to make critical comments on each chapter. We also thank the publisher Springer for recognizing the demand for such a book, and for realizing the increasing importance of the area of Blends and Interpenetrating Network.

P. M. Visakh
Sabu Thomas
Arup K. Chandra
Aji. P. Mathew

Contents

Advances in Elastomers: Their Composites and Nanocomposites: State of Art, New Challenges and Opportunities

P. Deepalekshmi, P. M. Visakh, Aji. P. Mathew, Arup K. Chandra and Sabu Thomas

Abstract The field of elastomers, their composites and nanocomposites has gained a lot of interest in recent years. These composite materials have great significance both from the fundamental and application point of view. Since this field is growing at a faster rate, it is always necessary to address the structure, properties and applicability of such materials. The present chapter gives a brief account on various elastomer systems, their composites and nanocomposites. Various topics such as elastomer based macrocomposites, nanocomposites, interphase modification, compatiblisation of rubber based nanocomposites, fully green elastomer nanocomposites, elastomeric micro and nanocomposites for tyre applications, elastomer based bionanocomposites, bio-medical applications of elastomeric composites and nanocomposites have been very briefly discussed. Finally the applications, new challenges and opportunities of these composites and nanocomposites are also discussed.

P. Deepalekshmi · P. M. Visakh (✉) · S. Thomas
School of Chemical Sciences, Mahatma Gandhi University, Kottayam, Kerala 686560, India
e-mail: visagam143@gmail.com

P. Deepalekshmi · P. M. Visakh · S. Thomas
Centre for Nanoscience and Nanotechnology, Mahatma Gandhi University, Kottayam, Kerala 686560, India

P. M. Visakh · A. P. Mathew
Department of Applied Physics and Mechanical Engineering Division of Wood and Bionanocomposites Luleå, University of Technology, 97187 Luleå, Sweden

A. K. Chandra
R&D Centre, Apollo Tyres Ltd., Limda, Waghodia, Vadodara, Gujarat 391760, India

P. M. Visakh et al. (eds.), *Advances in Elastomers II*,
Advanced Structured Materials 12, DOI: 10.1007/978-3-642-20928-4_1,

1 Introduction

Elastomers are one of the most applicable materials in science and technology. The high level of flexibility, mechanical strength, flexural properties, curing behavior etc. makes this class of polymers quite significant. In order to enhance the existing properties as well as to achieve new qualities such as conductivity, impermeability etc. filler particles are incorporated in elastomer matrices. In the past, the most interested property for the fabricated composites was mechanical strength. Improved mechanical properties and strength of composites are not a new idea, since there are quite a lot of naturally occurring composite materials. For example, tissues in the body, wood, bamboo, muscles, etc. exhibit high strength in addition to flexibility. In synthetic elastomer composites, the main aim is to achieve optimum mechanical properties required for specific applications. Depending on the dimensions of the filler particles, the newly obtained composites vary in its name and properties. The macro sized and micro sized fillers are added to rubbers to respectively fabricate the macro composites and micro composites. Such composites are used in technology from the time immemorial. Recently the invention of nanomaterials has opened a new area in composite technology, since the nanoparticles are required in small quantities only. These main three categories of composites—Macrocomposites, microcomposites and nanocomposites—are discussed in detail in this chapter. Effort has been made to emphasize on the filler properties and the applicability of the composite materials.

1.1 Macrocomposites

The nature and dimension of filler materials have a great role in regulating the matrix properties. In conventional macrocomposites the dispersed phase has dimensions in the order of tens of microns to tens of millimeters. The concept of macrocomposite has a long history back to the ancient age. The combination of two different materials—one dispersed in the other in macro scale can be called as macro composites. In ancient times the swords were made by combining tough, ductile steel with hard, brittle steel. The result was a combination of stiffness and toughness far superior than that could be obtained from any material available at that time. It is this ability to produce combinations of stiffness and toughness not obtainable from a single material gives great potential to the concept of macrocomposite as a means of using engineering plastics to replace steel and other metals. To make such a macrocomposite, one resin would be molded into which the second molded resin is inserted through a secondary molding operation. The two different resins would be chosen depending on the properties required and also the compatibility between the two [1, 2]. The macro composites and blends have similar meaning but the former varies in the fact that the dispersed phase can be not only a polymer but other inorganic filler materials in the macro dimension.

Whereas in the case of polymer blends, the two components are polymers. There are two types of macrocomposites: Large particle composites and fiber-reinforced composites [3–5]. In large particle macrocomposites big sized particles of one type of material is uniformly dispersed in a matrix material. The matrix may be metal, polymer or ceramic. The elastomeric matrices can be of natural or synthetic rubber with a variety of ingredients to confer the required properties.

1.2 Microcomposites

As the name indicates, micro sized particles are reinforced in elastomer matrices to attain superior properties, which the macro composites failed to achieve. Fillers such as carbon black are very good reinforcements for elastomers like natural rubber especially in tyre industry. Fibre reinforced composites are known for their enhanced mechanical strength and stiffness for long time [6, 7]. Fibres obtained from natural resources like coconut, rice husk, pineapple, banana etc. are well known fillers in rubber industry. In the composite the micro fibres carry the load whereas the matrix keeps them in good orientation and act as the load transfer medium. Such composites show better dimensional stability, high internal damping and better vibration energy absorption [8–10]. In addition to the natural fibres, glass fibre and carbon fibers are extensively used in high performance applications. Even though natural fiber filled composites have many advantages like biodegradability, combustibility, light weight, non toxicity, decreased environmental pollution, low cost, ease of recyclability etc. moisture absorption, less thermal stability, biodegradation, photodegradation etc. limits their usage in engineering applications [11].

1.3 Nanocomposites

In the present scenario, nanotechnology holds considerable promise in different technological areas. Many nanotechnological applications are based on novel as well as conventional materials deliberately engineered to nano structure. Nanofillers have superior qualities such as enhanced surface area and high aspect ratio. These make them highly efficient structural reinforcement in elastomers. Due to the very large surface area, the required quantity of such materials is quiet small. Carbon based nanofillers like carbon nanotubes, graphenes and graphitic derivatives, inorganic fillers like nanosilica, nanometal particles and nanoclays, bio nanofillers such as cellulose, starch and chitin etc. are a few of the fillers having nanodimension imparting huge reinforcement for the elastomer based systems. Rubber-clay nanocomposites exhibit outstanding properties, at low loading levels of clay, as compared with unfilled rubber compounds or conventional filled composites [12]. In general, the preparation methods for rubber-clay nanocomposite can be divided into four major groups, according to the processing technique which are in-situ polymerization, intercalation of rubber via solution blending, direct melt intercalation method, intercalation of rubber via latex compounding. Montmorillonite is the most commonly used layered silicate for preparation of nanocomposites because of its high aspect ratio, large surface area, and surface reactivity. Regarding the type of nanofiller, Sadhu and Bhowmick [12] investigated the effect of various nanoclays such as sodium montmorillonite, bentonite and potassium montmorillonite, and organic amines of varying lengths on the behavior of styrene butadiene rubber (SBR) and found enhanced properties upon the filler addition. Magaraphan et al. [13] reported that the montmorrilonite (MMT) clays treated with long primary amines led to much more improved mechanical properties when incorporated in NR matrices, than those treated with quaternary amines having same number of carbon atoms. They also observed that the cure time is independent of the length of hydrocarbon chain in the alkylamines, whereas in the case of clay nanocomposites prepared with long quaternary amine, the cure time enhanced sharply.

Carbon nanotubes and graphene nanoparticles are promising materials for rubber reinforcement as it can make rubber conducting. In the case of multi walled carbon nanotubes (MWCNTs), the addition of 1 phr leads to 45 % increase of modulus and 70 % increase of tensile strength for SBR based nanocomposites [14]. The addition of 0.44 vol% of graphene oxide nanosheets enhanced the tensile strength and modulus at 200 % elongation of carboxylated NBR by more than 50 and 100 % [15]. The main reason for reinforcing rubbers with nanofillers is to improve the mechanical and thermal properties as well as reducing cost, and sometimes weight of the obtained nanocomposites. The electrical properties of CNT's are considered to be very important in the tyre industry as a source of dissipating static electricity [16]. In addition to the reinforcement, nanoparticles also play the role of compatibilizer for mixtures of immiscible polymers.

Carbon black and silica are the most popular fillers in rubber industry. The selection of such fillers in tyre industry depends on the performance requirements, as they are different for the tread, sidewall and apex. These fillers improve the mechanical properties, such as tensile strength, modulus, tear resistance and abrasion resistance of rubbers [17]. Ingredients also come into play to aid the processing of the tyre and/or other rubber applications. Elastomers are relatively expensive materials, and it became common practice to add cheap mineral powders, e.g. silica or calcium carbonate to reduce the cost, the effect being to increase the elastic modulus of the cured compound with little effect on the strength or other properties. However, it was soon found that very fine powders, particularly carbon black, manufactured by burning oil or natural gas in controlled conditions, could result in a large increase in both strength and modulus, though with an increase in hysteresis [18, 19]. The effectiveness of fillers in composites and nanocomposites depends on its characteristics, such as particle size, shape, size of the agglomerates; concentration, and more significantly, on surface activity i.e. on the strength of polymer-filler interactions. To achieve maximum improvement, strong interaction with the polymeric matrix will be needed because the interface/interphase between filler and matrix has a large influence on the mechanical properties of the composite. These interactions may also increase the effective degree of crosslinking and such an effect is particularly strong if the particles have some reactive surface groups [20].

Increasing the area of contact between rubber matrix and filler particles seems to be the most important factor for ensuring a strong reinforcement effect and the surface area of the interface is controlled by the size of filler particles and their volume fraction. Furthermore, the magnitude of adhesive bonding between rubber matrix and filler is a key factor in determining the degree of reinforcement. The dispersion and distribution of carbon black in a rubber matrix are important factors to achieve optimum physical properties. Moreover, their structures, wettability by the rubber phase and the related interactions have all major influence on the properties of the rubber compound. Certain compatibilizers reduce interfacial tension and, therefore, are capable of promoting adhesion leading to a finely dispersed morphology that shows stability against gross segregation. For instance polyolefins functionalized with acrylic acid [21], maleic anhydride [22] or other thermoplastics with polar groups such as ionomers [23] act as compatibilizers. Silane coupling agents are widely used in the tyre industry to increase the dispersibility of silica into elastomeric matrices, thus promoting the formation of covalent bonds between the phases. Calcium carbonate ($CaCO_3$) is one of the most inexpensive and recyclable natural resources, with the highest reserves on earth and is widely used in polymer nanocomposites. $CaCO_3$ nanoparticles were used for reinforcement of PP/SEBS and PP/carboxylated SEBS (C-SEBS) blends [24].

There are many advantages of using fillers from plant sources as reinforcement in polymer matrices, e.g. low density, low cost and low energy consumption, high specific strength, renewability and biodegradability, abundance in a variety of forms throughout the world, flexibility, non-abrasive nature to processing equipment, non-toxicity, ease of handling, reactive surface, organic nature and the economic development opportunity for non-food farm products in rural areas [25]. Bio fillers such as cellulose, hemicelluloses and lignin are the major constituents of plant body. These molecules have been extensively studied in many applications, using them in natural form or as derivatives [26, 27]. In nanoscale these materials are excellent fillers to polymers particularly polyurethanes and rubbers leading to the production of blends, composites and nanocomposites. Another material of interest is chitin and chitosan, which are of animal origin [28–32]. Other molecules such as proteins, vegetable oils, polyhydroxyalkanoate, terpenes, tannins, starch and polylactic acid from renewable sources having the potential to produce new polymers, are also extensively studied [33].

For the production of fully green elastomer composites and nanocomposites the biggest challenge is the production of polymers from renewable sources, which can be used as a matrix in these composites. This is an area in which there are still many possibilities to explore. The combination of elastomeric polymers and filler from renewable resources to produce a fully green elastomer composite has seldom been explored when the elastomeric polymer matrix is a material other than natural rubber [34, 35]. Such bionanocomposites system, which is a sub class of bio-nanohybrid materials, has several advantages of long range of elasticity and viscoelasticity in addition to their inherent eco-friendly, biodegradable and renewable characteristics [35]. Wu et al. [36] have synthesized novel high strength elastomeric bio-nanocomposites using microcrystalline cellulose and PU. Significant improvement in tensile strength, stiffness and strain to failure properties has been observed. Due to extensive hydrogen bonding as well as covalent bonding between PU and cellulose network such improvement in properties has been observed.

Hybrid effect of different fillers is another topic of interest as it can impart superior properties of whole counterparts. Wang et al. [37] have embedded starch nano-crystals and cellulose whiskers in waterborne polyurethanes. It is evident that the Young's modulus and tensile strength of such elastomeric bio-nanocomposites along with thermal properties have been significantly improved as compared to pristine PU. Actually, polysaccharide molecules and whiskers forms a hydrogen bonded network, giving synergistic effect. They belong to the class of eco-friendly elastomeric bio-nanocomposites. Zuber et al. [38] have synthesized chitin based polyurethane/clay nanocomposites. It has been observed that interaction among the clay and the polymer chain improves their dispersion in the PU matrix.

1.4 Applications

Tyre manufacturing is the most significant application of rubber materials. The components of tyres include fiber, textile, and steel cord and so the manufacture of this highly engineered product is very complex. Rubber itself and the filler materials are the two major ingredients in a rubber compound and their combination depends on the nature of end products. In tyre industry, the main requirements to achieve are optimization of performance, maximizing the traction in both wet and dry conditions and superior rolling resistance. All desired objectives can be achieved through the careful selection of one or more types of rubber, along with the type and amount of filler to blend with the rubber [39]. It is found that 6 phr nano clay has exhibited equivalent reinforcement to that of 40 phr carbon loaded composite with lower heat generation and better dynamic properties [40]. Other nano—fillers such as nano silica, nano carbon black; carbon nano fiber; MWCNT'S, clay and POSS [41] are also applied in tyre industry.

Applications of elastomer composites include automotive and railway industries and numerous mechanical, civil, electronics and electrical engineering sectors. These unique materials act as couplings between stiff structures for avoiding or at least reducing transmission of vibrations. Examples of these components are: pipes, top mounts, bushings and hydro bushings for suspensions and shock absorbers, torsion axes, supports for stabilizing bars, compression blocks, seals and membranes. Blending of starch with polymers replaces other plastics for packaging applications [42–44]. The high interfacial surface area of nanofillers cause reduction of filler quantity and this reduces the specific gravity and heat generation, leading to the high performance of tyre and in better fuel efficiency.

Elastomeric bio-nanocomposites have emerged as unique advanced material with potential biological applications. Bio-elastomers with well optimized properties have several advantages including high purity, good processibility, and optimized physical, chemical and mechanical properties. Such composites are useful in tissue engineering scaffolds, which are capable of degrading chemically following hydrolytic or enzymatic degradation mechanism [45]. Other applications include wound dressing, drug delivery, packaging etc. as well as in medical devices like cardiac assist pumps and blood bags, to chronic implants such as heart valves and vascular grafts [46, 47] hip-joint and knee replacements where elastomeric blends are used to compliment traditional materials such as metals and ceramics [48, 49], cosmetic surgery [50, 51], dental implants, skin care and in other medical devices such as contact lenses, intraocular lenses and renal dialyzers.

Polyurethane and silicone elastomers are known to offer good biocompatibility and biostability [52]. In biomedical applications, silicone elastomers are commonly used as catheters, drains and shunts. These include medical devices fabricated with silicone extrusions and those with nonsilicone substrates that are silicone-coated to minimise host reaction [53]. In short, the bio composite materials exhibit combined biocompatibility, biostability and mechanical properties which eliminate the possibility of (i) severe allergic reactions in patients, (ii) device rejection by the host environment, and (iii) premature device failure.

2 Conclusion

This chapter summarizes the importance of elastomer composites, particularly nanocomposites by giving special emphasis to its applications. The composite properties vary depending on the nature, size and dimension of filler reinforcement. Some of the current scenario in developing the composite materials is described in detail. A clear understanding of the relationship between processing, morphology, structure and properties of various kinds of nanocomposites has not been made yet. We have to go a long way to explore this field by making use of sophisticated techniques.

References

1. Hertz Jr, D.L.: Handbook of elastomers. In: Bhowmick, A.K., Stephens, H.L. (eds.) Marcel Dekker Inc., New York (1988)
2. Bryant, C.L.: Acrylonitrile-Butadiene (Nitrile) rubbers. In: Blow, C.M. (ed.) Rubber Technology and Manufacture. Newnes-Butterworths, London (1977)

3. Hofmann, W.: Rubber Technology Handbook, Hanser Publishers, Munich (1989). Rubber Plast. News **14**, 2, 21 (1984)
4. Blow, C.M. (ed.): Rubber Technology and Manufacture. Newnes-Butterworths, London (1977)
5. Rigbi, Z.: Reinforcement of rubber by carbon black. Adv. Polym. Sci. **36**, 21–68 (1980)
6. Geethamma, V.G., Joseph, R., Thomas, S.: Short coir fiber-reinforced natural rubber composites: effects of fiber length, orientation, and alkali treatment. J. Appl. Polym. Sci. **55**, 583–594 (1995)
7. Geethamma, V.G., Pothen, L.A., Rao, B., Neelakantan, N.R., Thomas, S.: Tensile stress relaxation of short-coir-fiber-reinforced natural rubber composites. J. Appl. Polym. Sci. **94**, 96–104 (2004)
8. Geethamma, V.G., Kalaprasad, G., Groeninckx, G., Thomas, S.: Dynamic mechanical behavior of short coir fiber reinforced natural rubber composites. Compos. A **36**(11), 1499–1506 (2005)
9. Geethamma, V.G., Thomas, S.: Diffusion of water and artificial seawater through coir fiber reinforced natural rubber composites. Polym. Compos. **26**(2), 136–143 (2005)
10. Das, D., Datta, M., Chavan, R.B., Datta, S.K.: Coating of jute with natural rubber. J. Appl. Polym. Sci. **98**(5), 484–489 (2005)
11. Nashar, D.E.El.,Abd-El-Messieh, S.L.,Basta, A.H.: Newsprint paper waste as a fiber reinforcement in rubber composites. J. Appl. Polym. Sci. **91**, 3410–3420 (2004)
12. Sadhu, S., Bhowmick, A.K.: Rubber Chem. Technol. **76**, 860–875 (2003)
13. Magaraphan, R., Thaijaroen, W., Lim-Ochakun, R.: Rubber Chem. Technol. **76**, 406–418 (2003)
14. Bokobza, L.: Polymer **48**, 4907–4920 (2007)
15. Bai, X., Wan, C., Zhang, Y., Zhai, Y.: Carbon **49**, 1608–1613 (2011)
16. Boonstra, B.B.: Role of particulate fillers in elastomer reinforcement: a review. Polymer **20**(6), 691–704 (1979)
17. Pramanik, M., Srivastava, S.K., Samantaray, B.K., Bhowmick, A.K.: J. Polym. Sci. B **40**, 2065–2072 (2002)
18. Fröhlich, J., Niedermeier, W., Luginsland, H.D.: The effect of filler–filler and filler–elastomer interaction on rubber reinforcement. Compos. A Appl. Sci. Manuf. **36**(4), 449–460 (2005)
19. Park, Soo-Jin, Cho, Ki-Sook: Filler–elastomer interactions: influence of silane coupling agent on crosslink density and thermal stability of silica/rubber composites. J. Colloid Interface Sci. **267**(1), 86–91 (2003)
20. Bokobza, L.: J. Appl. Polym. Sci. **93**, 2095–2104 (2004)
21. Faisant, J.B., Ait-Kadi, A., Bousmina, M., Deschênes, L.: Polymer **39**, 533–545 (1998)
22. Willis, J.M., Caldas, V., Favis, B.D.: J. Mater. Sci. **26**, 4742–4750 (1991)
23. Dagli, S.S., Xanthos, M., Biesenberger, J.A.: Emerging technologies in plastics recycling. In: Andrews, G.D., Subramanian, P.M. (eds.) ACS Symposium Series, 513 Chapter 19 (1992)
24. Hikasa, S., Nagata, K., Miyahara, K., Izumi, T., Suda, T., Toyohara, A., Kato, A., Nakamura, Y.: J. Appl. Polym. Sci. **114**, 919–927 (2009)
25. Dufresne, A.: Cellulose-based composites and nanocomposites. In: Belgacem, M.N., Gandini, A. (eds.) Monomers, Polymers and Composites from Renewable Resources, pp. 401–418. Elsevier, Amsterdam (2008)
26. Zhou, Q., Zhang, L., Zhang, M., Wang, B., Wang, S.: Miscibility, free volume behavior and properties of blends from cellulose acetate and castor oil-based polyurethane. Polymer **44**, 1733–1739 (2003)
27. Yoshioka, M., Hagiwara, N., Shiraishi, N.: Thermoplasticization of cellulose acetates by grafting of cyclic esters. Cellulose **6**, 193–212 (1999)
28. Zia, K.M., Barikani, M., Zuber, M., Bhatti, I.A., Sheikh, M.A.: Molecular engineering of chitin based polyurethane elastomers. Carbohydr. Polym. **74**, 149–158 (2008)
29. Barikani, M., Honarkar, H., Barikani, M.: Synthesis and characterization of chitosan-based polyurethane elastomer dispersions. Monatshefte für Chemie/Chemical Monthly **141**, 653–659 (2010)

30. Rao, V., Johns, J.: Thermal behavior of chitosan/natural rubber latex blends: TG and DSC analysis. J. Therm. Anal. Calorim. **92**, 801–806 (2008)
31. Barikani, M., Honarkar, H., Barikani, M.: Synthesis and characterization of polyurethane elastomers based on chitosan and poly(ε-caprolactone). J. Appl. Polym. Sci. **112**, 3157–3165 (2009)
32. Ciobanu, C., Ungureanu, M., Ignat, L., Ungureanu, D., Popa, V.I.: Properties of lignin–polyurethane films prepared by casting method. Ind. Crops Prod. **20**, 231–241 (2004)
33. Belgacem, M.N., Gandini, A.: Monomers, Polymers and Composites from Renewable Resources. Elsevier, Amsterdam, p. 553 (2008)
34. Darder, M., Aranda, P., Hitzky, E.R.: Adv. Mater. **19**(10), 1309 (2007)
35. Darder, M., Colilla, M., Hitzky, E.R.: Chem. Mater. **15**, 3774 (2003)
36. Wu, W., Henrikson, M., Liu, X., Berglund, L.A.: Biomacromolecules **8**(12), 3687 (2007)
37. Wang, Y., Tian, H., Zhang, L.: Carbohydr. Polym. **80**, 665 (2010)
38. Zuber, M., Zia, K.M., Mahboob, S., Hassain, M.: Int. J. Biol. Macromol. **47**, 196 (2010)
39. Leblanc, J.L.: Rubber–filler interactions and rheological properties in filled compounds. Prog. Polym. Sci. **27**(4), 627–687 (2002)
40. Gopi, J.A., Patel, S.K., Chandra, A K., Tripathi, D.K.: J. polym. Res. **18**, 1625–1634 (2011)
41. Chandra, A K.: Nano materials for rubber/tyre application paper presented at the International conference on innovation in technologies for processing of rubber & elastomer, Ramada Palm Grove, Juhu, 26–27 Oct 2012
42. Avérous, L.: Biodegradable multiphase systems based on plasticized starch: a review. J. Macromol. Sci. C **C44**, 231–274 (2004)
43. Wang, X.L., Yang, K.K., Wang, Y.Z.: Properties of starch blends with biodegradable polymers. J. Macromol. Sci. C **C43**, 385–409 (2003)
44. Yu, L., Dean, K., Li, L.: Polymer blends and composites from renewable resources. Prog. Polym. Sci. **31**, 502–576 (2006)
45. Shi, R., Chen, D., Li, Q., Wu, Y., Xu, X., Zhang, L., Tian, W.: Int. J. Mol. Sci. **10**, 4223 (2009)
46. Christenson, E.M., Wiggins, M.J., Anderson, J.M., Hiltner, A.: Surface modification of poly(ether urethane urea with modified dehydroepiandrosterone for improved in vivo biostability). J. Biomed. Mater. Res. **73A**, 108–115 (2005)
47. Gunatillake, P.A., Martin, D.J., Meijs, G.F.: Designing biostable polyurethane elastomers for biomedical implants. Aust. J. Chem. **56**, 545–557 (2003)
48. Khorasani, M.T., Zaghiyan, M., Mirzadeh, H.: Ultra high molecular weight polyethylene and polydimethylsiloxane blend as acetabular cup material. Colloids Surf B **41**, 169–174 (2005)
49. Onatea, J.I., Comin, M., Bracerasa, I.: Wear reduction effect on ultra-high-molecular-weight polyethylene by application of hard coatings and ion implantation on cobalt chromium alloy, as measured in a knee wear simulation machine. Surf. Coat. Technol. **142–144**, 1056–1062 (2001)
50. Barr, S., Bayat, A.: Current implant surface technology: an examination of their nanostructure and their influence on fibroblast alignment and biocompatibility. ePlasty **9**, 22 (2009)
51. Barr, S., Hill, E., Bayat, A.: Patterning of novel breast implant surfaces by enhancing silicone biocompatibility, using biomimetic topographies. Eplasty **10**, 246–268 (2010)
52. Kanyanta, V., Ivankovic, A.: Mechanical characterisation of polyurethane elastomer for biomedical applications. J. Mech. Behav. Biomater. **3**, 51–62 (2010)
53. Colas, A., Curtis, J.: Biomaterials Science High Molecular Weight Polyethylene in Total Joint Replacement and Medical Devices. Academic Press, Elsevier (2009)

Elastomer Macrocomposites

N. K. Anifantis, S. K. Georgantzinos, G. I. Giannopoulos and P. A. Kakavas

Abstract This chapter summarizes many of the recent technical research accomplishments in the area of elastomer macrocomposites. Firstly, it explains the compounds that exist in elastomeric matrices as well as the types of filler used in order to reinforce them. Then, the various recent attempts reported on advances of elastomer based macrocomposites are discussed. In addition, an analytical description in their manufacturing techniques and processes is comprehensively reported. Moreover, the techniques used to structurally and mechanically characterize the elastomeric macrocomposites are covered. Their usage in commercial applications is described as a final point.

1 Introduction

Materials with synergistic properties are chosen to create composites with tailored properties; for example, brittle carbon fibers of high-modulus, are added to low-modulus polymers, to create stiff, lightweight composites with a reasonable degree of toughness. The properties achieved by conventional macrocomposites, however, usually involve compromises [1–10]. In conventional macrocomposites the dispersed phase has dimensions in the order of tens of microns to tens of

N. K. Anifantis (✉) · S. K. Georgantzinos · G. I. Giannopoulos · P. A. Kakavas
Machine Design Laboratory, Mechanical and Aeronautics Engineering Department, University of Patras, GR 26500 Patras, Greece
e-mail: nanif@mech.upatras.gr

S. K. Georgantzinos
e-mail: sgeor@mech.upatras.gr

G. I. Giannopoulos
e-mail: g_giann@mech.upatras.gr

P. A. Kakavas
e-mail: kakavas@upatras.gr

P. M. Visakh et al. (eds.), *Advances in Elastomers II*,
Advanced Structured Materials 12, DOI: 10.1007/978-3-642-20928-4_2,

millimeters. The mechanism of mechanical property reinforcement in these materials is well understood and the properties can be predicted using sum of properties of each component weighted by the volume fraction of that component. Because the size of individual component phases, in macrocomposite elastomers, cannot be reduced beyond a limit, the composite properties are linearly dependent on the component properties; non-linear or exponential improvement in composite properties therefore requires looking beyond conventional macrocomposites [11–14]. Macrocomposite is making a part from two or more different materials. In ancient times *swords* were made by combining tough, ductile steel with hard, brittle steel. The result was a combination of stiffness and toughness far superior than could be obtained from any material available at that time. It is this ability to produce combinations of stiffness and toughness not obtainable in a single material which gives the macrocomposite concept great potential as a means of using engineering plastics to replace steel and other metals. To make such a macrocomposite, one resin would be molded into a part which would then be used as an insert when a second resin was molded in a second molding operation. The two different resins would be chosen so that one would supply toughness where toughness was required, and the other would supply stiffness where stiffness was needed [15, 16].

Mostly the properties of interest in composites are mechanical properties. There are many naturally occurring composite materials. For example, tissues in the body, wood, bamboo, muscles, etc. They exhibit high strength in addition to flexibility. In synthetic composites, the aim is to achieve optimum mechanical properties required for specific applications. Three types of composites may be distinguished: Macrocomposites, Microcomposites and Nanocomposites.

There are two types of macrocomposites: Large particle composites and fiber-reinforced composites [17–20]. In large particle macrocomposites big sized particles of one type of material is uniformly dispersed in a matrix material. The matrix may be metal, polymer or ceramic. Concrete is a typical example of large particle composite in which the cement matrix is mixed with particulate of sand and gravel. Cermets are large particle composites in which ceramic particles are embedded in a metal matrix. Other examples of large particle composites are carbon particles added to vulcanized rubber matrix. This enhances the tensile strength and toughness of the rubber material. These materials are used for auto motive tires. For effective reinforcement, the particles should be small and must be uniformly distributed throughout the matrix. The particle should form strong adhesive bond with the matrix. Fiber reinforced composites are designed to increase the strength and elastic modulus of the material without increasing its specific gravity. High strength and low weight is the main characteristic aimed at in the design of fiber reinforced composites [21–23].

Carbon black has been used for the reinforcement of rubber compounds since the beginning of the twentieth century, and has a global production rate of approximately 10 million tonnes/year. Carbon black is produced by incomplete combustion of oil under specific conditions, with a long history of safe production and use. Several parameters are controlled in the process in order to achieve the specific characteristics of the finished carbon black products. Specifically, the oil is injected at high speed, approximately 900 m/s, into a reactor where it is pyrolysed at high temperature (approx. 1800 °C). The combustion reaction is controlled so that the oxidation remains incomplete and therefore carbon black is formed. The reaction is intentionally stopped by spraying water (quenching). The carbon black obtained is filtered and then compressed into pellets of millimetre dimension. During the first split second of the above combustion reaction carbon nodules are formed with dimensions from 20 to 100 nm, depending on the grade of carbon black. On the basis of the proposed definitions these nodules may be considered nano-objects. However the lifespan of these nodules is very short as they immediately and irreversibly cluster together to form aggregates of sizes between 100 to 500 nm. These aggregates then subsequently combine together to form very solid entities: agglomerates (Fig. 1).

These agglomerates measure between 1 and 50 microns. The aggregates and agglomerates however, may meet the proposed definition of nanostructured materials. In aggregates, covalent bonds exist between particles; in agglomerates the bonds between aggregates are electrostatic. Aggregates cannot deaggregate at all. Agglomerates cannot de-agglomerate during standard handling conditions. After the agglomeration reaction, the agglomerates pass through the pelletiser to compact the carbon black into a pellet form of millimetre dimension [24].

Amorphous precipitated silica apart from being used to produce rubber articles is used in many other applications, e.g. cosmetics, paper and many other applications related to nutrition and health. The world production of amorphous precipitated silica is 1.4 million tons of which 1/3 is used in tire production. Silica has been used in the treads of tires for more than twenty years in order to reduce the fuel consumption of vehicles, thus contributing to a reduction in vehicle emissions of greenhouse gases. Amorphous precipitated silica is produced by a two step process: obtaining silicate, and then the production of precipitated silica. To make the silicate, very pure sand is mixed thoroughly with sodium carbonate. The homogeneous mixture is transferred to a glass oven, heated above 1000 °C, so that

Fig. 1 Carbon black pellets [24]

glass, vitreous silicate, is formed. Then rapid cooling causes solidification and fractionation of the silicate into small granules, a few cubic centimeters in volume. The vitreous silicate is dissolved in water and transferred to a reactor in which, through acidification and under agitation, amorphous silica is precipitated out. During this precipitation, as is the case with carbon black, there is an instantaneous initial formation of elementary particles, from 5 to 40 nm, of a very short lifespan. These also may be considered *nano-objects* according to the above definition (Fig. 2).

These particles, however, immediately and irreversibly cluster to form non-dissociable aggregates, from 50 to 300 nm in size, which subsequently bind together to form agglomerates from 1 to 50 microns. The aggregates and agglomerates meet the above definition of nanostructured materials. In aggregates, covalent bonds exist between particles; in agglomerates the bonds between aggregates are electrostatic. Aggregates cannot de-aggregate at all. Agglomerates cannot deagglomerate during standard handling conditions [25, 26].

2 Elastomer Based Macrocomposites

2.1 Elastomeric Matrices

The Elastomeric Matrices (EM) consists of compounds of natural or synthetic rubber with a variety of ingredients to confer the required properties. The natural rubber (NR) forms as a latex, a suspension in water of rubber particles coated with a natural surfactant [27]. The synthetic rubbers (SR) have been developed for specific purposes, and are made by polymerization of the monomers under pressure and heat in the presence of a catalyst either in suspension, to form latex, or in solution. The elastomers may be defined as long-chain organics, often polyolefin molecules, consisting of repeating units of one or more monomers, which exhibit very high extensibility combined with an ability to return to the original

Fig. 2 Silica micro pearls

configuration. Analysis showed that their high extensibility was due to uncoiling of the linear polymer chains from an initially random configuration to one of general alignment. There are several natural limits to both the elastic extension and the permanent set in raw elastomers:

(i) Chain entanglements, which result in stiffening at very high elongations, an effect that increases with molecular weight [28–31].
(ii) Crystallinity, which in elastomers requires high purity and very regular chain structure, e.g., in polybutadiene better than 98 % cis content, percentage of covalent bonds in the main chain. This level is unrealizable in cross-linked material. Spherulitic crystallinity develops in unstrained polymers e.g., NR or polychloroprene at low temperatures, thus increasing their stiffness. Crystallinity can also develop at high extension levels as fibers align along the stretching direction, where it increases both stiffness and strength [32–34].
(iii) Second-order transition temperature [35, 36].

Since elastomers owe their elastic properties to thermal agitation, they are naturally sensitive to temperature variations. In general, their stiffness increases as the temperature is reduced, until a temperature is reached where the material assumes a glass-like consistency, with a corresponding increase in hysteresis, internal viscosity. The temperature at which this occurs, at the glass transition temperature, Tg, increases with the number and size of the side groups on the polymer chain, for example vinyl groups, benzene rings, chlorine, and the tendency is thus for high glass transition temperature to be associated with high hysteresis and incidentally low air permeability and high friction. Elastomers are usually mixed with several different ingredients to form a compound, the ingredients being chosen to optimize the physical properties of the compound in order to meet the specification for a particular application.

Studies have shown that the process causes sulfur cross-links between two and eight atoms long to form at irregular intervals between chains, by breaking the covalent bonds which are a feature of the polyolefin structure. Efficient vulcanization systems, having short cross-links, tend to have higher modulus and lower extension at break. The cross-link density can be deduced from swelling tests in solvents, and is usually defined in terms of the molecular weight between cross-links, M_c, typical values being around 100,000, compared with 78 for the polyisoprene monomer unit, or 26,106 for a natural rubber molecule. Where the main elastomer chain contains no covalent bonds, it is necessary either to provide such sites by copolymerization with an olefin, or to form cross-links directly between carbon atoms on the chains without using sulfur by radiation or organic peroxides. It was found that some materials, typically sulfenamides or thiazoles, can catalyze cross-linking reactions and are added as accelerators to reduce the reaction time. In some applications, the reaction rate needs to be reduced and a retarder (e.g., phthalic anhydride) is then added. All these additives comprise the curative system. The amounts of each ingredient vary, but rarely exceed 5 % by weight.

2.2 Fillers

Elastomers are relatively expensive materials, and it became common practice to add cheap mineral powders, e.g. silica or calcium carbonate to reduce the cost, the effect being to increase the elastic modulus of the cured compound with little effect on the strength or other properties. However, it was soon found that very fine powders, particularly carbon black, manufactured by burning oil or natural gas in controlled conditions, could result in a large increase in both strength and modulus, albeit with an increase in hysteresis [37–40]. The reinforcing effect increases with the amount added; up to about 30 % by volume, and also with reduction in particle size or increase in the structure the tendency of the particles to form chains. The particle size and structure are defined by an internationally accepted system, ASTM Standard, D-1765, thus N330 black ca. 30 mm, medium structure is made under a variety of trade names by various manufacturers. Reinforcing silicas having similar particle size, but with special surface treatments, have appeared, which appear advantageous in terms of hysteresis. Another means of reducing cost is to add oil to the compound, which may also improve processability, but usually with a small reduction in physical properties. Except in racing tire treads it is rarely used in quantities greater than 5 % by volume. By using highly aromatic oil, it is possible to raise the glass transition temperature, Tg, of the compound, which is found to improve the wet grip of tires, while aliphatic paraffinic oils tend to reduce it. Soon after the development of cross-linking systems, it was observed that rubber compounds deteriorate with time, an effect that is typically manifest as cracking or crazing of the surface. This is usually due to oxygen or ozone attack on the covalent bonds in the elastomer chains, a process perpetuated by the formation of free radicals. There are a variety of means of reducing the deterioration, e.g., the addition of waxes which bloom to the surface to form a protective skin, or the addition of small amounts of antioxidants or antiozonants which act as free radical acceptors, thus breaking the chain reaction. In addition to reducing the cracking, these materials may also affect the cross-linking reaction, hence necessitating an increase in the amount of accelerator used.

Where rubber is required to adhere to a polymeric or metal substrate e.g., cords, special compounding techniques are used. In some cases, one or more bonding layers such as latex resin mixtures are interposed between the substrate and the rubber, which bond well to both surfaces, while in other cases special compounding ingredients e.g., organic cobalt salts are added to promote adhesion. The cord reinforcement acts as the main load bearing component of the composite and enables it to withstand the variety of loads that it is subjected to in its service life while maintaining an acceptable degree of dimensional stability. Cords are made by twisting together a parallel array of fibers to form a singles yarn and then twisting together two or more of these yarns together to form the cord. Many different materials can be used as the cord reinforcement, the main requirement being that it is strong in a longitudinal direction and sufficiently flexible to withstand the forces imposed upon it in service. Initially cords were made from natural fibers such as cotton, flax, etc. The development of regenerated natural fibers such as rayon and manmade fibers such as nylon, which have a much greater tensile strength, enabled the performance of cord reinforced elastomers to be greatly improved. In addition to organic polymer materials, some inorganic materials such as glass and steel are also used for cord reinforcement. Rayon is produced from regenerated cellulose. Cellulose is the main structural material of many plants such as trees, grasses, and cotton. The main source of cellulose for rayon production is wood pulp derived from spruce wood. Nylon was the first commercially successful synthetic fiber, the raw materials being derived from coal or oil. Nylon is a generic name for polyamide fibers composed of linear polymers, the monomer units being linked by amide groups. Polyester is the generic name for polymers with the monomer units joined by ester linkages formed by reacting an acid with an alcohol, the raw materials being derived from petroleum. Aramid fibers are aromatic polyamides formed from an aromatic acid such as terephthalic acid or an aromatic dichloride such as terephthaloyl dichloride and para-phenylene diamine. The performance of cord reinforced elastomers has improved dramatically, particularly when used in tires, mainly due to the service requirements of car manufacturers. In order to enable these demands to be met there has been continual research and development in polymer chemistry to find new fibers with improved properties. Two fibers which are new to the marketplace are polyethylene naphthanate and polyolefin ketone. Steel wires of higher tensile strength have been developed; these wires have a slightly higher carbon content and are known as high tensile and superhigh tensile. Steel cords, which are made from a number of steel wires, were first used in radial car tires as the tread stabilizing layer. In a similar way that steel would not normally be considered as a textile material, glass when extruded as a filament and drawn down fine enough can be processed similarly to a normal textile material. Glass is spun into air from the molten state, it is then rapidly quenched and attenuated to prevent crystallization and coated with size to aid future processing. Carbon fiber is virtually 100 % carbon with a highly orientated graphitic structure. Its main properties of interest as tire reinforcement are similar to those of glass but it is lighter and has even higher modulus and strength.

Rubber is a widely used material having properties of flexibility, strength and elasticity. The basic raw material, either natural rubber or synthetic rubber, is in latex or solid form, and processed into many different products. The raw rubbers, composed essentially of long polymer chains, are joined together with "cross-links" in a process called vulcanization to give the final material its characteristic properties. Natural rubber hardens below 0 °C and softens and weakens above 80 °C, losing its strength and becoming tacky. In between these temperatures it can flow under stress and permanent deformation occurs under prolonged strain. These undesirable properties are reduced by vulcanization, in which the reactivity the double bonds impart to the molecule is utilized to make it react with added material to form crosslinks between the chains. The crosslinking increases the useful temperature range of the rubber and hardens the rubber so that it becomes much stronger and does not creep but returns to its original shape on release of stress. Its surface properties are improved and its solubility decreased. Sulphur is still the most important vulcanizing compound for natural rubber, but not the only one. The raw rubbers include butadiene rubber, butyl rubber (isobuteneisoprene copolymer), synthetic isoprene rubber, ethylene-propylene rubbers, chloroprene rubber (the methyl group of isoprene is replaced by chlorine), nitrile rubbers (acrylonitrile butadiene copolymer), styrene-butadiene rubber, and silicone rubbers which are polysiloxanes.

Compounds added to the raw latex must be in the form of emulsions or dispersions. They are prepared by milling the substances with distilled or softened water in ball or gravel mills which revolve for anything from a few hours up to several days. Gelatin, casein, glues, soaps etc. are used as wetting or dispersing agents. Substances added are softeners, fillers, pigments for color, the vulcanizing agents and antioxidants. The latex itself has to be stabilized with surface-active agents to prevent coagulation which can be irreversible. These agents act by imparting a charge to the surface of the minute rubber particles or by holding an envelope of water around the particle, thereby preventing any mixing. The compounding materials and the latex are mixed and are then ready for dipping, molding, foaming or spreading. The production sequence is mixing, forming and vulcanizing. The solid rubber and the other materials have to be mixed. This is done with two basic machines, a two-roll mill in which the material is passed between two heavy metal rollers mounted horizontally, and a Banbury mixer, an internal mixer in which the materials are sheared between the internal rollers and the inside of the casing. There are four basic methods of forming the material to the required shape; spreading onto fabric from solution, extruding, calendering and molding.

Sulphur vulcanization, the most common form, involves the formation of polysulphide crosslinks between the chains. If sulphur only is used, curing times of 8 h are necessary. Modern methods using "activators" and "accelerators" reduce the curing time, eliminate cyclic structures and can shorten the sulphur links down to one or two sulphur atoms. Shorter sulphur links give greater stability to heat and improve the ageing properties.

Fillers range from inert dilutants such as whiting, talc, clays, $CaCO_3$, etc. which "cheapen" the product and make it go further, to reinforcing fillers which increase tensile strength and abrasion resistance. Carbon black is a most important reinforcing agent in tires and tubes and is usually produced by burning oil or natural gas in a limited supply of oxygen. Protective agents prevent ageing or deterioration, which comes about mainly by oxidation. Softeners and lubricants, in the form of Stearic acid, waxes, mineral oils, tars, etc. are added to plasticise or increase self-adhesion of the rubber mix. Certain resins increase the tackiness of the rubber for use on adhesion tapes. Elastomers are polymers that will elongate when subjected to a tensile force. They will return to the original shape when the force is removed. Rubber is an elastomer. Natural rubber is composed of isoprene units. Isoprene is polymerized into polyisoprene. Natural rubber is soft and tacky when hot. Reacting it with sulfur cross-links the polyisoprene and makes the rubber harder. This process is known as vulcanization. Synthetic rubber is similar to polyisoprene. One example is polybutadiene. Neoprene is very similar to polybutadiene, but contains chlorine instead of one CH-group. It is more resistant to solvents like oil and gasoline. Another synthetic rubber is styrene-butadiene rubber (SBR). SBR is a copolymer of styrene (25 %) and butadiene (75 %). It is tougher and more resistant to oxidation than natural rubber, but its mechanical properties are less satisfactory [41–46].

Addition of filler increases hardness of the cured product. All fillers are not created equal, so that there is a range of reinforcement from very high to very low, corresponding to the primary size of the filler particle, from around 10 nm for very fine particle carbon blacks giving high reinforcement, to greater than 300 nm for some calcium carbonate which give low reinforcement. Use of the latter reduces compound cost. The shape and surface chemistry of the filler particle also play an important part in reinforcement. Some popular fillers are, in order of decreasing reinforcement, carbon blacks and silicas, clays and then whitings. Carbon black is a material of major significance to the rubber industry, so it is no surprise that most rubber products we see in the market place are black in color. There are two common methods of producing carbon black today. Heating natural gas in a silica brick furnace to form hydrogen and carbon produces a moderately reinforcing material called thermal black. Alternatively, if we incompletely burn heavy petroleum fractions, then furnace blacks are produced. These are the most important blacks in terms of quantity used and available types. Carbon black consists of extremely small particles from around 10 to 300 nm. This gives two primary properties allowing a whole range of grades designated by both a particular particle size (surface area) and a specific level of structure [47–50].

3 Recent Advances in Elastomer Macrocomposites

The effect of rubber wood on curing characteristics and mechanical properties of epoxidized natural rubber (ENR) has been investigated in the loading range of 0–50 phr. Results indicate that the scorch and cure times decrease with increasing rubberwood loading. Tensile modulus and hardness of the composites increase with increasing rubberwood loading whereas tensile strength and tear strength show a decrease. SEM studies and rubber-rubberwood adhesion measurements indicate that the increasing rubber wood loading has weakened the rubber-rubber wood interactions [51]. Properties of natural rubber (NR) filled with various loadings of ultra-fine vulcanized acrylate rubber powder (ACMP) have been investigated. ACMP loading was varied from 0 to 20 phr and, after compounding, the compound properties have been determined. Results reveal that increasing ACMP loading leads to improved processability, as evidenced by the reduction of both mixing energy and Mooney viscosity. ACMP, however, has negative effect on cure, that is, both scorch time and optimum cure time are prolonged while the state of cure is reduced with increasing ACMP loading. Due to the reinforcing effect of the fine ACMP particles, both modulus and hardness are found to increase consecutively with increasing ACMP loading. The tensile strength is also found to improve with increasing ACMP loading up to 10 phr. However, due to the cure retardation effect and the high thermoplastic nature, the presence of ACMP causes deterioration of elasticity. As ACMP is highly polar and fully saturated, the addition of ACMP enhances the resistance to oil and thermal aging of the NR vulcanizate. Significant improvement of thermal aging resistance is found when 10 phr or more of ACMP is added [52].

The puncture and burst properties of short-fiber reinforced polychloroprene rubber under various conditions has been investigated to yield the best mechanical properties. In addition to five types of interphase conditions, four types of fiber aspect ratios (length of fiber/diameter of fiber) and three different fiber contents have been studied for their roles in the puncture and burst properties of rubber products. Certain interphase conditions and higher fiber aspect rations have shown to provide higher puncture and burst stresses at a given fiber content. Since both testing methods measure biaxial properties of reinforced rubber, the relation between the two properties have studied. The discrepancy between regressed puncture and burst force is explained based upon the rubber stiffness due to reinforcing parameters and the stress concentration upon sharp edge. Overall, it was found that the interphase condition, fiber aspect ratio, and fiber content have an important effect upon puncture and burst properties [53].

Mica or carbon black was used as filler in composites of acrylic rubber. The fillers differ not only in nature, mica being a mineral material and carbon black being organic, but also in form and particle size. The content of filler varied from 0 to 50 phr and its influence on acrylic rubber was evaluated based on cure parameters, mechanical, and swelling properties. The cure parameters allow the conclusion that the presence of mica does not have a negative effect on the cure or processability; the swelling results indicated a weak interaction between acrylic rubber and mica even though the mechanical properties of acrylic rubber composition with 40 phr of mica were found to be similar to those of 20 phr carbon black. As a result of mica being less expensive than carbon black, is light colored and easily processible, these properties are of industrial importance. All properties analyzed have been compared with gum type composition without filler [54].

A networked silica prepared by interconnecting silica particles with polymeric methylene diphenyl diisocyanate has developed for use as a highly effective reinforcing material for rubber compounds without the need to add silane coupling agents. Methylene diphenyl diisocyanate incorporated onto the silica surface formed networks among neighboring silica particles with urethane linkages and produced networked silica at low cost. The TEM photographs illustrated the improved dispersion and formation of openings among the silica particles, which could allow easy intrusion of rubber molecules. The networked silica has showed a high reinforcing performance for styrene-butadiene rubber (SBR) compounds, suggesting the possibility of replacing the silica reinforcing systems with coupling agents. Due to the absence of any silane-containing coupling agents, the networked silica does not suffer the disadvantages associated with coupling agents. Since the networked silica reinforces rubber compounds by the physical entanglement between rubber molecules and the methylene diphenyl diisocyanate chains, high loading of the networked silica is more effective in enhancing the mechanical properties of rubber compounds [55].

Different elastomer-based composites for microwave absorbers are developed. The influence of chemical character and structure of the polymer matrix and chemical nature and concentration of fillers with high values of the imaginary part of the complex dielectric permittivity and magnetic permeability on the microwave properties of the absorbers is investigated. Natural rubber, styrene butadiene rubber, butadiene rubber, chloroprene rubber, nitrile butadiene rubber have tested as polymer matrix. Graphite, furnace carbon black, acetylene carbon black and active carbon are used as fillers with high dielectric losses; natural magnetite is used as filler with high magnetic losses in the experiments. Some more important microwave parameters of the absorbers as function of frequency and composition (mass ratio filler/rubber) are measured [56].

The production of hydrosilylated impact polypropylene (PP) copolymer has been demonstrated in a co-rotating twin-screw extruder. This has been accomplished through a two-step reactive extrusion process that involves: (i) the formation of terminal double bonds on a commodity polypropylene copolymer through peroxide initiated degradation reactions, and (ii) the melt-phase functionalization of these double bonds with hydride-terminated polydimethylsiloxane (PDMS). Spectroscopic analysis of the peroxide degraded polypropylene and the purified hydrosilylated polypropylene has been performed to confirm the attachment of PDMS onto the polypropylene chains. In addition, the hydrosilylated polypropylene has been characterized in terms of molecular, rheological, surface, and mechanical properties. It has been found that the two-step hydrosilylation reaction decreases molecular weight, imparts formation of branching/chain extension, decreases impact strength, and improves surface hydrophobicity. Finally, oscillatory shear measurements exhibit an unusual up-turn in the viscosity curve at low frequencies [57]. A theory for incompressible, rubber-like shells of arbitrary geometry undergoing finite rotations and finite strains, including transverse normal strains but neglecting transverse shear strains, is presented [58].

The effect of fillers on the mechanical properties such as stress–strain behavior, tensile strength, percentage strain at-break, Young's modulus and tear strength has been investigated. Different filled crosslinked thermoplastic elastomers of styrene butadiene rubber (SBR)/high density polyethylene blends (HDPE) blends were prepared using silica, HAF-carbon black, china clay and TiO_2. The reinforcement ability of the filler was increased in the order of silica > HAF-black > clay > TiO_2. The nature of the filler and filler loading has a dramatic effect on the mechanical properties of SBR/HDPE blends. Filled blends showed improved mechanical properties such as enhanced of strain at-break, when fillers are incorporated. The initial trend of properties for all filled system is the enhancement of properties. When HAF-black is used as the filler, at higher loading strain at- break is found to decrease due to the stiffness of the matrix. In the case of clay, there is a deterioration of properties occurs on higher loading, which is attributed to dilution effect and all TiO_2 filled system have lower elongation at break than the base polymer.

Theoretical models namely Guith's and Kerner's model have been compared with the experimental values of Young's modulus of filled system. The experimental values of modulus are found to be higher than the theoretical values indicating strong interaction between the filler and the matrix. SEM studies of the tensile and tear fractured surfaces of the filled blends have been carried out. The variation in properties was correlated with the morphology of the system [59].

The use of the sol–gel process on general-purpose grade rubbers has reviewed in the absence or presence of silane coupling agents. The sol–gel reactions of tetraethoxysilane in epoxidized natural rubber (ENR), styrene-butadiene rubber (SBR) or butadiene rubber (BR) vulcanizates produced silica generated in situ. This silica was found to be a good reinforcing agent by investigating tensile and dynamic mechanical properties and morphology observation by transmission electron microscopy (TEM). The amount of silica formed was limited by the degree of swelling of the rubber vulcanizate by of tetraethoxysilane which was the precursor of the silica. However, the dispersion of silica generated in situ was better than conventionally added silica due to its formation in place. Also, it was noted that the diameter distribution of in situ silica was monodispersed. Silane coupling agents, such as mercaptosilane, aminosilane, and bis (3-triethoxysilyl-propyl) tetrasulfide, were compounded in the vulcanizates and their effects on silica generated in situ were evaluated. Their effects were significant. The dispersion of the silica in the rubbery matrix became better and the particle size became smaller and monodispersed, as observed by transmission electron microscopy, which improved mechanical properties. The superior properties of silica generated in situ have studied further to elucidate the mechanism of reinforcement [60].

Fatigue life prediction and evaluation are the key technologies to assure the safety and reliability of automotive rubber components. Fatigue life prediction methodology of vulcanized natural rubber was proposed by incorporating the finite element analysis and fatigue damage parameter determined from fatigue test [61]. Heat-aging effects on the fatigue life prediction of natural rubber were experimentally investigated. Fatigue test were performed using the 3-dimensional dumbbell specimen, which were aged different amounts. The Green–Lagrange strain at the critical location determined from the finite element method used for evaluating the fatigue damage parameter. Fatigue life prediction equation effectively represented by a single function using the Green–Lagrange strain. According to fatigue life prediction equation, fatigue life ambient temperature was longer than at 70 °C. Predicted fatigue lives of the rubber component were in fairly good agreements with the experimental fatigue lives within factor of two [61].

The relationship between the variation of hydrogen bonding and macroscopic properties of composites composed of hydrogenated nitrile butadiene rubber (HNBR) and 3,9-bis {1,1-dimethyl-2 [β-(3-tert-butyl-4-hydroxy-5-methylphenyl) propionyloxy] ethyl}-2,4,8,10-tetraoxaspiro [5, 5]-undecane (AO-80). Hydrogen bonding of the composites was studied by Fourier-transform infrared (FT-IR) and ultraviolet (UV) spectroscopies. FT-IR spectra at room temperature revealed that the stretching vibration peak of O–H and $C = O$ of AO-80 red shifted with increasing AO-80 content, whereas that of C ≡ N of HNBR blue shifted only when the AO-80 content exceeded 10 parts per 100 resin (phr) [62].

Miscibility and dynamic mechanical properties of hydrogenated nitrile butadiene rubber (HNBR) and 3,9-bis{1,1-dimethyl-2-[β-(3-*tert*-butyl-4-hydroxy-5-methylphenyl) propionyloxy]ethyl}-2,4,8,10-tetraoxaspiro [5, 5]-undecane (AO-80) composites have investigated. The composites with low-AO-80 content have showed the incomplete miscibility, and a part of AO-80 dissolved into HNBR, while the rest was dispersed into HNBR in the form of deep-submicron-sized microspheres based on the micromorphological and thermal analyses. However, with the increasing AO-80 content, the system became completely miscible. When blending with AO-80, the resulting composite exhibited the remarkably improved dynamic mechanical property. The loss tangent peak of the composites gradually shifted to room temperature with the increasing AO-80 content [63].

4 Manufacturing of Elastomer Macrocomposites

4.1 Introduction of Processing Rubber Compounds

Most products are made from dry rubber. The successive operations for natural rubber being [64–71]:

(i) Tapping the rubber tree to collect the latex in cups, to be followed by screening to remove contaminants.
(ii) Coagulation, using acetic acid to neutralize the surfactant charge.
(iii) Consolidation squeezing out the excess water in a mill, followed by drying.
(iv) Shipping, usually in the form of 50 kg bales, during which crystallization may develop.
(v) Softening, by passing the rubber between the ridged rollers of a Breaker Mill, so reducing the molecular weight and crystallinity.
(vi) Internal mixing, where the rubber is sheared by rotors within a heated chamber while the compounding ingredients are added. A few minutes mixing are typically required to achieve good dispersion of the filler, which may be accomplished in two or three stages to avoid overheating and the risk of 'scorch' (premature cure). The cross-linking ingredients are naturally always added last.
(vii) Sheeting-out to a required dimension by passing the compound between the heated rollers of a two-roll mill or (with more precision) a three- or four-roll calendar.
(viii) Extrusion, which involves forcing the compound through a shaped die to achieve the desired profile by means of a ram or screw feed.

(ix) Vulcanization, by heating to 130 ± 180 8C for a period of 2 ± 120 min (depending on the thickness), to promote the cross-linking reaction. In normal rubber manufacture a slug of compound is usually placed in a mold of the required shape, and loaded into a press, which applies external load to prevent the mold opening or internal porosity developing in the rubber. Narrow vents (spew holes) are provided to allow the slight excess of rubber to escape.

Textile and glass cords are usually processed using the traditional spinning and weaving route to form a fabric prior to calendering but steel cords are usually fed from a creel situated in front of the calendar. After spinning the bundle of continuous filaments, these are brought together to form the yarn. This is then given a spin finish and a small amount of twist to aid processing and wound onto a cylindrical shaped center to form a cheese. The starting material of a steel cord is a hot-rolled wire rod, usually 5.5 mm in diameter, which has been specially cooled. The main production process consists of a series of drawing and heat-treatment stages during which the steel rod is reduced in diameter to give a flexible wire suitable for use as a textile material, and also having the desired crystalline structure to give a high-modulus, high-tensile wire. After the final wire drawing stage, several wires are brought together and twisted to form a strand, and for heavier cords, such as those used in truck tires, several of these strands are twisted together to form a cabled cord or further layers of wires are wrapped around a core strand to form a layered cord. The great majority of products made from cord reinforced elastomers, whether they be tires, hoses, or conveyor belts, are assembled from layers of uncured rubber/cord fabric to form the final article. There are several ways of making this fabric, the most common by far being calendering, but spreading and extruding methods can also be used. Another technique of making a cord/rubber fabric is that of extruding a sheet of cords and rubber through a die to produce a tape. The process for doing this with steel cords is called the Steelastic process, although a similar process can be used to produce textile cord fabric. The benefit of this technique is that it requires much less space and resources to set up, and produces much less waste material.

In the rubber industry, for the design and production of a commercial material that satisfies precise technical requirements, it is necessary to:

(a) Select the right raw materials;
(b) Blend them in the appropriate proportions in suitable equipment;
(c) Form the resultant blend into the desired shape; and
(d) Render the finished product dimensionally stable.

The first stage, known as compounding, refers to the formulation of a blend of rubber and various additives. These are first thoroughly mixed and then formed. All the operations inherent to the blending of the various ingredients and their forming constitute the processing. The physical properties of an unvulcanized elastomer do not allow a manufactured item to retain its dimensional and mechanical characteristics over time; therefore, it is necessary to generate a stable molecular network, making use of a chemical reaction capable of joining the polymer chains one to the other. This process is known as vulcanization. Finally, it should be recalled that the definition of the physical and chemical properties of elastomers has been the subject of standardization carried out by the American Society for Testing and Materials (ASTM). The standard methods of measurement are continually updated and adapted to new instrumentation and needs, verifying their reliability through cross-tests between different laboratories. Once the characteristics of the finished product have been established, the basic elastomer and the package of ingredients necessary to obtain the properties required are selected. The recipe of the blend consists of a list of the various ingredients, and instructions on how to mix them to prepare the blend. The unit of measurement used for the quantities by weight of the various constituents of the recipe is phr (per hundred rubber), which indicates the quantity of additive required per 100 parts of rubber.

4.2 The Materials Needed for the Preparation of a Blend

The materials needed for the preparation of a blend belong to the categories indicated below:

- Base polymer (raw elastomer, gum rubber). This is the main ingredient and can consist of natural or synthetic rubber, or thermoplastic elastomers.
- Vulcanizers. These are the substances needed to generate the three-dimensional network that gives the rubber its typical characteristics (except in the case of thermoplastic elastomers that do not need to be vulcanized); a typical vulcanizer is sulphur, used in quantities in the range of 0.5–3 phr.
- Accelerators. They are the substances that interact with the vulcanizer to reduce the time of vulcanization; they are used in quantities in the range of 0.5–1.5 phr.
- Activators. Activators are made up of metal oxides (such as zinc, lead and magnesium), of carbonates and of alkaline hydroxides; they are added in quantities of 2–3 phr, form chemical compounds with the accelerators and modify the speed of vulcanization and the number of links between the different macromolecules (density of cross-linking) unlike diamonds and graphite, is not found in nature. It is the generic term used to indicate a family of materials made up of elementary carbon in the form of aggregated spheroidal particles, obtained by the thermal decomposition of hydrocarbons in a shortage of air. Carbon blacks are produced using different technologies, from which the various types take their names: furnace blacks and thermal blacks make up 95 and 4 % respectively of the world market. These particles, which have an average diameter from 10 to 100 nm (the size depending on the preparation technology), are aggregate forming larger structures (60–200 nm); their surface area is as much high as 150 m^2/g, and on the surface there are phenol and quinone groups, carboxylic acids and lactones, in different numbers and quantities according to the type of carbon black.

- Vulcanization retardants. These are substances that interact with the vulcanizer-accelerator-activator system, creating a period of time during which vulcanization does not take place. They thus ensure the completion of the various transformation operations thereby avoiding pre-vulcanization; they are used in quantities in the range of 0.1–0.3 phr.
- Organic acids. Their reaction with the activators provides the cations necessary for the formation of chemical complexes with the accelerators. Monobasic acids with high molecular weights, such as stearic, oleic, lauric, palmitic, and myristic acids, and hydrogenated palm, castor and linseed oils are used in quantities of 1–3 phr.
- Antioxidants. Added in quantities of 1–2 phr, they protect the rubber from oxidation, accelerated by light and ozone, which generally causes a structural modification to a lesser or greater degree in the polymer chain with consequent variations to its mechanical properties. This category includes secondary amines, which exhibit a high tendency to coloring, phenols hindered by *t*-butyl groups in the ortho-position (primary antioxidants, they reduce the peroxide radicals) and organic phosphites (secondary antioxidants, they reduce the hydroperoxides). The primary and secondary antioxidants are used in synergic combinations.
- Fillers. Initially they were added to rubber in the form of small particles for economic reasons. The addition of carbon black to natural rubber had a reinforcing effect, improving some characteristics such as resistance to abrasion and teat, and increasing the values of the elastic modulus and of the tensile strength. Fillers are subdivided into two groups: the first includes reinforcing fillers and the second inert fillers. The inert fillers (kaolin, barytes, carbonates of calcium and of magnesium, of iron and of lead), powdered to dimensions of 0.1 mm, are used to modify some technological characteristics of the vulcanized elastomer such as its hardness, its density or its electrical properties. The reinforcing fillers (carbon black and silicas), instead, have a major effect on the mechanical and dynamic characteristics of the vulcanizate in that, by interacting with the macromolecules, they take part in the elastic network.
- Carbon black is an allotropic form of carbon that, unlike diamonds and graphite, is not found in nature. It is the generic term used to indicate a family of materials made up of elementary carbon in the form of aggregated spheroidal particles, obtained by the thermal decomposition of hydrocarbons in a shortage of air. Carbon blacks are produced using different technologies, from which the various types take their names: furnace blacks and thermal blacks make up 95 and 4 % respectively of the world market. These particles, which have an average diameter from 10 to 100 nm (the size depending on the preparation technology), are aggregate forming larger structures (60–200 nm); their surface area is as much high as 150 m^2/g, and on the surface there are phenol and quinone groups, carboxylic acids and lactones, in different numbers and quantities according to the type of carbon black.
- Plasticizers. These are materials capable of improving processability, of reducing the hardness of the vulcanizates and increasing their elasticity and cold

flexibility, without undesired effects on their more important physical properties. They elong to two main classes: extender oils, derived from the petroleum refining industry, which are suitable for diene-based rubbers (SBR, NR and BR), and esters, which are recommended for polar rubbers such as NBR rubbers. Extender oils are, in turn, classified (the paraffin content is indicated in brackets) as highly aromatic oils (5–15 %), aromatic oils (15–32 %), naphthenic oils (48–56 %) and paraffinic oils (60–75 %); the choice of the most suitable plasticizer is made based on a criterion of compatibility.
- Processing auxiliaries. These are additives different sorts introduced to facilitate the incorporation of other ingredients (peptizers, adhesion promoters, dispersants), to regulate the process (lubricants), or to facilitate the separation of the vulcanized article from the mould (release agents). Among the peptizers of natural rubber is thio-bnaphthol, which is used in quantities of 0.6 phr during the mastication phase, di-o-tolylguanidine and diphenylguanidine for polychloroprene and zinc laurate for SBR rubber.
- Various additives. They are substances of various types that are added in varying quantities and proportions—flame retardants (aluminium hydroxide, antimony oxide, zinc borate), antistatics (metal powders or fibres, carbon black), colorants (metal oxides) and substances that increase adhesion to metals during the molding or extrusion phase (resorcinol–formaldehyde resins and derivatives of isocyanates).

4.3 Mixing

The final properties of a vulcanized elastomers depend not only on its intrinsic properties, composition, primary structure, but also on the degree of homogeneity with which the various fillers have been dispersed within it. The mixing of solid particles with highly viscous material is a very complex operation, which can be split into three successive stages: incorporation, dispersion and distribution. During the incorporation stage, starting with the separate ingredients of the compound, a homogeneous mass capable of flowing is obtained. Within this stage there can be three further distinct phases:

- encapsulation, during which the free surface of the elastomer wraps around the fillers,
- subdivision, during which the reciprocal distances and dimensions of the encapsulated fillers are reduced, deformed in shear or elongation; and, finally,
- the immobilization of a considerable fraction of the rubber inside the voids contained in the filler aggregates, with the important consequence that shielded polymer plays no part in the flow behavior. If the compound is subjected to moderate strain, the whole filler agglomerate and the associated entrapped rubber behave as if they were a single unit of filler.

Since, when a polymer is mixed with a rigid additive, the viscosity of the compound increases with the volumetric fraction of the additive and its elastic memory diminishes, a *blend* with a less than optimal dispersion possesses an ever-increasing viscosity and an ever-decreasing post-extrusion swelling as compared to a material in which the filler has been effectively dispersed. The use of plasticizers and oils, which produce greater molecular mobility, allows the polymer to quickly exit the voids of the reinforcing filler aggregates, thereby diminishing the proportion of entrapped rubber, reducing the viscosity of the mass and favoring the subsequent phase of dispersion. A suitable parameter for describing the effectiveness of mixing is the amount of power applied, usually correlated linearly with the rate of dispersion. During mixing, the temperature rises notably and the viscosity decreases along with the amount of applied power; the mixing time should therefore be increased, but not beyond certain limits in order to avoid early vulcanization. The time interval, measured from the moment that the compound containing vulcanizing agents is heated to the moment at which the reaction of cross-linking starts, is called the scorch time. This time can be modified by using retardants as well as by the choice of vulcanizer and accelerator. From the technological point of view, while for thermoplastic materials broad use is made of continuous mixing, for the processing of rubber, batch mixers are mainly utilized, consisting typically of open mixers (two roll mills) and of internal mixers. The reasons for this derive from the lack of availability of rubber in free-flowing granules at a cost comparable to that of rubber in bales, from the difficulty of accurately feeding a large number of ingredients and from the impossibility of adapting continuous mixers to the processing of different types of rubber and formulations without heavy changes. In an open mixer, there are three zones: one is located between the high powered (internally cooled) rollers (a); another, the bench (b), acts as a reservoir to feed the region between the rollers where the process of encapsulation takes place; the third, the belt (c), carries the rubber from zone a to zone b. The rollers rotate in opposite directions at different speeds with a ratio that varies between 1 and 1.1 (Fig. 3).

Due to its versatility (the ability to mix a great variety of elastomers with the same setting), and due to its ability to accept rubber in bales, the internal mixer is the machine most used in the rubber industry. Figure 4 contains a diagram of the two counter-rotating rotors, of the piston, which allows for the introduction of the various ingredients of the blend into the mixing chamber when lifted, but which keeps the rubber in the mixing area when lowered, and of the discharge door. Cooling water passes through the rollers, the walls of the chamber and the discharge door. Observing the mixing of rubber in a roll mill makes it possible to identify two limiting situations: a dry behavior, which shows a critical factor relating to the shearing of the elastomers (the rubber breaks up); and a cheesy behavior, typical of materials characterized by poor elastic behavior and poor tensile properties, mainly related to the absence of high molecular weight fractions. Aside from these two limiting cases, the behavior considered to be good is that characterized by the formation of a continuous belt between the rolls. The scale of visual evaluation is associated to a behavior that can be related to the

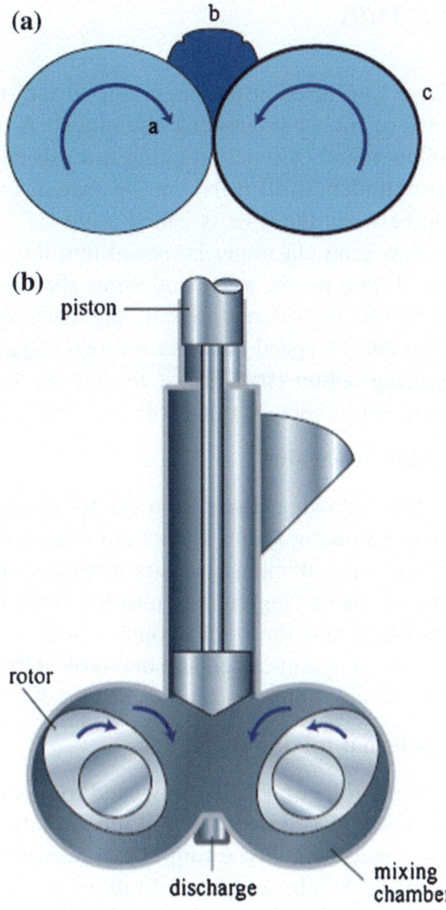

Fig. 3 Open mixer (**a**) and internal mixer (**b**)

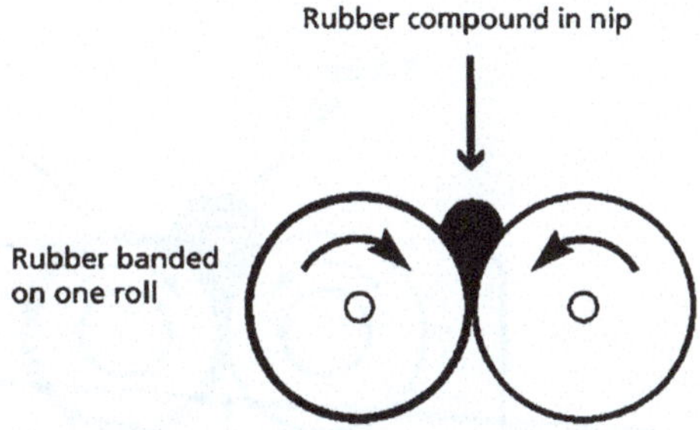

Fig. 4 Conceptual view of rubber mill rolls

extensional rheology of the unfilled polymer, which is particularly sensitive to its molecular structure. The principle stems from the theory of the elasticity of rubber at high deformation, and is based on the calculation of the elastic (recoverable) and viscous (non-recoverable) contributions of the polymer used in the blend.

4.4 Mills

These were used at the beginning of the rubber industry and are still an essential piece of rubber processing equipment. A mill consists of two horizontally placed hollow metal cylinders rotating towards each other (Fig. 4). The distance between the cylinders (mill rolls) can be varied, typically between 0.25 and 2.0 cm. This gap between the rolls is called a nip.

Raw gum elastomer is placed into the gap between the two mill rolls, the mill nip. It then bands, as a continuous sheet, onto one of the rolls. The speeds of the two rolls are often different, the back roll rotating faster than the front. The difference in speed between the two rolls is called the friction ratio and allows a shearing action (friction) at the nip to disperse the ingredients and to force the compound to stay on one roll, preferably the front one.

- **Mill processing**

The following description relates primarily to compounds which use sulfur as the crosslinking agent. The key to mixing (in a Banbury mixer or a mill) (Fig. 5) is to maintain sufficient viscosity to ensure an adequate shearing action, to distribute the non-rubber ingredients into the raw gum elastomer, or to force the raw gum elastomer into the microscopic spaces of each filler particle. Both mechanisms have been hypothesized and one typical mixing sequence might be as follows: The raw gum elastomer is placed into the nip and allowed to band onto the front roll

- **Internal mixing machines**

If the rolls of a mill are twisted to produce a corkscrew effect and then a block of steel is placed over the mill nip with the block connected to a steel rod above it, this would be called a ram. The ram would move up, to allow addition of ingredients to the nip, and it would move down to force the compound ingredients into the nip. If the whole thing is surrounded in a heavy metal jacket with a chute at the

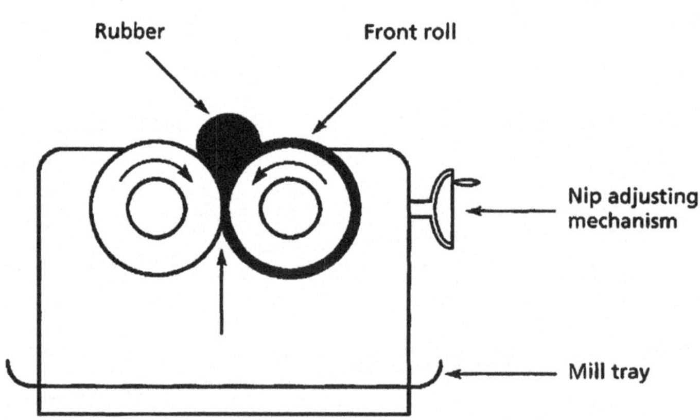

Fig. 5 Mixing in a mill

top to put ingredients in and a door at the bottom, to let the mixed material out, the result will be an internal mixing machine. Raw gum elastomer is dropped through the hopper into the mixing chamber where it is mixed by the rotors. The ram, pressing on to the rubber mixture, is forced down by a pneumatically or hydraulically controlled cylinder, whose pressure is adjusted to give the best control of the mixing process. Oil may be poured in from the hopper, or injected through a valve in the hopper wall just above the mixing chamber. Mixing can occur between the rotors (intermeshing rotors) or between the mixing chamber walls and the rotors.

- **Processing**

The mixing process will be discussed primarily with reference to unsaturated elastomers which are sulfur cured, unless otherwise stated. The mixing principles are similar to those for the mill. The raw gum elastomer is dropped into the hopper and the ram allowed to move down under pressure; the ram is raised for each addition of material and then lowered, to compact the mixture in the mixing chamber. Fillers are then added; large total amounts can be added incrementally and after most of the filler has been mixed in, any oil in the formulation may be then be introduced. If oil addition is delayed too long, the filler becomes totally 'encapsulated' by the elastomer and, the oil addition can cause a loss of shearing action, resulting in a slippery mess in the mixing chamber. During the mixing operation, feedback is received from the electrical power usage indicator, the temperature gauge, the time clock, and, for experienced operators, the sucking sound of the batch and the sound of the electrical motor driving the mixer.

- **Extruders**

Extruders (Fig. 6) are conceptually a pump, consisting of a screw to move the material forwards, a barrel around the screw to contain the material, help it move, and provide part of the temperature control. The back end has a hopper, sometimes with feed rollers, to put rubber into the screw, and the front end has a 'head' to hold a die, through which the rubber extrudes. An alternative to the screw extruder is the ram extruder, a well known trade name being Barwell. The ram extruder predates the screw extruder, but it is still used in certain specialized applications. Here, the screw is replaced by a ram, which forces the material through the die.

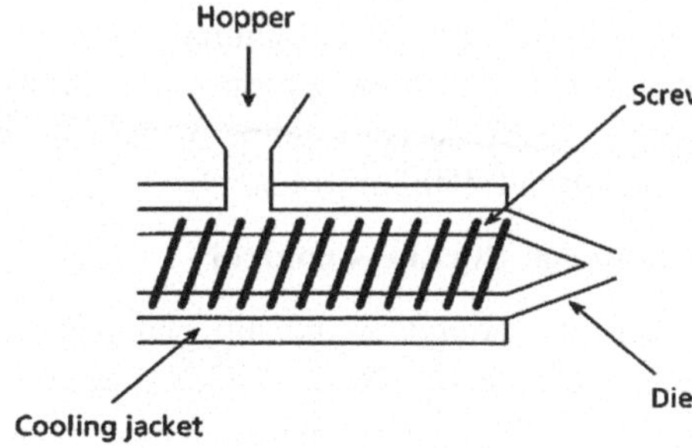

Fig. 6 Conceptual view of a basic extruder

Since the process is discontinuous, a slug of rubber is placed in the barrel, extruded, then another slug introduced, it is suited to making preforms for further use, such as placing into the cavities of molds.

- **Die swell**

The die is designed to avoid sudden discontinuities, as the compound moves through it and thus often has a contoured lead section. As the extrusion exits the die, the extrusion can shorten in length and increase in cross section. This is known as die swell, which is dependent on die design, screw speed, i.e., shear rate, temperature and the compound's viscosity and its elastic component. In practice, die swell can be quite complex and it might be necessary to modify the die a number of times, before the required extrusion shape is achieved. This recognizes that even uncured rubber has complex elastic and plastic behavior. Like an elastic band it can undergo elastic recovery on exiting the die.

- **Recent extruder design**

A problem with traditional extruders is the potential for reduced interblending of material as it moves along the screw. This causes uneven temperature distribution in the extrudate, which translates to a variable viscosity and therefore a continuously changing die swell. Layers of compound move along without intermingling, i.e., in laminar flow.

- **Calendars**

A calendar can be crudely thought of as a very high precision mill, with the rolls stacked on top of one another, with anything from two to four rolls in various configurations. The distance between the rolls can be varied to control calendared thickness.

As with the extruder, the calendar (Fig. 7) further processes the compound after it has been mixed in the internal mixer or on the mill. Calendering is a useful technique, if the final product is to be a roof or tank lining, fabricated hose, expansion joint or indeed any further product which needs accurately dimensioned continuous sheet. Calendering is also used for applying rubber compound to textiles. Sheet from a mill will have a thickness which is much too imprecise, can have a rough surface and may contain some trapped 'bubbles' of air.

A three roll calendar is very popular, where the middle roll is fixed, while the ones above and below it can be moved vertically to adjust the gap between the rolls. A four roll 'S' configuration might be considered more 'state of the art'. Calendars are extremely robust and solidly built machines, and may provide service for many decades. Some of the rolls can be a substantial size, i.e., 90 cm in diameter and 250 cm in length.

- **Material thickness control**

The compound in the nip can generate very large reaction forces pushing against the rolls. A force of 43,000 kg/m of nip is possible for a product gauge of 0.25 mm, which is enough to deflect the surface of the rolls. Since the middle of

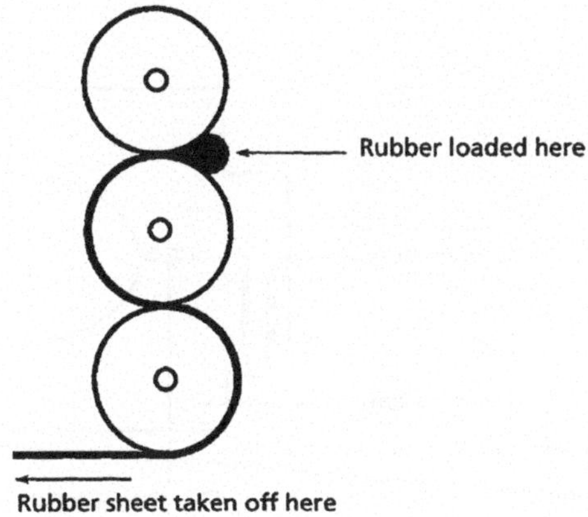

Fig. 7 Three roll vertical calendar

the roll length is furthest away from a solid support it will deflect the most, resulting in a curve shaped deflection with a maximum at the middle which can be 0.13 mm or more. Even this apparently small variation can result in significant material wastage and complications in further processing. To counteract this, an opposing curve called a crown is put into the roll by grinding it. This only goes some way to solving the problem, since deflection forces vary with the compound used, and the thickness of the rubber to be calendared.

- **Feeding the calendar**

If cold, at room temperature, compound were to be fed to the nip, it would heat up erratically and produce a variable viscosity. This would cause uncontrolled deflection forces on the rolls and hence an unacceptable thickness variation and surface quality of the rubber. The simplest way to feed the calendar, is to 'roll' small pieces of compound off a mill and immediately put them in the nip. The pieces then spread along the length of the nip and form a 'rolling bank'.

- **Curing equipment**

The rolls of sheeting have been calendared, the extrusions have been made, the *Barwell* has produced its preformed pieces, and shapes have been cut from milled sheet. The final step is to provide sufficient heat to change the uncured compound from a somewhat plastic state, to a dimensionally stable elastic substance, and additionally, in the case of molding, to achieve a final shape.

4.5 Molding

A mold might be described simplistically as at least two pieces of material (typically steel), which when fitted together form a cavity, resembling the shape of the product. This would be a very basic mold. Molding is by far the most important

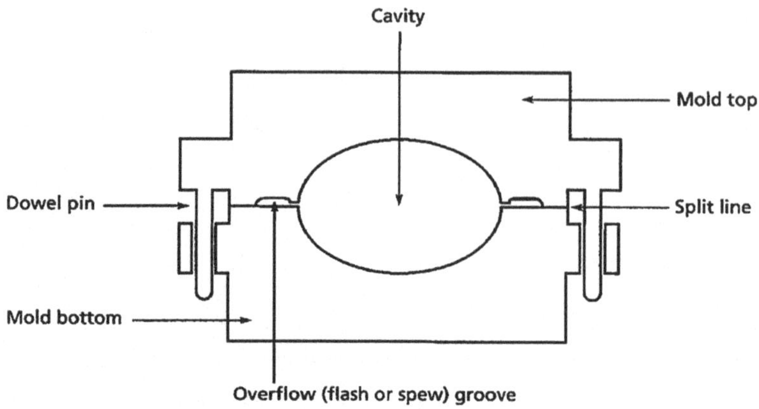

Fig. 8 Closed empty compression mold

curing process, where uncross-linked rubber is placed into a heated mold, which gives it the final product shape, and then vulcanizes the material.

- **Mold design**

A basic compression mold design is illustrated in Fig. 8 which shows a cross section through the center. It is very important that the two halves of the mold register (fit accurately together). In this case, pins built into the top section fit snugly into holes drilled into the bottom half. Any looseness between the pin and the hole may cause the top half of the product to be out of alignment with the bottom half. If the fit is too tight, attempts to manually open the mold may prove difficult.

There are different ways of introducing compound into the mold, some of which involve modifications to the basic design in Fig. 9. They each confer certain advantages not found in the others.

(a) In the most basic design, (Fig. 10), compression molding, pieces of rubber compound are placed in the bottom cavity and compressed using the top half of the mold.

(b) The first modification is transfer molding, which can be visualized as drilling holes through the outside of the top mold half of a compression mold through to the cavity. Thus the mold can stay closed while rubber compound is introduced through these holes into the cavity by using the force exerted by the press platen.

(c) If a separate device is used, not related to the press platen, which injects the compound through the holes, this would be injection molding.

- **Compression molding**

This is the simplest, cheapest, and probably the most widespread of the three basic molding techniques. It is ideally suited to small quantity production, say, from around fifty to a few thousand of each product annually. Figures 10, 11, 12

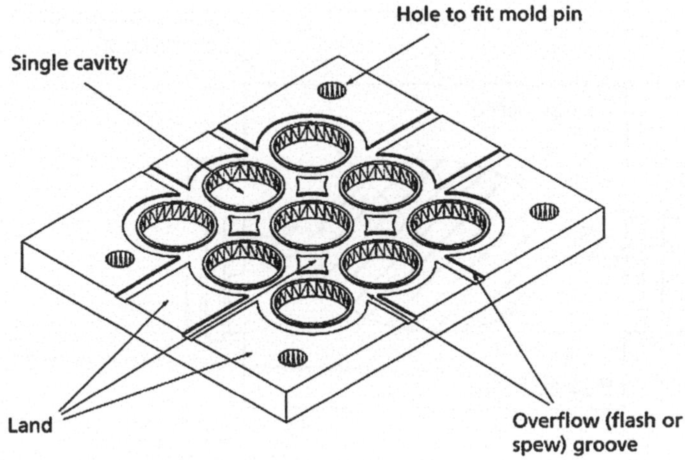

Fig. 9 The bottom half of a nine cavity mold

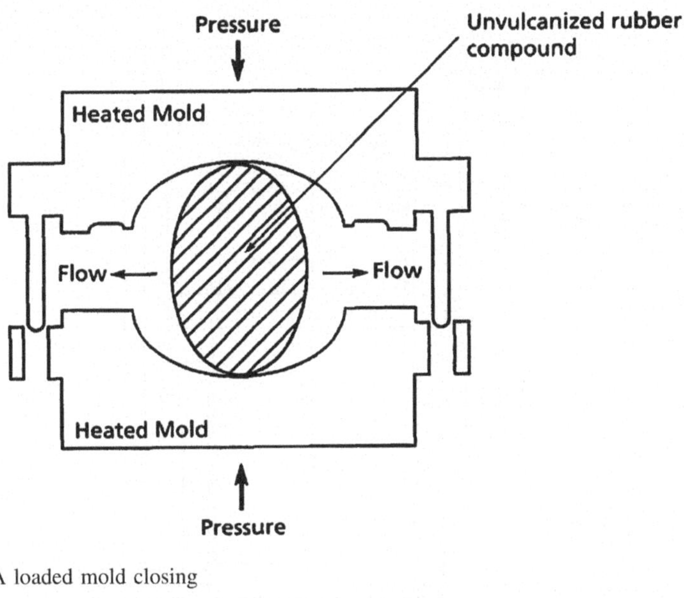

Fig. 10 A loaded mold closing

shows the various stages in the molding process. One of the keys to successful molding is adequate removal of air while the mold cavity is filling up with rubber.

The uncured pieces of compound placed in the mold are known-variously as preforms billets or load weights. For a ball, one might use an elliptically shaped extrusion, cut to an appropriate length from a *Barwell*. This shape is important and deliberately chosen so that air in the mold cavity will have a free path of escape when the mold begins to close.

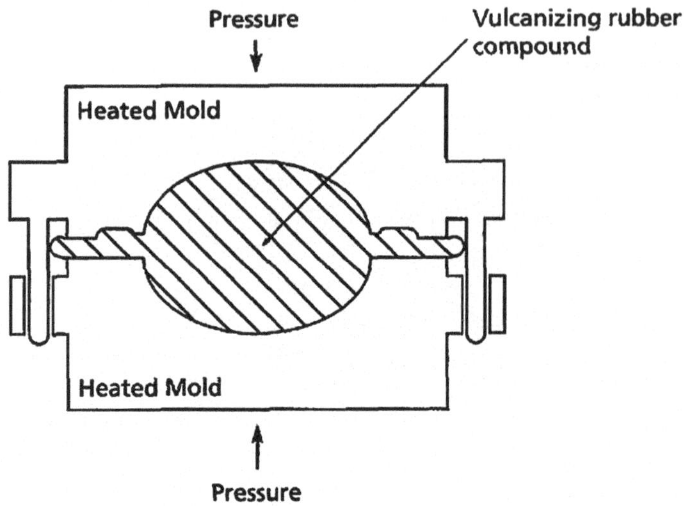

Fig. 11 A loaded mold, closed under heat and pressure

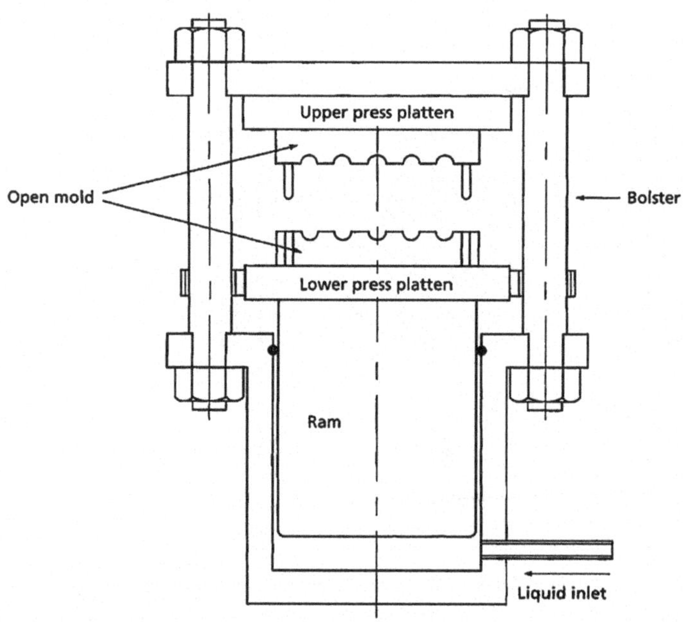

Fig. 12 A basic press for compression and transfer molding

As the press platens close the mold, excess compound begins to squeeze out into the flash grooves, taking air with it. Often, residual air remains and various methods have been devised to remove it. One method is to bring the mold pressure back down to zero and then return to full pressure by quickly lowering and raising the press platens a number of times. This 'shock' treatment is called 'bumping'. An additional line of attack is to find where air is being trapped in the final cured product and drill a small diameter hole through the mold cavity in the equivalent area; these are called bleeder holes. They permit an alternative escape route for the trapped air together with some rubber. The shape of the preform and also its placement in the mold is important. The uncured rubber, placed in the cavity, might be a single piece or a number of pieces.

- **Backrind**

Once the compression mold has closed, the compound continues to heat up and attempts to thermally expand. Its coefficient of expansion can be a least fifteen times greater.

- **Heat transfer**

How long should it take to cure a compound in a mold? The rubber laboratory uses a rheometer to help determine these using small samples of compound. The rheometer might indicate a cure time to be 25 min at 150 °C. If this is then applied to a shop floor molding, it must be remembered that the 25 min is based on the entire compound in the rheometer being at 150 °C at approximately the beginning of the 25 min period. Rubber can be very poor at transferring heat, so that for a large bulky part in a shop floor compression mold, it may take hours for heat to be transferred from the mold to the center of the part. The rheometer estimate of 25 min must now be revised, to take into account a constantly changing temperature throughout the part as the cure progresses. Carbon black is significantly better at transferring heat than a raw gum elastomer, thus for the same bulky part, a carbon black filled compound will vulcanize, through to its center, much sooner than a non-filled gum compound.

- **Transfer molding**

If we take the top half of a compression mold, then drill transfer holes through it and place a metal collar on the closed mold so as to surround all of the holes, we have in effect converted it into a transfer mold. All that is needed now is to put rubber compound into the pot and force it through the holes by placing a piston (plunger) into the pot and using the press platens to force the piston to push the compound down through the pot into the closed mold cavity. This conversion is used in the rubber industry. Alternatively, the transfer pot can be designed to be an integral part of the mold and the piston can be fixed to the upper press platen (see Figs. 13, 14, 15).

- **Injection molding**

Injection mold consists of a cylinder (injection barrel) with a ram or screw inside it, so that the rubber compound can be moved towards a nozzle at its end. The nozzle is then pressed against a hole made in the top half of a closed mold. This hole is then connected to smaller holes (gates and runners) which enter the cavities of the mold. The compound can be presented to the barrel as a continuous

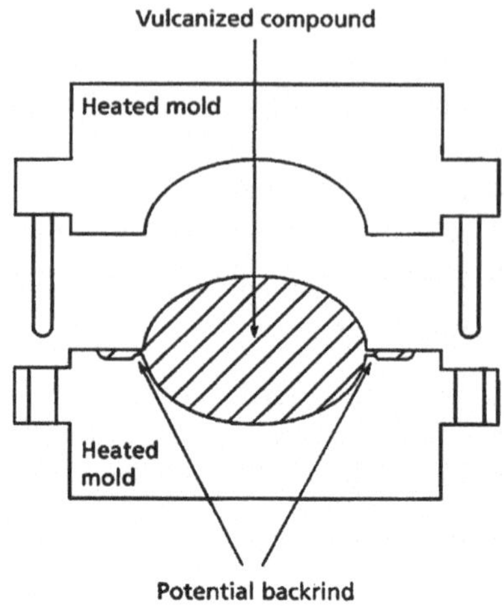

Fig. 13 An open mold after rubber curing

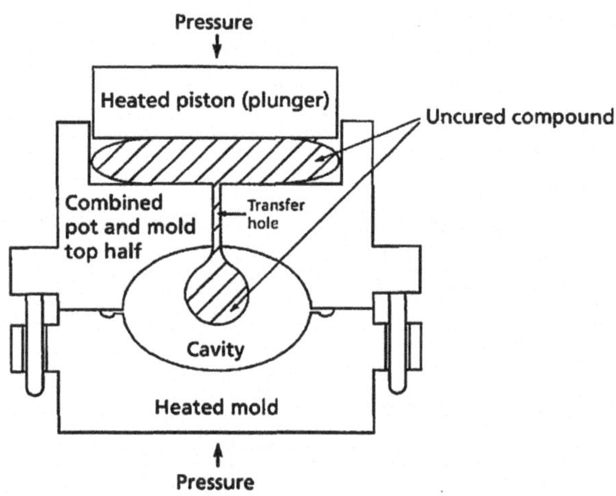

Fig. 14 Conceptual cross section through a transfer mold—compound moving from pot to cavity

strip, or in granulated form through a hopper, as in plastics injection molding. A ram has a tighter fit in the barrel than a screw and therefore there is less leakage backwards through the barrel; it is also cheaper than a screw. The screw 'mixes' the compound as it moves towards the nozzle, creating more frictional heat and therefore higher temperatures which translate to easier flow and shorter cure times. A combination of ram and screw is popular.

- **Autoclave curing**

An autoclave is a cylindrical steel pressure vessel, used to cure extrusions, sheeting and all manner of hand fabricated parts. They can be very small or huge, for example, 30 m long and 3 m or more in diameter. The heat needed to cure is commonly provided by wet steam, often at 0.3 MPa, although some curing

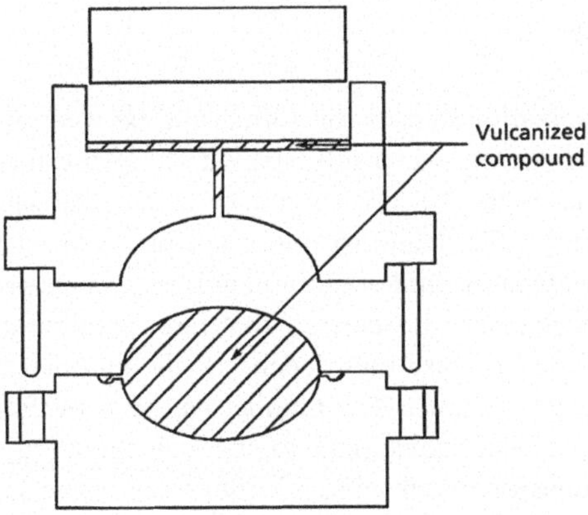

Fig. 15 Conceptual cross section through a transfer mold—transfer mold opened after cure. Since rubber can be considered a thixotropic non-Newtonian fluid, the shear between it and the walls of the transfer holes reduces its viscosity, thus allowing the compound to enter the mold cavity more easily. Shear also heats the compound which reduces viscosity and speeds up cure

processes might need pressures of 0.7 MPa or more. In special cases carbon dioxide or nitrogen might be used, either separately or in combination with wet steam, to provide higher pressures than the wet steam alone could produce at a given temperature. The pressure of wet steam is restrained by its temperature/pressure equivalence. Choosing 150 °C as the desired cure temperature would only generate 0.3 MPa from the wet steam. In cases where porosity in the product is a problem it might be advantageous to independently increase the pressure in the autoclave. If enough nitrogen is introduced to give an extra 0.6 MPa pressure then there is now a total of 0.9 MPa, which might significantly reduce any porosity problems encountered at the lower pressures.

The preparation of the blend is usually followed by a phase of molding and, subsequently, of vulcanization. The two phases can be simultaneous as in the case of compression molding, which consists in putting a part of the appropriately shaped blend into a mould. The closing of the mould and application of pressure bring about the adaptation of the rubber to the mould, while the elimination of excess material takes place through appropriate holes. Pneumatic tires are shaped and vulcanized by a similar technique. A technology for the shaping of thermoplastic elastomers, but also for vulcanizable elastomers, is that of injection molding, typical of the plastic industry, but adapted also for elastomers; in this case, the molding and vulcanization phases are also usually simultaneous. Instead, in the case of shaping by extrusion, which is used for the preparation of pipes or tapes, the vulcanization phase is separate and is effected continuously in high temperature baths (200–300 °C), in hot air tunnels or by means of continuous hot pressing systems (heated cylinders).

4.6 Vulcanization

The process of vulcanization causes the formation, between the macromolecules, of bonds that are statistically randomly spaced along the molecular axis with a frequency of one link every 50–100 monomer units. The introduction of bonds between the macromolecules leads to important physical variations of the elastomer, which changes from a fluid soluble in solvents to an insoluble elastic solid characterized by mechanical properties that are technologically useful. The properties of the vulcanized elastomer depend on the number and type of the links that connect the molecular chains. The number and type of the links are, in turn, a function of the degree of advancement of the vulcanization and of the type of accelerator. As the density of the network increases, there is an increase in the static and the dynamic elastic moduli, at high frequency, and in hardness, while the permanent deformation values after compression decrease. The tensile strength, the tear strength and the fatigue resistance reach their maximum. The nature and the characteristics of the intermolecular bonds, as well as their number, have a great influence on the final properties of the vulcanized elastomer: short bonds, the thermal stability and the dynamic properties; improve tensile strength, fatigue resistance and tear strength. During vulcanization with sulphur, an evolution of the type and number of intermolecular bonds is observed, with consequent variations over time of the properties of the vulcanizate: polysulphide bonds steadily decrease as the vulcanization time increases, the monosulphide bonds increase and the disulphides remain constant (about 20 %). As vulcanization proceeds, the viscosity of the material increases and this increase is measured by the rise in the torque needed to keep the amplitude of the oscillation constant. Since the measurement is carried out at high temperature, it is assumed that the viscous effect of the material is negligible and that the increase of the torque is proportional to the number of links per unit volume of rubber. Figure 16 contains a typical vulcanization curve, which shows the trend of the torque as a function of time.

The first period (induction), in which the viscosity does not change (the retardant systems that inhibit the formation of the activator-accelerator complex are active) is followed by a period of torque rise and, subsequently, of stabilization of the torque value. A decrease in this value is symptomatic of breakage of the intermolecular bonds caused by the temperature (reversion), while an increase is linked to a further cross-linking of the material.

4.6.1 Accelerated Sulphur Systems

For many years, the adjustment of the scorch time was carried out by means of salicylic or benzoic acid or N-nitrosodiphenylamine until the first vulcanization inhibitors were introduced in 1968: they represent an important class of substances that make it possible to reduce the risk of prevulcanization without substantially altering the rate of the process. The principal and most common one is N-(cyclohexylthio)phthalimide (CTP), which is used in quantities of the order of 0.1–0.3 phr.

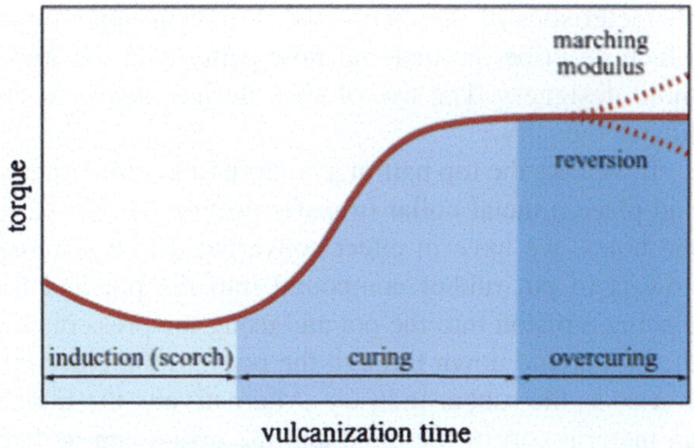

Fig. 16 Vulcanization plot for an elastomer

A mold might be described simplistically as at least two pieces of material, which when fitted together form a cavity, resembling the shape of the product. This would be a very basic mold. Molding is by far the most important curing process, where uncross-linked rubber is placed into a heated mold, which gives it the final product shape, and then vulcanizes the material. It can vary in size from a clenched fist to that of an automobile, and can have a single cavity to make one product at a time, or enough cavities to make a hundred or more.

Most rubber molding is based on introducing a solid compound into a mold, although urethanes and silicones can be introduced as solids or liquids. It takes a fairly high mechanical pressure, to close the mold, and thus form the product shape; this pressure is provided by a press. Thus the mold must be strong enough to avoid being crushed. Tool steel hardened to a Rockwell C hardness of about 60 might be needed. A basic compression mold design is illustrated in Figs. 10–12 which shows a cross section through the center. It is very important that the two halves of the mold register fit accurately together. In this case, pins built into the top section fit snugly into holes drilled into the bottom half. Any looseness between the pin and the hole may cause the top half of the product to be out of alignment with the bottom half. If the fit is too tight, attempts to manually open the mold may prove difficult. Since a number of compound materials expand with heat by at least an order of magnitude more than steel, they will also shrink correspondingly as they cool when taken from the steel mold. The chemist plays his part in achieving a smooth flow of material in the mold, by striving to control the uncured compound viscosity. This needs to be high enough to create the back-pressure required to expel air efficiently as the mold closes, and low enough to permit completion of flow into all parts of the cavity before vulcanization begins. If we look at a low cured-hardness rubber, it usually contains little or no filler (NR & CR), or alternatively fillers plus a large quantity of oil. This can often make its viscosity too low for successful compression molding and the compounder may strive to increase its viscosity, by choosing a raw gum elastomer grade with a high Mooney viscosity. The flow of material in a mold is a complex process, especially in compression molding. The rubber in the cavity is undergoing large temperature changes, which translate to viscosity variations thus continuously altering the flow characteristics of the compound. In recent years finite element analysis packages which describe the material flow patterns in the mold, have become available to mold designers. The use of such design aids is at an early stage in most of the rubber industry.

If we take the top half of a compression mold, then drill transfer holes through it and place a metal collar (transfer pot) on the closed mold so as to surround all of the holes, we have in effect converted it into a transfer mold. All that is needed now is to put rubber compound into the pot and force it through the holes by placing a piston into the pot and using the press platens to force the piston to push the compound down through the pot into the closed mold cavity. This conversion is used in the rubber industry. Alternatively, the transfer pot can be designed to be an integral part of the mold and the piston can be fixed to the upper press platen.

Raw gum elastomer is placed into the gap between the two mill rolls, the mill nip. It then bands, as a continuous sheet, onto one of the rolls. The speeds of the two rolls are often different, the back roll rotating faster than the front. The difference in speed between the two rolls is called the friction ratio and allows a shearing action at the nip to disperse the ingredients and to force the compound to stay on one roll, preferably the front one. A friction ratio of 1.25:l is common. Powders, liquids, etc., are then added to the nip in a specific way. The process produces friction which creates heat. This excess heat needs to be removed, either by spraying or flooding the inside of the roll with cooling water or by passing water through drilled channels in the wall of the roll. A device is necessary to prevent the rubber from moving past the end of the rollers. This is accomplished by a piece of metal called a guide, positioned at each end of the roll, so as to almost touch the surface.

Rubber equipment and its use fall off the mill rolls, so a tray (mill pan) is provided to catch them, to be swept up and returned to the rolls. When all the ingredients are completely blended rotating knives, in the shape of a disc, can be automatically applied to the rubber covered roll (this method can eventually cause scoring of the rolls), to take off one continuous sheet. Alternatively the operator can use a hand-held mill knife, and take off individual sheets. The following description relates primarily to compounds which use sulfur as the crosslinking agent. The key to mixing (in a Banbury mixer or a mill) is to maintain sufficient viscosity to ensure an adequate shearing action, to distribute the non-rubber ingredients into the raw gum elastomer, or to force the raw gum elastomer into the microscopic spaces of each filler particle. Both mechanisms have been hypothesized and one typical mixing sequence might be as follows: The raw gum elastomer is placed into the nip and allowed to band onto the front roll. In the case of NR, it needs to move though the nip quite a few times to reduce its nerve (elasticity) and to lower its high viscosity (low viscosity grades are available). It then forms a smooth, more plastic, band on the roll. Normally most powders (other than accelerators and sometimes sulfur) are then added. If significant heat is produced, then cross-linking agents and accelerator addition will be delayed to the last part of the mixing process. In some cases, when excessive heat is produced, it may be necessary to remove the compound from the mill before the accelerator is added, to avoid scorching (prevulcanization). The compound at this point is known as a *masterbatch*, defined in ASTM D 1566 as a homogeneous mixture of rubber and one or more materials in known proportions for use as a raw material in the preparation of the final compounds. The masterbatch is allowed to cool and subsequently returned to the mill for addition of the accelerator. If the rolls of a mill are twisted to produce a corkscrew effect (they would now be called rotors), and then a block of steel is placed over the mill nip with the block connected to a steel rod above it, this would be called a ram. The ram would move up, to allow addition of ingredients to the nip, and it would move down to force the compound ingredients into the nip. If the whole thing is surrounded in a heavy metal jacket with a chute at the top to put ingredients in and a door at the bottom (underneath the rotors), to let the mixed material out, the result will be an internal mixing machine. The Banbury mixer had modified rotors and the addition of a floating weight. The internal mixer rapidly became an essential part of the rubber industry. At the present time, mixers are available in sizes ranging from those capable of mixing a kg or so, to those that can mix more than 500 kg per load, equivalent to many large mills. The internal mixer is faster, cleaner, (produces less dust from powdery materials such as carbon black, silica and clay), uses less floor space, and is probably less operator sensitive. It has thus displaced the mill for most compounding operations. However, the variable nip opening on a mill, plus immediate visual feedback of the state of the mix, allows a good mill operator a high degree of control and consequently dispersion. The internal mixer has a fast mixing capability, from around two to ten minutes, and thus requires an efficient cooling system. This is provided by drilled channels in the walls of the mixing chamber, through which water passes to control the mix temperature. The rotors and discharge door can also be water cooled. The temperature of the compound being mixed is measured by a thermocouple in the side of the mixing chamber. Other parameters which can be measured and controlled during the mixing process are electrical power (amperage or watts) and time. The mixing principles are similar to those for the mill.

For natural rubber, the time taken to achieve coherence can be somewhat long, due to its high initial viscosity which needs to be reduced by mastication. Sometimes this involves a separate step, where the mixer is exclusively and completely filled with NR raw gum elastomer, which is worked to reduce its elasticity and increase its plasticity. It is then dumped from the mixer, and after resting, a portion of it is returned to be mixed with other compounding ingredients. Controlled, lower viscosity natural rubber is available which can eliminate this extra step. Compounds with larger quantities of reinforcing filler can often reach temperatures over 150 °C by the time they are mixed. They would therefore be dumped from the mixer, often without sulfur and definitely without any accelerators (called a first stage mix or masterbatch), as otherwise the vulcanization process could commence in the mixer. The masterbatch would then be cooled, prior to being returned to the mixer for the addition of these materials, allowing the batch to be dumped at a final temperature closer to 100 °C. Since the heat generated during mixing is often associated with reinforcing fillers, a compound without this raw material (or with some non-reinforcing filler) can reach full mixing (complete dispersion and distribution) at a much lower temperature than a compound with reinforcing fillers. This temperature will generally be comfortably below that needed to initiate vulcanization. Thus it may be feasible to experiment with adding the whole cure system in the first stage, i.e., as a single stage mix. This would be done with due regard to the required compound scorch time and plasticity of the mix for further processing.

4.7 Extruders

Extruders are conceptually a pump, consisting of a screw to move the material forwards, a barrel around the screw to contain the material, help it move, and provide part of the temperature control. The back end has a hopper, sometimes with feed rollers, to put rubber into the screw, and the front end has a 'head' to hold a die, through which the rubber extrudes. An alternative to the screw extruder is the ram extruder, a well known trade name being Barwell. The ram extruder predates the screw extruder, but it is still used in certain specialized applications. Here, the screw is replaced by a ram, which forces the material through the die. Since the process is discontinuous (a slug of rubber is placed in the barrel, extruded, then another slug introduced), it is suited to making preforms for further use, such as placing into the cavities of molds. Thus a rotating blade is fitted in front of the die to chop the extrusion into volumetrically accurate preforms. Extruders are used to make hose and general profiles such as window channels, coated wires, and preforms for further processing. They can also be used to produce sheet rubber, where a large extruder makes a tube, which is immediately slit, producing a continuous sheet.

The die is designed to avoid sudden discontinuities, as the compound moves through it and thus often has a contoured lead (entrance) section. As the extrusion exits the die, the extrusion can shorten in length and increase in cross section. This is known as die swell, which is dependent on die design, screw speed (i.e., shear rate), temperature and the compound's viscosity and its elastic component. In practice, die swell can be quite complex and it might be necessary to modify the die a number of times, before the required extrusion shape is achieved. This recognizes that even uncured rubber has complex elastic and plastic behavior. Like an elastic band it can undergo elastic recovery on exiting the die. A problem with traditional extruders is the potential for reduced interblending of material as it moves along the screw. This causes uneven temperature distribution in the extrudate, which translates to a variable viscosity and therefore a continuously changing die swell. Layers of compound move along without intermingling, i.e., in laminar flow. A relatively recent idea, introduced in the 1970s, is to introduce pins protruding from the inside of the barrel towards the screw. This breaks up the layers, mixes them, reducing thermal variation and increasing homogenization. Such a machine is known as a pin barrel extruder. An extension of this concept is called a Pinconvert extruder. After a conventional pin section, there is a portion which has hydraulically adjustable pins protruding into a helically grooved liner on the inside of the barrel. This gives it a high degree of flexibility for controlling temperature and output. The device describes has an L/D ratio of 8:1, making it quite compact for a cold feed extruder. An alternative concept is to introduce small bowl shaped cavities into the end section of the screw and the inside of the barrel. This creates turbulent flow in the cavities and therefore increased physical and thermal blending. Such an arrangement would be added onto the end of a standard extruder, and is called a cavity transfer mixer.

4.8 Pneumatic Vechicle Tires

The initial patent for pneumatic bicycle tires dates from the middle of the nineteenth century, and had the objective of improving comfort by reducing the shock loads transmitted from the road to the rider, essentially by reducing the radial stiffness of the tire. The solid rubber tire, which carries load by compression, is replaced by an inextensible but flexible shell made of cord-reinforced elastomer. On inflation, the radial stiffness is much lower than the solid tire. It is an example of an invention which had to await the development of new materials to become viable. The concept was independently reinvented 40 years later, and has been much developed in the following 100 years, and applied to a variety of vehicles, ranging from bicycles and wheelbarrows to cars, trucks, aircraft, and earthmoving vehicles. The main requirements of most pneumatic tires are described in the following:

- Tire shape. On inflation, the tire must assume the desired shape without generating excessive internal stress, and any subsequent changes e.g., due to creep, should be minimal. It may be noted that the inflated shape of the tire is determined as much by the construction (the arrangement of the various cord-reinforced layers) as by the profile of the mold in which it was made.
- Structural integrity. Since a burst tire can lead to a major accident, and in some cases fatalities, this is a safety–critical requirement. The tire must survive repeated (several million) cycles of deflection, and in the case of truck tires one or more retreading processes, and continue to withstand the service conditions it is exposed to. These conditions can involve high values of speed, temperature, and deflection, inappropriate inflation pressure (high or low) and road surface roughness, all of which increase the technical demands on the tire.
- Comfort. Although the original objective of pneumatic tires was to reduce radial stiffness to improve comfort, it later became apparent that low longitudinal (i.e., circumferential) stiffness was also desirable. Tire uniformity is also important in minimizing the generation of vibrations.
- Durability. Apart from structural failure, tire life in service is likely to be limited either by penetration of the structure by foreign objects or by abrasion of the tread material. Since tires are relatively expensive to replace, maximizing the tread life is a major objective.

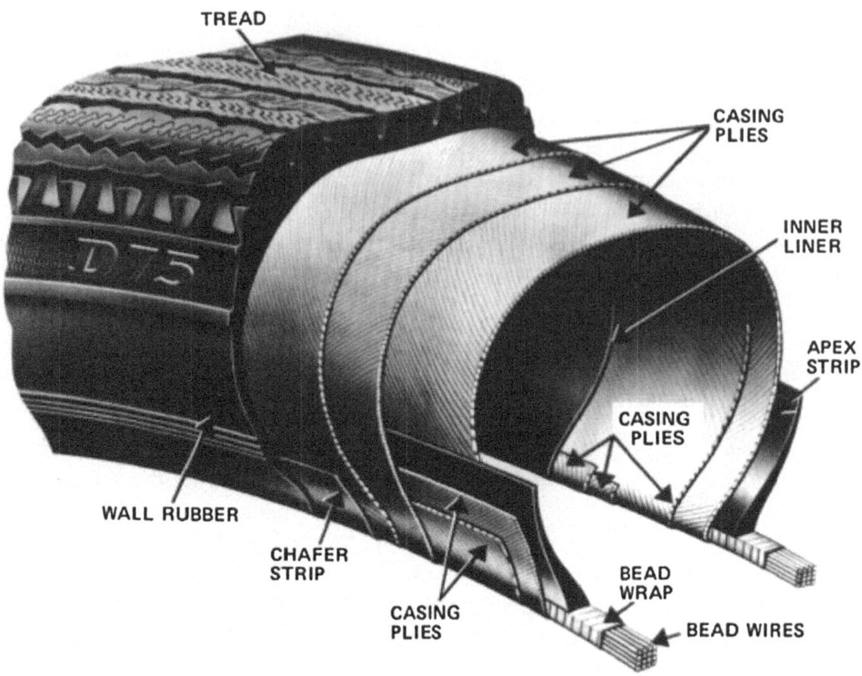

Fig. 17 Various components of tires

- Controllability. During cornering or braking, frictional forces are developed at the road surface, whose limiting values are dependent on the road surface and also the tread pattern and material. These forces distort the structure of the tire, causing internal stresses. It is important that these forces increase progressively with steering or braking input, and that (for example) the slip angle for a required level of cornering force should not be excessive (Fig. 17).

4.8.1 Compact Rubber Extruder with Gear Pump

Gear pumps with herringbone gearing in combination with compact extruders for precision technical profile production. Influenced only by its rpm, the gear pump generates an even, constant volume flow, a pre-condition for optimum production of hi-quality profiles.

- Functional principle. Filling of the tooth gaps, transport in the tooth gaps, discharging of tooth gaps (Fig. 18).
- Extrusion Blow Molding. The melt processible rubber can be processed on a range of extrusion blow molding machines to produce a variety of hollow rubber goods at low unit cost. The Extrusion Blow Molding Process Blow molding enables the production of hollow parts of complex design and variable internal

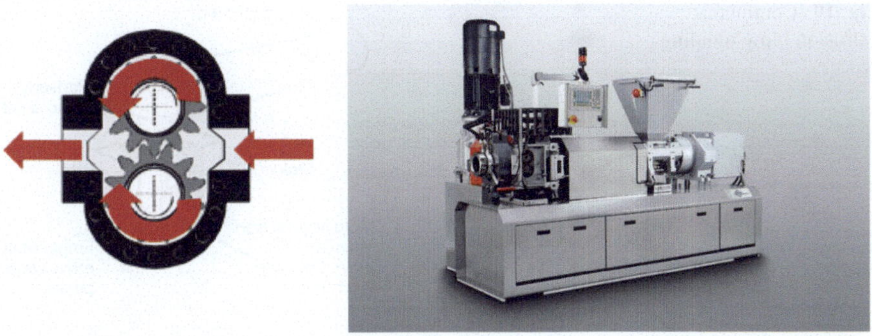

Fig. 18 Compact rubber extruder with gear pump

configuration from thermoplastic resins. The process involves forming a hot tube of molten or softened resin called a parison, and clamping it inside a cooled mold. As the mold closes, the parison is inflated with compressed air until it conforms to the mold cavity configurations. Thus, convolutions, texturing, and other fine details can be produced in the part. The finished part is ejected from the mold when it has developed enough strength (by cooling) to be removed without distortion. Blow-molded parts are prepared commercially by three processes:

- Continuous extrusion blow molding
- Intermittent extrusion blow molding
- Injection blow molding

• Continuous Extrusion Blow. In this process (Fig. 19), the melt is continually fluxed and extruded vertically downward through the parison die. The parison is formed, the mold closes and moves automatically to another station where the parison is blown and the finished piece is cooled and ejected. Meanwhile, an empty mold moves into position beneath the parison die head, a new parison is extruded, and the cycle is repeated. Continuous extrusion blow molding yields high production rates and is suitable for heavy weight parts. It is the preferred process for blow molding, because there is little opportunity for the melt to hold up in the machine and degrade.
• Intermittent Extrusion Blow. In this process the melt is fluxed and delivered to a reservoir, or accumulator. When enough melt has accumulated a ram or plunger forces it through the parison die. When the parison is completely formed, the mold closes and the parison is blown. Meanwhile, the ram retracts and the extruder plasticates more melt and delivers it to the accumulator for the next shot. There are three variations of the intermittent process:

- The accumulator heat system (Fig. 20).
- The ram-accumulator system (Fig. 21).
- The reciprocating screw system (Fig. 22).

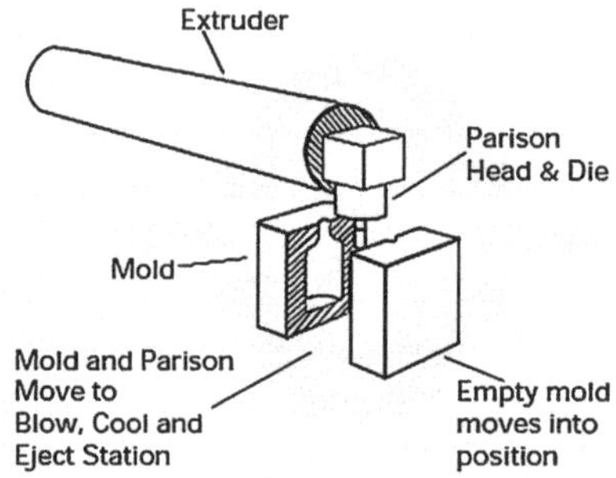

Fig. 19 Continuous extrusion blow molding process

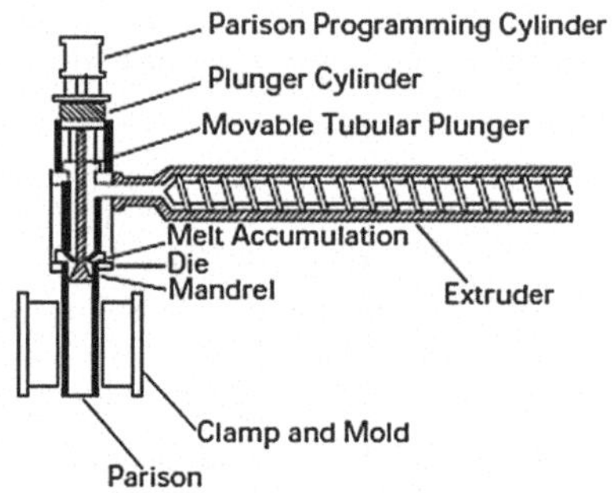

Fig. 20 The Accumulator heat system

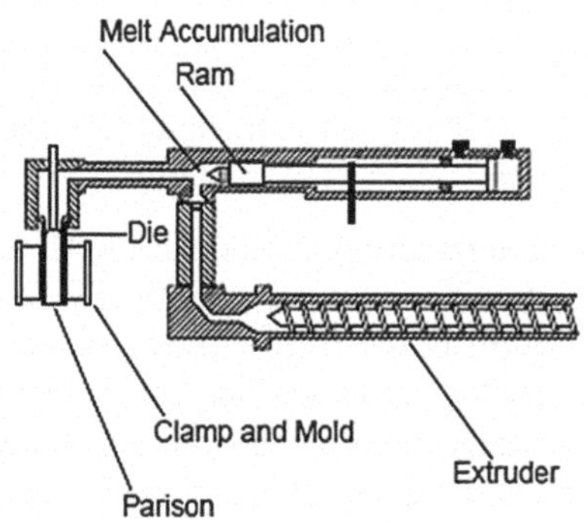

Fig. 21 The Ram-accumulator system

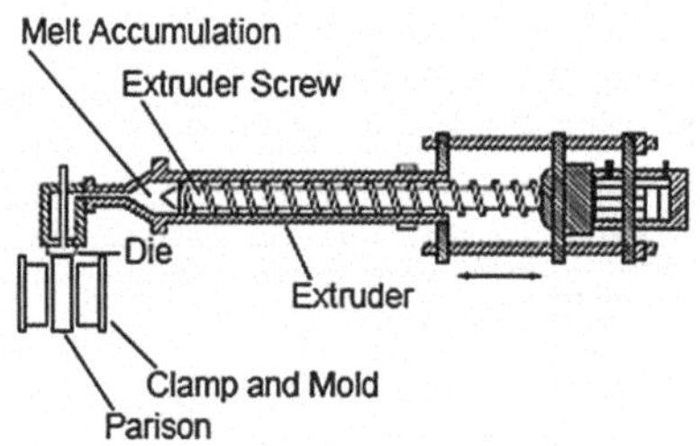

Fig. 22 The Reciprocating screw system

Intermittent extrusion blow molding is suitable for large parts and is economical for short production runs. It is best used with non-heat sensitive resins.

4.9 Injection Molding

The injection molding technique has to meet the increasing demand for a high quality product, which should still be economically priced. This is feasible only if the molder can adequately control the molding process, if the configuration of the part is adapted to the characteristics of the molding material and the respective conversion technique, and a mold is available which satisfies the requirements for reproducible dimensional accuracy and surface quality. Therefore injection molds have to be made with the highest precision. They are expected to provide reliable and fully repeatable function in spite of being under extreme loads during the molding process, and a long service life to offset the high capital investment. Injection molding is the most widespread technique for making 3-D configurations. In the simplest although most frequent case, the injection mold consists of two halves, which are directly mounted to the plates of a molding machine. These two basic elements, the stationary injection and the movable ejection half can be found in every mold regardless of its design. They are called the male and female half. Material under high pressure is forced to flow through a distribution system in the mold to the mold cavity. Pressure is applied until the mold is full and the part has cooled enough to prevent warpage. The pressure may be built up by an extruder equipped with a ram to force the polymer melt into the mold at the appropriate time in the cycle. The gate controls the flow of polymer melt into the cavity and prevents backflow of melt out of the cavity as the pressure is released and the melt is cooling. The runners distribute the melt to several different cavities and must be balanced to prevent some cavities from not getting enough melt to fill them. The factors, which dictate the dimensions of sprue, runner, and gate systems for injection molds, namely, pressure drop, temperature rise, and injection rate, are interrelated. The calculation of these dimensions is difficult, due to the need to solve coupled flow and heat transfer equations, and most mold manufacturers adopt a past experience approach. However, valuable guidance can be obtained from simplified flow analyses. In addition to the pressure drops and temperature rises in channels of constant or smoothly changing cross section, the transitions between them must also be taken into consideration, particularly when the change in cross section is considerable, as in the case for runners and gates. In the scheme of mold design, it is necessary to make an initial estimate of the number of cavities which may be included in the mold, the layout of the sprue and runner system required to feed them, and the volumetric flow rate for a desired injection time. The length of the flow path to each cavity should be the same wherever possible, to give an equal and simultaneous fill in each cavity. After setting the preceding parameters an initial selection of the sprue and runner cross sections is required. The dimensions of the nozzle and gates should also be included at this stage, since they are important parts of the total flow path during injection. The gate type and cross section is often fixed by the requirements of the product. The die, the male and female half, is split to allow removal of the product. It must be kept firmly shut during injection, with the aid of a large hydraulic cylinder, or hydraulically actuated mechanical clamps, or a mechanical clamp combined with a shortstroke hydraulic cylinder. The clamping force is calculated from the projected area of the moldings and the recommended injection pressure. Shrinkage is the term used to describe the difference in the dimensions of the mold and the article produced from it, when both are measured at ambient temperature. Shrinkage factors must be determined for individual compounds if strict dimensional accuracy is required, and can be obtained by curing a standard test specimen at production temperatures. In general, linear shrinkage figures fall within the range 1.5–3.0 %, depending on polymer type and filler loading. Molding is the operation of shaping and vulcanizing the plastic rubber compound, by means of heat and pressure, in a mold of appropriate form. Fundamentally, all processes of molding are similar, the ways of introducing the material into the mold distinguishes one technique from another. The basic processes are compression, transfer, and injection molding.

In compression molding, chunks or sheets of material are placed in an open mold, which is then closed around it. After the part is formed, the mold is opened and the part removed. In injection molding, material is injected under pressure into a closed mold. After the part is formed and vulcanized, the mold is again opened and the part is removed. Transfer molding represents a combination of injection and compression molding in which rubber is injected from a reservoir into a compression mold. The main feature of injection molding, in which it differs from compression molding, is in the presentation of rubber to the mold at or near molding temperature. A large part of the cure time necessary to allow the whole of the rubber mass in the cavity to reach the vulcanizing temperature by conduction and convection, as in the case of compression molding, is eliminated. With compression and transfer molding, it is necessary to place a rubber preform into a compression mold cavity or into a transfer pot. A strip of rubber, or granulated rubber, automatically supplies an injection molding machine. Another difference is that systems for injection molding are much more complex than those for compression or transfer molding. There are several controls to adjust temperature, pressure and other variables during injection molding. These controls are not normally a part of compression or transfer molding systems. With compression and transfer molding, presses provide a force to close a mold. In injection molding, a press is referred to as a clamp. A clamp is an integral part of an injection molding machine. Injection molds are normally attached to the clamp and thus open and close with the clamp. One reason for attachment between mold and clamp is the need for accurate alignment between an injection molding machine and its mold. Injection molds must be capable of withstanding extremely high pressures without mold distortion. The pressure reached in an injection mold is about ten times greater than that for compression and transfer, i.e. injection pressures up to about 200 MPa (29,000 psi) are used. Therefore, high quality steel such as 4140 hardened to a Rockwell C of 28–32, is recommended for injection molds. The mold plates should be thick enough to resist bending at the high pressures encountered during injection molding. Injection molding can be combined with compression or transfer molding. With injection/compression molding, an injection machine pumps hot rubber into an open compression mold. Then the mold is closed, generally using much less force to close it than is normally used in injection molding. This method was developed to protect shoe uppers from being torn apart or wrinkled by the strong forces typically associated with injection molding. With injection/transfer molding, hot rubber from an injection nozzle is placed in the pot of a transfer mold. Then a plunger transfers hot rubber from the pot to the mold cavities. With both injection/compression and injection/transfer molding, cycle times are shorter than for regular compression or transfer molding. Rubber entering the mold is preheated and this rubber therefore crosslinks in a shorter time.

- **Types of injaction molding machines**

The types of injection machinery available differ basically in the method of heating and pre plasticisation of the compounded rubber. The main difference usually arises between vertically oriented machines, which include both ram and screw-ram presses, and horizontally oriented machines, which are of the reciprocating screw type. In the former the mold temperature is maintained by conduction from the heated platens of the press, whereas the heaters are integral with the mold in the latter case. The configuration of molds for horizontal machines often result in large surface areas for potential convective heat loss and long paths for conductive heat flow from the press platens. The horizontal machine generates its own preplasticising heat by work on the rubber, which is fed into it in strip form, between the screw and the barrel. This heat is governed by a temperature-controlled fluid medium (water, glycol, or oil), which circulates around the barrel and, in larger machines, through the center of the screw. Progression of the rubber up the screw both heats and preplasticises it. The rubber, collected in front of the screw, pushes the screw back until a trip switch is operated, and the preplasticisation phase is completed. Final heating of the stock takes place during passage through a small die into the runner system of the mold. The second type of injection press uses the principle of vertical injection. This type of machine has a separate screw and injection piston in a V-head configuration situated near a vertical upstroking press. The preplasticising screw feeds into the injection chamber through a non-return valve where the compound is injected into the mold. There are many independently adjustable parameters of both the injection head and press unit. Molds for vertically oriented machines are usually larger in plan area, giving improved conductive heating, and also expose smaller areas for convective heat loss. A horizontal injection machine is shown in Fig. 23.

An injection mold consists of a cylinder (injection barrel) with a ram or screw inside it, so that the rubber compound can be moved towards a nozzle at its end. The nozzle is then pressed against a hole made in the top half of a closed mold. This hole is then connected to smaller holes (gates and runners) which enter the cavities of the mold. The compound can be presented to the barrel as a continuous strip, or in granulated form through a hopper, as in plastics injection molding. A ram has a tighter fit in the barrel than a screw and therefore there is less leakage backwards through the barrel; it is also cheaper than a screw. The screw 'mixes' the compound as it moves towards the nozzle, creating more frictional heat and therefore higher temperatures which translate to easier flow and shorter cure times.

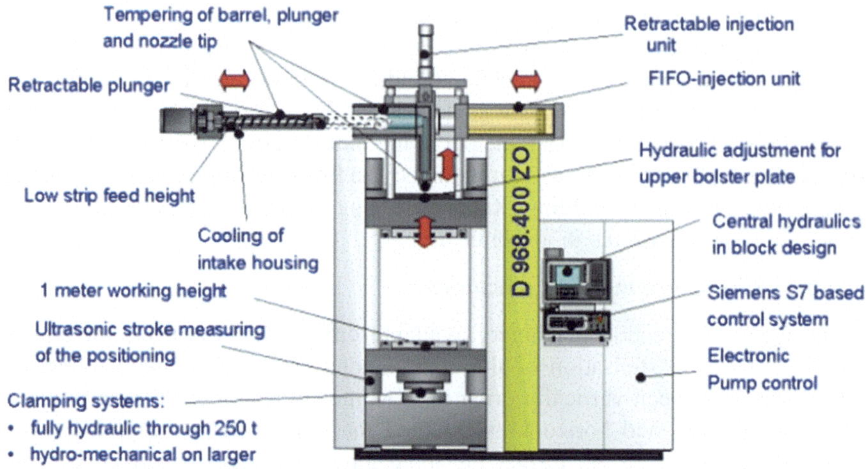

Fig. 23 Horizontal injection machine

5 Characterization of Elastomer Macrocomposites

Early structural characterization of polymers focused on solution properties and their relationship to molecular weight. Subsequently spectroscopic and chromatographic techniques were developed, and reviews are widely available. The chemical *elemental analysis* of polymers can often be carried out by the methods used for low molecular weight organic compounds [72–88]. This is particularly true when combustion of the sample is involved. Thus, C, H, and N can be determined on milligram samples by complete combustion followed by gas chromatographic analysis of the gases evolved. Accuracy is about 0.3 %. Sulfur and halogens are also easily determined after combustion, by titration of sulfate or SO_2 for S, and by potentiometric titration with $AgNO_3$ for halogens after treatment of the gases with NaOH and hydrazine sulfate, for example. Quantities as low as ppm metals can be determined quantitatively and quickly by x-ray techniques. For basic information, *infrared spectroscopy* (invariably Fourier transform infrared [FTIR]) is a straightforward technique, accessible to the nonspecialist. Thin films of elastomers can be measured directly, and obtained by casting or molding 20–30 mg of sample between polyester or aluminum foil (preferably Teflon-coated). Care must be taken to minimize oxidation if the molding is done at elevated temperature. Most analyses make use on the mid-infrared region at 4000 to 400 cm^{-1}, with sample identification made through comparison, using widely available spectral libraries. If the polymer is crosslinked, sample-forming is more restricted. Raman spectroscopy yields analogous information, but is complementary to infrared absorption, in that vibrations which are infrared inactive are generally Raman active, and vice versa. For example, carbon–sulfur bonds are easily detectable via Raman measurements. Since the detected light is scattered from the sample, spectra are readily obtained on crosslinked specimens.

Interferences due to fluorescence are avoided by using a longer wavelength Source Pyrolysis of samples can lead to the production of characteristic fragments, which may be analyzed by gas chromatography (GC) or mass spectrometry (MS). Since the relationship between fragments and the original polymer is often complex, this technique is a last recourse, for insoluble polymers or samples not amenable to more facile and reliable characterization methods. Combined GC/MS has been used to analyze the volatile components in natural rubber. In ozonolysis, an unsaturated sample is reacted to form an unstable intermediate, ozonide, which is then further reacted for chemical identification. Ozonolysis of rubber is usually combined with GC analysis. In secondary ion mass spectrometry (SIMS), the sample surface is irradiated with an ion beam, followed by mass spectrometry of the emitted secondary ions. SIMS has found various applications in rubber, including surface analysis and studies of carbon black interaction. Pyrolysis can also serve as a fingerprinting technique for routine analyses. In thermogravimetric analysis (TGA), the polymer degradation by-products volatize, whereby the residue provides a measure of the carbon black or other filler content.

Electron paramagnetic resonance (EPR), or electron spin resonance (ESR), can be used to detect types and quantities of free radicals. Such information is of value in studying the chemistry occurring during degradation and fracture of polymeric materials. EPR can also be applied to study carbon black and other fillers in polymers. The technique of temperature rising elution fractionation (TREF) has been developed to measure the compositional distribution of semicrystalline polymers. Polymer is dissolved off a substrate as temperature is raised through the melting region, so that discrimination is based on differences in crystallizability of the fractions. A similar method uses supercritical fluids. TREF can also provide information about the sequence distribution, since longer sequences of a monomer unit are more crystallizable. Different arrangements of the monomer units give rise to different chemical shifts and scalar couplings (splittings) in the NMR spectra. Using selection rules and empirical knowledge of chemical shifts, chemical structures can be assigned. Since chemical shifts in 13C NMR spectra are larger than in proton spectra, subtle structural differences can be seen for carbon atoms separated by up to five bonds from the point of reference. The viscoelastic response of amorphous polymers at elevated temperatures is governed to a significant extent by the average molecular weight, Mw, the presence of any long chain branching, and the MWD. The chain length distribution is usually presented as a plot of the mole fraction or weight average of molecules versus molecular weight. The various average molecular weights represent the moments of the chain length distribution. For low molecular weight samples, the preferred method is by vapor pressure osmometry (VPO). This technique is based on the decrease of vapor pressure of a solvent due to the presence of dissolved polymer. The different equilibrium vapor pressures cause a difference in condensation rate on two matched thermistors, contained in a chamber saturated with solvent vapor. One thermistor is coated with solvent and the other with a solution of the polymer. Membrane osmometry relies on the lowering of the activity (free energy) of the solvent by dissolution of a solute, to yield an direct determination of Mn.

When a polymer solution is brought in contact with pure solvent, the concentration gradient induces mixing by diffusion. If a semipermeable membrane is placed between the pure solvent and solution, the polymer is trapped but solvent can pass. Light scattering is another absolute technique for the determination of molecular weights (>103 g/mol). (We are only concerned with static light scattering for structural information; however, inelastic light scattering is a powerful technique for studying polymer dynamics. Structural information can also be obtained from measurements of small angle neutron scattering (SANS). Scattering of neutrons is due to their interaction with nuclei. It differs from light scattering, in that contrast arises from differences in neutron scattering length, rather than refractive index differences. Neutron cross-sections are commonly expressed in terms of the relevant correlation function. Elastic, coherent scattering is proportional to the spatial Fourier transform of the pair-correlation function. The classical method of solvent-nonsolvent fractionation according to MWD and compositional distribution depends on solubility differences among the various species. The method is empirical and tedious, involving characterization of phase-separated "cuts" as they are recovered. The term field flow fractionation (FFF) refers to a family of one-phase chromatographic techniques, carried out in thin flow channels. In principle, FFF yield absolute molecular weights, although in practice calibrations similar to SEC are used. Long chain branching (LCB), defined as branches having molecular weights of at least a few times the entanglement molecular weight, is common in rubbers. Its most important effect is increasing the viscosity; LCB is present in some commercial rubbers in order to reduce cold flow (i.e., the room temperature creep of rubber during storage). Substantial LCB may require use of a low *Mn* polymer, in order to retain a viscosity low enough for processing; however, the consequent plethora of chain ends may entail sacrifice of cured properties, especially those relating to heat buildup or strength. Branching affects other properties, affording a means to characterize the degree of branching.

With light scattering measurement of the radius of gyration, direct comparisons of g values can be made. In Q solvents, the experimental g values generally agree with calculations, with some notable exceptions. Advanced SEC analytical techniques take advantage of online light scattering and viscometry [IV A]. With the SEC calibrated for the product of [h] times M versus elution time for a particular polymer species, together with [h] Br or MW Br measurements, one can calculate g factors as the ratio of the measured intrinsic viscosity to the value of [h] calculated for the corresponding linear polymer (i.e., linear polymer that would have eluted at the same time). When a measurement of polymer dynamics is made (using, for example, mechanical or dielectric spectroscopy, or dynamic light or neutron scattering), a maximum in the susceptibility (absorption peak) is observed for molecular motions transpiring on the timescale of the experimental variable (e.g., frequency of the applied stress or voltage). In the limit of small perturbations (for which the material response is linear), the relaxation directly reflects the equilibrium Brownian motion. Such correspondence follows from fluctuation dissipation theory, originally developed to explain Johnson noise in electrical conductors. The gigantic size of polymer molecules provides for an enormous number of degrees of freedom, and thereby motion encompassing many decades of time overly a broad range of length scales.

The local segmental dynamics are associated with the rubber-glass transition, occurring at Tg. It is only in the glass transition zone of the viscoelastic spectrum that both chain modes and local segmental modes can be measured simultaneously at the same temperature, using conventional techniques. Since the chain modes and the local segmental dynamics have different temperature dependences, a breakdown of the time–temperature superposition principle is observed in the glass transition zone. Thus, master curves for the chain dynamics can be constructed, which extend from the end of the softening zone through the terminal relaxation, and master curves of the local dynamics are possible. However, in the softening zone, the shape of the viscoelastic spectrum changes with temperature (Fig. 24).

The most common means to assess Tg is via heat capacity (Cp) measurements, usually carried out in a differential scanning calorimeters (DSC). The temperature of the sample changes at a fixed rate, with the heat flow monitored. An older, and less useful, variation is differential thermal analysis (DTA), in which heat flow is programmed, and the consequent rate of temperature change is measured. ge in heat capacity. This is nearly equal to an alternative definition, the temperature of the intersection of the extrapolated baseline with the tangent of the maximum slope (Fig. 25).

Analyzing the morphology of an elastomer includes not only the characterization of the molecular structure of the polymer itself and of the compounding ingredients, but may also extend to the super molecular scale. Rubber elasticity arises from the orientation of chain segments, and the degree of this orientation underlies the mechanical properties. The most facile way to quantify the orientation is from the (optical) birefringence, defined as the difference in refractive indices for two perpendicular directions. When a material is semicrystalline, the observed birefringence is the sum of the Dn from the oriented amorphous phase and the crystalline regions. There is also a contribution due to form birefringence, arising from distortion of the light wave transverse the refractive index boundary between the amorphous and crystalline phases. Another method to characterize orientation is deuterium NMR. Deuterons have a spin quantum number equal to unity and therefore a nuclear quadrupole moment. Coupling between the quadrupole moment and the two possible NMR transitions (resulting from the three Zeeman levels of the spin = 1 nucleus) yields differences in the energy levels of the two transitions, giving rise to a doublet. The separation of this doublet is proportional to $f_2(f)$, which depends on the orientation of the chemical bond axis of the deuteron relative to the applied external magnetic field; thus, the magnitude of the observed splitting yields directly the orientation. Although many rubbers can be mixed to form an ostensibly homogeneous blend, the overwhelming majority of such mixtures have phase-separated morphologies. To form a useful elastomer, the components need only be sufficiently compatible that a satisfactory dispersion is attained. Actual thermodynamic miscibility, implying segmental mixing of the components, is rare, although it is important to recognize when is achieved. The phase size in a heterogeneous blend will depend on the compatibility of the

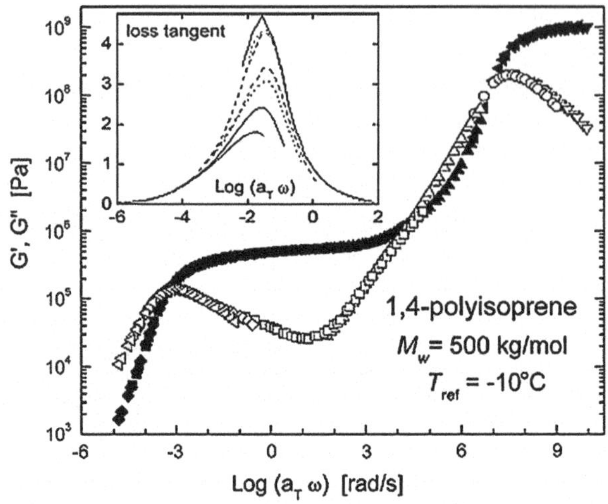

Fig. 24 Apparent master curves (reference temperature = −10 °C) for the storage (solid symbols) and loss (hollow symbols) moduli of cis-1,4-polyisoprene (Mw = 500,000 g/mol). The breakdown of time-temperature superpositioning, barely evident in the softening zone, is seen clearly in the loss tangent peak, shown in the *inset* for temperatures from −66 °C to −48 °C

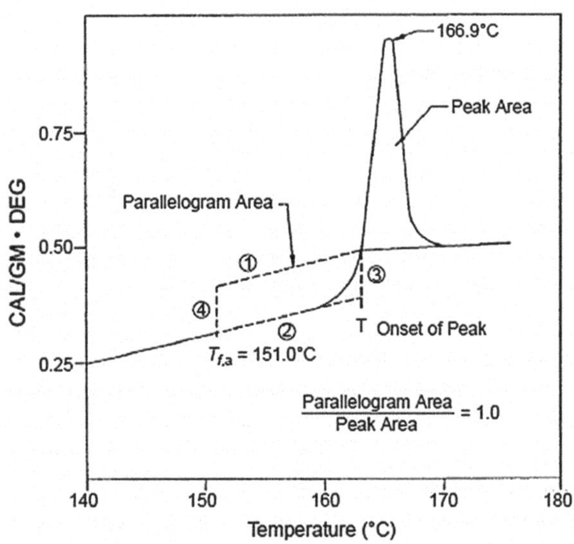

Fig. 25 DSC of a cured epoxy resin. To determine the fictive temperature, a parallelogram is constructed to the low temperature side of the enthalpy peak, with horizontal sides parallel to the measured heat capacity. The onset of the peak defines the high temperature boundary, while the low temperature boundary (i.e., T_f) is chosen such that the area of the parallelogram equals the peak area

components and, to a lesser extent, on the mixing conditions. The simplest method of determining whether the phase morphology is homogeneous is by calorimetry. The observation of two transitions, corresponding to the respective Tg of each component, indicates a phase-separated morphology. However, a single transition does not guarantee thermodynamic miscibility, especially if the component Tg's are close. NMR of swollen rubber has been used to determine crosslink distributions in blends. Swelling enhances chain mobility, and the isotropic motion of nuclei averages local fields, thereby narrowing the spectral lines. This allows individual resonances to be characterized. Rubbery behavior-large, reversible extensibility—implies an absence of crystallinity, and this is usually the case for undeformed elastomers. However, small extents of crystallization may be present at ambient temperature in some elastomers, including EPDM with high ethylene content, epichlorohydrin rubber, and polypropylene oxide rubber. The crystallites in these materials can act as reinforcing agents. Many thermoplastic elastomers have crystalline domains which function as reversible crosslinks It is well known that elastomers, like virtually all solid materials, have preexisting, "naturally occurring", flaws. By intensifying local stresses, such flaws exert an influence on the failure properties of elastomers. More recently, interest in these flaws has increased, due to concerns about their potential for reducing the barrier performance of rubber films. The elastic behavior of rubber for large strains reflects the effect of topological interactions known as entanglements. Entanglements constrain the chains, suppressing lateral motions. The pseudo-network of entanglements gives rise to the characteristic plateau in the time-dependence of the mechanical response of uncrosslinked rubber. While the length (extent over time or frequency) of the rubbery plateau is determined by the molecular weight, its height, G_0N, reflects the concentration and effectiveness of the entanglements. The storage modulus varies only weakly with frequency, and is approximately proportional to the entanglement concentration. Such proportionality in the melt is purely entropic, and is not affected by energy differences between the conformers. The molecular weight between entanglements, depends on chemical structure, and thus is characteristic of the polymer species. The entanglement interactions govern the rheology of uncrosslinked polymers, influencing the viscosity, the dynamic modulus, and the recoverable compliance.

6 Applications of Elastomer Macrocomposites

6.1 Tire Manufacturing Process

A tire is a black donut made of rubber, placed on a vehicle so the driver can transport himself and his cargo from point A to point B. The tire has to allow for easy steering, braking and cornering. It must provide for a comfortable, safe ride. It needs to be durable. Actually, a tire is an advanced engineering product made of a lot more than rubber. Fiber, textile, and steel cord are just some of the components that go into the tire's innerliner, body plies, bead assembly, belts, sidewalls, and tread. As you can imagine, the manufacture of this complex product is, well, complex. It requires the latest technology, heavy equipment, precision instruments and-most importantly-qualified people. Some of the activities that go on in the tire factory are the mixing of the rubber compound; preparation of the fabric cord, steel cord, and bead wire; "calendering" of the innerliner, steel belt and ply cord;

extrusion, or shaping, of the tire's sidewall and tread; and the actual building, curing, and inspection of the tires. Read on for a more detailed explanation and refer to the manufacturing flowchart for a visual reference to each process [89].

The two major ingredients in a rubber compound are the rubber itself and the filler, combined in such a way as to achieve different objectives. Depending on the intended use of the tire, the objective may be to optimize performance, to maximize traction in both wet and dry conditions, or to achieve superior rolling resistance. The desired objective can be achieved through the careful selection of one or more types of rubber, along with the type and amount of filler to blend with the rubber. In general, there are four major rubbers used: natural rubber, styrene-butadiene rubber (SBR), polybutadiene rubber (BR), and butyl rubber, along with halogenated butyl rubber. The first three are primarily used as tread and sidewall compounds, while butyl rubber and halogenated butyl rubber are primarily used for the innerliner, or the inside portion that holds the compressed air inside the tire [90, 91].

The most popular fillers are carbon black and silica, and there are several types of each. The selection depends on the performance requirements, as they are different for the tread, sidewall, and apex. Other ingredients also come into play to aid in the processing of the tire or to function as anti-oxidants, anti-ozonants, and anti-aging agents. In addition, the "cure package"—a combination of curatives and accelerators—is used to form the tire and give it its elasticity. Once the compound is determined, the next challenge is how to mix it all together. The mixing operation is typically a batch operation, with each batch producing more than 200 kg of rubber compound in less than 3–5 min. The mixer is a sophisticated piece of heavy equipment with a mixing chamber that has rotors inside. Its main function is to break down the rubber bale, fillers, and chemicals and mix them with other ingredients. The sequence in which the ingredients are added is critical, as is the mixing temperature, which can rise as high as 160–170 °C. The compound can be damaged, if the temperature is too high, so the mixing operation is typically accomplished in two stages. The curative package is normally added in the final stage of mixing, and the final mixing temperature cannot exceed 100–110 °C or scorching may occur [92, 93].

Once the mixing is completed, the batch is dumped out of the mixer and sent through a series of machines to form it into a continuous sheet called a "slap." The slap is then transferred to other areas for bead wire assembly preparation, innerliner calendering, steel and/or fabric belt/ply cord calendering, tire sidewall extrusion, and tire tread extrusion. Because tires have to carry heavy loads, steel and fabric cords are used in the construction to reinforce the rubber compound and provide strength. Among those materials suitable for the tire application: cotton, rayon, polyester, steel, fiberglass, and Aramid [94, 95].

- Fabric cord. Fabric cord quality is based on its strength, stretch, shrinkage, and elasticity. The yarn used is first twisted, and then two or more spools of yarn are twisted into a cord. Before shipping the cord to the tire factory, the manufacturer pre-treats the cord and applies an adhesive to promote good bonding with the rubber. The temperature, humidity, and tension control are critical before the fabric cords are calendared with rubber compound. For this reason, fabric cord is kept in a temperature-and-humidity-controlled room once it arrives at the factory.

- Steel cord. Steel wire cord quality is based on tensile strength, elongation, and stiffness. It is manufactured from steel rod with high carbon content; and while the steel wires used have different configurations, all are brass-coated strands twisted together into cords. If the wire is used in a multi-ply tire rather than a belted tire, the fatigue performance will be important. If used in belted tires, then stiffness is of primary concern. Since the steel wire is brass coated, storage conditions are important to maintain the steel wire to rubber bonding properties. Therefore, the steel wires are also kept in a temperature and humidity controlled room once they arrive at the factory.

To produce fabric or steel belts, the fabric or steel cord must go through a calendering process—an operation in which the rubber compound is pressed on and into cords. Because the bonding of fabric to rubber or steel to rubber is critical to performance, the calendering process is an important step. The calendar is a heavy-duty machine equipped with three or more chrome-plated steel rolls which revolve in opposite directions. The roller temperature is controlled via steam and water. In this process, the rubber compound is applied to the cords [96, 97].

First, a pre-set number of fabric or steel cords under proper tension are continuously pressed through two steel rollers, and rubber compound is added to the opening area between the rollers. Then the rubber compound is pressed into, on top of and on the bottom of the fabric or steel cords. A continuous sheet of cord-rubber composite goes through several more rollers to ensure good penetration and bonding between the rubber and cords. Quality is measured by the thickness of the sheet, spacing between cords, the number of cords and the penetration of rubber into the composite sheet. The composite sheet is then cut into appropriate sizes, shapes, and angles depending on the desired contour of the tire.

Finally, the tire is ready to be built by a highly robotized machine which ensures quality and efficiency. All components-bead assemblies, calendared plies, belts and innerliner, tread and sidewall sections—are assembled and the building process begins. A typical radial tire is built on a flat drum in a two-stage process. In the first stage, the innerliner is wrapped around a drum and the first body ply is wrapped on top, followed by the second body ply. The bead assemblies are then positioned, and a bladder on the drum is inflated and pushed in from both ends of the drum, forcing the body plies to turn up to cover the bead assemblies. The sidewall sections then are pressed onto both sides. In the second stage of the tire building process, another machine is used to apply the belts, nylon cap, and tread on top of the first stage. At this point, the tire still needs curing because there is no tread pattern on it. A tire is essentially a cord–rubber composite. Tires have plies of reinforcing cords extending transversely from bead to bead, on top of which is a belt located below the tread. The belt cords have low extensibility and are made of steel and fabric depending on the tire application. The belt cords are at a relatively low angle, between 12° and 25°, and serve as restrictions to the 90° casing plies.

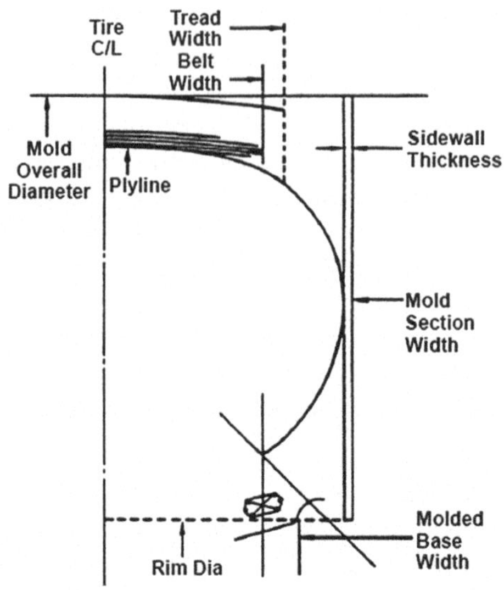

Fig. 26 Ply line boundaries of the tire

Tire mold design initially begins with determination of the inflated dimensions of the required tire size. By use of inflated tire and growth characteristics of the tire, preliminary ply line and mold dimensions are computed (Fig. 26). Once the mold boundary dimensions, location of the ply line, and tread width and depth are known, the contours of the tread, shoulder, sidewall, and bead components can be established. These dimensions and contours are developed using computer-aided engineering techniques Structural mechanical calculations such as finite-element analysis are used to analyze both the inflated and loaded deflected shapes of a tire cross-section and the resulting stress–strain relationships in the belt area. Such studies permit both quantitative analysis and qualitative comparisons of the range of belt configuration options. Figure 27 shows a heavy-duty truck tire in the loaded and unloaded states. The density of grids is designed so as to preserve the essential features of the tire cross-section geometry while maintaining the total number of grid points.

Steel wire used in tires are of various configurations, but all are brass-coated wire strands wrapped together to give cords of different characteristics, depending on the application. Steel tire cord is manufactured from high-carbon-steel rod which is first drawn down to a diameter of approximately 1.2 mm. A brass plating is then added to the wire before a final drawing to 0.15–0.40 mm. These filaments are next stranded to form a cord construction which is designed and optimized for a specific service requirement. Steel tire cord is manufactured from high-quality steel which is necessary because of the performance demands to which tires are subjected. The composition of a typical steel cord is illustrated in Table 1.

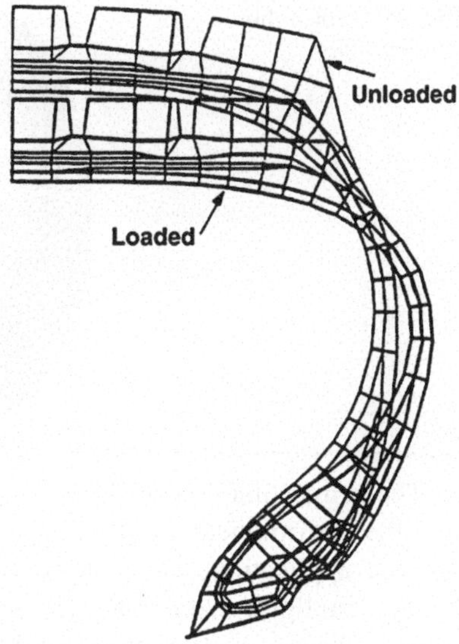

Fig. 27 Finite-element structure of a heavy-duty truck tire

Table 1 Composition of steel tire cord

Element	Composition (%)	Function
Carbon	0.65	Strength
Chromium	0.05	Strength
Copper	0.02	Strength
Manganese	0.60	Deoxidation
Silicon	0.25	Deoxidation
Sulfur	0.03	Machinability

6.2 Hoses

A hose is a reinforced, flexible conduit used to move materials from one point to another or to transmit energy. It is flexible to accommodate motion, alignment, vibration, thermal expansion and contraction, portability, ease of routing, and ease of installation. Most hoses are made up of three elements: (1) a tube, (2) reinforcement, and (3) an outer cover. Each of these components is usually adhered to the adjacent components by bonding agents or thin layers of specially compounded rubber. The tube is the innermost rubber or plastic element of the hose. Reinforcement can be textile, plastic, or metal, alone or in combination, built into the body of the hose to withstand internal pressures, external forces, or a combination of both. The type and amount of reinforcing material used depends on the method of manufacture and on the service requirements. For example, a residential garden hose does not need the same level of reinforcement as required for high pressure air hose used in construction and mining applications. The cover is the outer element and can be made [98].

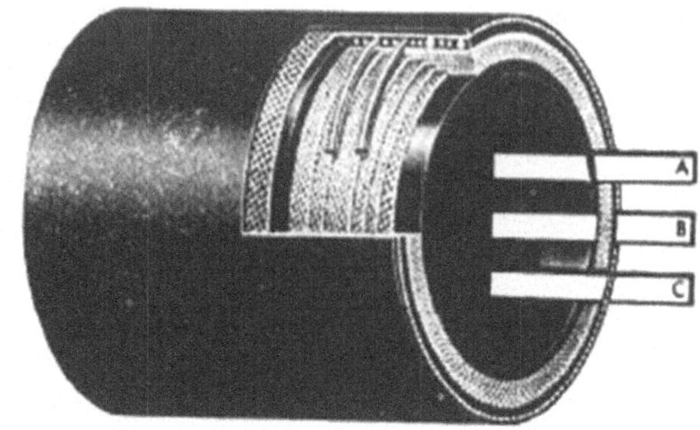

Fig. 28 Typical hose construction

The prime function of the cover is to protect the reinforcement from damage and the environment in which the hose will be used. Covers are designed for specific applications and can be made to be resistant to oils, acids, abrasion, flexing, sunlight, ozone, etc. The basic materials in the manufacture of hose are rubber, plastics, textile yarns, textile fabrics, and metal in the form of wires and cables. Throughout this book, the term "rubber" will be used in its broadest sense. This will include all elastomeric materials that are compounds of natural or synthetic elastomers or combinations of these materials. To provide a wide range of physical properties for specific service needs, elastomers are mixed with various chemicals. Space does not permit discussion of the compounding ingredients or compounding methods, so only the basic elastomers will be discussed. There are many of these available to the hose manufacturer. In addition, many types may be blended in almost unlimited combinations to obtain the most desirable properties. A typical hose is shown in Fig. 28.

Textile fabrics used as reinforcement in hose construction provide the strength to achieve the desired resistance to internal pressure or to provide resistance to collapse, or both. The properties of a fabric depend on the construction and the material from which the yarn is made and on the type of weave used. One common hose fabric is woven from warp yarns, which run lengthwise, and filling yarns, which run cross-wise. Usually they are woven at right angles to each other. The most common weave is known as "plain weave", Fig. 29.

Yarns are used in hose for reinforcement of the tube material to provide the strength to achieve the desired resistance to internal pressure or to provide resistance to collapse, or both. The basic yarn properties required for hose reinforcement are: adequate strength, acceptable heat resistance, dynamic fatigue resistance, and satisfactory processability for the various methods of reinforcing hose. Other special properties such as stiffness, adhesion, conductivity, etc., may be developed depending upon the specific hose application. Yarn is available in two basic forms: staple (sometimes referred to as spun yarn) and filament.

Staple yarn is made by twisting bundles of short fibers to form a continuous yarn. The staple obtains its strength from the binding effect of the twist imparted to the individual fibers. The base staple yarn is called a singles. It is made from fiber

Fig. 29 Plain weave

bundles twisted together in one direction to form a singles strand. If two or more single yarns are twisted together, usually in a direction opposite that of the singles yarn, the result is a plied yarn.

Filament yarn is produced by extruding synthetic material through a spinnerette containing hundreds of orifices. The monofilaments from each of the orifices are brought together to form a multifilament yarn. Filament yarns have higher tenacity (strength per unit of weight -grams per denier), in the range of 2–3 times that of staple yarn on the same material type and size. Steel wire has strength, high modulus for dimensional stability, fatigue resistance, and low cost, and is the major reinforcement used in high pressure hose and in most suction hose. Under certain service conditions, carbon steel wire is not suitable. An *alloy wire* is used instead. One of the most commonly used is stainless steel which offers exceptional resistance to corrosion and heat. Where light weight is essential, alloys of aluminum are used. *Static wires* and other conductive materials are used in hose to prevent static electricity buildup. Wires can be made from many metals including copper, steel, monel, aluminum and tin-coated copper. Static wires may be solid, stranded, or braided.

Most hose is used for pressure service; however, some applications require the hose to resist collapse in suction and vacuum service. Such hose is subjected to crushing forces because the atmospheric pressure outside the hose is greater than the internal pressure. The hose can collapse and restrict the flow unless the hose is constructed to resist these pressure differentials. The most common method of preventing hose collapse is to build a helical wire reinforcement into the hose body. The size and spacing of the wire reinforcement depends on the size of the hose and the expected pressure differential for the application. In suction applications approaching a perfect vacuum, most of the carcass plies are applied over the wire reinforcement. The hose is constructed with high adhesion between the tube and the carcass to prevent tube separation. Suction hose must be specifically designed for the service for which it is used.

Static wires and conductive rubber components are used in hose to help prevent static electricity build-up and subsequent a discharge as a spark. Electrical engineers differ in opinion on the effects of static electricity and the means of dissipating it. In handling gasoline and other petroleum-based liquids, recognized national associations and companies have conflicting opinions on the need for conductive hoses. In some specific applications, specially around high voltage electrical lines, it is imperative for safety that the hose be nonconductive. Unless the hose is designed particularly to be non-conductive and is so branded, one dare not conclude that it is nonconductive. Many black rubber compounds are inherently and inadvertently conductive. Nonconductive hose is usually made to a qualifying standard that requires it to be tested to verify the desired electrical properties.

Thermoplastic products such as vacuum cleaner hoses, used for very low pressure applications are often manufactured with blow molded or tape forming techniques. Blow molded products are shaped into a circumferentially corrugated profile at the tube extruder when the thermoplastic material is still in the molten state. The corrugations provide a tremendous improvement in product flexibility and stretch characteristics. The profiling is accomplished by injecting air into the tube pushing it into a series of metal die blocks corrugated with the intended profile. As the tube cools while traveling along the die block track, the tube becomes permanently corrugated circumferentially. A similar process, vacuum forming, uses the same technique of corrugated die blocks at the extruder, but instead of blowing air in the tube, a vacuum is drawn through the blocks pulling the molten tube into the corrugations. The appearance of the final product from each method is quite similar. However the vacuum forming process generally yields superior corrugation uniformity. The corrugated tube may be the final product or used in conjunction with other hose components. For instance, for higher pressure applications an adequate reinforcement may be applied and then a smooth cover extrusion. Combinations of rubber and plastic layers may provide the best appearance for a specific application.

References

1. ASTM D 1566-98, Standard Terminology Relating to Rubber (1991)
2. Suzuki, D.: The tree that changed the world, (videotape). Canadian Broadcasting Corporation, Ottawa, Canada, Rubber Developments, **44**(1), 11 (1991)
3. Subramaniam.: Natural rubber, in rubber technology. Morton, M. (ed.). Van Nostrand Reinhold, New York (1987)
4. Stern, H.J.: History, In rubber technology and manufacture. Blow, C.M. (ed.). p. 2, Newnes-Butterworths, London (1977)
5. Buist, J.M.: Proceedings of the Institute of Materials International Rubber Conference, IRC 96, Manchester, UK, Paper No.1 (1996)

6. Duerden, E.: Plastics and rubber international. **11**(3), 22 (1986)
7. Stern, H.J.: History, in Rubber technology and manufacture. Blow, C.M. (ed.). Newnes-Butterworths, London (1977)
8. White, J.L.: Rubber processing: technology, materials and principles. Hanser Publishers, Munich (1995)
9. Stern, H.J.: History, in Rubber technology and manufacture. Blow, C. M. (ed.), Newnes-Butterworths, London (1977)
10. Famulok, T., Roch, P.: Proceedings of the Institute of Materials International Rubber Conference, IRC 96, Manchester, UK, Paper No.3, (1996)
11. Stern, H.J.: History, in rubber technology and manufacture. Blow, C. M. (ed.). Newnes-Butterworths, London (1977)
12. Kuzma, L.J.: Rubber technology, 3rd edn. Morton, M. (ed.). Van Nostrand Reinhold, New York (1987)
13. Stern, H.J.: History, in rubber technology and manufacture Blow, C. M. (ed.). Newnes-Butterworths, London (1977)
14. White, J.L.: Rubber processing: technology materials and principles. Hanser Publishers, Munich (1995)
15. Hertz, D.L.: Jr., Handbook of elastomers. Bhowmick A.K. and Stephens H.L. (ed.). Marcel Dekker Inc., New York (1988)
16. Bryant, C.L.: Acrylonitrile-Butadiene (Nitrile) Rubbers, in rubber technology and manufacture. Blow, C.M. (ed.). Newnes-Butterworths, London (1977)
17. Hofmann, W.: Rubber technology handbook. Hanser Publishers, Munich (1989)
18. Rubber and Plastics News. **14**(2), 21 (1984)
19. Blow, C.M.: Rubber technology and manufacture. Blow, C.M. (ed.). Newnes-Butterworths, London (1977)
20. Rigbi, Z.: Reinforcement of rubber by carbon black. Adv. Polym. Sci. **36**, 21–68 (1980)
21. Yan, L.: Mullins effect recovery of a nanoparticle-filled polymer. J. Polym. Sci. Part B Polym. Phys. (2010)
22. Mark, J.E.: Monte carlo simulations on nanoparticles in elastomers. effects of the particles on the dimensions of the polymer chains and the mechanical properties of the networks. Macromol. Symp. **256**(1), 40–47 (2007)
23. Mdarhri, A.: Microwave dielectric properties of carbon black filled polymers under uniaxial tension. J. Appl. Phys. **101**(8), 084111–084122 (2007)
24. http://www.etrma.org/pdf/
25. http://www.cb4reach.eu/
26. http://corporate.evonik.com/en/
27. Treloar, L.R.G: The physics of rubber elasticity. Oxford University Press (2005)
28. Tsenoglou, C.: Rubber elasticity of cross-linked networks with trapped entanglements and dangling chains. Macromolecules. **22**(1), 284–289 (1989)
29. Adolf D.: Origins of entanglement effects in rubber elasticity. Macromolecules **21**(1), 228–230 (1988)
30. Brereton, M.G., Filbrandt, M.: The contribution to rubber elasticity of topological entanglements. Polymer **26**(8), 1134–1140 (1985)
31. Goppel, J.M.: On the degree on crystallinity in natural rubber II. The orientation of the rubber crystallites in stretched samples. Appl. Sci. Res. **1**(1), 18–26 (1949)
32. Negahban, M.: Modeling the thermomechanical effects of crystallization in natural rubber: III Mechanical properties. Int. J. Solid. Struct. **37**, 2811–2824 (2000)
33. Nielsen, L.E., Stockton, F.D. : Theory of the modulus of crystalline polymers. J. Polym. Sci. Part A: General Papers, **1**(6), 1995–2002 (1963)
34. DiBenedetto, T.: Prediction of the glass transition temperature of polymers: A model based on the principle of corresponding states. J. Polym. Sci. Part B: Polym. Phys. **25**(9), 1949–1969 (1987)
35. Sperling, L.H.: Introduction to physical polymer science, 4th edn. John Wiley (2006)

36. Yatsuyanagi, F., Suzuki, N., Ito, M., Kaidou, H.: Effects of secondary structure of fillers on the mechanical properties of silica filled rubber systems. Polymer **42**(23), 9523–9952 (2001)
37. Boonstra, B.B.: Role of particulate fillers in elastomer reinforcement: a review. Polymer **20**(6), 691–704 (1979)
38. Fröhlich, J., Niedermeier, W., Luginsland, H.-D.: The effect of filler–filler and filler–elastomer interaction on rubber reinforcement. Compos. A Appl. Sci. Manuf. **36**(4), 449–460 (2005)
39. Park, S-J., Cho, K-S.: Filler–elastomer interactions: influence of silane coupling agent on crosslink density and thermal stability of silica/rubber composites. J. Colloid Interface Sci. **267**(1), 86–91 (2003)
40. Bekaert N.V.: Steelcord Catalogue. (1991)
41. Clark S K.: Chemical Technology. **41**, 482 (1968)
42. Clark, S.K.: Mechanics of Pneumatic Tires, 2nd edn. US Department of Transport NHTSA, Washington DC (1981)
43. Evans, L.R., et al.: Use of precipitated silica to improvebrass-coated wire-to-rubber adhesion. 147th Rubber Division A.C.S. Paper No. 16, Spring (1995)
44. Goodyear, Tire and Rubber Co., Eur Pat. 902 046-A2 (1999)
45. Gough V.E.: Mechanics of pneumatic tires. US Dept. Transport. NHTSA. p. 204 Washington (1981)
46. Gough, V.E.: Stiffness of cord and rubber constructions. Rubber Chem. Technol. **41**, 1021 (1968)
47. Hartman, D.R. Greenwood, M.E., Miller, D.M.: High strength glass fibers. Technical paper, Owens-Corning Fiberglas Corp. (1994)
48. van Ooij, W.J.: Mechanism and theories of rubber adhesion to steel tire cords. Rubber Chem. Technol. **57**, 421–456 (1984)
49. Williams M.L., Landell R. F., Ferry, J.D.: The temperature dependence of relaxation mechanisms in amorphous polymers and other glass-forming liquids. J. Am. Chem. Soc. **77**, 3701 (1955)
50. Ismail, H.: The potential of rubberwood as a filler in epoxidized natural rubber compounds. J. Elastomers Plast. **33**, 34–46 (2010)
51. Pongdhorn, S-O., Sirisinha, C., Kannika H.,: Properties of natural rubber filled with ultra fine acrylate rubber powder. J. Elastomers Plast. **42**, 139–150 (2010)
52. Ryu, S.R., Lee, D.J.: Effects of interphase and short fiber on puncture and burst properties of short-fiber reinforced chloroprene rubber. J. Elastomers Plast. **42**, 181–197 (2010)
53. Renata M.B. Fernandes, Leila L.Y. Visconte, and Regina C.R. Nunes: Characteristics of acrylic rubber composites with mica and carbon black. J. Elastomers Plast. **42**, 65–74 (2010)
54. Jeong, D.S., Hong, C.K., Lim, G.T., Seo, G., and Ryu, C.S.: Networked silica with exceptional reinforcing performance for SBR compounds: interconnected by Methylene Diphenyl Diisocyanate. J. Elastomers Plast. **41**, 353–368 (2009)
55. Shtarkova R. and Dishovsky N.: Elastomer-based microwave absorbing materials. J. Elastomers Plast. **41**, 163–174 (2009)
56. Bulsari, P.M., Tzoganakis, C., Penlidis, A.: Hydrosilylation of Impact Polypropylene Co-polymer in a Twin-screw Extruder. J. Elastomers Plast. **40**, 365–380 (2008)
57. Faruk Yókseler, R.: A theory for rubber-like shells. J. Elastomers Plast. **40**, 39–60 (2008)
58. Jayasree, T.K., Predeep, P.: Effect of fillers on mechanical properties of dynamically crosslinked styrene butadiene rubber/high density polyethylene blends. J. Elastomers Plast. **40**, 127–146 (2008)
59. Shinzo J., Yuko, I: Reinforcement of general-purpose grade rubbers by silica generated in situ. Rubber Chem.Technol. **73**(3), 534–550 (2000)
60. Woo, C-S., Kim, W-D., Lee, S-H., Choi, B-I., Park, H-S.: Fatigue life prediction of vulcanized natural rubber subjected to heat-aging. Procedia Eng. **1**(1), 9–12 (2009)
61. Cao, Y., Mou, H., Shen, F., Xu, H., Hu, G-H., Wu, C.: Hydrogenated nitrile butadiene rubber and hindered phenol composite. II. Characterization of hydrogen bonding. Polym. Eng. Sci. **51**(1), 201–208 (2011)

62. Cao, Y., Shen, F., Mou, H., Cao, D., Xu, H., Wu, C.: Hydrogenated nitrile butadiene rubber and hindered phenol composite. I. Miscibility and dynamic mechanical property. Polym. Eng. Sci. **50**(12), 2375–2381 (2010)
63. Blow, C.M.: Rubber technology and manufacture, Butterworth Scientific, London (1982)
64. Crawford, R.J.: Rotational molding of plastics, 2nd edn. p. 260. Research Studies Press, London (1996)
65. Dieter, G.E.: ASM Handbook, materials selection and design, vol 20. ASM International (1997)
66. Freakley, Philip K.: Rubber processing and production organization. Plenum Press, London (1985)
67. Grulke, Eric A.: Polymer process engineering, PTR Prentice Hall, Englewood Cliffs, NJ (1994)
68. Kresta, Jiri E.: Reaction injection molding. Am. Chem. Soc. (1985)
69. Harry, L: Basic compounding and processing of rubber. Am. Chem. Soc. (1985)
70. Mark, James E.: Science and technology of rubber. Academic Press, San Diego (1994)
71. Dawkins, J.V.: Developments in polymer characterization, vol. 1–5. Elsevier, New York (1986)
72. Booth, Price, C.: Comprehensive polymer science. Polymer characterization, vol. 1. Pergamon, New York (1989)
73. Yamakawa, H.: Modern theory of polymer solutions. Harper, New York (1971)
74. Flory, P.J.: Statistical mechanics of chain molecules. Oxford Univ. Press, New York (1969)
75. Tanaka, Y.: Rubber Chem. Technol. **64**, 325 (1991)
76. Campbell, White, J.: Polymer characterization. Chapman and Hall, New York (1989)
77. Baldwin, F.P., Ver Strate, G.: Rubber Chem. Technol. **44**, 709 (1972)
78. Hsu, S.L.: Handbook of vibrational spectroscopy: A companion for polymer scientists. Wiley, New York (2004)
79. Stuart, B.H.: Polymer Analysis. Wiley, New York (2002)
80. Koenig, J.L.: Spectroscopy of Polymers, 2nd edn. Elsevier, New York (1999)
81. Braun, Simple: Methods for Identification of Plastics, 3rd edn. Hanser, New York (1996)
82. Mitchell, J.: Applied polymer analysis and characterization, vol. 2. Hanser, New York (1992)
83. Collins., Bares, J., Billmeyer, F.: Experiments in polymer science, Wiley, New York (1973)
84. Tyler, W.: Rubber Chem. Technol. **40**, 238 (1967)
85. Ishida, H.: Rubber Chemical Technology **60**, 497 (1987)
86. Messerschmidt, R., Harthcock, M. (eds.): Infrared microspectroscopy, practical spectroscopy series, vol. 6. Marcel Dekker, New York (1988)
87. http://www.spectroscopynow.com/
88. Groover, M.P.: Fundamentals of modern manufacturing: materials, processes and systems, 3rd edn, Accédez directement à la nouvelle édition (2006)
89. Campbell, F.C.: Manufacturing process for advanced composites. p. 513. Elsevier, New York (2004)
90. Freakley, P.K.: Rubber processing and production organization. Adv. Mater. Process. **1**(1), 53 (1985)
91. Leblanc, J.L.: Rubber–filler interactions and rheological properties in filled compounds. Prog. Polym. Sci. **27**(4), 627–687 (2002)
92. Thiruvarudchelvan, S.: Elastomers in metal forming: A review. J. Mater. Process. Technol. **39**(1–2), 55–82 (1993)
93. Michaeli, W.: Extrusion dies for plastics and rubber, design and engineering computations, 3rd edn, Hanser Gardner Publications Inc., (2003)
94. Brzoskowski, R., White, J.L., Szydlowski, W.: Air lubricated die for extrusion of rubber compounds, Gummi Fasern Kunstst, **42**(7) 312, 314–317 (1989)
95. Song, H.J., White, J.L., Min, K., Nakajima, N., Weissert, F.C.: Rheological properties, extrudate swell, and die entry extrusion flow marker experiments for rubber-carbon black compounds. Adv. Polym. Technol. **8**(4), 431–449 (1988)

96. Isayev, A.I., Huang, Y.-H.: Unsteady flow of rubber compounds at injection molding conditions. Adv. Polym. Technol. **9**(3), 167–180 (1989)
97. Haberstroh, E., Wehr, H.: Rubber processing with gas-assisted injection moulding (R-GAIM). Macromol. Mater. Eng. **284–285**(1), 76–80 (2000)
98. Goettler, L.A.: The extrusion and performance of plasticized poly(vinyl chloride) hose reinforced with short cellulose fibers. Polym. Compos. **4**, 249–255 (1983)

Rubber Nanocomposites: Latest Trends and Concepts

Deepalekshmi Ponnamma, Hanna J. Maria, Arup K. Chandra and Sabu Thomas

Abstract Rubber nanocomposites have a unique position both in academic and industrial point of view and extensive research studies are progressing in this area. Due to their ever increasing significance, a thorough investigation is necessary especially when the application side is considered. The enhanced surface area and high aspect ratio of nano materials lead to superior matrix/filler interaction and this results in the versatile properties and wide range of applications for the obtained nanocomposites. Nano fillers like layered silicates, carbon nanotubes (CNTs), fullerenes, silica, metallic nanoparticles, metal oxides, polyhedral oligomeric silsesquioxane (POSS), biomaterials, nanodiamonds etc. are used extensively in rubber composites fabrication. In this chapter, attempt has been made to briefly explain the recent advances in the area of nanofillers and their rubber nanocomposites. A thorough survey has been made by giving special emphasis to the filler geometry and composite morphology on one side and the composite properties on the other. Finally the study ends up with novel applications of rubber nanocomposites and the future perspectives in this area.

Abbreviations

A_f	Aspect ratio
AFM	Atomic force microscopy
CB	Carbon black
CNT	Carbon nanotube
CVD	Chemical vapor deposition

D. Ponnamma (✉) · H. J. Maria · S. Thomas
School of Chemical Sciences, Mahatma Gandhi University, Kottayam, Kerala 686560, India
e-mail: lekshmi.deepa09@gmail.com

S. Thomas
Centre for Nanoscience and Nanotechnology, Mahatma Gandhi University, Kottayam, Kerala 686560, India

A. K. Chandra
R&D Centre, Apollo Tyres Ltd., Limda, Waghodia, Vadodara, Gujarat 391760, India

P. M. Visakh et al. (eds.), *Advances in Elastomers II*,
Advanced Structured Materials 12, DOI: 10.1007/978-3-642-20928-4_3,

CRG or GE	Chemically reduced graphene
EG	Expanded graphite
GO	Graphene oxide
HR-TEM	High resolution-transmission electron microscopy
MMT	Montmorillonite
NR	Natural rubber (poly isoprene)
NBR	Nitrile–butadiene rubber
XNBR	Carboxylated nitrile–butadiene rubber
PDMS	Polydimethylsiloxane
RTVSR	Room temperature vulcanizing silicone rubber
RGO	Reduced graphene oxide
SWCNT	Single wall carbon nanotube
SEM	Scanning electron microscopy
SDS	Sodium dodecyl sulfonate
SR	Silicone rubber
SBR	Styrene–butadiene rubber
TEM	Transmission electron microscopy
TRG	Thermally reduced graphene
TPU	Thermoplastic polyurethane

1 Introduction

Rubbers an important class of polymers, generally known as elastomers are useful materials from the time immemorial. It is a very essential material in industry due to their unique properties particularly viscoelasticity. For making them applicable in various fields, fillers are added to it resulting in the manufacture of varieties of composites. This will impart high elastic moduli and durability for the elastomers, which were always limiting their practical usage. Mineral fillers are known to improve the strength and stiffness of rubbers. However the extent of property enhancement depends on various factors such as the size and shape of the particles, filler aspect ratio, degree of dispersion and orientation of particles in the matrix and the interfacial adhesion between filler and polymer chains [1]. In general, it is said that the filler–filler [2–4] and the filler-rubber interactions [5] cause better reinforcement. The nano materials with average particle size in the range of 1–100 nm are extremely useful polymeric reinforcements. Compared to traditional fillers, these particles can enhance the composite properties in relatively small concentrations. In the nano-scale regime, some materials exhibit additional or different features or properties as compared to coarser materials. These materials are now used in a wide range of innovative technological applications and products. Nano sized fillers have the unique ability to form very fine and homogenous dispersion in the elastomer matrices and occupy substantially greater number of sites in it. In general, nanocomposite manufacturing processes include in situ polymerization, solution mixing and melt compounding [6–9].

Among the nonblack fillers, layered silicates are very useful due to the low cost, versatility, availability and stiffening properties. Clays contain hexagonal platelets having thickness in the nanometer range, and length and breadth in the micro range. But, when compared to the carbon black fillers, clay filled composites exhibit less mechanical properties since the hydrophilicity of clay causes poor filler/polymer interaction. This as well as the slow curing effect observed in clay systems can be avoided by the organic modification of clay platelets. Apart from the mechanical and thermal applications, clay composites are used in tyre inner liners where they are helpful in reducing the air permeability. The reason for this behavior is the formation of tortuous paths by the gas molecules due to orientation of filler platelets, and thus causing the air permeability of the composite to decrease. More recently, extensive studies using fillers like carbon nanotubes, graphene etc. are going on other than clay [10].

Carbon nanotubes (CNTs) are made of cylindrical graphitic sheets with fullerene end-caps [11], diameter ranging from 1 to 100 nm and length up to several millimetres. Its density varies between 1 and 2 g/cm^3 and modulus is greater than 1 TPa [12]. Based on the structure, there are single walled (SWCNT) and multi walled (MWCNT) nanotubes and both are widely used as nanofillers. Recently graphite and its derivatives like graphite oxide (GO) are noted for their high dispersive capacity, long coherence length and superior barrier properties [13, 14]. High conductivity of these fillers make them applicable in electronics for instance in sensor skins [15, 16], flexible display [17–19], and in dielectric actuators [20–22] instead of the brittle material based systems currently used. Polyhedral oligomeric silsesquioxanes (POSS) having combined organic/inorganic material properties are excellent lightweight, high performance hybrid materials used to modify many polymer properties [23]. Having 1–3 nm diameter, POSS can enhance the service temperatures, decomposition temperatures, oxidation resistance, surface hardening, mechanical properties, flame retardancy, heat evolution [24] etc. of several polymeric materials to a great extent. Very recently, bionanofibers and their nano crystals obtained from natural resources as reinforcing agents in several polymeric matrices have attracted the attention of researchers. Bionanofibers include nanocellulose, nanochitin and nanostarch and their nanocrystals. This chapter deals with fillers having various morphologies such as layered sheets (montmorillonite, hectorite, and graphene), cylindrical tubes (carbon nanotubes,) and spherical particles (such as POSS) and their rubber nanocomposites.

Amidst of the superior properties imparted by the nano fillers, their incompatibility due to poor dispersion in the polymer matrix cannot be neglected. This is the most important problem faced by the material scientists. The interfacial strength between the filler and polymer is another important aspect which should be given due consideration. The poor filler-polymer adhesion cause the formation of nanoparticle aggregates during the composite fabrication, which is responsible for failure at the interface and for the decreased physical as well as chemical properties of the final composites. The nano particle aggregation can be minimized to a great extent by modifying the filler surface either chemically or physically. This can regulate the reaction rate [25] as well. In addition to the information about various kinds of fillers special emphasis has been given in discussing about the composite manufacturing techniques, properties of composites and their versatile applications in this survey, by addressing filler dispersion problems and their modification methods.

2 Nanofillers

Generally, fillers are solid particulate materials (inorganic, organic) that have irregular, spherical, fibrous, or platelet-like shape. Whereas a nano-filler (exist in defined singular form that have at least one dimension in the nano scale, <100 nm) impart properties as desired and can be tailor made according to the individual needs. As already mentioned, such nanofillers at small percentages (<10 %) can improve polymer properties such as heat resistance, barrier properties, strength, stiffness, flame retardancy etc. The small particle size (size range 1–100 nm), high specific surface area and hence the high surface area-to-volume ratio of nano fillers change the reactivity and physical properties of the elastomers without affecting its bulk properties like density or light transmission. The enhanced surface area to-volume ratio also helps nanoparticles to interact with one another in different ways. Nano fillers improve mechanical or physical properties of elastomers significantly, reduce the material cost and increase the processability [26]. One of the biggest rubber industries, the tyre manufacturing units, use fillers like nano-silica, nano-zinc oxide, nano-black, etc. for performance improvement. This increase in properties may be attributed to the chemical bond formation, disruption of the elastomer conformational position and orientation of polymer chains and the immobilization of adjacent polymer chains.

Based on the number of dimensions in nano scale, nano fillers can be classified into three types—One-dimensional (one dimension in nanometer scale, usually the thickness, e.g. plates or laminas layered silicates, layered double hydroxides (LDHs)), two dimensional (two dimensions in the nanometer scale and the third a little longer (nanotubes and nano fibres) and three-dimensional (nanogranules, nanocrystals and spherical for instance POSS, nano silica)—[28, 29] as illustrated in Fig. 1. All these kinds of nano fillers possess different surface properties than bulk due to their large surface area facilitating large volume of interfacial matrix material. As we move from macro to micro to nano-materials, nano sized materials have new and surprising important properties related to solubility, reactivity, selectivity, optical, electrical, magnetic and mechanical properties. Among nanomaterials, nano-clays are the most commonly used commercial additive for the preparation of nano-composites, accounting for nearly 80 % of the volume used. Since the interface between fillers and rubber is one of the factors determining the composite properties, a comparative study has been done on the variation of the

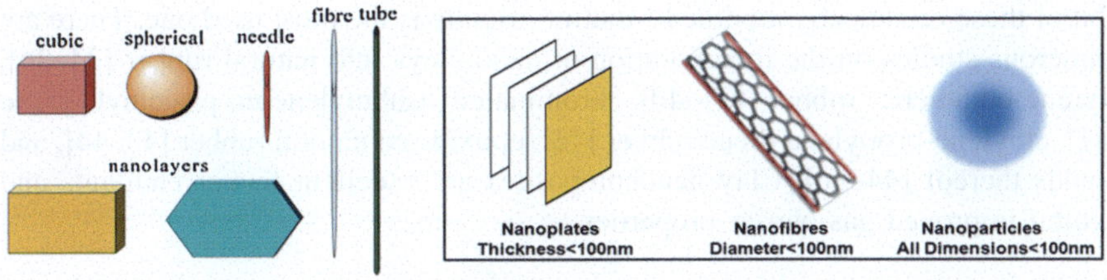

Fig. 1 Various types of nanoscale materials [27]

Fig. 2 Plot of volume of interfacial polymer based on 10 nm thick interfacial region surrounding each nanoparticle against volume fraction [31]

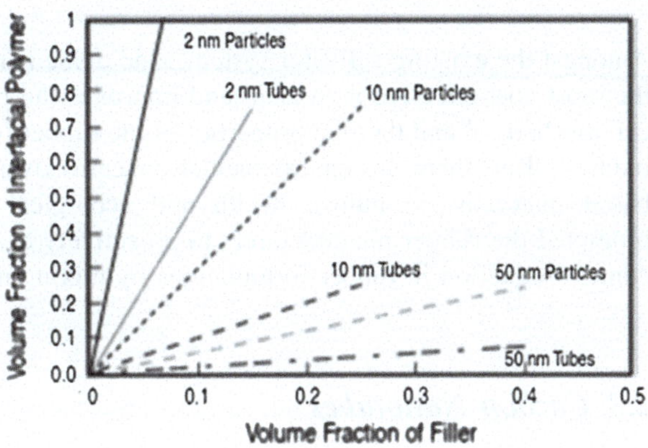

interfacial volume of polymers with filler dimensions [30]. The three general categories of nano fillers—tube like (diameter <100 nm and an aspect ratio of at least 100), plate-like (thickness on the order of 1 nm, aspect ratio in the other two dimensions of at least 25) and sphere like (<100 nm dimension)—changes the interfacial interaction significantly as shown in Fig. 2. An increase in interfacial volume with decreasing nanoparticle size is clearly evidenced from the plot.

A few of the numerous nano materials useful in rubber reinforcement are discussed below:

2.1 Nano-Clays

Layered silicates or clays come under the class of hybrid organic–inorganic nanocomposites [32, 33]. These are the most widely used reinforcement due to their natural abundance and high aspect ratio. They contain two structural units—a silica tetrahedral sheet fused to an aluminum octahedron—by sharing oxygen atoms. Since the clays have surface charges, they are highly hydrophilic, making it less compatible with a wide range of non polar elastomers. This can be solved by adding certain surfactants or covalently modifying the surface of the clay sheets, which leads to the formation of the so called organically modified clays (OMMTs).

Out of these organically modified Montmorillonite is the most used one. There are numerous studies on the incorporation of nano-clays into natural rubber [34–36], styrene-butadiene rubber [37–40], brominated isobutylene-co-p-methylstyrene [41], ethylene-propylene-diene rubber [42], epoxidised natural rubber [43, 44], and blends thereof [44–46]. Clay nanocomposites are excellent flame retardants and exhibit improved gas barrier properties.

2.2 *Nano-Oxides*

Amongst the existing activator system, zinc oxide (ZnO) has been widely used as the most cost-effective ingredient and improves the processability and the physical, mechanical and thermal properties of the rubber [47, 48] especially in the tyre industry. But, there are environmental concerns over the harsh influence of Zn-based materials on human health and ecological systems. This reason has prompted the rubber manufacturers to proactively evaluate the strategies for zinc content reduction in rubber formulations by mixing nano-ZnO [49].

2.3 *Carbon Nanotubes*

Carbon nanotubes (CNTs) also been considered as an attractive candidates for imparting several properties to elastomers. A carbon nanotube is a tube-shaped material, made of carbon, having a diameter measuring on the nano-meter scale. Carbon nanotubes have many structures, differing in length, thickness, and in the type of helicity and number of layers. Commonly used nanotubes are single-walled nanotubes (SWCNTs) and multi-walled nanotubes (MWCNTs). SWNTs contain one cylinder formed by wrapping one layer of graphene sheet and MWNTs by several layers. The diameter of SWCNTs is close to 1 nm, whereas for MWNTs, it varies up to 100 nm. Structure of CNTs is well understood from the TEM images given in Fig. 3. Although they are formed from essentially the same graphitic sheet, their electrical characteristics differ depending on these variations, acting either as metals or as semiconductors. CNTs improve the electrical as well as thermal conductivity of elastomers. The electrical properties of CNTs are significant and considered to be very important in the tyre industry as a source for dissipating static charge. Replacing the carbon black by carbon nanotubes improved skid resistance and reduced abrasion of the tire [50]. Carbon nanotubes may provide a safer, faster, and eventually cheaper transportation in the future [51]. CNTs themselves are superior conductors but they may not exhibit the same level of conductivity, when integrated into other materials [52].

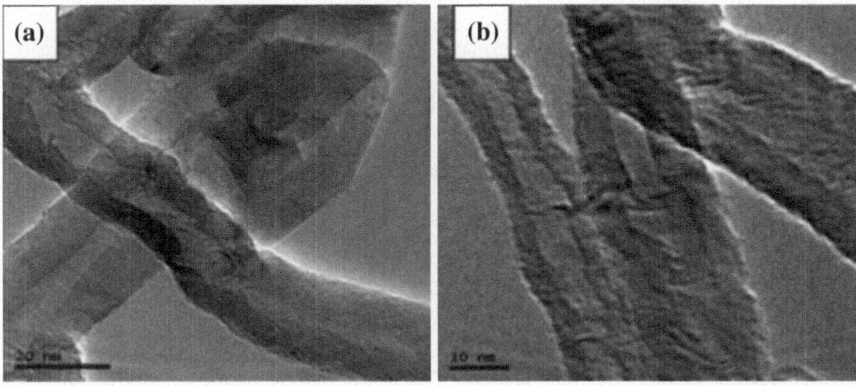

Fig. 3 TEM Images of carbon nanotubes **a** at low resolution **b** at high resolution [53]

2.4 Graphene

In addition to CNTs, the one atom thick two-dimensional graphene sheet can also provide effective reinforcement to elastomers. In graphene, the sp2-hybridized carbons are arranged like a honeycomb structure, which can provide huge surface area (experimentally 1850 $m^2\ g^{-1}$) since because of the availability of both sides of the sheets [54]. Graphenes have the unique capacity of enhancing the thermal, electrical and gas barrier properties. The two dimensional sheets of graphite, expanded graphite, graphene oxides, graphite oxides etc. are also widely used in regulating the polymer properties.

2.5 Polyhedral Oligomeric Silsesquioxane

Polyhedral oligomeric silsesquioxane (POSS) is the recent development in polymer science and technology. The three dimensional POSS molecules are considered as the smallest particles possible for silica and they consists of an inorganic silica $Rn(SiO_{1.5})n$ core cage structure where n is the number of silicon atoms of the cage (n = 8, 10, 12) with diameter less than 3 nm and mass about 1000D. This enables POSS to get nearly equivalent dimensions of polymers, with less radius of gyration than polymer chains and thereby applicable in tremendous polymer nanocomposites. Several structural representations of silsesquioxanes with the empirical formula $RSiO_{1.5}$ are possible, with the two most common representations being a ladder-type structure and a Si-O cage framework (Fig. 4). There are reports of using POSS in tyres. In a patent by Crutchley [55] a rubber composition comprising one or more POSS included in the rubber composition as part of a rubber compound master batch or added as an additive to a rubber compound master batch. In another patent [56]; Puhala claimed a rubber composition having POSS as reinforcement.

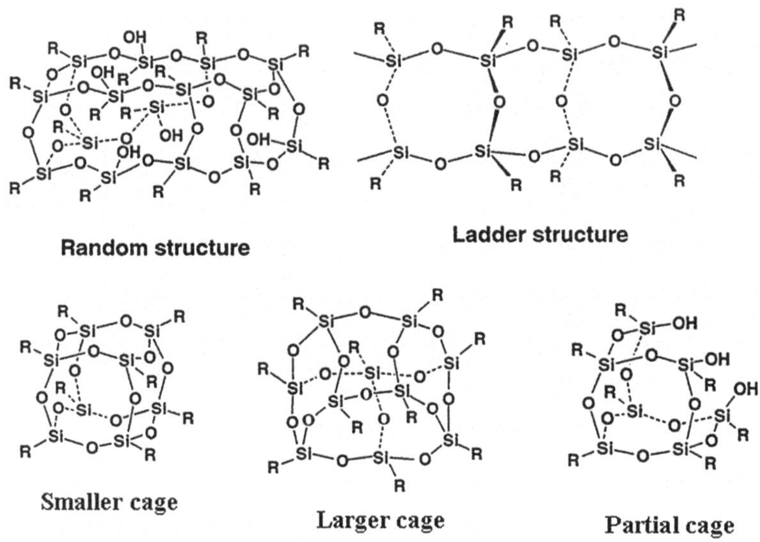

Fig. 4 Structural representations of silsesquioxanes [57]

2.6 Nano Carbon Black and Nano Silica

The nano-structures carbon black and nano silica are the new category of fillers in tyre applications. Nano-structured black is a family of new carbon black characterized by a rough surface and enhanced filler-polymer interaction. It hinders the slippage of polymer molecules along the rough nano-structured surface and reduces the hysteresis significantly. This type of black ideally meets truck tyre requirements, as it provides improved tread wear in addition to low hysteresis [58]. Synthesis of nano silica has gained much attention due to its superior properties and most widely being used as filler in rubber. These silica nano-particles show superior physico-mechanical properties and improvement in the processing behaviour.

2.7 Bionanofillers

Over the past two decades much effort have been devoted to the use of microcrystals obtained from natural resources as reinforcing agents in several polymeric matrices. Environmental pollution, depleting oil reserves, oil barrel price evolution etc. have triggered the need to find bio-based solutions. The advantages of natural fillers are their low density, renewable character, and biodegradability associated with the highly specific properties of nanoparticles. Nanosized derivatives of polysaccharides like starch, chitin and cellulose can be synthesized in bulk and can be used for the development of bionanocomposites. They can be promising substitutes for environment pollutant carbon black, for reinforcement of rubbers even at higher loadings (up to 50 phr) via commercially viable process. The combined effect of size reduction and organic modification improves filler–matrix adhesion and in turn the performance of bionanofillers. Cellulose, chitin and starch are abundant, natural, renewable and biodegradable polymers from where we can prepare cellulosic nanofibers, cellulose nanowhiskers, starch nanocrystals and chitin nanowhiskers [59–70].

2.7.1 Starch Nanoparticles

The use of polysaccharides as reinforcing agents for polymer materials are due to their properties like renewable nature, availability, diversity of sources, low density, low energy consumption, low cost, high mechanical properties, comparatively easy processability, nonabrasive nature, relatively reactive surface, which can be used for grafting specific groups etc. Native starch granules contain more or less concentric "growth rings" that are readily visible by optical or electronic microscopy. The starch structure has been under research for years, and because of its complex structure, a universally accepted model still need to be proposed [71]. The structure of starch forms a concentric semicrystalline multiscale structure that allows the production of new nanoelements like:

(i) starch nanocrystals from the disruption of amorphous domains from semicrystalline granules by acid hydrolysis
(ii) starch nanoparticles produced from gelatinized starch.

These starch nanoparticles are proved to improve mechanical and barrier properties of bionanocomposites when used as fillers. Because of their use in industrial packaging, researchers are continuously looking for innovative solutions for efficient and sustainable systems. Therefore, starch nanoparticles have been the focus of an exponentially increasing number of works devoted to develop bionanocomposites by blending starch nanoparticles with different biopolymeric matrices.

2.7.2 Cellulosic Nanoparticles

Being one of the most abundant biomass materials in nature, cellulose allows different kinds of nanoscale cellulosic fillers—called cellulose nanocrystals or microfibrillated cellulose (MFC) [72]. Cellulose nanoparticles have been the focus of an exponentially increasing number of works or reviews devoted to understanding such materials and their applications. Over the last decades researchers have shown that cellulose nanoparticles could be used as fillers to improve mechanical and barrier properties of biocomposites. The microfibrils, was found to possess width in the range of 5–30 nm, [72] and are highly crystalline materials formed by lateral packing of long cellulose molecules with hydrogen bonding.

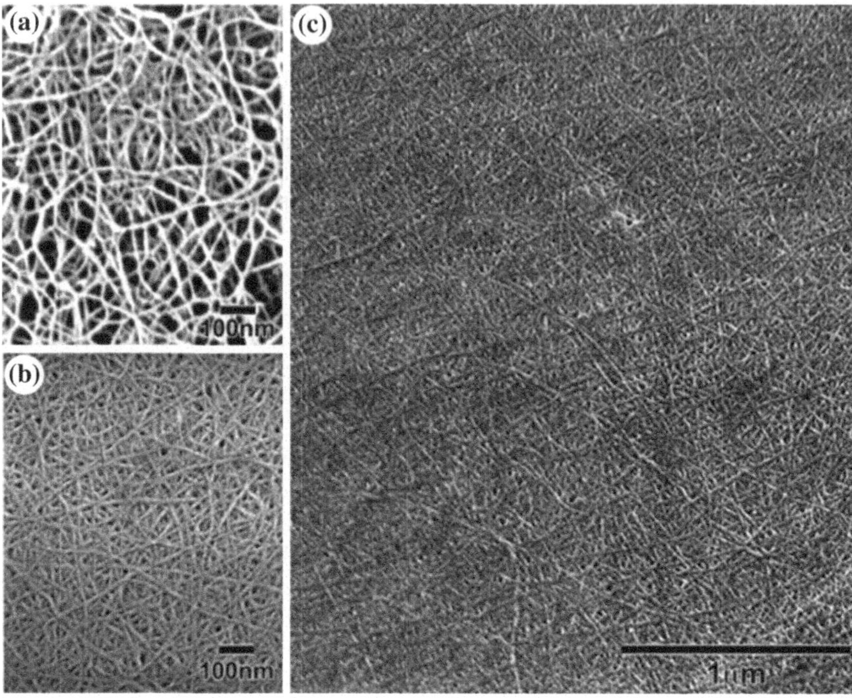

Fig. 5 FE-SEM micrographs of nanofibers [79]

Figure 5 shows the FE-SEM micrographs of cellulose nanofibers. The outstanding mechanical properties, with a high Young's modulus of 138 GPa in the crystal region along the longitudinal direction [73] and a very low coefficient of thermal expansion along the longitudinal direction [74] was shown by a stable structure. Therefore, cellulose whiskers and fibrils have great potential for use as reinforcement in nanocomposites and have attracted a great deal of interest recently [75–79]. Various conditions and steps involving general synthesis of starch and cellulosic nanoparticles are represented in Fig. 6 [80].

2.7.3 Chitin Nanowhiskers

Chitin, the animal counter part of cellulose is the second most available material in nature. In recent years considerable interest has been devoted to biomaterials based on chitin and on its amino-derivative chitosan. It is estimated that about 1010–1011 tons of this polymer are synthesized each year [81]. Chemically, chitin molecules consist of N-acetyl-D-glucosamine units. Chitin is known to be non-toxic, odorless, biocompatible with living tissues, and biodegradable [82] and hence the application of this as the reinforcing agent is boundless. The advantage of chitosan over other polysaccharides (cellulose, starch etc.) is that, its chemical

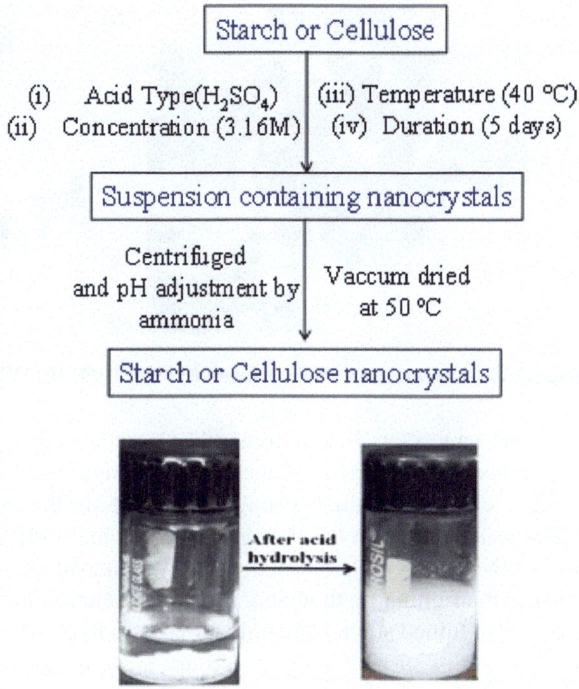

Fig. 6 Different steps involving general synthesis of starch and cellulosic nanoparticles [80]

structure allows specific modifications without too many difficulties at the C-2 position. Specific groups can be introduced to design polymers for selected applications [83]. Regardless of the chitin sources, the hydrolytic condition is commonly-used for obtaining chitin whiskers [84–86] Chitin whiskers obtained from squid pen [87] and Riftiatubes [88] was found to have good reinforcing effect in copolymer of styrene and butyl acrylate [87] or poly(caprolactone) [88].

2.8 Other Nano-Particles

There are several other nano-particulate materials which are getting attention from compounders such as nano-calcium carbonate [89], nano-carbon fiber [90], nano-cellulose [91], nano-mica [92], nano-talc [93], etc. In recent years, there is development of new nano-particles from different polymers known as core–shell nano-particles. These nano-particles have a crosslinked core and a crosslinked shell. Conducting polymers like polyanilines (PAni) are also notable fillers due to its low cost, good conductivity, stability and easy synthesis. Conducting elastomer blends of polyanilines and rubbers such as polychloroprene, ethylene-propylene-diene (EPDM), styrene-butadiene (SBR) rubber, and nitrile rubber (NBR) have been reported [94–101]. These composites showed conductivity values in the range of 10^{-6} to 10^{-9} S cm^{-1}.

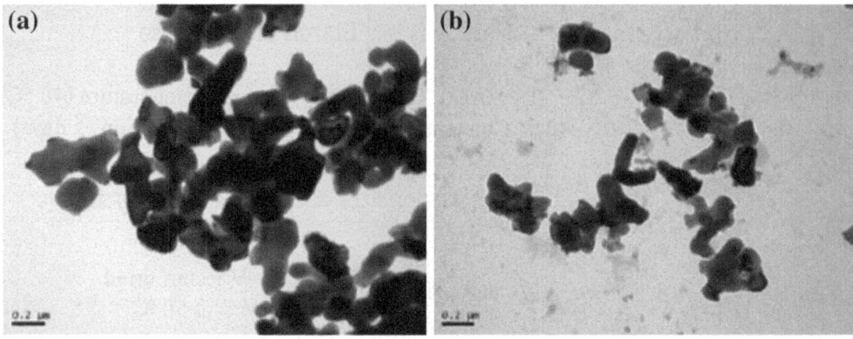

Fig. 7 TEM of untreated Al_2O_3 (**a**) and KH570-Al_2O_3 (**b**) [103]

Other nanofillers like alumina (Al_2O_3), silica (SiO_2) and titania (TiO_2) improve the dielectric properties of the polymers significantly and modification of their surface can impart much stronger influence on the properties. Nanoparticles like silica usually exist in the powdered or colloidal [94] form and is obtained by sol–gel or the microemulsion method [102]. Nanosilica can also be produced by following the fuming method and/or the precipitation method. The white, amorphous powder of fumed silica is manufactured by a hydrothermal process and it possesses three kinds of surface hydroxyl groups such as isolated free, SiOH, geminal free, $Si(OH)_2$ and vicinal, OH groups bound together through hydrogen bond. This can help in the better grafting of filler particles on polymer chains. Figure 7 shows the difference between modified and unmodified alumina and it is clear that modification significantly avoids agglomeration of particles and enhances the dispersion.

Filler modification includes treating it with suitable surfactants or chemical modification on the filler surfaces. Surfactants cause a wrapping effect on the filler particles thereby reducing their agglomeration tendency. Surfactants can also decrease the surface tension of the suspension which also helps in particle segregation and well dispersion. Chemical modification always leads to the addition of functional groups on filler surfaces and helps in making filler polymer covalent bonds. Even though this method offers better bonding, chemical treatments can cause filler defects which sometimes cause a negative impact on composite properties. Nanocomposites are synthesized by incorporating all these kinds of fillers in suitable matrices (Table 1).

3 Rubber Nanocomposites

Properties of rubber nanocomposites have been greatly influenced by the nature of filler and rubber, interfacial filler-elastomer interactions, compatibilisation and mode of dispersion. Among the various fillers used in the fabrication of rubber nanocomposites, nano silica, carbon nanotubes, graphene, bionanofibers and

Table 1 Structural characterization of EG, Graphite nanoplatelets, clay, CNT and graphene

Material	Parameters	Surface area (m^2/g)	Electrical conductivity
Clay	100–150 nm [27] Interlayer spacing	700–900 [27]	–
GNP	2–150 nm [104] Thickness	15–2630 [104]	$5.98 \pm 0.11 \times 10^4$ (S/m) [105]
Graphene	0.34 nm [106] Thickness	~2630 [106]	$1.28 \pm 0.04 \times 10^2$ (S/m) [105]
POSS	1–1.5 nm [107] Diameter	500 [107]	6000 S/cm [105]
CNT	~1 nm [108] (Diameter)	10–20 [108]	5–50 μΩ.cm [109]
PAni	50–100 nm [110]	40–50 [110]	3–300 (S/m) [110]

nanoclays have unique importance in regulating the properties and are mainly focused in this survey.

3.1 Preparation Methods

A lot of preparation methods are employed for fabricating the nanocomposites, out of which the most commonly used techniques are melt mixing, solution casting and in situ polymerization. Apart from describing these three conventional methods, effort has been made to point out other methods also. Table 2 illustrates a few of the various preparation methods applied in the rubber composite manufacture.

Solvent casting: This technique is based on dispersing the filler as well as the matrix in suitable solvents separately and then mixing both together. The solvent evaporates from the casted suspensions to get dried composite samples. Using ethanol as dispersing agent, Das et al. [111] got good dispersion of CNTs and thus

Table 2 A few preparation methods of rubber nanocomposites

Preparation method	Composite system	Reference
Solvent casting	NR latex and clay rectorite	[136]
Solution mixing and melt mixing	Isobutylene/isoprene rubber/organic modified clay	[137]
Masterbatch method	NR or SBR/clay and carbon black	[138]
Latex stage compounding	NR latex/CNT	[120]
Hetero coagulation approach	SBR/carbon black, CNTs	[139]
Melt blending	NR/graphene	[134]
Solution mixing	PU/POSS	[140]
Emulsion in situ polymerization	Styrene and butyl acrylate and laponite	[141]
Melt blending	TPU/clay	[142]
Electrospinning	Quantum dots/elastomer	[143]

Fig. 8 Illustration of preparation of RGO–NBR composite [119]

excellent mechanical properties for the composites. Khalid et al. [112] also dispersed CNTs in NR via solvent casting method using toluene. Ganter et al. prepared rubber nanocomposites based on styrene–butadiene rubber (SBR) and butadiene rubber (BR) containing organophilic-layered silicates [113, 114] by following solution mixing process. Graphene/rubber nanocomposites are also reported via solution blending [115–117]. Haiqing et al. fabricated [118] silicon rubber/graphene (and CNTs) composites by solution mixing in THF solvent and obtained excellent reinforcement. Figure 8 shows a schematic representation of the fabrication of reduced graphene oxide (RGO) dispersed in NBR composite using the solvent xylene [119]. The use of solvents like toluene is considered to be the disadvantage of this technique even if good dispersion can be achieved through this.

Latex stage compounding: Since most of the rubbers also exist in latex form, this compounding technique is of great significance. This is done by adding the filler and the curing agents like sulphur, ZnO and stearic acid directly to the latex and then following the mixing, casting steps as in the solution mixing. This method has been reported for the fabrication of NR latex/MWCNTs [120, 121] and NR/10 wt% natural and synthetic layered silicates nanocomposites [120]. Rubber/Ca-montmorillonite (Ca-MMT) nanocomposites containing well exfoliated Ca-MMT layers were prepared by a combined latex compounding and melt mixing [121] technique. NBR/clay composites are synthesized by following continuous dynamic latex compounding [122]. Graphene/rubber nanocomposites were also reported by this method [123]. The interaction of CNTs with the latex particles in

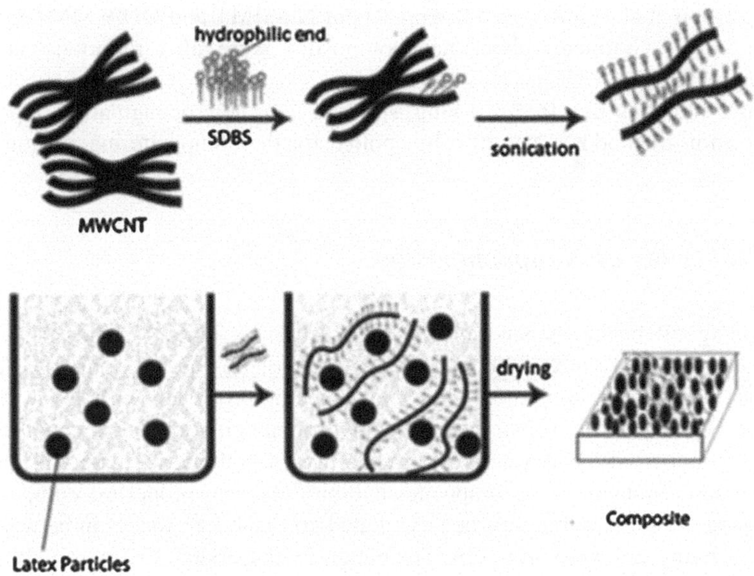

Fig. 9 Schematic representation of the surfactant interaction on the surface of CNTs after sonication, interaction of CNT with latex and subsequent drying of the latex film [124]

the presence of surfactants are described in the Fig. 9. After casting, well exfoliated nanotube bundles in the matrix are observed.

In-situ polymerization: In situ polymerization is an effective method to attain uniform dispersion of fillers in elastomers. Here, the filler is mixed with the monomer and essential reagents to trigger the polymerization reaction in the presence or absence of a solvent. During the polymerization reaction, a uniform dispersion of fillers inside the matrix can be obtained. This will lead to the greater reinforcement effect through the strong interfacial bonds formed [125, 126]. Kaminsky et al. used bis (cyclopentadienyl) zirconium compounds in the presence of Zeigler catalysts to synthesize atactic poly propylene and further reacted it with ethylene and diolefines to manufacture EPDM rubber [127]. They also prepared EPDM terpolymers with ethylidene norbornene as diene monomer using a zirconium based Ziegler catalyst [128, 129].

Melt blending/Extrusion: Absence of solvent is considered to be the major advantage of this method. This process is industrially important and offers mixing at high temperature and at high shear force. Extruders, internal mixers etc. are used for this purpose. Melt mixing can result in good dispersion of CNTs in styrene butadiene (SBR) and butadiene rubber (BR) blend as reported by Das et al. [130]. Vu et al. [131] reported that the compounds of epoxidised natural rubber (ENR) and sodium MMT clay prepared by melt mixing exhibit higher tensile properties than those compounds of precipitated silica. Sulfur-cured ENR recipes containing layered silicates produced by melt compounding were studied by Varghese et al. [132]. EPDM/nano ZnO composites were fabricated following the same process

[133]. Arroyo et al. reinforced NR with 10 phr Na-MMT and ODA-MMT by melt mixing [134]. Graphene/rubber nanocomposites were also reported via melt blending [115, 134–135].

Simple methods like freeze drying, spray drying, heterocoagulation approach, pulverization method etc. can also be applied for the composite manufacture.

3.2 Properties of Nanocomposites

Reinforcing fillers for rubber typically have relatively large particle sizes. Such fillers are commonly inert without any reactive moieties on their surfaces. These fillers are added to the polymers to promote specific properties and also to reduce the cost. However, to improve the surface hydrophobicity, electrical conductivity, thermal conductivity and relative permeability of polymeric materials, micro/nano-fillers are added. Nanocomposites exhibit enhanced properties compared to the pristine elastomers. Fillers when exfoliated go to the free spaces in between the polymer chains and wrap around it. The better the dispersion, the more will be the reinforcement. This reinforcing effect of clay platelets can be well explained by the following Fig. 10. Styrene butadiene rubber is the elastomeric co polymer used here and it is found that the force between the fillers and the matrix balances the polar, steric and electrostatic contributions due to the nice dispersion of fillers. Thus, the segments of the polymer chains (butadiene part) are adsorbed on the kaolin surfaces and pores of the clay aggregates. This enhances the hard part (styrene) in the matrix and accordingly the elastic constant of rubber enhances

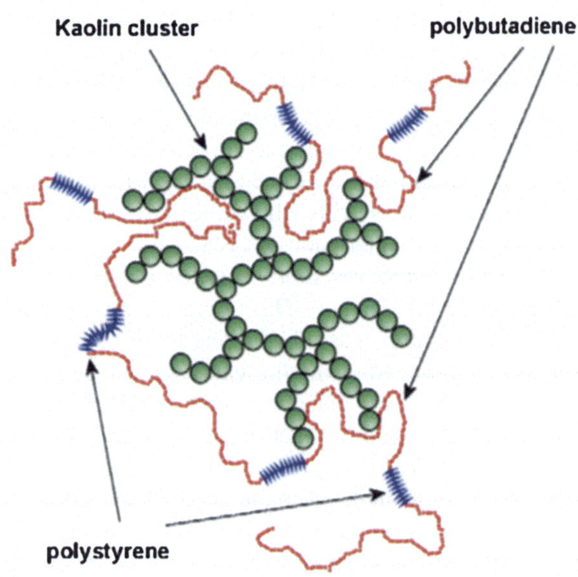

Fig. 10 Schematic representation of structure of styrene–butadiene block copolymer filled with kaolin powder [144]

dramatically when nanoclays are added. The usually addressed properties of all kinds of nanocomposites such as mechanical strength, rheology, electrical conductivity and gas permeability are discussed here in detail.

3.2.1 Mechanical Properties

The mechanical behavior of elastomer nanocomposites depends on several factors, including the filler dispersion, degree of interfacial adhesion between elastomers and fillers and crosslink density of the composite. Mechanical properties of a few nanocomposites are compared in Table 3.

The tensile strength and elongation at break of NR/SBR/organo clay (OC) nanocomposites is demonstrated in Fig. 11a, b. The authors correlated tensile strength with the microstructure of the prepared nanocomposites and found the influence of compatibilisers on the properties. The compatibilisers, epoxidised natural rubber (ENR50) and EPDM-g-MAH enhance the strength and stress at 100 and 300 % elongation to a great extent. They obtained highest reinforcement effect when the compatibilizer to clay ratio was 3 (B4EP3 and B4EN3).

The stress strain curves of Al_2O_3/NR nanocomposites with change in filler loading are shown in Fig. 12. The unique property of strain-induced crystallization of natural rubber is clearly shown in this curve. The modulus increases with

Table 3 Mechanical Properties of Elastomer/nano filler derivative composites

Elastomer	Filler	Content	Processing	Modulus (MPa) (composite/ neat polymer)	Tensile strength (MPa) (composite/ neat polymer)	Elongation at break (%) (composite/ neat polymer)
NR latex [123]	CRG In situ reduction	2 wt%	Solution	M 300 6.6/2.4	25.2/17.1	564/579
			Melt	2.47/2.4	18.8/17.1	600/579
NR [108]	MWCNT:Silica)/ 0 phr	3:40 phr	Melt	4.8/1 Stress at M100	29/12	470/480
EPDM [108]				5.5/1.1 M100	17.2/2.5	299/299
EPDM [132]	Nano Zno	3 phr	Melt	–	15.87	286
RTV [145]	Silica/POSS	Constant/		25 wt%	–	–
1.1	100					
NR65/ SR20/ BR15 [146]	Clay	10 phr	Melt	1.70	11.12	480

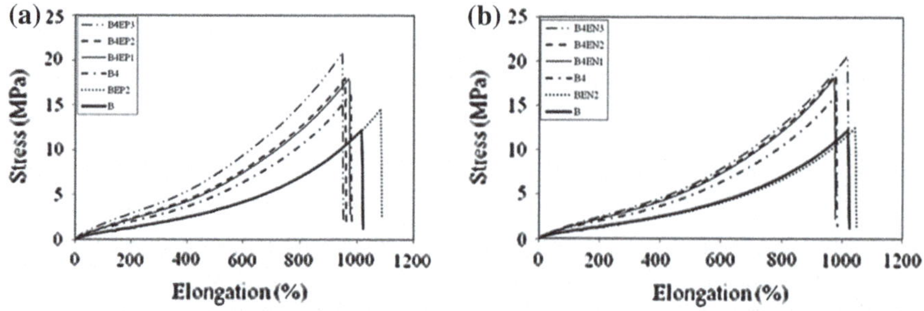

Fig. 11 Tensile strength and elongation at break for the NR/SBR simple blend (*B*), unfilled NR/SBR blend but containing compatibilizer (*BEP2, BEN2*), uncompatibilized NR/SBR/OC sample (*B4*), and NR/SBR/OC Nanocomposites compatibilized by: **a** EPDM-g-MAH and **b** ENR50 with compatibilizer to clay ratio of 1, 2 and 3 [147]

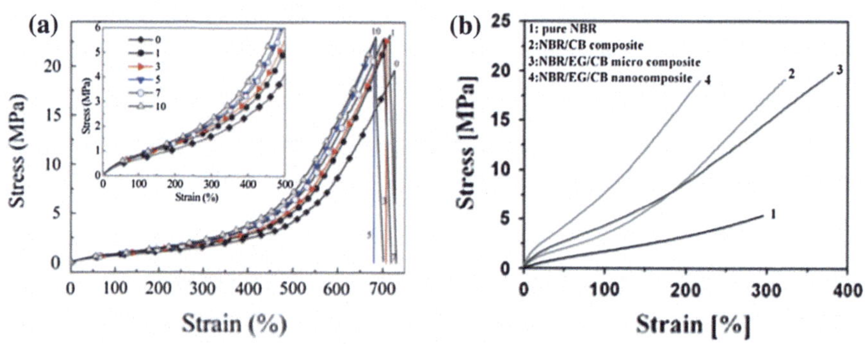

Fig. 12 Stress–strain curves of **a** KH570-Al_2O_3/NR Nanocomposites with different filler content [103] **b** NBR and its composites [148]

increase in strain at lower strain region and then sharply increases due to the development of induced crystallization. When the nanocomposites are considered, the tensile strength and modulus values are higher than that of pure NR, but the strain induced crystallization weakens. This is due to the better dispersion of Al_2O_3 nanoparticles, and the increased force between the nano layers and the NR chain. Figure 12b shows the tensile properties of NBR composites and it is observed that the elongation at break and modulus values of NBR/EG/CB (expanded graphite and carbon black composites) composites is higher than NBR/CB composites. Carbon black (CB) mainly contributes to rubber reinforcement and along with this, the higher dispersion of thin graphite sheets lead to significant enhancement in tensile modulus (at 100 % elongation) and hardness.

The steady increase in the tensile strength with strain is reported in the case of NBR–nano calcium carbonate (NBR-NCC) composites [149] which are typical for all synthetic elastomers. The addition of NCC increases stress in the composite for the same rate of strain and the strength increases with NCC amount. Here, the tensile

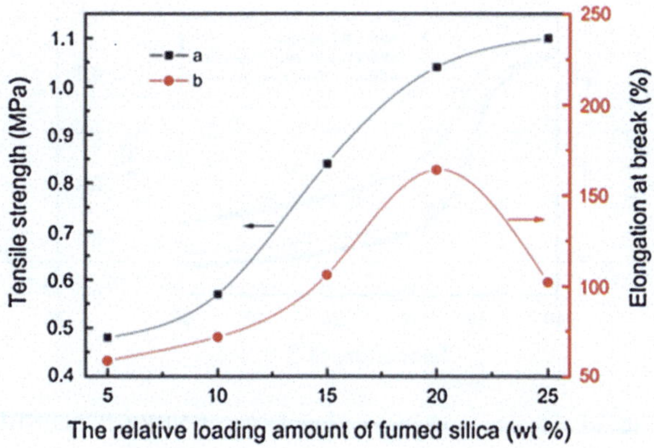

Fig. 13 Effects of the relative loading amount of fumed silica on mechanical properties of the novel RTV silicone rubbers: **a** tensile strength **b** elongation at break [145]

strength increases by 24 and 42 % respectively for 5 and 10 phr of NCC filler loading, which is equivalent to that attained with two times of commercially precipitated $CaCO_3$. The synergistic effect of fumed silica and POSS nanofillers on the mechanical properties of RTV silicone rubbers are reported [145] as shown in Fig. 13. The tensile strength of the neat elastomer increases from 0.48 to 1.10 MPa with the amount of filler loading (Fig. 13a), whereas the elongation at break increase from 58.6 to 164.1 % in the beginning and thereafter decreases with fumed silica (Fig. 13b). Both POSS-rich domains and fumed silica phases significantly enhance the thermal and mechanical properties of RTV silicone rubbers. The tensile properties of NBR/PAni DBSA blends containing 15 and 27 % of PAni are compared [150] and good conclusions were drawn. Even though PAni. DBSA exhibit poor mechanical strength, the presence of 15 and 27 % of it in blends of NBR fabricated by physical mixing and in situ polymerization resulted in an increase of both tensile strength and elongation at break values compared to pure NBR.

The dynamic mechanical property analysis reveals the storage modulus (E′) and loss factor (tan δ) of composites. These values for NBR/CB and NBR/EG/CB composites as a function of temperature are shown in Fig. 14. The E′ values for NBR/EG/CB nanocomposites are observed to be higher than that of NBR/CB and NBR/EG/CB microcomposites, similar to the tensile observation. For nanocomposites, the energy dissipation behavior is particular, for instance the hysterisis loss peak first decreases, broadens, and then increases, indicating the enhanced filler volume fraction and good nano-size dispersed graphite—rubber interfacial interaction [151]. This clearly illustrates that all nano particles have the unique ability of giving a positive effect on the mechanical strength in their composites, which again increases with the filler concentration and nature of dispersion. Similar trend has been observed in the case of natural rubber carbon nanotube composites as well [152].

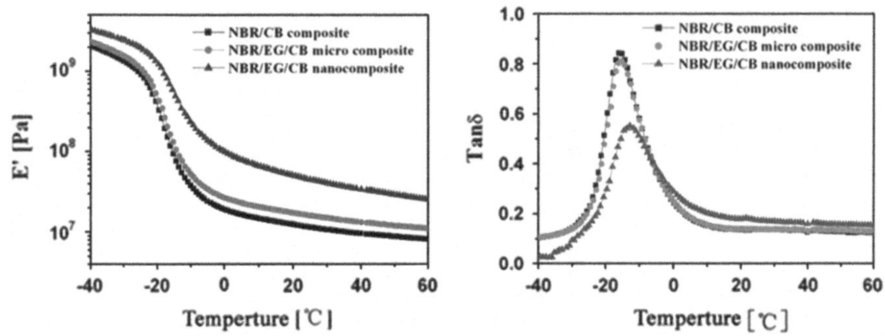

Fig. 14 Storage modulus and loss factor of NBR/CB and NBR/EG/CB composites [148]

3.2.2 Rheological Properties

Generally fillers are added into rubber to influence its physical properties. The effectiveness of a composite depends on the physical and chemical nature of the filler, which in turn depends on the amount of filler, filler–rubber and/or filler–filler interactions. This includes the nature and morphology of the filler, surface chemistry, concentration and the rate of dispersion in the matrices. Rheology is a very significant tool to explain the viscoelastic behavior of the material based on the nature of filler particles incorporated in the nanocomposites and understanding of the rheology of filled elastomers is significant in the design, processability and development of nanocomposites. This technique not only determines the final products properties but also controls the fluid and heat transfer characteristics during the processing. Thus it is very important to examine the dynamic viscosity as well as moduli based on the filler loading as well as the experimental factors like temperature, frequency, and strain rate [153]. Fillers in viscoelastic polymers can reduce melt elasticity and hence influence phenomena such as die swell [154]. At high filler loading yield stress is observed in composite systems, which is associated with particle interactions within the matrix. At this point, viscosity of the material enhances highly, showing a solid like behaviour. The difficulties addressed for rheological measurement of composites include the significantly smaller dimensions of suspended particles than the characteristic dimensions of the measuring equipment and the wall effect (formation of a melt layer at the wall surface) caused by a nonhomogeneous distribution of the dispersed phase. This layer of relatively low viscosity existing at the melt boundary gives rise to lubrication effects or apparent wall slippage [155]. Additionally, inertial and gravitational effects should be negligible.

The variation in storage modulus (G′) with frequency for different NR and SBR blend nanocomposites is shown in Fig. 15. Higher complex viscosity and storage modulus is observed for SBR when compared to NR, particularly at low frequency region. But the clay platelets have more influence in increasing these parameters for NR than that of SBR, suggesting greater ability of NR for intercalation process.

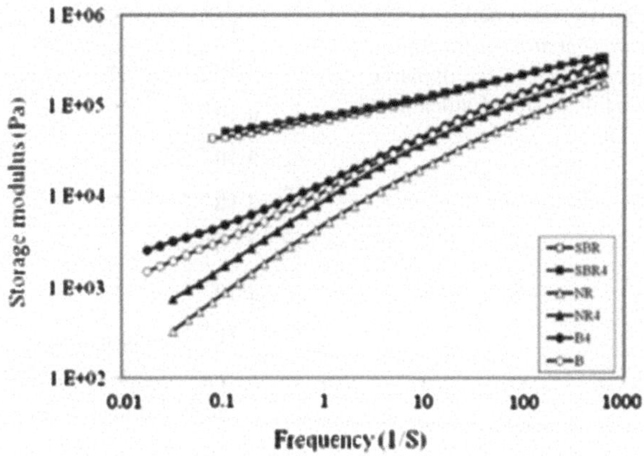

Fig. 15 Melt shear storage modulus (G′) as a function of frequency for unfilled NR, SBR, simple blend of NR/SBR, and filled counterparts [148]

On considering the polarity of the phases, the filler is expected to be present in both NR and SBR non polar constituents. The more intercalation effect of nanoclay on NR than SBR is also evidenced from the melt viscoelastic behavior of NR/SBR/clay composite (Fig. 15).

One of the important experimental parameters in determining the reinforcing action of elastomer properties is **Payne Effect**. It is known that the dynamic performance of tyres such as resistance to friction depend on the nonlinear viscoelastic behavior of tyre material, which is a vulcanized rubber composite. This kind of studies is very important as far as the fuel consumption and safety is considered and a lot of studies are focusing on the Payne effect. The effect is defined as the variation of storage modulus with strain amplitude and is related to the filler/rubber molecular interactions. Obviously the filled matrix shows higher storage (G′) and loss modulus (G″) values than neat rubber in the rubbery region depending on the filler dimensions.

Figure 16 illustrates respectively the loss and storage (G″ and G′ values) modulus, at small and large strain amplitudes. The Payne effect is generally demonstrated through the analysis of the low strain dynamic mechanical properties which describe the viscoelastic response of the rubbery material to periodic deformation. Filled elastomers display a unique dynamic viscoelastic behavior characterised by an amplitude dependence of the dynamic and loss moduli. The difference between the two limits (G″–G′) has been shown to depend on several parameters. It increases with increasing concentration of filler and with the specific surface area of the particles while it decreases with increasing temperatures and with improved dispersion. The contributions of the filler to the mechanical behavior are strongly strain dependent. Considering the influence of rubber occlusion by the filler particles, hydrodynamic effects account very well for the rheology of filled rubber at moderate strain levels [157–160]. At higher strains, debonding of the polymer chains from the filler, orientation of the latter phase, and microscopic crack deviation have been involved to explain the various physical properties.

Fig. 16 Schematic representation of the strain dependence of the dynamic moduli for filled vulcanizates [156]

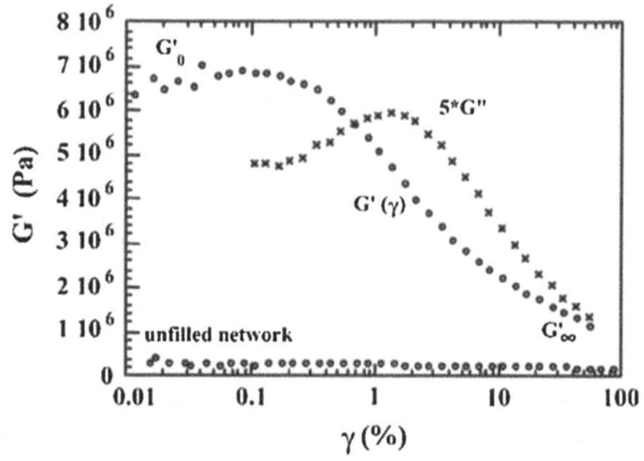

The magneto-rheological properties of nitrile rubber reinforced irregular, micrometer iron particles are examined within the frequency range 100–1250 Hz applying an external magnetic field. This material has good application in vibration isolators [161]. The shear modulus of this composite depends strongly on both frequency and magnetic field, while the loss factor is almost independent of those factors. In this technique, the instrumental set up increases the magnetic induction on one side while reducing the magnetic cross transfer via another part. The magnetic field independent loss factor and the frequency independent relative shear modulus magnitude of the material have important implications such as statistical elastic and quasi—statistical shear modulus modelling and frequency split. Of these, one factor considers magnetic field dependent, elastic and static properties while the other factor the frequency dependent, viscoelastic rubber properties within the audible frequency range.

The storage modulus (G′) as a function of frequency for a blend/CNT nanocomposite is shown in Fig. 17 with filler loading from 1 to 20 wt%. As expected, the influence of the nanofillers is more pronounced in the low-frequency region where G′ increases at a fixed frequency, and the slope decreases and tends rapidly

Fig. 17 Storage modulus (G′) as a function of frequency at T ¼ 120 °C for different blend/CNT composites [162]

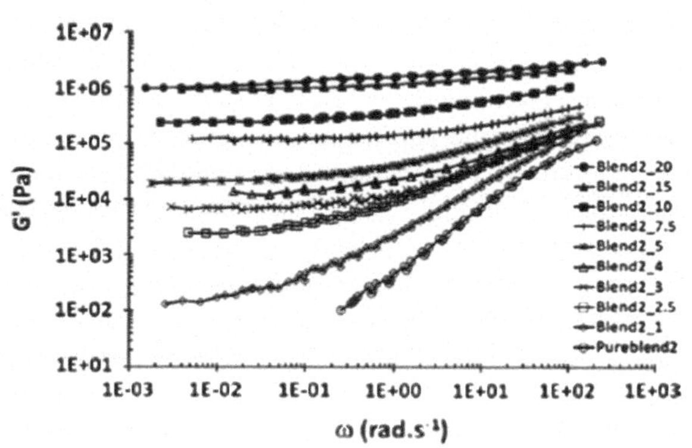

to zero when the filler content increases. This phenomenon appears from 1 wt% of CNT and could be explained by the existence of a percolated network. Above this percolation threshold, G′ was independent of the frequency, which is characteristic of a viscoelastic solid. Many studies report this phenomenon [163–165].

3.2.3 Electrical Properties

Conducting nanoparticles are added to various elastomer matrices to impart conductivity and this paves the way to flexible electronics. Fillers like CNTs, graphene and other graphitic fillers have the ability to reduce the volume resistivity to a great extent; the values for various nanocomposites are represented in Table 4.

The variation of volume resistivity with the nanotube content is given in Fig. 18. The percolation threshold for the three elastomer matrices, NR, SBR and EPDM is found to be 0.5 phr (volume fraction = 0.002), which is much lower than the already reported values [161, 167] for these elastomers as well as composites containing conventional microscale conducting fillers [168]. This property of CNTs to form a conducting network at low level of filler loading is due to their intrinsically high conductivity and high aspect ratio. The improvement in electrical conductivity obtained here indicates better dispersion of CNTs because of the processing method (sonication) evidenced from the microscopic technique.

When the carbon black and PAni fillers dispersed in a given polymer matrix are considered, the carbon black containing systems offer no solubility effects, and structuring of carbon black particles occurs during the melt processing step. But in the doped PAni polymer systems the shorter chains may partially dissolve in the melt and upon cooling partially precipitate out to form fine, discretely dispersed, doped PAni particles resulting in a conducting network. This phenomenon explains the conductivity increase with PAni-DBSA loading (Fig. 19). EPDM/PAni-DBSA-R crosslinked by electron beam irradiation produces highly conductive materials. Additional electron beam irradiation also enhances the conductivity.

Figure 20a shows the variation of log DC electrical conductivity of the composites against loading. Conductivity of pure XNBR in the order of 10^{-7} increases with CNT loading and becomes 10^{-2} orders of magnitude. At low filler concentration the conductivity remains very close to that of the pure, electrically

Table 4 Electrical conductivity of elastomer/conductive fillers nanocomposites

Elastomer	Concentration of filler	Electrical conductivity
NR [108]	3:40 phr MWCNT:Silica)	3.2×10^{-3} (S/m)
EPDM [108]	3:40 phr (MWCNT:CB)	2.1×10^{-1} (S/m)
XNBR [134]	3.9 phr MWCNT	0.07 (S/cm)
NBR [150]	PAni 48 wt%	2×10^{-2} (S/m)
SBR [166]	Graphene 5 %	4.56×10^{-7} (S/m)

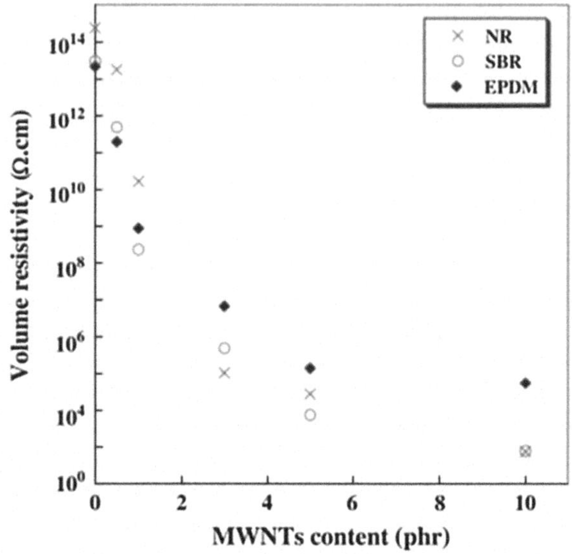

Fig. 18 Dependence of volume resistivity on CNT loading for NR, SBR, and EPDM composites [169]

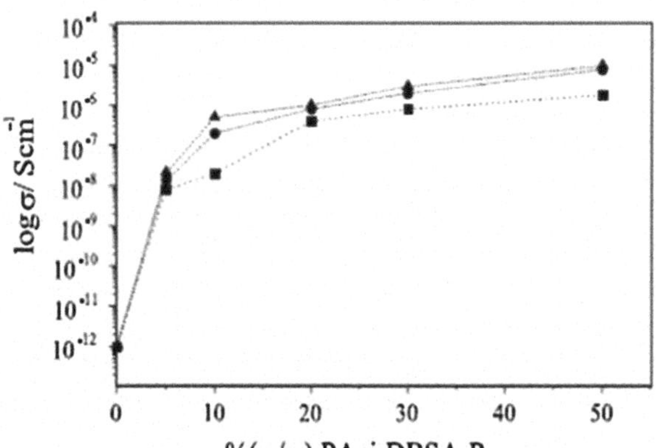

Fig. 19 Conductivity for EPDM/PAni-DBSA-R crosslinked with resin (*square shaped*) and electron beam using (*circle shaped*) 75 kGy and (*triangle shape*) 150 kGy [124]

insulating polymer matrix, since the fillers occur only individually or in small clusters throughout the matrix [170]. Percolation threshold and drastic increase in electric conductivity is a result of continuous network of CNT in the polymer matrix. The conducting CNTs are either making physical contact between them or are separated by small gaps [171].

The same effect of decrease in volume resistivity of silicone rubber with the increase of filler content is represented in Fig. 21. The percolation threshold for graphene is about 2 wt% with volume resistivity of 10^5 Ω cm whereas for MWNTs it is about 5 wt% with volume resistivity of 4.21×10^6 Ω cm. The two dimensional planar structure of graphene facilitates its better dispersion in the matrix compared to MWNTs. NTs tend to aggregate in silicone rubber, giving rise to a larger percolation value.

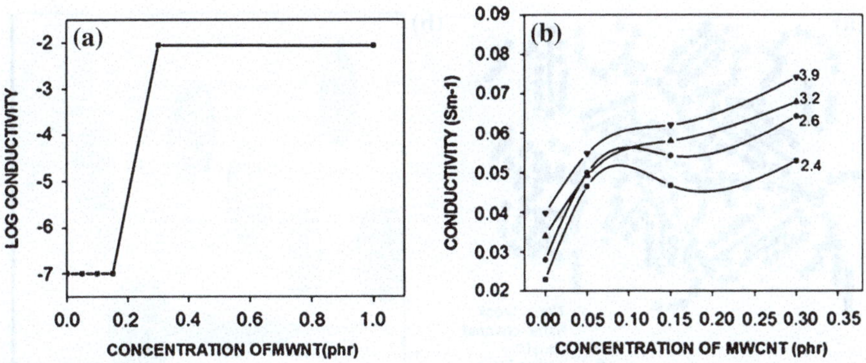

Fig. 20 Variation of log DC conductivity (**a**) and AC conductivity at different frequencies (**b**) against concentration [118]

Fig. 21 The resistivity of silicone rubber composites [118]

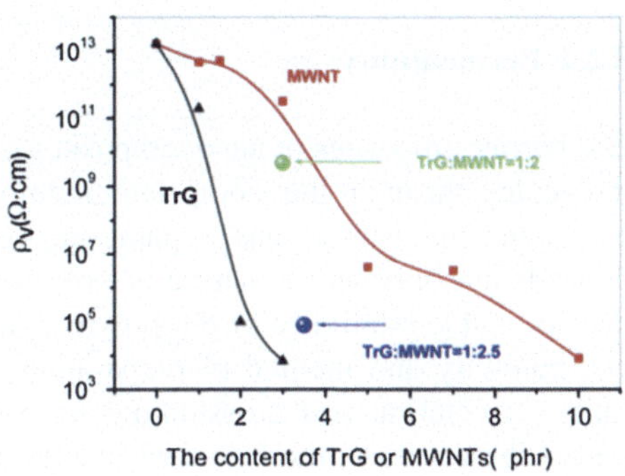

When both graphene and carbon nanotubes are filled in silicone rubber composites, carbon nanotube get well dispersed in presence of graphene and good conductive network is formed. Carbon nanotubes act as conductive bridges between graphene layers and the matrix. The synergistic effect of carbon black, carbon nanotube and graphene in forming the conductive network is schematically represented in the Fig. 22a, where as Fig. 22b shows the same effect between only CNTs and graphene flakes.

The large surface area of graphene and high aspect ratio of carbon nanotubes contribute towards the synergistic effect. These one dimensional and two dimensional fillers together form a "conductive bridge" in the silicone rubber (Fig. 22). MWNTs are difficult to disperse in SR alone, whereas it is much easier with the aid of graphene. The strong interaction between graphene and SR as well as between graphene and MWNTs account for the good dispersion of NTs. Graphene behaves as a compatibilizer in this case. Both physical isolation and hydrogen bonding between graphene and MWNTs during the dispersion lead to the better dispersion of NTs.

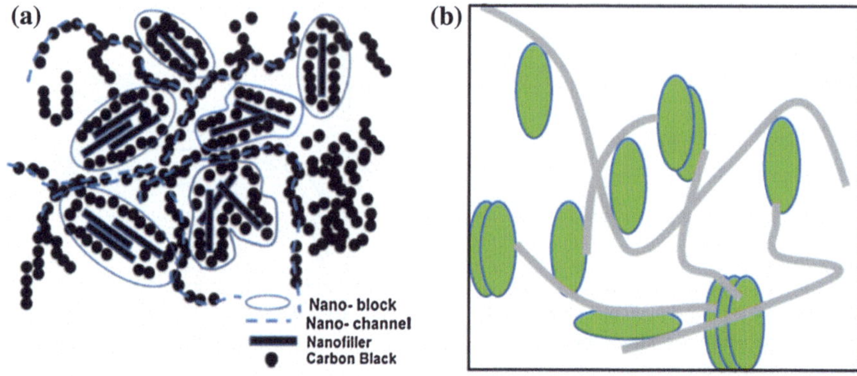

Fig. 22 **a** Microstructural developments in nanofiber based nanocomposites in the presence of CB [172] **b** Schematic of the synergistic interaction between graphene and carbon nanotubes [118]

3.2.4 Permeability

Gas barrier properties of nanocomposites is related to the transmission of penetrant molecules through the composite membranes and are very important in the packaging industries, gas separation, solvent studies etc. This phenomenon depends not only on the nature of polymer and diffusant but also varies with the density and crystallinity of the polymer, filler orientation, crosslinking, plasticizers, humidity and method of preparation of the composite. The density, crystallinity, orientation and crosslinking have a negative effect on permeability, the higher these values, the lower will be the gas permeability. Thickness of the film does not affect the permeability much, though different values are obtained from films of variable thickness. Inorganic fillers decrease permeability based on the type, shape and amount of filler and its interaction with polymer. Humidity and presence of plasticizers increases permittivity, being the effect more pronounced for hydrophilic polymers in the former case. Specific mechanisms by which diffusion occurs in polymeric systems have been reported [173]. Permeation does not depend on pressure, whereas an increase in temperature leads to decrease in penetrant solubility.

Nanoparticles having plate like structure have the unique ability to enhance the barrier properties of polymer films due to the formation of tortuous path [174] and by reduction of polymer mobilities. These interact with the elastomers, reduces its free volume and act as impermeable obstacles to gas molecules. It is important to develop barrier properties of rubber nanocomposites with superior mechanical and dynamic mechanical properties. The need for packaging and adhesive applications paves way for the scope of technological developments in this area. Recently a lot of scientists have done studies on gas transport in various elastomer nanocomposites. Kojima et al. [175] synthesized nitrile rubber/clay hybrid with 70 % of the permeability of neat NBR. Gatos et al. [176] have investigated the effect of the aspect ratio of silicate platelets on barrier properties of hydrogenated acrylonitrile butadiene rubber (HNBR)/layered silicate nanocomposites and found good correlation with Nielsen's model. Similar studies with clay [177–179] and graphite [180] were also reported on acrylonitrile butadiene rubber (NBR). Natural rubber latex based nanocomposites with an aqueous dispersion of layered clay have also reported for their high barrier properties [181]. Influence of processing conditions in regulating the transport properties of polyurethane/clay nanocomposites has established by Herrera-Alonso et al. [182]. Meneghetti et al. have reported styrene-butadiene rubber-clay nanocomposites with enhanced mechanical and gas barrier properties [183]. They found that surfactant chain length and functional groups determine the extent of dispersion and the modified nanoclays exhibit the best results. Generally the presence of platelet-like silicate layers greatly reduces the permeability for most of the elastomers due to their impermeability to diffusing molecules [184]. This enhances the path length required for the diffusant to travel through the composite [185]. The permeability values for various nanocomposites are given in Table 5.

Figure 23a, b illustrate the oxygen transmission rate (OTR) values of cloisite 15A/NR nanocomposites at different loadings of cloisite 15A respectively at 45 °C and at various temperatures. The enhancement in barrier properties for the composites over NR has been shown by plotting the relative permeabilities of these nanocomposites at same temperatures. It was observed that both response time and time to equilibration decreased with increase in temperature, which is explained in terms of increased membrane mobility and easier adsorption of permeants onto the membranes.

As described above, increasing in filler loading can provide longer diffusion path, thus improving the gas barrier property. Lu and Mai [190] and Bharadwaj [191] and Nielson models are generally used to explain the permeability effect based on the filler aspect ratio (diameter/thickness for silicate layer). Since exfoliated clay layers have much larger aspect ratios than intercalated structures, they have better gas barrier property. The dependence of relative gas permeability on the inorganic filler volume fraction of epichlorohydrin with predictions based on simple Nielsen model for different aspect ratios is showed in Fig. 24. The

Table 5 Gas permeability of rubber nanocomposites

Elastomer	Filler	Content	Processing	Permeant	Relative reduction
NR [186]	TRG	1.7 (vol%)	Solution /melt	Air	60
IIR [136]	Clay	8 (vol%)	Solution	Oxygen	20
Epichlorohydrin rubber (ECO) [187]	Clay	40 %	Melt	Nitrogen	70
SR [188]	FGS	0.43 (vol%)	–	Nitrogen	7.3
		1.31 (vol%)	–	Nitrogen	49
EPDM [187]	Clay	40 %	Melt	Nitrogen	30
SBR [187]	Clay	40 %	Melt	Nitrogen	50

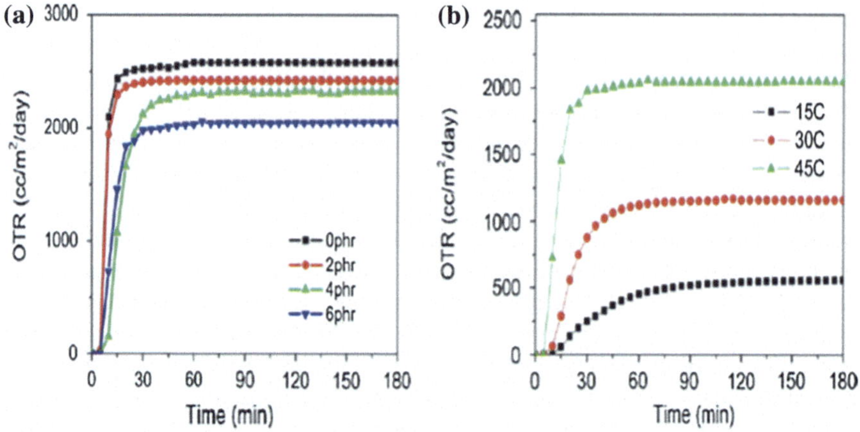

Fig. 23 OTR values of **a** clay/NR nanocomposites at different loadings of cloisite 15A at 45 °C and **b** of 6 phr clay filled NR nanocomposites at different temperatures [189]

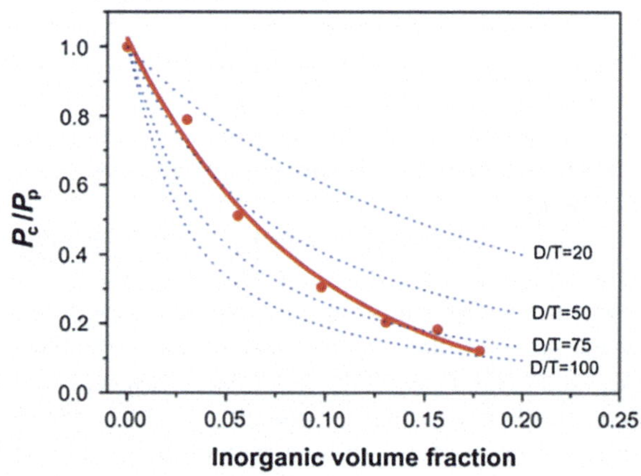

Fig. 24 Relationship between relative gas permeability (Pc/Pp) of ECOCNs (epichlorohydrin) and inorganic volume fraction. The *dotted curves* are numerical predictions for disk-shaped inclusions with aspect ratios 20, 50, 75, 100 (diameter/thickness) according to Nielson model [187]

experimental values are in good correlation with the theoretical predictions at lower filler loadings for fillers having aspect ratio 50 and at higher loading for the fillers with aspect ratio 75.

3.2.5 Morphology

Morphological analysis has very significance as far as the nanocomposite properties are considered. Morphology is an essential tool for the characterisation of the filler dispersion inside the polymer matrix and also the nature and type of interfaces. Figure 25 illustrates the transmission electron micrographs of the calcium carbonate reinforced composites of nitrile rubber. Lower filler loading

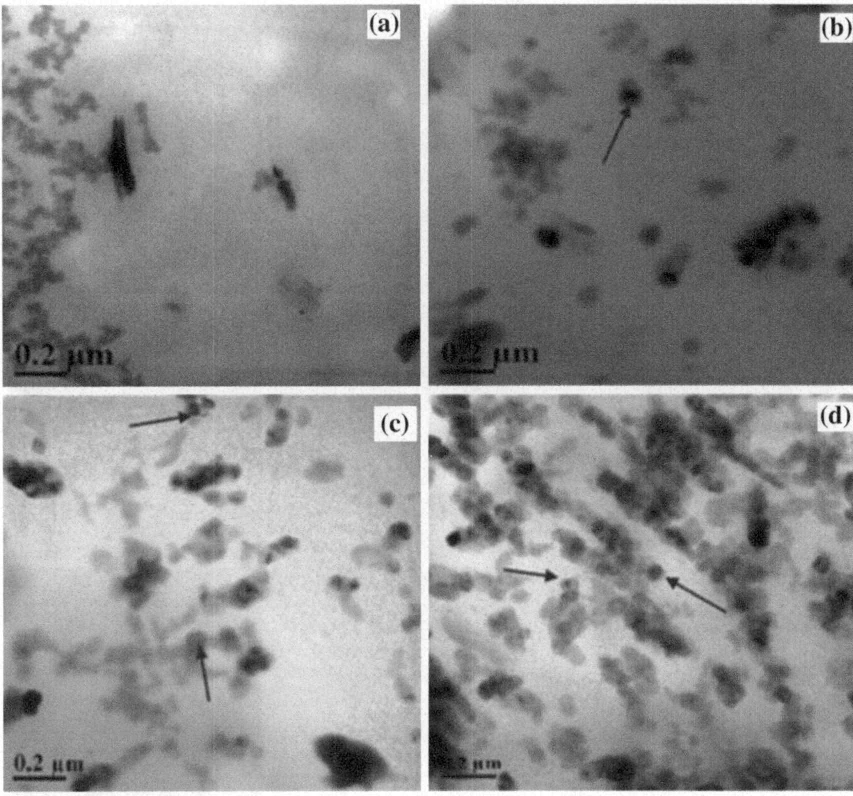

Fig. 25 TEM micrographs of NBR–nanocalcium carbonate composites **a** 0 phr, **b** 2 phr, **c** 5 phr, and **d** 10 phr [192]

essentially gives the idea of filler dispersion whereas at higher loadings the nanoparticles tend to have an agglomerating tendency which is inappropriate during the composite manufacture.

Morphology is really helpful for the characterization of carbon nanotube reinforced composites since the nanotubes have very strong tendency to form bundles due to their polar nature. Apart from the level of dispersion, the nature of interfacial interaction and the type of bonding in composites can also be found out from the micrographs. This is clear from Fig. 26, where three different kinds of nanocomposites are studied. Figure 26a–c shows typical TEM images of produced MWNT1s in rubber matrix containing 5, 9, and 16 wt% filler. MWNT2-filled rubber nanocomposites (Fig. 26d–f) depict the tortuous configurations of MWNTs after milling and their homogeneous dispersion in the composites. To further investigate the 3D morphology, the configurations of the MWNT2-filled rubber composite have also been investigated using 3D images with detailed information, as shown in Fig. 26g–i.

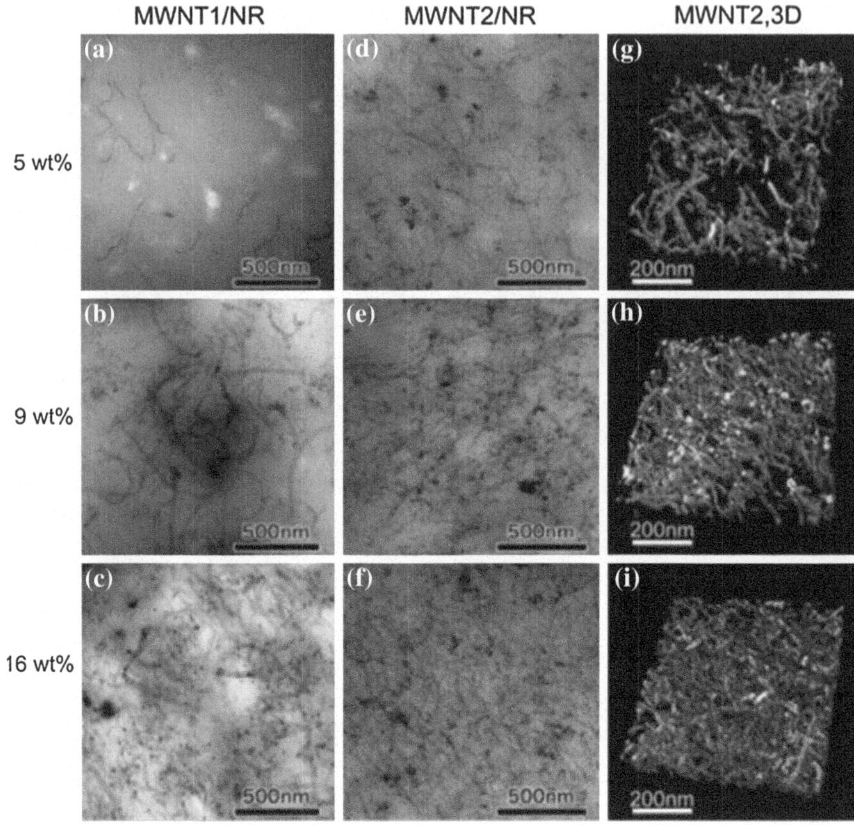

Fig. 26 TEM images of the produced composites containing 5, 9, and 16 wt% of fillers. **a–c** MWNT1s/rubber, **d–f** MWNT2s/rubber, **g–i** corresponding 3D images of MWNT2s/rubber. The thickness of each sample is 130 nm [192]

The modification of nanotubes and hence the enhanced filler dispersion is also evidenced by using TEM [187]. The high resolution transmission electron microscopy image of natural rubber composite of expanded graphite and carbon black is also reported [186]. The synergistic effect of two fillers in terms of good filler–filler and filler-polymer interactions can be well explained by analyzing the TEM image. The dispersion of reduced graphene oxide (RG-O) into natural rubber (NR) was also observed from morphology and is correlated with the dramatic enhancement in the mechanical, electrical, and thermal properties of composites [185]. Co-coagulating a stable RG-O suspension with NR latex afforded a web like morphology consisting of platelet networks between the latex particles, while two-roll mill processing broke down this structure, yielding a homogeneous and improved dispersion. The physical properties of composites with both morphologies were compared and found that the network morphology was highly beneficial for thermal and electrical conductivity properties and greatly increased

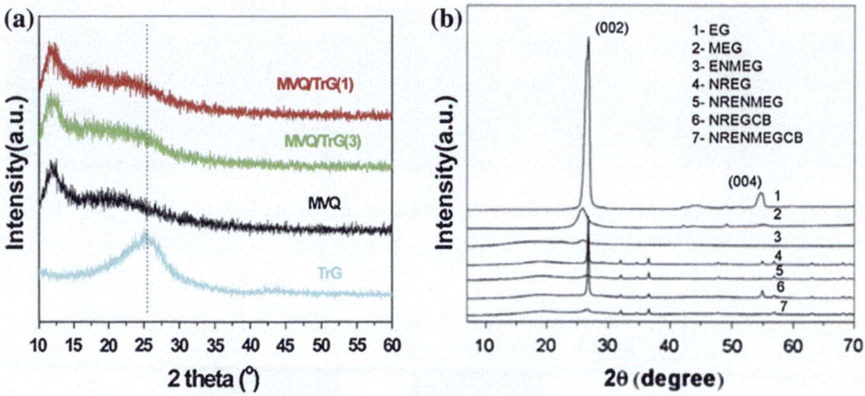

Fig. 27 XRD profiles of **a** graphene, silicone rubber and silicone rubber/graphene composite [124] and **b** EG, modified EG and EG/MEG filled NR based nanocomposites in presence and absence of CB [186]

stiffness but was detrimental to elongation. In addition to the microscopic techniques such as TEM, SEM and AFM, other techniques like XRD, Raman etc. are also used to characterise the nature of filler and composite morphology. Figure 27 illustrates the X-ray diffractogram of various composites of clay and graphitic fillers. The intensity and shift in the 2θ values gives clear picture about the filler exfoliation [193]. In the CNT and graphene case Raman spectroscopy plays an essential role for the structural information.

The importance of bio fillers in fabricating nanocomposites is increasing to a wider extent nowadays. Bio nanocomposites are composite materials comprising one or more phase(s) derived from a biological origin and the size is of nano range. Polysaccharides such as starch and cellulose have been used as reinforcing agents in many polymer matrix like natural rubber. The various advantages like renewability, non-toxicity and biocompatibility, of the bionanocomposites prepared from nanofillers like cellulose fibers, chitin whiskers, and starch nanocrystals are made use in a wide range of applications, like medicines, coatings, food products and packaging. Both solution blending and dry mixing methods have been employed for the development of bionanocomposites and they offer potential eco-friendly substitutes for the conventional filler carbon black [88]. It is found that the cellulose nanowhiskers prepared from bio sources are good reinforcement for natural rubber system [194]. Figure 28 explains the random orientation of nanowhiskers in the NR matrix and their homogeneous distribution. SEM also points out the dispersion of fillers without microscale aggregation, especially at 5 wt% CNW concentration.

Morphology is again used in finding out the nature of distribution of starch nanocrystals in rubber medium by Bouthegourd et al. [195]. The AFM images shown in Fig. 29 illustrates the uniform distribution of the nanocrystals at low filler content (<15 %), while a tendency for agglomeration is observed for content greater than 15 % w/w. The nanocrystals were prepared from potato base medium

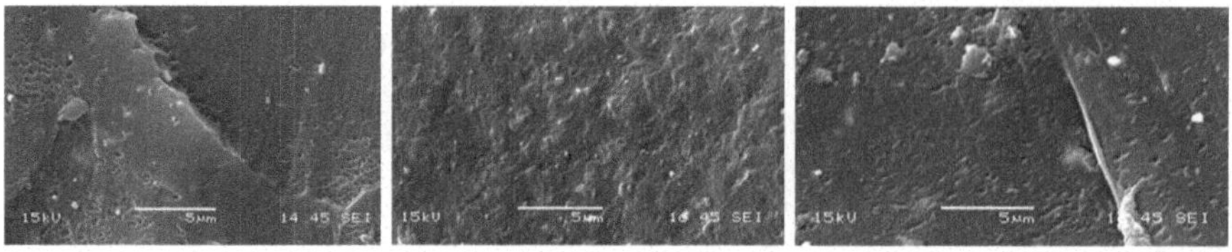

Fig. 28 SEM images of NR/Cellulose nanowhiskers nanocomposites: **a** neat NR, **b** NR-CNW2.5, **c** NR-CNW10 [194]

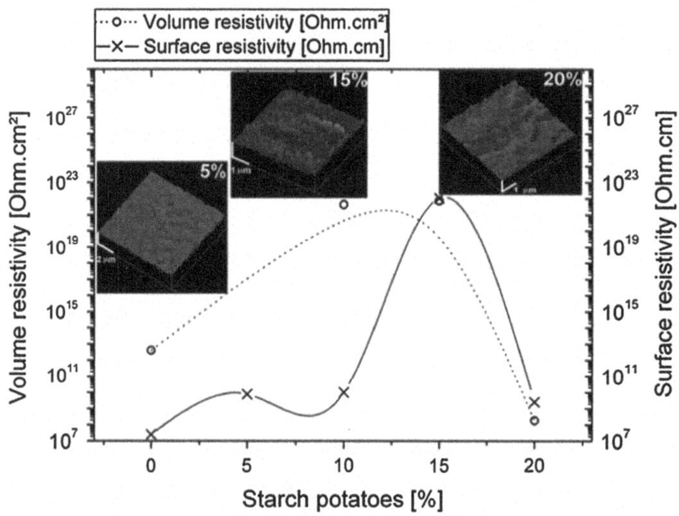

Fig. 29 Effect of starch potato nanocrystal composition on the volume and surface electrical resistivity of the nanocomposites. In this figure the AFM images for 5, 15 and 20 % of starch are also displayed. *Lines* are drawn as a guide for the eyes [195]

by following the method of hydrolysis. The electrical resistance value determined from the dielectric studies also supports the uniform distribution of crystals throughout the matrix. The increase of the electrical resistance at 15 wt% indicates that nanoparticles act as a barrier to the electrical charge movements.

All the above discussed properties give a detailed explanation of the improvement in composite properties while filling the matrix with nano particles.

3.2.6 Applications

A lot of applications have been envisaged for the nanocomposites developed. The synthesized composite membranes can be sandwiched between two electrodes to produce membrane electrode assembly (MEA) using hot press method at constant conditions of temperature, pressure and time. The performance of the fabricated

MEA was tested in a single PEM fuel cell using hydrogen as the fuel gas and oxygen as oxidant at room temperature [109]. Synthesis and characterization of nanosilver based antimicrobial natural rubber latex foam can strongly inhibit the growth of bacteria and fungus [132]. Elastomer nanocomposite with high electromagnetic dissipation and shielding properties is useful in electromagnetic interference (EMI) shielding devices. Combining the good electrical properties, flexibility and fluid resistance, nanocomposites yield very versatile and light weight materials for a variety of electromagnetic static discharge and EMI applications [196]. Porous nano composite materials have numerous applications in areas such as catalysis, chromatography, and separation, where control over pore structure and pore size strongly influences the efficiency of the material. POSS is noted for fabricating such composite materials [197]. Two dimensional fillers impart superior barrier properties to the composite membranes. Several nanocomposites of graphene and carbon nanotubes are reported as good dielectric materials. In addition to composites formed from single filler, multiple fillers also contributes towards better application due to their synergistic effect.

4 Conclusion

A brief account of the recent researches on the synthesis, properties and applications of rubber nanocomposites has been done. Various nano fillers like layered silicates, carbon nanotubes, graphene, POSS, PAni, nanosilica, biofillers, graphene oxide etc. were addressed. It is found that conductive nanofillers like CNT and graphene have tremendous applications in electric and dielectric fields as they can impart conductivity to the rubber insulator. The barrier properties of rubbers have superior applications in tyre industry and fillers with two dimensional plate like morphology are useful in this case. The incorporation of nano-fillers into tyre components is driven by potentially higher overall performance, especially focused on fuel efficiency through reduced weight and energy dissipation. Because of the enhanced surface area and high aspect ratio all kinds of nano fillers can result in superior mechanical strength. Though the nanocomposites offer vast applications, more studies should be focused on it to completely exploit their properties. Dispersion, modification of fillers, improved interfacial interactions are some of the problems yet to be addressed. Effort has been made to explain the important aspects of this area emphasizing the challenges. Further researches and more enhancement in properties and superior applications are anticipated.

Acknowledgments The authors would like to acknowledge the University Grants Commission and the Department of Atomic Energy Consortium for giving enough funding to carry out the research works on Carbon nanotube. Thanks are also due to Nanomission of DST, New Delhi for their support. We also acknowledge the financial support from the Ministry of Higher Education, Science and Technology of the Republic of Slovenia through the contract No. 3211-10-000057 (Center of Excellence Polymer Materials and Technologies).

References

1. Bokobza, L.: Polymer **48**(17), 4907–4920 (2007)
2. Payne, A.R., Whittaker, R.E.: Rubber Chem. Technol. **44**, 440 (1971)
3. Waddell, W.H., Beauregard, P.A., Evans, L.R.: Tire Technol. Int. **1995**, 24 (1995)
4. Wang, M.J.: Rubber Chem. Technol. **72**, 430 (1999)
5. Pliskin, I., Tokita, N.: J. Appl. Polym. Sci. **16**, 173 (1972)
6. Herrera-Alonso, J.M., Sedlakova, Z., Marand, E.: J. Membr. Sci. **349**, 251 (2010)
7. Ansari, S., Giannelis, E.P.: J. Polym. Sci., Part B: Polym. Phys. **47**, 888 (2009)
8. Usuki, A., Tukigase, A., Kato, M.: Polymer **43**, 2185 (2002)
9. Zhang, L.Q., Wang, Y.Z., Wang, Y.Q., Sui, Y., Yu, D.S.: J. Appl. Polym. Sci. **2000**, 78 (1873)
10. De Falco, A., Goyanes, S., Rubiolo, G.H., Mondragon, I., Marzocca, M.A.: Appl. Surf. Sci. **254**(1), 262–265 (2007)
11. Ebbesen, T.W.: Annu. Rev. Mater. Sci. **24**, 235–264 (1994)
12. Treacy, M.M.J., Ebbesen, T.W., Gibson, J.M.: Nature **381**, 678–680 (1996)
13. Lian, Y., Liu, Y., Jiang, T., Shu, J., Lian, H., Cao, M.: J. Phys. Chem. C **114**, 21 (2010)
14. Kalaitzidou, K., Fukushima, H., Drzal, L.T.: Compos. Sci. Technol. **67**, 2045 (2007)
15. Someya, T., Kato, Y., Sekitani, T., Iba, S., Noguchi, Y., Murase, Y.: Proc. Natl. Acad. Sci. U.S.A. **102**, 12321–12325 (2005)
16. Mannsfeld, S.C.B., Tee, B.C.K., Stoltenberg, R.M., Chen, C.V.H.H., Barman, S., Muir, B.V.O.: Nat. Mater. **9**, 859–864 (2010)
17. Kim, M.J., Shin, D.W., Kim, J.Y., Park, S.H., Han, I.T., Yoo, J.B.: Carbon **47**, 3461–3465 (2009)
18. Sekitani, T., Nakajima, H., Maeda, H., Fukushima, T., Aida, T., Hata, K.: Nat. Mater. **8**, 494–499 (2009)
19. Chang, H.X., Wang, G.F., Yang, A., Tao, X.M., Liu, X.Q., Shen, Y.D.: Adv. Funct. Mater. **20**, 2893–2902 (2010)
20. Hu, L.B., Pasta, M., La Mantia, F., Cui, L.F., Jeong, S., Deshazer, H.D.: Nano Lett. **10**, 708–714 (2010)
21. Pelrine, R., Kornbluh, R., Pei, Q.B., Joseph, J.: Science **287**, 836–839 (2000)
22. Chen, I.W.P., Liang, Z.Y., Wang, B., Zhang, C.: Carbon **48**, 1064–1069 (2010)
23. Joshi, M., Butolaa, B.S.: J. Macromol. Sci. Part C: Polym. Rev. **44**, 389–410 (2004)
24. Li, G., Wang, L, Ni Jr, H.: J. Inorg. Organomet. Polym. **11**, 123–154 (2001)
25. McAllister, M.J., Li, J.L., Adamson, D.H., Schniepp, H.C., Abdala, A.A., Liu, J., Herrera-Alonso, A.M., Milius, D.L., Car, R., Prud'homme, R.K., Aksay, I.A.: Chem. Mater. **19**, 4396–4404 (2007)
26. Xanthos, M.: Polymers and polymer composites (chapter 1). In: Xanthos, M. (ed.) Functional Fillers for Plastics, Second, updated and enlarged edition, pp. 9–14. WILEY-VCH Verlag GmbH & Co., Weinheim (2010)
27. Chrissafisa, K., Bikiaris, D.: Thermochim. Acta **523**, 1–24 (2011)
28. Kumar, A.P., Depan, D., Tomer, N.S., Singh, R.P.: Prog. Polym. Sci. **34**, 479–515 (2009)
29. Rozenberg, B.A., Tenne, R.: Prog. Polym. Sci. **33**, 40 (2008)
30. Marquis, D.M., Guillaume É., Joly C.C.: Nanocomposites and Polymers with Analytical Methods. In: Cuppoletti, J. (ed.) Nanocomposites and Polymers with Analytical Methods. Intech publishers. 978-953-307-352-1 (2011)
31. Schadler, L.S., Brinson, L.C., Sawyer, W.G.: Nanocomposite Mater. 50–58 (2007)
32. Giannelis, E.P.: Appl. Organomet. Chem. **12**, 675–680 (1998)
33. Zou, H., Wu, S., Shen, J.: Chem. Rev. **108**, 3893–3957 (2008)
34. Bhattacharya, M., Maiti, M., Bhowmick, A.K.: Rubber Chem. Technol. **81**, 782–808 (2008)
35. Ramorino, G., Bignotti, F., Pandini, S., Ricco, T.: Compos. Sci. Technol. **69**, 1206–1211 (2009)

36. Carretero-Gonzalez, J., Valentin, J.L., Arroyo, M., Saalwachter, K., Lopez-Manchado, M.A.: Eur. Polymer J. **44**, 3493–3500 (2008)
37. Ayippadath Gopi J., Patel S.K., Chandra A.K., Tripathy D.K.: J Polym Res **18**, 1625–1634 (2011)
38. Bhattacharya M., Biswas S., Bhowmick A.K.: Polymer **52**, 1562–1576 (2011)
39. Kim W-S., Yi J., Lee D-H., Kim I-J., Son W-J.,. Bae J-W, Kim W.: J Appl Polym Sci **116**, 3373–3387 (2010)
40. Kim W-S, Paik H-J., Bae J-W., Kim W.: J Appl Polym Sci **122**, 1766–1777 (2011)
41. Kumar, K.D., Tsoub, A.H., Bhowmick, A.K.: J. Adhes. Sci. Technol. **24**, 789–809 (2010)
42. Ahmadi, S.J., Yudong, H., Li, W.: Iran. Polym. J. **13**, 415–422 (2004)
43. Varghese, S., Karger-Kocsis, J., Gatos, K.G.: Polymer **44**, 3977–3983 (2003)
44. Pal, K., Rajasekar, R., Pal, S.K., Kim, J.K., Das, C.K.: J. Nanosci. Nanotechnol. **10**, 3022–3033 (2010)
45. Alipour A.: Study the morphology, microstructure. International Conference on Nanotechnology and Biosensors IPCBEE, vol. 25, pp. 44–48 (2011)
46. Ranimol, S., Rosamma, A., Treesa, C., Varghese, S., Kuruvilla, J., Sabu, T.: J. Appl. Polym. Sci. **101**, 2355–2362 (2006)
47. Janotti, A., Van de Walle, C.G.: Rep. Prog. Phys. **72**, 126501 (1–29) (2009)
48. Xu, S., Wang, Z.L.: Nano Res **11**, 1013–1098 (2011)
49. Henning S.K.: Reduced zinc loading: Using zinc monomethacrylate to activate accelerated sulfur vulcanization. Fall 172nd Technical Meeting of the Rubber Division, American Chemical Society, Cleveland, OH; 16–18 Oct 2007. ISSN: 1547-1977
50. Smith J.: Slicing it extra thin: Next big breakthrough will be measured in billionths, not billions. Tire Review (2005). Accessed http://www.tirereview.com
51. Holister P., Harper T.E., Vas C.R.: NANOTUBES white paper. CMP Científica (2003). Accessed www.cmp-cientifica.com
52. Rensselaer Polytechnic Institute.: Thermal superconductivity in carbon nanotubes not so 'super' when added to certain materials. Sci Daily (2003) Accessed http://www.sciencedaily.com/releases/2003/11/031112072719.htm
53. Abdullateef, A.A., Thomas, S.P., Al-Harthi, M.A., De, S.K., Bandyopadhyay, S., Basfar, A.A., Atie, M.A.: J. Appl. Polym. Sci. **125**, E76–E84 (2012)
54. Ray, S.S., Okamoto, M.: Prog. Polym. Sci. **28**, 1539–1641 (2003)
55. Crutchley G.S.: WIPO patent application WO/2006/027618. (2005)
56. Puhala A.S., Zanzig D.J., Holden B.D., Maly N.A.: US Patent 6852794, 2005
57. Li G., Wang L., Ni H., Pittman Jr C.U.. Polyhedral oligomeric silsesquioxane (POSS) polymers and copolymers: A review. J Inorg Organomet Polym **11**, 123–154 (2001)
58. Chandra A.K.: Tire technology—recent advances and future trends (chapter 32). In: Bhowmick, A.K. (ed.) Current Topics in Elastomers Research, pp. 919–933. CRC Press, New York (2008)
59. Favier, V., Canova, G.R., Cavaille′, J.Y., Chanzy, H., Dufresne, A., Gauthier, C.: Polym. AdV. Tech. **6**, 351 (1995)
60. Favier, V., Cavaille ′, J.Y., Chanzy, H.: Macromolecules **28**, 6365 (1995)
61. Helbert, W., Cavaille′, J.Y., Dufresne, A.: Polym. Compos. **17**, 604 (1996)
62. Dufresne, A., Cavaille ′, J.Y., Helbert, W.: Polym. Compos **18**, 198 (1997)
63. Favier, V., Canova, G.R., Shrivastava, S.C., Cavaille, J.Y.: Polym. Eng. Sci. **37**, 1732 (1997)
64. Chazeau, L., Paillet, M., Cavaille′, J.Y.: J. Polym. Sci. Polym. Phys. **37**, 2151 (1999)
65. Dubief, D., Samain, E., Dufresne, A.: Macromolecules **32**, 5765 (1999)
66. Dufresne, A., Kellerhals, M.B., Witholt, B.: Macromolecules **32**, 7396 (1999)
67. Dufresne, A.: Compos. Interfaces **7**, 53 (2000)
68. Angle's, M.N., Dufresne, A.: Macromolecules **33**, 8344 (2000)
69. Angle's, M.N., Dufresne, A.: Macromolecules **34**, 2921 (2001)
70. Mathew, A.P., Dufresne, A.: Biomacromolecules **3**, 609 (2002)
71. Buleon, A.: Int. J. Biol. Macromol. **23**, 85–112 (1998)

72. Siqueira G., Bras J,, Dufresne A.: Cellulosic bionanocomposites: A review of preparation, properties and applications polymers 01/2010. Polymers **2**(4), 728–765 (2010)
73. Sakurada, I., Nukushima, Y., Ito, I.J.: Polym. Sci. **57**, 651–660 (1962)
74. Nishino, T., Matsuda, I., Hirao, K.: Macromolecules **37**, 7683 (2004)
75. Favier, V., Canova, G.R., Cavaile, J.Y., Chanzy, H., Dufresne, A., Gauthier, C.: Polym. AdV. Technol. **6**, 351–355 (1995)
76. Azizi Samir, M.A.S., Alloin, F., Dufresne, A.: Biomacromolecules **6**, 612–626 (2005)
77. Nakagaito, A.N., Yano, H.: Appl. Phys. A **78**, 547–552 (2004)
78. Zimmerman, T., Pohler, E., Schwaller, P.: Adv. Eng. Mater. **7**, 1156–1161 (2005)
79. Wood K.A., Shinichiro I., Hiroyuki, Y.: Biomacromolecules **8**, 3276–3278 (2007)
80. Corre, D.L., Bras, J., Dufresne, A.: Biomacromolecules **4**(657), 665 (2003)
81. Gopalan, N.K., Alain, D.: Biomacromolecules **4**, 657–665 (2003)
82. Kumar, M.N.V.R.: React. Funct. Polym. **46**, 1 (2000)
83. Rinaudo, M.: Prog. Polym. Sci. **31**, 603–632 (2006)
84. Marchessault, R.H., Morehead, R.R., Walter, N.M.: Nature **184**, 632 (1959)
85. Nair, K.G., Dufresne, A.: Biomacromolecules **4**, 666 (2003)
86. Morin, A., Dufresne, A.: Macromolecules **35**, 2190 (2002)
87. Paillet, M., Dufresne, A.: Macromolecules **34**, 6527 (2001)
88. Thakore, S.: Role of Biopolymers in Green Nanotechnology. InTechOpen, Publishers. 2012-03-07
89. Mishra, S., Shimpi, N.G.: J. Sci. Ind. Res. **64**, 744–751 (2005)
90. Bhattacharya, M., Bhowmick, A.K.: J. Mater. Sci. **45**, 6126–6138 (2010)
91. Xu S.H., Gu J., Luo Y.F., Jia D.M. eXPRESS Polym. Lett. **6**, 14–25 (2012)
92. Mouri, H., Akutagawa, K.: Rubber Chem. Technol. **72**, 960–968 (1999)
93. Meli G., Europe L., Tan A., Asia L.: Nippon Gomu Kyokaishi (J. Soc. Rubber Ind. Jpn) **79**, 160 (2006)
94. Pinho, M.S., Gorelova, M.M., Dezzotti, M., Soares, B.G., Pertsin, A.J.: J. Appl. Polym. Sci. **70**, 1543 (1998)
95. Faez, R., De Paoli, M.A.: J. Appl. Polym. Sci. **82**, 1768 (2001)
96. Faez, R., De Paoli, M.A.: Eur. Polym. J. **37**, 1139 (2001)
97. Faez, R., Gazotti, W.A., De Paoli, M.A.: Polymer **40**, 5497 (1999)
98. Tsanov, T., Ditcheva-Kortchakova, M., Terlmezyan, L.: Polym. Polym. Comp. **8**, 115 (2000)
99. Faez, R., Schuster, R.H., De Paoli, M.A.: Eur. Polym. J. **38**, 2459–2463 (2002)
100. Schmidt, V., Domenech, S.C., Soldi, M.S., Pinheiro, E.A., Soldi, V.: Polym. Degrad. Stab. **83**, 519 (2004)
101. Vallim, M.R., Felisberti, M.I., De Paoli, M.A.: J. Appl. Polym. Sci. **75**, 677 (2000)
102. Bikiaris D.N., Vassiliou A.A.: Fumed silica reinforced nanocomposites. In: Wang, B.Y. (ed.) Current Status and Promises in Environmental Biodegradation Research Focus, pp. 189–215. Nova Publishers, New York (Chapter 7) (2008)
103. Fu, J.F., Chen, L.Y., Yang, H., Zhong, Q.D., Shi, L.Y., Deng, W., Dong, X., Chen, Y., Zhao, G.Z.: Polym. Compos. **33**(3), 404–411 (2012)
104. Viculis, L.M., Mack, J.J., Mayer, O.M., Hahn, H.T., Kaner, B.: J. Mater. Chem. **15**, 974 (2005)
105. Lian, H., Li, S., Liu, K., Xu, L., Wang, K., Guo, W.: Polym. Eng. Sci. **51**(11), 2254–2260 (2011)
106. Geng, Y., Wang, S.J., Kim, J.: J. Colloid Inter. Sci. **336**, 592–598 (2009)
107. Qin, Y., Zhu, F., Luo, M., Zhang, L.: J. Appl. Polym. Sci. **121**, 97–101 (2011)
108. Lorenz, H., Fritzsche, J., Das, A., Stöckelhuber, K.W., Jurk, R., Heinrich, G., Klüppel, M.: Compos. Sci. Technol. **69**, 2135–2143 (2009)
109. Abdulkareem, A.S., Afolabi, A.S., Idibie, C.A., Iyuke, S.E., Pienaar, H.C.vZ: Energy Procedia **14**, 2026–2037 (2012)
110. Blighe, F.M., Diamond, D., Coleman, J.N., Lahiff, E.: Carbon **50**, 1447–1454 (2012)

111. Das, A., Stockelhuber, K.W., Jurk, R., Fritzsche, J., Kluppel, M., Heinrich, G.: Carbon **47**, 3313–3321 (2009)
112. Khalid, M., Ismail, A.F., Ratnam, C.T., Faridah, Y., Rashmi, W., Al Khatib, M.F.: Radiat. Phys. Chem. **79**, 1279–1285 (2010)
113. Ganter, M., Gronski, W., Reichert, P., Mulhaupt, R.: Rubber Chem. Technol. **74**, 221 (2001)
114. Ganter, M., Gronski, W., Semke, H., Zilg, T., Thomann, C., Mulhaupt, R.: Kautsch. Gummi Kunstst. **54**, 166 (2001)
115. Kim, H., Miura, Y., Macosko, C.W.: Chem. Mater. **22**, 3441–3450 (2010)
116. Bai, X., Wan, C., Zhang, Y., Zhai, Y.: Carbon **49**(5), 1608–1613 (2011)
117. Lian, H., Li, S., Liu, K., Xu, L., Wang, K., Guo, W.: Polym. Eng. Sci. **51**(11), 2254–2260 (2011)
118. Hua, H., Zhao, L., Liu, J., Liu, Y., Cheng, J., Luo, J., Liang, Y., Tao, Y., Wang, X., Zhao, J.: Polymer **53**, 3378–3385 (2012)
119. Singh, V.K., Shukla, A., Patra, M.K., Saini, L., Jani, R.K., Vadera, S.R., Kumar, N.: Carbon **50**, 2202–2208 (2012)
120. Anoop, A.K., Sunil, J.T., Rosamma, A., Rani, J.: Inter. J. Polym. Mater. **59**, 33–44 (2010)
121. Peng, Z., Feng, C., Luo, Y., Li, Y., Kong, L.X.: Carbon **48**, 4497–4503 (2010)
122. Tan, J., Wang, X., Luo, Y., Jia, D.: Mater. Des. **34**, 825–831 (2012)
123. Zhan, Y., Wu, J., Xia, H., Yan, N., Fei, G., Yuan, G.: Macromol. Mater. Eng. **296**(7), 590–602 (2011)
124. Nair, K.P., Thomas, P., Joseph, R.: Mater. Des. **41**, 23–30 (2012)
125. Tang, B.Z., Xu, H.Y.: Macromolecules **32**(8), 2569–2576 (1999)
126. Fan, J.H., Wan, M.X., Zhu, D.B., Chang, B.H., Pan, Z.W., Xie, S.S.: J. Appl. Polym. Sci. **74**, 2605–2610 (1999)
127. Star, A., Stoddart, J.F., Steuerman, D., Diehl, M., Boukai, A., Wong, E.W., Yang, X., Chung, S.W., Choi, H., Heath, J.R.: Angew. Chem. Int. Ed. **40**, 1721–1725 (2001)
128. Kaminsky, W., Schlobohm, M.: Macromol Symp (2011). dOI:10.1002/masy.19860040114
129. Kaminsky W., Miri M.: *J. Polym. Sci.: Polym. Chem. Ed.*, 2003, DOI: 10.1002/pol.1985.170230807
130. Das A., Stöckelhuber K.W., Jurk R., Saphiannikova M., Fritzsche J., Lorenz H., Klü ppel M., Heinrich G.: Polymer, **49**, 5276–83 (2008)
131. Vu, Y.T., Mark, J.E., Pham, L.H., Engelhardt, M.: J. Appl. Polym. Sci. **82**, 1391 (2001)
132. Varghese, S., Karger-Kocsis, J., Gatos, K.G.: Polymer **44**, 3977 (2003)
133. Akhlaghi, S., Kalaee, M., Mazinani, S., Jowdar, E., Nouri, A., Sharif, A., Sedaghat, N.: Thermochim. Acta **527**, 91–98 (2012)
134. Arroyo, M., Lopez-Manchado, M.A., Herrero, B.: Polymer **44**, 2447 (2003)
135. Prud'homme R.K., Ozbas B., Aksay I.A., Register R.A., Adamson D. H.: W.O. Patent 2008045778 A1, 2008
136. Wang, Y., Zhang, H., Wu, Y., Yang, J., Zhang, L.: Eur. Polym. J. **41**, 2776–2783 (2005)
137. Lianga, Y., Wang, Y., Wu, Y., Lu, Y., Zhang, H., Zhang, L.: Polym. Test. **24**, 12–17 (2005)
138. Tan, J.H., Wang, X.P., Luo, Y.F., Jia, D.M.: Appl Mech Mater **110–116**, 3810 (2011)
139. Zhou, X.W., Zhu, Y.F., Liang, J.: Mater. Res. Bull. **42**, 456–464 (2007)
140. Zhang, Q., He, H., Xi, K., Huang, X., Yu, X., Jia, X.: Macromolecules **44**(3), 550–557 (2011)
141. Herrera, N.N., Putaux, J.L., David, L., Lami, E.B.: Macromolecules **39**(26), 9177–9184 (2006)
142. Hwang, S., Liu, S., Hsu, P.P., Yeh, J., Yang, J., Chang, K., Chu, S.: Int. Commun. Heat Mass Transf. **38**(5), 597–606 (2011)
143. Zhu, J., Wei, S., Patil, R., Rutman, D., Kucknoor, A.S., Wang, A., Guo, Z.: Polymer **52**(9), 1954–1962 (2011)
144. Lagazzo, A., Lenzi, S., Botter, R., Cirillo, P., Demicheli, F., Beruto, D.T.: Particuology **8**, 245–250 (2010)
145. Chen, D., Liu, Y., Hu, C.: Polym. Degrad. Stab. **97**, 308–315 (2012)

146. Zarei, M., Naderi, G., Bakhshandeh, G.R., Shokoohi, S.: J. Appl. Polym. Sci. (2012). doi:10.1002/APP.37687
147. Tavakoli, M., Katbab, A.A., Nazockdast, H.: J. Appl. Polym. Sci. **123**, 1853–1864 (2012)
148. Wang, L.L., Zhang, L.Q., Tian, M.: Mater. Des. **39**, 450–457 (2012)
149. Balachandran, M., Bhagawan, S.S.: J. Appl. Polym. Sci. **000**, 000 (2012)
150. Soares, B.G., Amorim, G.S., Oliveira, M.G., Pereira da Silva, J.E.: Macromol. Symp. **233**, 95–101 (2006)
151. Yang, J., Tian, M., Jia, Q.X., Shi, J.H., Zhang, L.Q., Lim, S.H., Yu, Z.Z., Mai, Y.W.: Acta Mater. **55**, 6372–6382 (2007)
152. Bhattacharyya, S., Sinturel, C., Bahloul, O., Saboungi, M.L., Thomas, S., Salvetat, J.P.: Carbon **2008**, 46 (1037)
153. Malkin, A.Y.: Adv. Polym. Sci. **96**, 69 (1990)
154. Carreau, P.J., Lavoie, P.A.: Macromol. Symp. **108**, 111 (1996)
155. Nah, C., Ryu, H.J., Kim, W.D., Choi, S.S.: Polym. Adv. Technol. **13**(9), 649–652 (2002)
156. Bokobza, L.: Macromol Symp **171**(1), 163–170 (2001)
157. Medalia, J.: Colloid. Interface Sci. **32**, 115 (1970)
158. Kraus, G., Polym, I.: Sci. **I3**(8), 601 (1970)
159. Medalia, A.: Rubber Chem. Technol. **47**, 411 (1974)
160. Ulmer, J.D., Hess, W.M., Chirico, V.E.: Rubber Chem. Technol. **47**, 729 (1974)
161. Kari, L., Lokander, M., Stenberg, B., Stockholm: Kautschuk Gummi Kunststoffe **55**, Nr. 12/2002
162. Charman, M., Le′ onardi, F., Dominguez, S., Bissuel,C., Derail, C.: J. Polym. Sci. Part B: Polym. **49**, 1597–1604 (2011)
163. Zhao, J., Morgan, A.B., Harris, J.D.: Polymer **46**, 8641–8660 (2005)
164. Zhang, Q., Rastogi, S., Chen, D., Lippits, D., Lemstra, P.J.: Carbon **44**, 778–785 (2006)
165. Pö tschke, P., Fornes, T.D., Paul, D.R.: Polymer **43**, 3247–3255 (2002)
166. Kim, J.S., Yun, J.H., Kim, I., Shim, S.E.: J. Ind. Eng. Chem. **17**, 325–330 (2011)
167. Bokobza, L., Belin, C.: J. Appl. Polym. Sci. **105**, 2054 (2007)
168. Perez, L.D., Zuluaga, M.A., Kyu, T., Mark, J.E., Lopez, B.L.: Polym. Eng. Sci. **49**, 866 (2009)
169. Bokobza, L.: Polym. Adv. Technol. (2012). doi:10.1002/pat.3027
170. Bose, S., Khare, R.A., Moldenaers, P.: Polymer **51**, 975–993 (2010)
171. Reffaee, A.S.A., El Nashar, D.E., Abd-EI-Messieh, S.L., Abd-El Nour, K.N.: Mater. Des. **30**, 3760–3769 (2009)
172. Bhattacharya, M., Bhowmick, A.K.: J. Mater. Sci. **45**, 6126–6138 (2010)
173. Schlotter, N.E., Furlan, P.Y.: Polymer **33**, 3323–3342 (1992)
174. Qutubuddin, S., Fu, X.: Polymer-clay nanocomposites: synthesis and properties. In: Rosoff M. (ed.) Nano-surface Chemistry, pp. 653–673. Marcel Dekker, New York (2002)
175. Kojima, Y., Fukumori, K., Usuki, A., Okada, A., Kurauchi, T.: J. Mater. Sci. Lett. **12**, 889–890 (1993)
176. Gatos, K.G., Karger-Kocsis, J.: Eur. Polym. J. **43**(4), 1097–1104 (2007)
177. Kim, J., Oh, T., Lee, D.: Polym. Int. **52**, 1058–1063 (2003)
178. Hwang, W.G., Wei, K.H.: Polym. Eng. Sci. **44**(11), 2117–2124 (2004)
179. Smoug D.: Mod. Plast. **2**, 28–34 (1998)
180. Tian, M., Wang, Y., Zou, H., Yang, J., Zhang, L.Q.: PMSE Prepr. **98**, 910–912 (2008)
181. Varghese, S.: Indian Patent, IN 2005CH00562 A 20070727, 2007
182. Herrera-Alonso, J.M., Marand, E., Little, J.C., Cox, S.S.: J. Membr. Sci. **337**, 208–214 (2009)
183. Meneghetti, P., Shaikh, S., Qutubuddin, S., Nazarenko, S.: Rubb. Chem. Technol. **81**(5), 821–841 (2008)
184. Yano, K., Usuki, A., Okada, A., Kurauchi, T., Kamigaito, O.: J. Polym. Sci. Polym. Chem. Ed. **31**, 2493 (1993)
185. Potts, J.R., Shankar, O., Du, L., Ruoff, R.S.: Macromolecules (2012). doi:10.1021/ma300706k

186. Malas, A., Das, C.K., Das, A., Heinrich, G.: Mater. Des. **39**, 410–417 (2012)
187. Lu, Y.L., Li, Z., Yu, Z.Z., Tian, M., Zhang, L.Q., Mai, Y.W.: Comp. Sci. Tech. **67**, 2903–2913 (2007)
188. Shuyang, P., Aksay, I.A., Prud'homme, R.K.: US Patent 20110178224 A1, 2011
189. Bhattacharyaa, M., Biswasa, S., Bandyopadhyay, S., Bhowmick, A.K.: Polym. Adv. Technol. **23**(3), 596–610 (2012)
190. Lu, C.S., Mai, Y.W.: Phys. Rev. Lett. **95**, 088303 (2005)
191. Bharadwaj, R.K.: Macromolecules **34**, 9189–9192 (2001)
192. Deng, F., Ito, M., Noguchi, T., Wang, L., Ueki, H., Niihara, K., Kim, Y.A., Endo, M., Zheng, Q.: ACS Nano **5**, 3858–3866 (2011)
193. Zhang, Q., Liu, Q., Zhang, Y., Cheng, H., Lu, Y.: Appl. Clay Sci. **65–66**, 134–138 (2012)
194. Visakh, P.M., Thomas S, Oksman, K, Mathew, A.P.: Composites: Part A **43**, 735–741 (2012)
195. Bouthegourd, E, Rajisha, K.R., Kalarikkal, N, Saiter, J.M, Thomas, S.: Mater. Lett. **65**, 3615–3617 (2011)
196. Fletcher, A., Gupta, M.C., Dudley, K.L., Vedeler, E.: Comp. Sci. Technol. **70**(6), 953–958 (2010)
197. Normatov, J., Silverstein, M.S.: Macromolecules **40**(23), 8329–8335 (2007)

Reinforced Elastomers: Interphase Modification and Compatibilization in Rubber-Based Nanocomposites

Petroula A. Tarantili

Abstract An extended review is presented on the structure and properties of filler-matrix interface in reinforced elastomeric materials, since the above characteristics are critical for the overall performance of the related products. The current trends for interphase modification in rubber systems containing various fillers, such as Carbon black, Silica, Calcium Carbonate, Clays with emphasis on clay nanofillers, as well as Graphene is discussed. The use of fibrilar reinforcements is also reported, including traditional materials, such as Natural fibres, as well as Aramids and Carbon Nanotubes. On the other hand, the concept of hybrid composites, i.e., those composed of a mixture of matrices or reinforcements, further enhances the versatility of those materials, since it provides new possibilities of extending the area of rubber applications. In fact, the above products combine the property improvement attributed to each one of the system's components and, moreover, they can usually take advantage of their synergistic action. In the same context the role of some other additives, necessary to adjust mechanical properties (e.g. plasticizers) or to promote phase miscibility in complex systems, such as compatibilizers, was investigated. The vulcanization of elastomeric materials is a critical step, with high impact on the properties of final products. In fact, the extent of this reaction, the cross-links density along with the other network parameters, are some important factors controlling the overall behavior of the vulcanized rubber and, therefore, monitoring, control, and modeling of rubber vulcanization are also examined in this work. The above review showed that the main reason for reinforcing rubbers is to improve their mechanical and thermal properties, as well as to reduce cost and sometimes the weight of a construction. It seemed that recent advances in nanoparticles have attracted much attention in manufacturing of rubber nanocomposites, because of the small size of filler and the corresponding increase in the surface area, which leads to the required mechanical properties at low filler loading. Carbon nanotubes and graphene nanoparticles are promising

P. A. Tarantili (✉)
Polymer Technology Lab., School of Chemical Engineering, National Technical University of Athens, 9 Heroon Polytechniou Str GR 15780 Athens, Greece
e-mail: taran@chemeng.ntua.gr

P. M. Visakh et al. (eds.), *Advances in Elastomers II*,
Advanced Structured Materials 12, DOI: 10.1007/978-3-642-20928-4_4,

materials, offering good electrical properties. Surface treatment of the filler particles with the appropriate coupling agents is often vital, in order to promote proper dispersion and adequate filler/matrix interactions. Also, efficient dispersion of the reinforcement into rubber matrices usually needs the assistance of functionalized polymers, i.e., compatibilizers. Among the different modifying agents, maleic anhydride is the most commonly used and seems to ensure best results at relatively low cost. Finally, the cross-linking parameters must be controlled for an optimal network formation. The newly developed polyblends, based on mixtures of rubbers with polyolefins, require the suitable compatibilization in order to reveal their unique properties. Nanoparticles, being very efficient reinforcing agents even at low concentrations, were also found to play the role of compatibilizer for these mixtures of immiscible polymers.

Abbreviations

3-APE	3-aminopropyl-triethoxy silane
IIR	Butyl rubber
ACM	Acrylic rubber
APTES	Aminopropyltriethoxysilane
ATO	Sb doped tin dioxide
BIMMS	Brominated polyisobutylene-co-paramethylstyrene
BPDA	Benzophenone-3,3′,4,4′-tetracarboxylic dianhydride
BR	Butadiene rubber
Bt	Bentonite
C-8	Octylamine
CB	Carbon black
CNT's	Carbon nanotubes
C-SEBS	Carboxylated SEBS
DDA	Dodecylamine
DFT	Density functional theory
DSC	Differential scanning calorimetry
EAR	Ethylene acrylate rubber
ENR	Epoxidized natural rubber
EOC	Ethylene-octene copolymer
EPDM	Ethylene-propylene-diene rubber
EPR	Ethylene-propylene rubber
EPR-g-MA	Ethylene-propylene rubber grafted with maleic anhydride
EVA	Ethyl-vinyl acetate copolymer
FGS	Functionalized graphene sheets
FKM	Fluoroelastomer
GIC	Graphite intercalated compound
GF	Glass fibres
GMA	Glycidyl methacrylate
GN	Graphite nanosheets
GO	Graphene oxide
HDPE	High density polyethylene

HDS	Hexadecyltrimethoxy-silanes
HNBR	Hydrogenated acrylonitrile butadiene rubber
HTV-SR	High temperature vulcanized silicone rubber
HXNBR	Carboxylated NBR
Ipp	Isotactic polypropylene
KF	Kenaf fibres
LB	Liquid polybutadiene
LDPE	Low-density polyethylene
MA-g-EPDM	Maleic anhydride grafted EPDM
MA-g-PB	Maleic anhydride grafted 1,2 polybutadiene
MB	Master batch
MDI	Methylene-bis-diphenyl diisocyanate
MG	Modified graphene
MH	Magnesium hydroxide
MMT	Montmorillonite
MPPB	Maleic anhydride grafted propylene-butadiene copolymer
MPTS	3-mercaptopropyltribis(triethoxysilylpropyl)tetrasulfide
MPS	γ-ethacrylopropyltriethoxysilane
MPS	γ-methacryloxypropyltrimethoxy
MRPS	γ-mercaptoproyltrimethoxy
MVMQ	Methyl vinyl silicone rubber
MWNTs	Multiwall nanotubes
NBR	Acrylonitrile-butadiene rubber
NG	Natural graphite
NR	Natural rubber
NR-g-MA	Maleic anhydride grafted natural rubber
NXT, NXT Z	3-octanoylthio-1-propyltriethoxysilane
ODA	Octadecylamine
OMEC	Online measured electrical conductance
OMMT	Organophilic montmorillonite
PMDA	Pyrromellitic dianhydride
PR	Petroleum resin
PU	Polyurethane
PUR	Polyurethane rubber
PA	Poly(amide), nylon
PB	Polybutadiene
PDMS	Poly(dimethyl siloxane)
PE-g-MA	Polyethylene grafted maleic anhydride
Phr	Parts per hundred
PPEAA	Poly(propylnene-eyhylene acrylic acid)
PP-g-MA	Polypropylene grafted maleic anhydride
PSA	Polysulfonamide
RBFMs	Rubber-based friction materials
RFL	Resorcinol fromaldehayde latex

RGO	Reduced GO
RTV	Room temperature vulcanized
RP	Red phosphorus
SACP	Surface-acetylated cellulose powder
SBR	Styrene-dutadiene rubber
SBS	Styrene-butadiene-styrene
SEBS	Styrene-ethylene- butylene- styrene
SEBS-g-MA	Styrene-ethylene- butylene- styrene grafted maleic anhydride
SEM	Scanning electron microscopy
SEP	Poly(styrene-b-ethylene-co-propylene) diblock copolymer
Si69	Bis(triethoxysilylpropyl)tetrasulfide
SRBC	Styrene rubber block copolymer blends
SWNTs	Single wall nanotubes
TEM	Transmittance electron microscopy
TEOS	Tetraethoxysilane
TESPT	1,4-phenylene diisocyanate (PPDI), methylene-bis-diphenyl
T_g	Glass transition temperature
TPNT	Thermoplastic natural rubber
ULM	Ultrasonically assisted latex mixing process
VPR	Butadiene–styrene–vinyl pyridine rubber
WHRA	White rice husk ash
XRD	X-ray diffraction

1 Introduction to Interphase Modification and Compatibilization in Composites

The structure and properties of filler-matrix interface play a major role on the mechanical and physical performance of composite materials. The quality of interface is a dominant factor for fracture toughness properties of composites as well as for their response to aqueous and corrosive environments. In fact, the interfacial area between filler and matrix, is believed to constitute a separate phase, otherwise called interphase or mesophase, with its own characteristics. Composite materials with poor interfacial properties have relatively low strength and stiffness but they display high resistance to fracture, whereas materials with strong interfaces ensure high strength and stiffness, often accompanied with brittleness. In a simple system bonding at an interface is due to adhesion between reinforcement and matrix. Adhesion may be attributed to five main mechanisms which can occur at the interface, either in isolation or in combination to produce the bond: (i) absorption and wetting (ii) diffusion of the polymer molecules on one surface into molecular network of the other surface (interdiffusion) (iii) electrostatic attraction and (iv) mechanical adhesion [1].

The effectiveness of a filler depends on its characteristics, such as particle size and shape and, more significantly, on the strength of polymer-filler interactions. To achieve maximum improvement, strong fibre interaction with the polymeric matrix will be needed because the interface/interphase between fibre and matrix has a large influence on the mechanical properties of the composite. These interactions may also increase the effective degree of crosslinking and such an effect is particularly strong if the particles have some reactive surface groups [2].

A great deal of effort has been directed towards ensuring that the fibre-matrix interface can transfer the stresses efficiently, so that the properties achieved by reinforcing are maintained and not degraded by the environment. The use of coupling agents, i.e. substances chemically reactive with both matrix and reinforcement, and/or chemical modification of the surface of one or more constituents, have been the most successful means of providing reasonably well controlled bond between matrix and the encapsulated reinforcement.

The rapid growth of using multiphase polymer systems (blends and composites) is undoubtedly related to the availability of methods of controlling the physical and chemical interactions at the interface. Compounds acting as interfacial agents are commonly known as "compatibilizers" in blends, or "coupling agents" in composites, and their function is to promote interfacial adhesion and enhance overall properties [3].

In general, compatibilizers reduce interfacial tension and, therefore, are capable of promoting adhesion leading to a finely dispersed morphology that shows stability against gross segregation. Compatibilizers are often based on polyolefins functionalized with acrylic acid [4], maleic anhydride [5] or other thermoplastics with polar groups such as ionomers [6].

2 Current Trends in Interphase Modification and Compatibilization of Rubbers

Rubbers, also called elastomers, are high molecular weight materials which possess very low interchain force, low glass transition temperature, T_g, and predominantly amorphous nature with low strength. The common characteristics of rubbers are their elasticity, flexibility and toughness and their commercial grades can be classified into two types: crosslinked systems and thermoplastic elastomers. Most of the commonly used rubbers are polymeric materials with long chains, which are chemically crosslinked during the curing process. These materials, once formed, cannot be reshaped, softened, melted nor reprocessed by a simple subsequent reheating. Thermoplastic elastomers, on the other hand, are rubbers which act at room temperature in a manner similar to crosslinking materials but are copolymers, with one phase being rubbery and the other crystallizable. Therefore, they can repeatedly shaped upon heating [7]. Rubbers are often compounded with reinforcing fillers to improve their mechanical properties, such as tensile strength, modulus, tear resistance and abrasion resistance and the most usual filler for this reason are carbon black and silica [8].

The performance of a filler in the rubber matrix is controlled by its characteristics, such as particle size and shape, concentration, surface activity, degree of interactions with rubber matrix and structure of the particle agglomerates. Increasing the area of contact between rubber matrix and filler particles seems to be the most important factor for ensuring a strong reinforcement effect and this surface area of the interface is controlled by the size of filler particles and their volume fraction. Furthermore, the magnitude of adhesive bonding between rubber matrix and filler is a key factor in determining the degree of reinforcing.

2.1 Carbon Black/Rubber Composites

Carbon black (CB) is an essential component in many rubber formulations. It is composed of carbon particles solidly fused together by covalent bonds, thus forming aggregates which cannot be broken into smaller sizes during the ordinary processing conditions of the material. The incorporation of CB with high surface area into an elastomer results in a high reinforcing level and gives higher tensile strength, tear strength and abrasion resistance. There are likely both chemical and physical interactions between CB and rubber, resulting in property improvement, but the understanding of the nature of this type of reinforcement is still growing. The close interfacial contact of a rigid solid phase (CB) and a soft phase (rubber) results in chemical or physical absorption of the rubber molecules into the CB surface. The consequence of this absorption is the formation of "bound rubber" on the CB surface. "Bound rubber" is the portion of elastomer which can not be separated from the filler surface when the rubber mix is extracted by one of its good solvents, such as toluene, over a specific period of time at room temperature. The formation of bound rubber structures, is believed to enhance the mechanical and physical properties of CB-filled rubber compounds [9].

The dispersion and distribution of CB in a rubber matrix are important factors to achieve optimum physical properties. Moreover, its structure, wettability by the rubber phase and the related interactions have all major influence on the properties of the rubber compound. It is known that the wettability of a filler particle by a polymer chain can be quantified in terms of surface energy and similar values of this parameters ensure good compatibility.

Non-uniform distribution of CB in a rubber blend is a major problem, especially in the case of dissimilar rubber blends, such as natural rubber/ethylene-propylene-diene rubber (NR/EPDM) or styrene-butadiene rubber/ethylene-propylene-diene rubber (SBR/EPDM). In such blends, CB tends to partition into the highly unsaturated phase leaving the saturated phase less occupied. For this filler, the surface energy is high compared to that of elastomers like SBR, NR and EPDM. Therefore, in order to get an appreciable influence on the dispersion and distribution properties of CB into a rubber blend, a significant reduction of its surface energy is needed. For instance, surface modification of a fullerenic CB by plasma polymerization was successfully carried out by Mathew et al. [10], in an attempt to reduce its surface energy and further compatibilize the filler with various elastomers. The plasma-coated CB shows low filler interaction in the different rubber systems compared to the untreated version. This is due to the low surface energy of plasma-coated carbon black and it was found that the filler/polymer interaction remains unaffected in unsaturated SBR and acrylonitrile-butadiene rubber (NBR) matrices, while in saturated EPDM rubber it decreases. EPDM rubber showed a decrease of tensile properties due to a substantially reduced interaction between the plasma-polymer film and the elastomer because of its saturated nature. In SBR and NBR the properties remained almost unaffected due to possible interactions between some residual unsaturations left in the plasma-polymer layer and the double bonds present in backbone of these rubbers [10].

2.2 *Silica/Rubber Composites*

Synthetic silicon dioxide (silica) can be produced either by precipitation or by pyrogenic (thermal) process and, therefore, this filler can be classified into two groups; namely precipitated and pyrogenic (fumed) silica. It is the most important filler competing CB in the area of rubber technology. However, in the early stages of silica usage, their uses as reinforcing filler instead of CB were limited due to a number of problems:

- Silica-filled compounds show higher viscosities
- They are more difficult to mix and process
- There is a concomitant increase in vulcanization time
- Silica-filled compounds often show lower crosslinking density

Surface silanol groups of silica have received keen interest in recent decades, because of their important role of interfacial interactions in several applications, such as the reinforcement of synthetic or natural elastomers. The silanol groups show strong filler–filler interactions and cause the absorption of the polar materials such as curatives on the surface. Such absorption of curatives results in a reduction in the crosslinking density and delay of the scorch time of the silica filled rubber compounds.

The main concern in using silica as a reinforcing filler for elastomeric matrices is the difficulty of obtaining a very efficient dispersion of silica particles, due to the filler's surface behavior, dominated by the chemistry of the terminal silanol groups, which can be isolated, vicinal or germinal. Silanols make the surface of hydroxylated silicas hydrophilic, inducing strong interactions between silica particles and leading to a poor dispersion of the filler in the conventional natural or synthetic non-polar type elastomer. Dispersion of silica can be enhanced by either, reducing the polarity of the filler particles or sterically avoiding their aggregation. For example, silane coupling agents containing sulfur are widely used in the tire industry to increase dispersibility of silica into elastomeric matrices, thus promoting the formation of covalent bonds between the phases.

The strength of chemical interactions between silica of different chemical characteristics and NBR, was studied by infrared spectroscopy measuring the increase of C≡N fundamental vibration frequency and by the density functional theory (DFT) [11]. The results showed that functional groups on the silica's surface (after modification with electron donors, alkaline and acidic groups), play an important role on the compatibility of immiscible blends of NBR copolymers, as they strongly interact with the rubber phase.

In the field of saturated elastomers, Hui et al. [12] explored the effect of pristine silica nanoparticles on a model low-density polyethylene (LDPE)-ethyl-vinyl acetate copolymer (EVA) thermoplastic rubber blend. It was observed that the blend properties were strongly affected by the sequence of addition of nanofillers during the composite preparation. The addition of a silane coupling agent (Si-69) to the formulation of silica-filled polymer blends leads to significant increase of tensile strength, modulus, extensibility and dynamic properties of the blends. Also, morphological studies clearly indicated that the differential properties of these mixtures primarily stemmed from the extent of dispersion and alternation of crystalline morphology, which in turn was a strong function of preferential incorporation in the LDPE or EVA matrix and the agglomeration tendency of nanofillers.

It has been suggested that partial or complete chemical modification of silica surface, which can be achieved by the reaction of the surface hydroxyl groups with coupling agents such as organosilanes, affords more or less significant changes in the thermodynamic properties of surface. This, in turn, will influence the silica/silica interactions and, furthermore, its dispersibility into a polymeric matrix.

Some silanes, such as bis(triethoxysilylpropyl)tetrasulfide (TESPT) and 3-octanoylthio-1-propyltriethoxysilane (NXT and NXT Z silanes), have been proposed for use as coupling agents. TESPT displays bifunctionality, which permits it to react both with silanol OH groups present on the silica surface and with the elastomer during vulcanization by sulfur.

Bertona et al. [13] prepared a new type of filler starting from precipitated silica modified with alkoxysilane coupling agent (sb1), obtained by grafting 3-mercaptopropyltrimethoxysilane (MPTS) onto polybutadiene oligomer chains. Styrene-butadiene copolymer was used as matrix and TESPT was the reference modifier of silica. The above authors found that the best tensile properties were obtained with TESPT-modified silica. The properties of the compound loaded with sb1-pregrafted silica were scantily lower than those corresponding to the other compounds and this was attributed to higher crosslinking density given by the sulfur deriving from TESPT decomposition during vulcanization process.

A better understanding of the impact of physical characteristics of the filler-rubber interface and the resulting interphase on the macroscopic mechanical performance of rubber composites was attempted by Stöckelhuber et al. [14]. The rubber matrix consisted of a solution of styrene-butadiene copolymer, filled with 20 or 40 phr pyrogenic or precipitated silicas, with different surface modifications by silanes, as well as with a CB sample as a reference. The "layered fibre model" was proposed, based on the hypothesis that during deformation of composites, the polymer chains slip off from the polymer interphase around the filler particles into the gaps between filler particles, where uniaxially oriented, high-strength polymer fibres are formed. The results of experiments with cured vulcanizates at its operating temperature, support the assumption of an immobilized layer of absorbed polymer molecules in the vicinity of the filler particle interface.

Some more sophisticated processing techniques have also been adopted for special product requirements. Thus, sol–gel techniques generating in situ silica from the sequential hydrolysis and condensation reactions of tetraethoxysilane (TEOS) have been adopted to synthesize transparent rubber-silica hybrid nanocomposites from polar rubbers, like acrylic rubber (ACM) and epoxidized natural rubber (ENR) [15]. It is mentioned that polarity of the host rubber matrix is significant in determining the structure distribution of the silica particles.

2.3 Calcium Carbonate/Rubber Composites

Calcium carbonate ($CaCO_3$) is one of the most inexpensive and recyclable natural resources, with the highest reserves on earth and is widely used in polymer processing. $CaCO_3$ nanoparticles were used for reinforcement of PP/SEBS and PP/carboxylated SEBS (C-SEBS) blends [16]. The composite with SEBS showed morphology with the SEBS domains and $CaCO_3$ particles independently dispersed within the PP matrix. On the other hand, the composite with C-SEBS showed a core-shell structure.

To use $CaCO_3$ as reinforcing agent, surface treatment with coupling agents or organic acids is absolutely necessary. Song et al. [17] studied the factors affecting the mechanical properties of styrene-butadiene-styrene (SBS) block copolymer composites filled with $CaCO_3$, after its surface treatment with liquid polybutadiene (LB). The above authors, concluded that the mechanical properties of SBS can be improved remarkably through reinforcing with $CaCO_3$ subjected to the above surface treatment, proving the filler surface with high content of 1,2-double bonds. When SBS, $CaCO_3$ and LB were directly blended, secondary aggregation of $CaCO_3$ took place and the mechanical properties of the composite were significantly lower. In the integral blend method, LB was shown to function as a plasticizer.

Coating of $CaCO_3$ with stearic acid leads to significant decrease of its surface tension, which results in decreased interactions and, hopefully, contributes to limited aggregation [18]. The phase separation of ternary phase composites is influenced by the melt rheology of the system, the compounding techniques, as well as by the surface characteristics and mutual wettability of the fillers and polymer components. In PP/EPDM/$CaCO_3$ composites, the surface treatment of filler was found to result in separate dispersion of phases (EPDM and $CaCO_3$), whereas encapsulation occurred when untreated calcium carbonate was used.

To promote the adhesive bonding between polymer and filler particles, functionalization of polymer phases was carried out. In Polypropylene/Ethylene-propylene rubber (EPR)/filler systems, it was shown that incorporating PP functionalized with maleic anhydride resulted in separate dispersions of elastomer filler. The use maleated EPR and non-functionalized PP was said to give a filler encapsulation structure [19, 20].

In another related work [21], the properties of PP/EPR/coated nano-$CaCO_3$ composites, with and without compatibilizers, were studied. The results indicated that good dispersion of nanoparticles in PP/EPR depends on their surface coatings of stearic and other fatty acids. In both cases, the final morphology is the core-shell structure, in which EPR acts as the shell part encapsulating coated $CaCO_3$. The use EPR-g-MA copolymer does not improve the interface between PP/EPR and nanoparticles but propylene ethylene (PEP) copolymer should be preferentially localized at the interface of PP and EPR/nano-$CaCO_3$ phases generating an improved adherence, which will ensure a better cohesion of the whole material. It is believed that the synergistic effect of both, EPR elastomer and $CaCO_3$ nanoparticles, should account for the balanced performance of the ternary composites.

2.4 Clay/Rubber Composites

Rubber-clay nanocomposites exhibit outstanding properties, at low loading levels of clay, as compared with unfilled rubber compounds or conventional filled composites. In general, the preparation methods for rubber-clay nanocomposite can be divided into four major groups, according to the processing technique:

- In situ polymerization
- Intercalation of rubber via solution blending
- Direct melt intercalation method
- Intercalation of rubber via latex compounding

The main objective of preparing organoclay nanocomposites is to achieve high degree of dispersion of organoclay aggregates within the polymer matrix, which can yield very large surface area. The efficient dispersion of organoclay platelets in the polymer matrix leads to improvement of the overall properties of the polymer.

Clay minerals have hydrophilic character and hence a nanometric dispersion into a hydrophobic polymer is hard to achieve, particularly through solution and melt blending methods. To obtain a homogeneous dispersion in a hydrocarbon matrix, the alkaline-earth cations of pristine clay minerals are replaced by hydrophobic organomodifiers, typically ammonium cations bearing long alkyl chains a compensating cations. The role of the alkyl ammonium cations is to lower the host's surface energy and thereby improve wetting-out by the polymers. The obvious outcome of the polymer-organoclay interactions is improved strength characteristics because of the strong interfacial interactions between silicate layers and the polymer chains. By appropriate choice of the alkyl ammonium cation, the interactions between the host layers and the intercalated polymer chains can be tailored to synthesize new organic–inorganic hybrids.

Montmorillonite is the most commonly used layered silicate for preparation of nanocomposites because of its high aspect ratio, large surface area, and surface reactivity. Some typical nanocomposite systems owing rubber matrix are the following:

2.4.1 Clay/SEBS Nanocomposites

SEBS block copolymer is one of the widely used thermoplastic elastomers, which exhibits elasticity and easy processability, good thermal stability etc. To further improve rubber properties clay nanoparticles were studied as reinforcing additive. Chang et al. [22] recorded increased tensile strength of SEBS/clay hybrids, at 10 phr clay concentration with 5–20 phr compatibilizer of SEBS-g-MA.

The effect of blending sequence on the microstructure of ternary nanocomposites composed of nylon 66/organoclay/(SEBS-g-MA) was studied by Dasari et al. [23], who also studied the microscale and nanoscale deformation behavior of nylon 66/organoclay/SEBS-g-MA ternary nanocomposites.

Lai et al. [24] prepared SEBS block copolymer/clay (Cloisite 20A) nanocomposites via melt mixing technique, using two types of maleated compatibilizers: SEBS-g-MA and polypropylene grafted maleic anhydride (PP-g-MA) in order to improve dispersion of organoclay. PP-g-MA compatibilizer system conferred higher tensile strength and tear strength than SEBS-g-MA compatibilized system, mainly due to its semi-crystalline nature. At fixed content of compatibilizers, the above mechanical properties were improved with increasing clay content as well.

Martín et al. [25] studied the intercalation capability of SEBS in nanocomposite systems prepared by melt blending, consisted of isotactic PP with 5 % organically modified montmorillonite (Cloisite 20A). They found that organoclay is not in direct contact with the PP phase because the clay is always located inside the elastomer domains. The elastomer surrounds the nanoclay, hindering the clay exfoliation and preventing its dispersion in the PP matrix. On the other hand, the presence of nanoclay causes a decrease in the coalescence of rubbery phase. There is a reduction of the size of SEBS domains in comparison with binary PP/SEBS systems and these domains are better distributed in the PP matrix.

2.4.2 Clay/Propylene-Butadiene and SBR Nanocomposites

Maleic anhydride grafted propylene-butadiene copolymer (MPPB) was investigated as compatibilizer for a composite of PP/SBS/organophilic montmorillonite (OMMT) [26]. In the presence of 10 phr MPPB, the impact strength of the composite was improved by about 20 %. TEM images showed that the OMMT was better dispersed in the matrix upon inclusion of MPPB. A better distribution of the rubber phase and a rugged fracture surface were observed in SEM images as the MPPB content increased.

In another work, SBR and OMMT were directly mixed to obtain nanocomposites, which have shown intercalated and partly exfoliated structure with improved viscoelastic and mechanical properties [27]. Also, SBS/MMT nanocomposites were successfully prepared by in situ living anionic polymerization with n-BuLi initiator [28]. The results from the kinetic study and ^{1}H NMR indicated that the addition of OMMT did not change the living copolymerization as well as the components of copolymer, when OMMT content was lower than 3 wt%. The result from TEM and X-ray diffraction revealed that a completely exfoliated structure existed in the nanocomposite with 25 wt% styrene units and OMMT content from 1 to 4 wt%, and styrene played an important role in expanding the OMMT layers. Moreover, nanocomposites possessed higher glass-transition temperature, thermal stability, tensile strength and elongation at break than SBR when the OMMT content ranged from 2.5 to 4 %. Regarding the type of nanofiller, Sadhu and Bhowmick [29] studied the preparation and properties of different nanoclays based on sodium montmorillonite, bentonite and potassium montmorillonite, and organic amines of varying lengths in the SBR matrix. The tensile strength of SBR nanocomposites increased with increasing chain length of amine. Among the examined clays, modified Na^+-MMT exhibited better mechanical properties than either bentonite or K^+-MMT in accordance with X-ray data. The incorporation of amine governs the organophilicity of the clay. This in turn leads to better polymer-filler interaction and hence better strength of nanocomposites. The higher polymer-filler interaction explained the shift of the glass transition temperature (T_g) in the positive direction as measured from DSC experiments.

2.4.3 Clay/NR Nanocomposites

Magaraphan et al. [30] mentioned that MMT clays treated with long primary amines led to much more improved mechanical properties when incorporated into NR matrices, than those treated with quaternary amines of the same number of carbon atoms. The length of hydrocarbon in the alkylamines had no effect on the curing time, whereas the nanocomposites prepared with long quaternary amine showed comparatively faster cure time.

Varghese et al. [31] prepared NR, polyurethane rubber (PUR) and NR/PUR-based nanocomposites from the related lattices by adding 10 phr pristine synthetic sodium fluorohectorite. It was observed that, in blends composed of polar PUR and apolar NR, the silicate layers were preferentially embedded in the polar PUR phase in well a intercalated/exfoliated stage. The properties of the PUR/NR-based nanocomposites were similar to those containing plain PUR. This result was claimed to be of great economic significance as NR latex is cheaper than PUR latex.

In an attempt to produce rubber nanocomposites based on NR or SBR, an innovative organoclay containing polybutadiene chains was prepared. It was obtained by reacting a maleinized polybutadiene oligomer with the $-CH_2CH_2OH$ moieties present in the ammonium cation of a commercially available organoclay (Cloisite 30B). A second organoclay was prepared in situ during melt blending by reacting Na^{+}Mt with dialkyl dimethylammonium chloride (Mt-2HT). Satisfactory exfoliation and dispersion of the innovative organoclay in both rubber matrices were obtained, not far from those observed in systems containing Mt-2HT.

In both NR and SBR, Mt-PB4 promoted a fast crosslinking reaction and a lower crosslinking density, with respect to Mt-2HT. It was observed that Mt-PB4 seems recommended to promote reinforcement of rubber compounds, without exceedingly increasing their rigidity and sacrificing their ultimate properties. It was confirmed that the enlargement of the basal spacing in a layered clay mineral, thanks to the use of a long chain hydrophobic substituent of the ammonium cation, is a positive strategy to achieve both exfoliation and dispersion of the organoclay [32].

The organically modified material is polar, hence it may not ensure an efficient dispersion upon direct incorporation in non-polar rubbers like SBR and EPR. Therefore, a polar rubber, which is compatible with the matrix polymer can be used as a compatibilizer.

Epoxidized natural rubber (ENR), obtained by epoxidation of 1,4-polyisoprene, has a higher glass transition temperature, polarity and superior compatibility with SBR and EPR. Rajasekar et al. [33] studied the effect of dual fillers on the properties of SBR and EPR compounds prepared using ENR as a compatibilizer. ENR-organically modified nanoclay (Cloisite 20A) composites were prepared by solution mixing and the obtained composite was incorporated in SBR and EPR matrices along with carbon black. The morphological studies proved the intercalation of nanoclay platelets in ENR and further incorporation of EC in SBR and EPR matrices leads to partial exfoliation of nanoclay platelets. A curing study demonstrated faster scorch time, cure time and increased maximum torque for the compatibilized SBR and EPR nanocomposites containing a dual filler system compared to control. Dynamic mechanical thermal analysis showed an increase of storage modulus for SBR and EPR compounds containing dual fillers as compared to compounds containing pure and single filler. The same compounds showed substantial improvement of mechanical properties.

2.4.4 Clay/EPDM Rubber Nanocomposites

EPDM rubber is a widely used engineering elastomer with outstanding resistance to ozone degradation, combined with high thermal stability and, therefore, EPDM/clay nanocomposite materials have attracted special attention. However, this elastomer does not have any polar groups in the backbone and, therefore, strong interactions between EPDM chains and clay layers can hardly be achieved, unless a suitable compatibilizer is used.

EPDM/montmorillonite (MMT) composites were prepared through a melt process and three kinds of surfactants with different ammonium cations were used to modify MMT and thus, to alter the morphology of composites [34]. In fact, the morphological configuration is dependent on the alkyl ammonium salt length, that is, the hydrophobicity of the organic surfactants. Organophilic montmorillonite (OMMT), modified by octadecyltrimethyl ammonium salt and distearyldimethyl ammonium salt, was intercalated and partially exfoliated in the EPDM matrix, whereas OMMT modified by hexadecyltrimethyl ammonium chloride exhibited a morphology with the clay nanoparticles existing as a common filler. Ethylene-propylene-diene rubber grafted with maleic anhydride (MA-g-EPDM) was used as compatibilizer and greatly affected the dispersion of OMMT. When OMMTs were modified by octadecyltrimethyl ammonium chloride and distearydimethyl ammonium chloride, the EPDM/OMMT/MA-g-EPDM composites (100/15/5) had an exfoliated structure and they showed good mechanical properties and high dynamic moduli.

Gatos et al. [35] studied the characteristics of EPDM/organoclay nanocomposites, resulting from different processing conditions and formulations. The change in the rubber polarity was attained by replacing half of EPDM rubber by maleic anhydride (MA)—and glycidyl methacrylate (GMA)-grafted EPDM. It was shown that incorporation of grafted EPDMs strongly improves the strength and stiffness of nanocomposites containing 10 phr organoclay.

EPDM-g-MA was used as compatibilizer in EPDM/OMMT nanocomposites prepared by melt compounding [36]. XRD analysis indicated that matrix chains were intercalated into the gallery space of OMMT effectively. An almost complete dispersion was obtained when the compatibilizer-filler ratio was three. In the samples containing less EPDM-g-MA, strong flocculation of the stacked dispersed clay particles was evidenced. Better dispersion of silicate layers was obtained in EPDM matrices with a higher Mooney viscosity, as they exert higher shearing during mixing. Also, the linear polymer chains can be more intercalated into OMMT galleries than long-branch EPDM chains.

2.4.5 Clay/NBR Nanocomposites

Rubber-clay nanocomposites have been prepared by modification of the naturally occurring Na montmorillonite clay followed by mixing with rubber in solution. Intercalation and/or exfoliation of the clay layers has been observed in SBR, NBR, butadiene rubber (BR), brominated polyisobutylene-co-paramethylstyrene (BIMS), EOC and thermoplastic elastomers.

The resulting structure and orientation of clay particles in the matrix depend on the nature of both rubber and clays. Usually nonpolar rubbers are better compatible with the organoclays where long aliphatic amine chains are present, whereas some polar rubbers have better attraction towards unmodified clay. When rubber and clay are compatible, more uniform dispersions of clay are obtained when the width of the clay layers is thinner than in the case of lower compatibility [37].

Kim et al. [38] prepared nanocomposites of octadecylamine-modified (ODA) MMT and NBR in an internal mixer using 3(-mercaptopropyl)trimethoxysilane as a coupling agent. No change in the maximum torque values were observed with the addition of coupling agent thereby proving that the coupling agent took no part in the crosslinking, at 180 °C with a sulfur curing system. The above authors reported enhanced thermal stability of the organo-MMT/NBR nanocomposite in the presence of the coupling agent, which was attributed not only due to the difference in chemical structure but also due to the restricted thermal motion of macromolecules in the silicate layer. It was observed that incorporation of below 5 phr coupling agent in the organoclay/NBR nanocomposites enhanced the tensile strength and modulus while above 5 phr content these properties were decreased. The improvement was explained by the authors due to a better interaction between organoclay and NBR whereas the decrease was attributed to the excessively added silane compound existing as a nonreactant after curing which form multimolecular layers between the boundaries of organoclay and NBR, thus causing a defect in the nanocomposites.

The effect of clay modification on organo-MMT/NBR nanocomposites was further studied by Kim et al. [39]. The alkylamines used for modification were octylamine (C8), dodecylamine DDA (C12), and ODA (C18). It was found that the mechanical properties increased in the order C8-MMT< C12-MMT< C18-MMT, depending on the length of the alkyl chain in the alkyl ammonium. With increasing organo-MMT content, the organo-MMT/NBR nanocomposites showed substantial improvement of mechanical properties compared with unfilled NBR. This high reinforcement effect implied a strong interaction between the matrix and the clay interface, which was attributed to the nanoscale and uniform dispersion of the silicate layers in the NBR matrix.

Study of the vulcanization reaction showed that as the chain length of the modifier increased the rheometric torque decreased and the optimum curing time were shortened [40]. They also observed the extent of swelling in methyl ethyl ketone (MEK) clearly decreased with increase in organo-MMT loading. Thus the organo-MMT/NBR nanocomposites have excellent barrier properties compared with vulcanized NBR.

2.5 Carbon Nanotubes/Rubber Composites

Carbon nanotubes (CNT) can be visualized as graphene layers rolled into cylinders consisting of planar hexagonal arrangement of carbon–carbon bonds. During their growth, depending on the synthesis methods, they can assemble either as concentric tubes (multiwall nanotubes, MWNTs) or as individual cylinders (single wall nanotubes, SWNTs). They attracted enormous attention for their fundamental behavior and for their use in a wide variety of applications in nanoelectronic devices, probe tips for scanning probe microscopes or in the automotive and aerospace industries for the dissipation of electronic charges. The extensive interest in CNTs arises from their unique structural and physical properties: their small size in the nanometer scale, their unique electronic behavior, their exceptional properties of ballistic transport, their extremely high thermal conductivity and high optical polarizability, as well as their unparalleled mechanical properties such as high elastic modulus and tensile strength.

One of the biggest challenges in the preparation of polymer composites, is to obtain a homogeneous dispersion of carbon nanotubes in a polymer matrix because van der Waals interactions between individual tubes often lead to significant aggregation or agglomeration, thus reducing the expected property improvement of the resulting composite.

Different techniques intended to optimize the nanotube dispersion within the polymeric medium have been used. These include oxidation [41] or chemical functionalization [42] of the tube surface and use of surfactants [43]. The above methods intend to promote adhesion between nanotubes and polymer, thus enabling effective stress transfer at the polymer–filler interface.

CNTs treated with 3-aminopropyltriethoxysilane improved the mechanical properties of NR composites, through the stronger polymer-filler interactions between reinforcement and rubber matrix [44].

Composites based on MWCNT and block copolymer SBS with two different contents of styrene have been investigated and their electrical conductivity and mechanical properties have been evaluated [45]. According to the results of dynamic mechanical analysis, the MWCNTs interact with both phases of the copolymer acting as reinforcing filler, whereas the dispersed agent acts as plasticizer. However, it was shown that the reinforcing effect of MWCNTs is dominant, resulting in an overall improvement of mechanical properties of the composites.

MWCNTs were used in SBR-based rubber compound for partial replacement of carbon black and with high content of plasticizers [46]. By the increase of their apparent shape factor it was shown that the oil permits better dispersion and alignment in the rubber matrix. The reinforcing effect of MWCNTs has been studied for both, sulphur-vulcanized and radiation cured nanocomposites. In those cases, a strong reinforcing effect has been recorded at low elongations, but the extra-reinforcement tends to disappear at high elongations due to the poor interaction between the nanotube surface and rubber matrix, The incorporation of plasticizer permits to partially reduce the fast heat build up and, hence, the strong mechanical hysteresis normally observed in nanocomposites prepared with MWCNTs.

In another work [47], it was found that the addition of 1 phr MWCNT to SBR leads to 45 % increase of modulus and 70 % increase of tensile strength. On the other hand, TEM analysis of composites has revealed poor dispersion. The large decrease of equilibrium swelling in toluene with MWNT content was not ascribed to filler-matrix interfacial interactions but to the occlusion of rubber into the aggregates.

2.6 Graphene/Rubber Composites

Graphene possesses similar mechanical properties as CNTs, but has superior electrical and thermal properties and larger surface area because of its 2-dimensional crystal structure. This material as well as its derivatives as fillers for polymer matrix composites, have shown a great potential for various important applications. The interface between the dispersion layers and the matrix plays a crucial role in determining the structure and properties of nanocomposites. Surface functionalization, including non-covalent or covalent modification of nanoparticles, is an effective method for improving the interfacial interactions between nanoparticles and polymer matrix [48].

Graphene oxide (GO) is oxygenated graphene possessing abundant oxygen-containing functional groups including epoxide, hydrpoxyl and carboxyl groups on the edges and surfaces. It has attracted increasing interest as a filler for polymer nanocomposites due to its high dispersive capacity, long coherence length and the barrier property. The improvement of mechanical properties of GO/polymer composites was attributed to its high elastic modulus ($\sim$650 GPa) and the efficient stress transfer through the polymer matrices to GO. This enhancement of mechanical properties mainly depends on two factors: filler dispersion and interfacial adhesion. It also should be noted that the oxygen-containing groups of GO assist its dispersion into polymer matrices and form strong hydrogen-bonding interactions especially with polar polymer matrices [49].

Bai et al. [50] successfully prepared, by a modified Hummers method, chemical exfoliated GO nanosheets with high aspect ratios. Such nanosheets were well mixed with carboxylated NBR (HXNBR) by a simple solution-mixing method and produced GO/HXNBR composites. The addition of 0.44 vol.% of GO nanosheets enhanced the tensile strength and modulus at 200 % elongation of HXNBR by more than 50 and 100 %. This was claimed to be due to strong interfacial interactions between the oxygen-containing functional groups of nanosheets and the carboxyl groups present in HXNBR. This speculation is further supported by the increase of the glass transition temperature of HXNBR from −23.2 to −21.6 °C, at a GO content of 1.3 vol.%. The results indicated that GO efficiently reinforced HXNBR due to the good dispersion and strong interfacial interactions.

In another work, butyl rubber (IIR) nanocomposites based on modified graphene (MG) sheets, were fabricated by solution processing followed by compression molding. MG was prepared from natural graphite (NG) through graphite oxide route [51]. X-ray diffraction showed that the exfoliated MG was homogeneously dispersed in the IIR matrix with doping levels of 1–10 wt% as evidenced by the lack of the characteristic graphite reflection in the composites. In contrast, the graphite retained its stacking order and showed the sharp characteristic peak in the NG-IIR composites. Scanning electron microscopy images of the fracture surfaces of the IIR matrix showed that MG nanofillers exhibited better compatibility than NG. The mechanical properties of MG-IIR nanocomposites were significantly improved due to the efficient distribution of the large surface area MG sheet.

Well-dispersed GO/EPDM/petroleum resin (PR) composites were successfully prepared by Chen et al. [52], via the combination of solution mixing and two-roll mill processing. The matched surface energy as well as low interfacial energy of EPDM and GO probably contributed to the good dispersion of EPDM/PR composites. These results would provide valuable guidance for evaluating the dispersion of GO in polymer matrices through surface energy and interfacial energy. Furthermore, the addition of 0.5 wt% GO increased the tensile modulus, tensile strength and elongation at break of the GO/EPDM composite. Low GO loading improved the damping properties of EPDM/PR matrices, which seemed to provide a novel approach to prepare damping materials. However, the thermal stability of EPDM/PR mildly decreased with low GO loading.

Ozbas et al. [53] demonstrated the use of functionalized graphene sheets (FGS) as multifunctional nanofillers to improve mechanical properties, lower gas permeability, and impart conductivity of several elastomeric matrices. Sixteen times higher loadings of traditional carbon black (CB) fillers are required to obtain the same stiffness as FGS. All these elastomers showed similar and significant improvements of mechanical properties with FGS, indicating that the mechanism of property improvement is inherent to the FGS and not simply a function of chemical crosslinking. FGS slightly accelerates the curing rate of NR, but the insensitivity of the increase of modulus to the polymer type suggests that the major mechanism of reinforcement does not arise from increased crosslinks density but from the FGS nanofiller geometry. The decrease of gas permeability was attributed to the high aspect ratio of the FGS sheets. This creates a tortuous path mechanism of gas diffusion; fitting the permeability data to the Nielsen model yields an aspect ratio of $\sim$1000 for the FGS. Electrical conductivity is demonstrated at FGS loadings as low as 0.08 % in poly(dimehtyl siloxane) (PDMS) and reaches 0.3 S/m at 4 wt% loading in NR. This combination of functionalities imparted by FGS is shown to result from its high aspect ratio and carbon-based structure.

An interesting technique of preparing nanocomposites of reduced GO (RGO) platelets with NR, was based on the co-coagulation of NR latex and RGO followed by compounding in order to incorporate the curing system, by one of two methods: solution treatment or two-roll milling [54]. It was demonstrated that property improvements were strongly dependent upon the processing history and nanocomposite morphology. Solution treatment (implantation of peroxide curing agent by swelling the NR in toluene) preserved the segregated filler network morphology produced by the co-coagulation procedure, whereas the milling process destroyed this network and generated a homogeneous dispersion of RG-O platelets in the NR matrix. The segregated network morphology was shown to be advantageous for conductivity properties and greatly increased the stiffness of the composite versus neat NR, but also significantly reduced the elongation to break of the samples. The milled nanocomposites also exhibited enhanced stiffness and strength while maintaining a high elongation to break. On the other hand, reinforcing effect in the solution-treated nanocomposites was attributed to the formation of a sample-spanning network of strongly interacting RGO platelets, located in the interstitial regions between latex particles, whereas in the milled nanocomposites, reinforcement is due to simple mechanical restraint along with promotion of strain-induced crystallization by the high aspect ratio RGO platelets.

2.7 Aramid Fibre/Rubber Composites

Natural rubber (NR) is one of the most important elastomers characterized by excellent elasticity and flexibility. This type of elastomer contains small amounts of fatty acids and proteinaceous residues, acting as accelerators to sulfur curing, which is the main route used in industry to vulcanize natural rubber. Because of the double bonds in its macromolecular backbone, it displays poor oil and ozone-resistance due to oxidative or thermal degradation. Moreover, loss of mechanical properties, such as low tensile strength and poor tear resistance, are also some drawbacks of NR. Many techniques were investigated in order to improve these disadvantages, such as reinforcement with various particulate fillers (e.g. carbon black, ultra-fine calcium carbonate and modified montmorillonite).

Aramid fibres were also used as reinforcement for rubbers, because they combine high specific strength and modulus, high thermal resistance, chemical inertness and low electrical conductivity compared with metallic fibres. There is, however, a recognized problem of poor interfacial adhesion between aramids and the usual rubber matrices and, therefore, various methods have been used to modify their surface. The application of chemical treatment like N-alkylation, RFL dipping, plasma treatment and polymer grafting to improve surface roughness and to bring various functional groups on the fibre surface is highly required for these kinds of fibres to improve adhesion with the elastomeric matrix [55–58].

In order to improve aramid-rubber interactions, these fibres have been treated with resorcinol formaldehyde latex (RFL). Previous studies have shown that the aramid-NR interaction was improved by the introduction of RFL treatment on the fibre surface [59].

Similar attempts in the past, have used di-isocyanate treatment of the aramid fibre surface, for its subsequent incorporation into polyurethane elastomers [60].

The use of a reactive compatibilizer like maleic anhydride grafted polymer with a non polar part having a structure similar to polymer matrix is also used to promote fibre-matrix interactions in aramid fibre reinforced composite [61].

Thermoplastic compounds of natural rubber and high density polyethylene, at a ratio of 70/30, were reinforced with aramid fibres [62]. The incorporation of graft-co-poly(ethylene/maleic anhydride) (PE-g-MA) as compatibilizer significantly improves the mechanical properties of composites with an optimum corresponding to 20 % fibre loading.

Short aramid fibres coated with RFL were studied by Shibulal and Naskar [63] as reinforcement for ethylene-octene copolymer (EOC), a new generation metallocene catalyzed thermoplastic elastomer which has been implemented successfully as an impact modifier of PP. Significant improvement was observed in the low strain (10 %) modulus and modulus at 100 % elongation by tensile tests, as well as in the storage modulus at the low strain region by rheological analysis. Improvement in tensile strength coupled with elongation at break and good fibre dispersion particularly at high fibre loaded composite were achieved with the incorporation of low molecular weight maleic anhydride grafted 1,2 polybutadiene (MA-g-PB), which indicates that it acts as an interface modifier through compatibilization between the fibre and the EOC matrix, being a good dispersing agent at the same time. SEM photomicrographs of the fractured surfaces of the composite with MA-g-PB revealed sticking of polymer traces on the pulled out fibre surface and further support the evidence of compatibilizing action of MA-g-PB.

Another simple chemical treatment reported in the literature for aramids is surface hydrolysis. Improvement of the interfacial adhesion of aramid fibre to styrene-ethylene-butylene-styrene (SEBS), a thermoplastic block terpolymer, was obtained by slightly hydrolyzing the fibre with sodium hydroxide solution to increase the number of reactive amino end groups and then mixing with the matrix and maleic anhydride grafted SEBS (MA-g-SEBS), acting as compatibilizer [64]. The tensile strength of the compatibilized composite was found to increase for compatibilizer concentration up to 5 % wt.

2.8 Natural Fibre/Rubber Composites

Over the past decade there has been growing interest in the use of lignocellulosic fibres, as reinforcing elements in elastomers. Compared with other fillers, cellulose owns a series of advantages, such as its renewable nature, low cost, low density, ease of chemical modification and biodegradability. Natural-fibre reinforced polymer composites are used as an alternative low cost sustainable material for structural and non-structural applications, such as building products, automotive applications, packaging, furniture and consumer goods [65–67].

It is well known that different surface properties between fibre and matrix, i.e. the former is highly polar and hydrophilic while the latter is, generally, non-polar and relatively hydrophobic, make necessary the suitable modification, in order to improve fibre/polymer compatibility and their interfacial adhesion.

Several strategies of changing the surface profile were developed, including chemical or physical treatment, the use of compatibilizers and surface modifications techniques. Alkali treatments, acetylation, graft co-polymerization or the use of maleic-anhydride-propylene co-polymer are some techniques appropriate to adjust the surface polarities between the natural fibre and polymer matrix [68].

In addition, the chemical modification using coupling agents bearing two reactive groups, one of which being likely to react with the OH function at the fibre surface, whereas the other one is left to copolymerize with the matrix, constitutes an interesting way allowing the establishment of covalent bonding between fibres and matrix, thus leading to materials with high mechanical properties.

For instance silane coupling chemicals present three main advantages: (i) they are commercially available in a large scale (ii) at one end, they bear alkoxysilane groups capable of reacting with OH-rich surface, and (iii) at the second end, they have a large number of functional groups which can be tailored as a function of the matrix to be used. The last feature ensures, at least, a good compatibility between reinforcing element and the polymer matrix or even covalent bonds between them.

Regarding the study of elastomeric composite systems, cellulose fibres were incorporated in low density polyethylene and natural rubber after chemical modification involving three silane coupling agents, namely γ-methacryloxypropyltrimethoxy (MPS), γ-mercaptoproyltrimethoxy (MRPS) and hexadecyltrimethoxysilanes (HDS). Composites reinforced with cellulose fibres treated with MPS and MRPS displayed good mechanical performance [69]. On the other hand, with HDS treated fibres only a modest enhancement on composite properties was observed.

Varghese et al. [70] reported that the surface of sisal fibres modified by acetylation treatment could increase its adhesion properties to natural rubber. Also, Martins and Joekes [71] observed an improved performance of rubber composites after surface acetylation of cellulose fillers.

In another work [72], wood fibres were subjected to mechanochemical reaction with acetic anhydride to prepare surface-acetylated cellulose powder (SACP) through pan-milling in the solid state at ambient temperature. Curing characteristics indicated that NR compounds filled with SACP possessed better processability and scorch safety. The mechanical properties of vulcanizates, i.e. tensile strength, tear strength and modulus at 100, 200 and 300 % elongation, increased with SACP, but elongation at break and hardness were less as compared to those of pristine cellulose fibres filled composites. They attributed this improvement in better filler dispersion and increase filler-matrix interfacial adhesion.

The use of particulate cellulose fillers, obtained from natural fibres, was also investigated. In fact, cellulose particles were successfully modified using different coupling agents, such as pyrromellitic dianhydride (PMDA), benzophenone-3,3′,4,4′-tetracarboxylic dianhydride (BPDA), 1,4-phenylene diisocyanate (PPDI), methylene-bis-diphenyl diisocyanate (MDI), γ-mercaptopropyltriethoxysilane (MRPS) and γ-ethacrylopropyltriethoxysilane (MPS) [73]. The stiff backbone of these grafted agents ensures that the one reactive group is preserved and directed outwards from the fibre surface. Grafting and preservation of the second function were proven by several techniques. These functionalized fibres were exploited in the preparation of cellulose-reinforced composites based on OH- or C=C rich matrices such as NR.

Similarly, the treatment of cellulosic agro-based fillers (sorghum stalk) for rubber reinforcement with alkali, glyceride oil, and cationic agents indicated that glyceride oil further inhibited adhesion between filler and matrix [74]. The alkali treatment was aimed at cleaning the surface of the cellulosic filler by removing its natural admixtures, such as lignin, pectines and some fats and oils. However, it is believed that some surface roughening will also be imparted. Improvement in tensile strength, hardness and density was observed in comparison with untreated fibre composites. The cationic treatment was designed to offer a positive potential to the filler as a prelude to trigger electrostatic attraction with rubber. It was found that this objective was fulfilled, the cationic filler composite was dense and for some low loadings it had the highest strength and hardness.

Sisal fibre/NR composites were prepared with filler loading 10–30 phr [75]. fibre treatment (alkalization) and addition maleic anhydride grafted natural rubber (NR-g-MA) were followed to improve interfacial adhesion of fibres to NR. Composites of alkali treated sisal fibre and NR exhibited higher tensile properties and hardness than those containing untreated fibres at all fibre concentrations, due to the improved adhesion between treated fibre and NR matrix through the mechanical interlocking mechanism. The addition of NR-g-MA into composites increased their mechanical properties but prolonged scorch time and cure time. NR-g-MA provided to the composite a more effective improvement of mechanical properties when compared to fibre alkalization.

2.9 Rubber Reinforcements for Special Uses

As already stated, EPDM, an elastomeric terpolymer consisting of ethylene, propylene and unsaturated diene, has become one of the most popular synthetic rubbers, because of its unique properties. However, due to the copolymerization of ethylene with propylene, EPDM possesses low crystallinity and cannot reinforce itself like NR or polychloroprene. There are two traditional methods to reinforce EPDM: incorporation of filler (carbon black, silica, clay minerals) and blending with crystal resin.

Sammarium borate ($SmBO_3$) and Sb-doped SnO_2 (ATO), due to their special electric and magnetic properties, were used to reinforce EPDM composites. However the tensile strength of those EPDM composites is not satisfactory, therefore blending with crystalline polymers, such as polyolefins, was proposed. Su et al. [76] used three types of ethylene-olefin copolymer (POE) in order to reinforce EPDM/$SmBO_3$ and EPDM/Sb-doped SnO_2 (ATO) compounds.

EPDM composites reinforced with polysulfonamide (PSA) short fibres were prepared and compared with those contaning aramid fibres [77]. It was found that the ablation rate and thermal conductivity of PSA/EPDM composites were lower, whereas their thermal degradation temperature and limiting oxygen index were higher. Dynamic mechanical thermal testing and morphology observations revealed that strong interfacial bonding between fibre and matrix was developed and this was beneficial for improving the ablation properties of the composites.

Manjhi and Sarkhel [78] reported an enhancement of filler-matrix bonding and filler dispersion in EPDM/kaolin systems with the incorporation of MA-g-EPDM as compatibilizer.

Maleic anhydride grafted EPDM (MA-g-EPDM) was used as compatibilizer to enhance miscibility in EPDM/bentonite (Bt) composites [79]. The addition of 10 parts per hundred (phr) of MA-g-EPDM into EPDM/Bt composites results in significantly improved tensile properties and solvent resistance, as compared with composites without compatibilizer. The cure and scorch time, maximum torque and minimum torque of compatibilized composites were increased upon addition of MA-g-EPDM. Finally, SEM micrographs of tensile fractured surfaces of compatibilized EPDM/Bt composites showed better dispersion of Bt and interfacial adhesion between EPDM and Bt as compared with uncompatibilized composites.

Pasbakhsh et al. [80] reported similar results, claiming that the incorporation of MA-g-EPDM promotes compatibilization by creating an interphase between the EPDM matrix and halloysite nanotubes, which produces stronger interfacial interactions and also improves the dispersion of nanotubes in the rubber matrix.

In the field of rubber composites suitable for special applications, poly(dimethyl siloxane) (PDMS) elastomers are a class of materials of prime importance. In fact, PDMS is a highly recommended elastomer for a variety of applications including those of medical, electronic, automotive and aerospace industries. However, these grades of elastomers are characterized by poor mechanical properties and, therefore, reinforcing fillers must be added to overcome this drawback. PDMS is commonly reinforced by fumed silica in a significant proportion to form high performance composites. The use of such large amounts of silica significantly increase the opacity of the resulting materials as well as their specific weight. Alternatively, fillers with large specific area and aspect ratio, such clay nanoplatelets, could bring better mechanical performance, improved barrier properties and thermal stability, keeping the material transparent and lightweight. In spite of their similarities in structure, montmorillonite does not naturally delaminate in PDMS matrix. In order to achieve the expected properties of nanocomposites, stacked clay platelets must be dispersed and strong interactions between the polymer and mineral phase must be promoted. Due to its hydrophilic nature, clay is generally modified by quaternary ammonium surfactants to increase the intergallery spacing and to achieve enough hydrophobicity, which will readily promote miscibility with the polymer matrix.

Successful treatments of the reinforcement include that of Burnside and Giannelis [81], who studied PDMS/OMMT nanocomposite systems, using melt intercalation. The organosilicate was prepared by ion exchanging Na^+-MMT with dimethyl ditallow ammonium bromide. The prepared nanocomposites exhibit decreased solvent uptake and increased thermal stability.

In another work, organosilicate was also prepared by ion exchange Na^+-MMT with hexadecyltrimethylammonium bromide and the silicone rubber/OMMT hybrids were prepared by simple mechanical mixing [82]. It was observed that the mechanical properties and thermal stability of hybrids were very close to those of aerosilica-filled silicone rubber.

A novel kind of OMMT was successfully prepared by Wang et al. who used *N,N*-di(2-hydroxyethyl)-*N*-dodecyl-*N*-methylammonium chloride as intercalation agent [83, 84]. Exfoliated nanocomposites were prepared using addition type silicone rubber via solution intercalation. The enhanced mechanical and physical properties demonstrated the efficient reinforcing and thermal stability properties of OMMT.

In a more recent work, Kim et al. [85] used bis(3-triethoxysilylpropyl)tetrasulfide (TESPT) to functionalize two commercially available clays, sodium montmorillonite and Cloisite 25A. Incorporation of the tetrasulfide group-containing clays was found to be effective for the enhancement of interfacial interaction between PDMS and clay, by possible chemical reaction of tetrasulfide groups with vinyl-terminated PDMS. According to the results obtained by swelling in toluene, the crosslinking density was lower than that of neat PDMS, indicating that the observed improvement in mechanical properties arise from enhanced compatibility between the constituents and not from increased crosslinking density.

Poly(dimethylsiloxane) (PDMS)/clay nanocomposites have also been produced by Labruyère et al. [86, 87] by the use of ω-ammonium functionalized oligo-PDMS surfactant [PDMS-$N^+(CH_3)_3$]. The clay ion-exchanged by PDMS-$N^+(CH_3)_3$ has been compared to nanocompositions prepared with PDMS and either non-exchanged sodium MMT and to two organoclays modified by using either alkyl ammonium cations ($C_{38}H_{80}N^+$) or hydroxyalkyl ammonium ($C_{22}H_{48}ON^+$) cations. The authors observed that nanocomposites based on PDMS-$N^+(CH_3)_3$ exhibit the best filler dispersion and significantly higher viscosity due to better intercalation of PDMS chains into clay galleries. Increase of stiffness with increasing the content of treated clay in PDMS was also recorded, but no significant differences between the different organo-modifiers were detected. Regarding the transport properties for organic vapours, the sorption is not influenced by the presence of organically modified clay, whereas the diffusion parameter is significantly affected in the range of sorbed vapour 0–4 % for acetone and 0–7 % for *n*-hexane [87].

Regarding the characteristics of the elastomeric matrix, recent studies try to investigate the related effects. Takeuchi and Cohen [88] published on systems of organo-montmorillonite and PDMS networks prepared from hydroxyl- or vinyl-terminated precursors. No improvement in the modulus of networks synthesized from vinyl-terminated precursors was obtained. The authors concluded that enhancement of the modulus was obtained only for non-optimal networks formed with hydroxyl-terminated precursor chains in contrast with vinyl-terminated chains. Their results indicate that reinforcing effect of these elastomers can be attributed to the anchoring of the hydroxyl end group to the silicate filler that dramatically reduces the soluble fraction and binds pendent chain ends.

The preparation procedure is also very critical for proper filler dispersion. A two step process was reported for the preparation of exfoliated/intercalated polymer/MMT nanocomposites, which included preparation of MMT solution via in situ polymerization of dimethyldichlorosilane inside the galleries of layered silicate hosts and, then, after separating most of the PDMS amount, the treated–MMT solution was blended with several polymers [89]. Bokobza [2] made a series of silicone/nanoclay composites from organically modified montmorillonite and addition cured PDMS. Nanoclay platelets were mixed with hydride terminated PDMS, subsequently cross-linked by 1,3,5,7-tetravynil-1,3,5,7-tetramethylcyclotetrasiloxane via a chemical reaction catalyzed by platinum divinyl-tetramethylsiloxane. Unfortunately, the process demands cure time of 12 h, making impractical the application of this method in a continuous molding process. Moreover, only minor improvements in tensile properties were reported. A new strategy to prepare disorderly PDMS nanocomposites was developed by Ma et al. [90]. More specifically, a quaternary ammonium-containing polysiloxane surfactant was adopted to modify the clay. The slurry of modified clay was then mixed with commercial RTV silicone rubber by hand and exfoliation was achieved as it was confirmed by TEM and XRD analysis. The resulting silicone/nanoclay composites presented significant improvement of their mechanical properties.

In order to overcome difficulties during preparation of high temperature vulcanized silicone rubber (HTV-SR) nanocomposites, Wang and Chen prepared a master batch of the above modified montmorillonite (OMMT-MB) by solution intercalation [91].

From the study of a multi-system, it has been shown that for nanocomposites based on silanol-terminated PDMS and alkylammonium modified layered-silicate fillers, there are two factors controlling silicate dispersion:

(i) the presence of the appropriate number of long ammonium-bound alkyl chains and
(ii) the presence of the appropriate number of polar functional groups. Therefore, dispersion is not a function of the molecular weight or of the differences in processing viscosity induced but it is rather influenced by the silanol end-group concentration [92].

Masterbatches of MMT and OMMT with siloxane-polyether surfactant were used for the preparation of high molar mass poly(dimethylsiloxane)-gum nanocomposites [93]. The incorporation of 5 phr OMMT into PDMS matrix, via masterbatch compounding, improved the tensile strength as much as that obtained with composites loaded with 3 phr OMMT clay by direct addition of the clay to PDMS.

Organoclay nanocomposites show increased versatility regarding their performance properties. Yang et al. [94] prepared flame-retardant methyl vinyl silicone rubber (MVMQ)/MMT nanocomposites by the solution intercalation method, using magnesium hydroxide (MH) and red phosphorus (RP) as synergistic flame-retardant additives, accompanied by aero silica as synergistic reinforcing filler. The above researchers claim that this kind of silicone rubber nanocomposites may be a promising flame-retardant polymeric material.

Liu et al. [95] examined the changes of diffusive properties, due to the incorporation of organo-modified MMT clay into a silanol terminated PDMS matrix, cross-linked with TEOS using 2-ethylhexanote. A cure time of 12 h was required and the study focused on the effect of deformation on diffusion rates. However, a notable reduction in the gas permeability of the silicone/OMMT samples with increasing filler loading was observed in the absence of external deformation.

Siloxane modified MMT nanoclays and commercially available organo-modified nanoclays were combined with liquid silicone rubber (LSR) matrix [96]. The morphologies of the LSR/nanoclay composites were analyzed by SEM and TEM, and partial exfoliation of the clay platelets was observed. Up to 20 % reduction of water permeability was achieved, as well as a 24 % improvement of tear strength and 40 % improvement of compression.

HTV-SR/OMMT-MB-20 % nanocomposites demonstrate enhanced tensile and thermal properties in comparison with HTV-SR/OMMT-20 %. In a recent work [97], this type of hyperbranched OMMT was directly added in HTV-SR system. The enhanced tensile properties suggest an efficient reinforcing agent because of the good dispersion of nanosilicate layers and the resulting "anchor" effect of the hyperbranched macromolecules in the composite.

Incorporation of aminopropyltriethoxysilane (APTES) to silicone rubber by copolymerization enhanced the adhesion properties of silicone rubber [98]. Stability of the silicone rubber against thermal degradation was examined through Thermogravimetric analysis. The tests revealed that compounding with Cloisite 30B boosted up the thermal stability of silicone rubber, much more efficiently than carbon black.

In an attempt to correlate silicate dispersion with the resulting mechanical properties in the PDMS/layered silicate nanocomposites, a systematic study was performed of the mechanical properties, equilibrium swelling and sol fraction measurements of cross-linked silanol-terminated PDMS networks as a function of the type and content of nanofiller as well as of the composition of base resin [99]. In montmorillonite systems, in particular, it was observed that equilibrium solvent uptake and mechanical properties are independent of dispersion state, which suggests that edge interactions play a more significant role than the degree of exfoliation.

The effect of intercalation agent for organic modification of MMT in the mechanical properties of condensation type polysiloxane was studies by Tarantili and Vasilakos [100]. It was observed that the incorporation of Cloisite 20A showed higher reinforcing capacity in PDMS in comparison with Cloisite 30B hybrids due to its higher content organic modification, which results in increased interlayer distance of clay platelets and, therefore, facilitates intercalation and further exfoliation processes.

2.10 Hybrids

Another alternative to conventionally reinforced elastomers is the use of a mixture of two or more fillers of different morphologies, which produces a composite called "hybrid". Wang [101] showed that vulcanizates filled with CB and silica, without coupling agent, display a weaker Payne effect than that resulting from direct addition of two single fillers. This suggests a less developed filler network in the filler blend compounds due to a weaker interaction between silica and CB aggregates than between two aggregates of any single filler.

One of the most important commercial hybrid fillers is carbon–silica phase dual filler, produced by pyrolizing petroleum based feedstocks with silicon containing feedstocks, by a unique co-fuming technology developed by Cabot Corporation [102]. This material, which consists of individual composite aggregates containing carbon and silica phases, was shown, when added to hydrocarbon rubber, to have stronger polymer-filler interaction compared to a blend of silica and carbon black of equivalent rubber.

Bokobza et al. [103] used spherical silica particles, generated in situ via sol-gel process and nanofibres of sepiolite for reinforcing NR. In a more recent work, composites of SBR filled with mixtures of pyrogenic silica combined with a silane agent and fibres of organophilic sepiolite were studied. It has been demonstrated that composites filled with mixtures of these fillers exhibit improved characteristics with regard to single filler materials. The morphologies, mechanical and swelling properties reveal synergistic effects arising by the incorporation of these two types of fillers.

As mentioned above, the major problem during preparation CNTs-elastomer composites is poor distribution and dispersion of CNTs in the elastomer phase. Many researchers have reported the use of hybrid systems instead of CNT alone. Bokobza et al. [104] used this approach for the investigation of reinforcing SBR with mixtures of CB and MWCNTs. The results revealed an improvement of mechanical properties and electrical conductivity with lower percolation threshold than that obtained with composites filled only with CNTs. Through a combination of the benefits of each type of filler, these hybrid composites could potentially exhibit improved characteristics with respect to single-filler systems. On the other hand, the dilution of expensive CNTs with a cheaper reinforcing component, without compromise of the mechanical properties of the resulting material, would have an economic impact as it would reduce the cost of nanotube-based composites [47].

Similarly, MWCNTs and silica were used as hybrid fillers in NR nanocomposites [105]. The results of this study showed that scorch and curing time decreased as the MWCNT content increased in the silica/MWCNT mixture and the maximum torque was increased about 30 %. Furthermore, the tensile strength, elongation at break and fatigue also increased up to a certain filler content and then decreased because of the agglomeration of both fillers, which reasonably results in reduction of their reinforcing capabilities. As the concentration of MWCNT increased, agglomeration became dominant and reduced the combined reinforcing effect. These results imply that silica/MWCNT hybrid presented better properties and created a synergistic effect for the nanocomposites.

Many researchers investigated the synergistic effect of different fillers, in order to achieve a balance between properties and cost. Although graphene and CNTs are similar in chemical composition, different topology distinguishes carbon nanotubes from graphene and offers them unique properties. One major advantage of graphene over CNTs lies in the precursor of graphene-graphite, which is natural abundant [106]. This is essential for industrial applications in the field of composites, since high yield production as well as cost effective and scalable process is allowed [107].

In comparison to graphene, the dispersion of MWCNTs into silicone rubber matrices was found extremely difficult, although both the above possess similar physical structure, whereas it is much easier for MWCNTs to be dispersed into the matrix with the aid of graphene. The strong interactions between graphene and silicone rubber, as well as between graphene and MWNTs, account for the good dispersion of MWNTs. Graphene also behaves as a compatibilizer, as it was observed that the addition of graphene provides a simple route to disperse carbon nanotubes and the synergistic effect facilitates fabrication of high performance polymer composites [108].

A new hybrid nanofiller (MWNTs/OMMT) has also been used as reinforcement for thermoplastic natural rubber (TPNR) composites, by incorporation into the rubber system via the melt blending method. TPNR is a blend of PP, liquid natural rubber (LNR) as a compatibilizer, and NR at the percentage of volume ratio 70:10:20, respectively [109]. The total reinforcement content was fixed at 4 wt%, so a compound of TPNR with 4 wt% OMMT, 4 wt% MWNTs, and hybrid 2 wt% OMMT 2 wt% MWNTs was compounded using the internal mixer. The results showed that the tensile strength and Young's modulus of the composite were improved by using the hybrid nanofillers. However, the elongation at break considerably decreased with the above hybrid fillers. The effect of MWNTs, OMMT, and hybrid MWNTs–OMMT was also confirmed by DMA. The results suggest that the storage modulus, loss modulus, and glass transition temperature (T_g) also increased for all samples reinforced with nanofillers. TEM micrographs revealed the interfacial adhesion between the fillers and the matrix, which contributed significantly to the improvement of the properties.

Elastomeric hybrids consisting of graphene oxide (GO) sheets were fabricated by utilizing butadiene–styrene–vinyl pyridine rubber (VPR) as the host through co-coagulation process and in situ formation of an ionic bonding interface [110]. The VPR/GO composites with a normal hydrogen bonding interface were also prepared. The mechanical properties and gas permeability of these hybrids with an ionic bonding interface were obviously superior to those of the composites with a hydrogen bonding interface. With the ionic interfacial bonding, inclusion of 3.6 % v/v of GO in VPR generates a 21-fold increase in glassy modulus, 7.5-fold increase in rubbery modulus, and 3.5-fold increase in tensile strength. The very fine dispersion of GO and the strong ionic interface in the hybrids are responsible for such unprecedented reinforcing efficiency of GO towards VPR.

Recently, studies of the synergistic effect of natural fibres and particulate fillers compounded with different polymeric matrices on the physicochemical properties of hybrid composites have been reported [111]. These studies revealed interesting results, showing the improvement of specific properties together with additional environmental and cost benefits.

Anuar et al. [112] prepared hybrid composites of thermoplastic natural rubber (TPNR) reinforced with *Hibiscus cannabinus* L fibre (kenaf fibre: KF) and short glass fibre (GF) via melt blending using internal mixer. The effects of coupling agents, silane and maleic anhydride grafted polypropylene (PP-g-MA) on the tensile properties were also investigated. The results on tensile strength have shown that the increasing in kenaf fibre content substantially reduced the tensile strength and modulus. Incorporation of silane coupling and PP-g-MA has enhanced the mechanical properties of the systems and the fibres were coated by the TPNR, subsequently reduced the formation of void as observed in SEM micrographs for untreated systems.

It is anticipated that the combined use of both short natural fibres and silica will combine the beneficial effects of individual reinforcements for the development of materials with desirable properties. NR based hybrid composites were prepared by incorporating cellulose fibres and silica in a laboratory two-roll mill and, in order to modify the surface properties of the fillers, a silane coupling agent [bis(triethoxysilylpropyl)tetrasulfide, Si69] was used [113]. It was found that silane treatment of both fillers had significant effect on the processability but offered little benefit as rubber reinforcement. It was also found that treated fillers were better dispersed into the composite matrix and their compatibility with NR increased with silane treatment. The above study showed that these hybrid fillers when combined with Si69 silane coupling agent, could be used as substitutes for silica to improve the processability of the products with the environmental advantage and acceptable performance for rubber applications.

Shojaei et al. [114] studied the effect of carbon and cellulose fibres on the tribological characteristics of rubber-based friction materials (RBFMs). They observed that the maximum attainable drum temperature caused by the friction-induced heat containing carbon fibres is lower than that of composites reinforced with cellulose fibres. CB influence slightly the coefficient of friction, due to the lubricating role of this material as well, but considerably improve wear resistance and fade behavior. Cellulose fibres offered high friction coefficient but it was found to be destructive for fade behavior and wear rate point, due to its weak thermal stability.

3 Crosslinking

The vulcanization of elastomeric materials is a complex process, which has a high impact on the properties of the final product. In fact, the extent of this reaction, the crosslinks density along with the other network parameters are some critical factors controlling the overall behaviour of the vulcanized rubber. Therefore, it is important to monitor and control this crosslinking reaction. Several articles on the kinetics of vulcanization of various rubbers are available in the literature and especially in silicone rubbers, which is the material with very interesting curing paths. However, there are only a few reports on the vulcanization kinetics of rubber nanocomposites and the changes in rheological and thermal behaviors during the network formation by the addition of nanofillers. It has been reported that curing behavior of organomodified rubbers conflicts as strong acceleration, no change or retardation effects.

Mathew et al. reported that the addition of organoclay accelerated the cure rate of ethylene acrylate rubber (EAR) and lowered the activation energy for curing reaction calculated from the autocatalytic model [115]. The pristine clay-filled EAR composite showed a similar trend, but much smaller effect.

Monsanto rheological measurements on NR reinforced with octadecylamine-modified bentonite have shown that the organoclay accelerates vulcanization reaction and gives rise to a marked increase of torque, indicating that the elastomer becomes more densely cross-linked [116]. These results are in agreement with those from swelling experiments and from the study of vulcanization reaction with thermal analysis, where an appreciable increase of the involved heat from curing reactions has been observed. Moreover, thermodynamic parameters have shown an increase in the structural order of the nanocomposite [116]. Also, the effect of incorporation of bentonite on the vulcanization kinetics of NR was investigated by means of both cure-meter and DSC under dynamic and isothermal conditions [117]. A marked decrease in the induction time and optimum cure time of the elastomer were observed in the presence of organoclay. Although octadecylamine itself accelerated the vulcanization process, the octadecylamine-modified clay gave rise to a further significant increase of vulcanization rate, due to a synergistic effect between filler and amine. Moreover, in the presence of organoclay, a dramatic increase of the torque value was recorded, because of the formation of higher number of cross-links, which could be attributed to the confinement of the elastomer chains within the silicate galleries and consequently, to better interactions between filler and the rubber.

In another work, the influence of inorganic nanoparticles on the cross-linking mechanism of NR nanocomposites has been evaluated by applying the tube model on equilibrium statistical mechanics. The obtained results have shown that a highly ordered structure with a huge amount of entanglements, wherein the polymer is nanoscopically confined, is formed by the addition of nanoparticles [118].

Carretero-González et al. studied the influence of nanoclay on the morphological and microstructural changes of NR network based on the results from broadband dielectric spectroscopy and in situ synchrotron wide-angle X-ray diffraction [119]. It was found that nanoclay introduces a dual crystallization mechanism, due to the alignment of nanoparticles during stretching. Moreover, the strong interfacial adhesion between nanoparticles-rubber matrix can induce an early promotion and enhancement of overall crystallization of NR chains under uniaxial stretching. The above authors reported that the organoclay does not affect the extent of curing and the observed increase in modulus was attributed to: (a) the hydrodynamic effect of the clay and (b) the formation of physical cross-links due to the presence of clay.

Cataldo observed that the addition of organo-modified montmorillonite in a NR/SBR mixtures causes an increase in the cure rate and this effect was explained by the high concentration of ammonium cations present in the exfoliated clay [120].

Very interestingly, differential scanning calorimetry was established as a useful analytical tool for the determination of the energy required during rubber vulcanization. Analysis through DSC is based on the assumption that the heat of reaction is only due to a single curing reaction and is proportional to the extent of the reaction [117, 121–124]. Therefore, in isothermal curing conditions, the degree of curing (a) can be calculated from the heat-flow peak of a DSC curve by the following equation:

$$a = \frac{\Delta H_t}{\Delta H_\infty} \tag{1}$$

where ΔH_t is the accumulated heat evolved to time t and during vulcanization, and ΔH_∞ is the total amount of heat generated during the reaction.

The dynamics of the vulcanization reaction are modeled by means of a differential equation with regard to time, where the influence of the temperature and vulcanization rate are separated. Thus, the equation expressing the vulcanization rate may be written as follows [125]:

$$\frac{da}{dt} = K(T) * f(a) \tag{2}$$

where t is the time, T is the temperature, and K is the kinetic constant.

The function K(T) is related to the activation energy according to the Arrhenius expression:

$$K(T) = K_0 * \exp\left(-\frac{E_a}{RT}\right) \tag{3}$$

where K_0 is the preexponential factor, E_a is the activation energy and R is the universal gas constant.

For solid-state transformations, combination of Eqs. (2) and (3) gives:

$$da/dt = K_0 e^{-Ea/RT} f(a) \tag{4}$$

In curing reaction the function f(a) may get different forms depending on the reaction mechanism. In our study, a model given for autocatalytic reaction with two reaction orders, known as Šesták-Berggren equation was adopted to simulate vulcanization process [126]:

$$f(a) = a^m * (1 - a)^n \tag{5}$$

where m and n denotes reaction order.

The choice of model depends on the type of chemical reaction. In the autocatalytic reaction, the maximum rate of cure is given for a conversion degree other than zero, because the reaction is promoted by the same products of the reaction.

By substitution of Eq. (5) in (4) the change of degree of conversion da/dt with temperature is given by [124]:

$$\frac{da}{dt} = K_0 * \exp\left(-\frac{E_a}{RT}\right) * a^m * (1 - a)^n \tag{6}$$

Parameters of curing process such as K(T), m, n were calculated by non-linear regression analysis plots of da/dt versus a.

Activation energy was calculated graphically, form the Arrhenius expression (Eq. 3), from the slope ($-E_a/R$) of linear regression of lnK versus 1/T:

$$\ln(K(T)) = \ln(K_o) + \left(-\frac{E_a}{R}\right) * \frac{1}{T} \tag{7}$$

In the case of non-isothermal experiments, i.e. when the sample temperature changes at a desired and constant heating rate, $\beta = dT/dt$, the calculation of activation energy of vulcanization reaction can be made from plots of the logarithm of the heating rate versus the inverse of the temperature at maximum reaction rate T_{max} while d(da/dt)dt = 0, by the Kissinger [127] method using the equation:

$$E_a = -R\frac{d\ln\left(\beta/T^2_{max}\right)}{d(1/T_{max})} \tag{8}$$

From a plot of $\ln\left(\beta/T^2_{max}\right)$ versus $1/T_{max}$, and fitting to a straight line, the activation energy E_a can be calculated from the slope.

The Ozawa method [128] can also be used through the following equation:

$$E_a = -R\frac{d\ln\beta}{d(1/T_{max})} \tag{9}$$

The activation energy for different heating rates can be determined from $\ln\beta$ versus $1/T_{max}$ plot.

In a recent paper [129] the curing mechanism and the related kinetics of montmorillonite/vinyl terminated polysiloxane nanocomposites was followed by the above described method and the effect of commercial unmodified montmorillonite as well as two different types of organically modified products, namely: Nanofil 116, Cloisite 20A and Cloisite 30B was recorded. It was found that Cloisite 20A/PDMS systems showed increased reaction rate at the early stages followed by a retardation of rate and extension of the curing time, in comparison with pure PDMS. On the other hand, Cloisite 30B seemed to decrease the curing rate of PDMS during the whole process, whereas a significant increase of this parameter was recorded in the case of unmodified clay/PDMS nanocomposites. This effect was attributed to physicochemical interactions between the cross-linking system and organoclay particles. In the attempt to further evaluate the obtained results, it was found that the autocatalytic model showed good fitting with the experimental data.

More specifically, for isothermal experiments, the increase of temperature increases, also the curing rate, as shown by the shifting of the exothermic peak to lower time values. Furthermore, the peak becomes narrow and vulcanization is completed faster. Also, the shape of conversion curve is dependent on the temperature. At low temperatures, the change in conversion is more gradual, but as the temperature increases the peak height of the conversion rate curve also increases with a shift of the peak position to higher values. The reaction rate was found to increase rapidly and after reaching a maximum it gradually slows down toward zero and this trend was observed for all the examined PDMS systems.

The incorporation of organoclay nanoparticles into the PDMS matrix affects the vulcanization reaction mechanism of the elastomer. The obtained data show that with increasing organoclay load the time required for completion of vulcanization slightly increases, in comparison with that necessary for unfilled PDMS. The reaction rate for the Cloisite 20A/PDMS systems is high at the beginning of vulcanization. A second stage is observed, where cross-linking proceeds slower, in comparison with the pure elastomer. Therefore, an inhibition during the second stage of the process, might be possible due to organoclay nanoparticles and leading to longer vulcanization time.

A stronger effect on the curing time was observed with PDMS nanocomposites reinforced with Cloisite 30B organoclay, where a significant delay of the vulcanization rate was observed, especially for hybrids reinforced with 5 and 8 phr loading. Therefore it seems that this type of reinforcement, containing hydroxy groups in the intercalation agent, increases to a higher extent the curing reaction time.

Further investigation, in order to elucidate the effect of the intercalation agent, included the study of the effect of another type of reinforcement, i.e. the unmodified montmorillonite Nanofil 116.

From the comparative study of effect of the examined montmorillonite reinforcements in the vulcanization of elastomeric matrix at 60 °C, it was observed that Cloisite 30B increases significantly the vulcanization time of vinyl terminated PDMS, followed by Cloisite 20A. However, the opposite effect was observed for the unmodified reinforcement, where the reaction was completed in shorter time and the maximum rate was achieved at shorter time, than that corresponding to pure elastomer. Curing retardation can be attributed to interactions between the catalyst system and organic moiety (quaternary ammonium compound) of the modified clay, which inhibits catalytic action [130].

In addition to the above, the consumption of a part of the cross-linker for chemical reactions with the active sites of the intercalation agent might be possible and this, would lead to a decrease of the extent of curing reaction especially at high loadings. Moreover, absorption of the cross-linking system into the montmorillonite nanoparticles would reduce the active concentration required for the cure reaction. The enhanced inhibiting effect, observed for hybrids of Cloisite 30B can also be related with the action of radical scavenging of OH groups of the intercalation agent of this type of organoclay. However, the unmodified montmorillonite seems to have a synergistic effect on the vulcanization process of the examined vinyl terminated PDMS leading to faster cure.

Contradictory results were reported in the literature regarding the effect of inorganic nanoparticles on elastomer's curing reaction. The chemistry of vulcanization reaction and the chemical activity of incorporated nanoparticles play a major part on the reaction mechanism.

Investigations of high temperature vulcanized silicone elastomers, filled with superconductive carbon blacks, have shown curing retardation, because of the radical scavenging effect of phenol-OH groups existing in filler surface [131].

On the other hand, Mathew et al. [123] reported an accelerating effect on the curing of ethylene acrylate rubber, introduced by quaternary ammonium ions, present in the organoclay modification medium. The same effect was observed during the vulcanization process of filled fluoroelastomer nanocomposites [124].

Secondary reactions with silane groups of the cross-linker have also been reported [117]. More specifically, it was claimed that the degree of cross-linking was strongly dependent on the amount of silane groups present. This effect was attributed to competitive oxidation/hydrolysis secondary reactions of the silane groups, leading to the formation of Si–OH groups, via reaction with other compounds such as moisture and O_2.

In addition to the above mentioned secondary reactions, in case of polysiloxane nanocomposites the introduction of clay nanoparticles must be taken into account. According to Takeuchi and Cohen [132] silicate oxygen groups on the surface of clays and hydrogen groups along the edge of these clay particles may enhance these secondary reactions.

In the macroscopic level, the properties of nanocomposites prepared with room temperature-vulcanized silicone rubber (RTV-SR) reinforced with MMT modified with di(2-oxyethyl)12-alkane-3 methyl-amine chloride, were explained taking into consideration incomplete curing of the whole system due to reactions of alkoxyl silicane catalyst not only with the hydroxyl containing base but also with the intercalation agent [133].

A decrease of the gel fraction (GF) with increasing clay content was observed by Kong et al. in silicone rubber nanocomposites reinforced with Fe-MMT and Na-MMT, which were modified with cetylthimethyl ammonium bromide [134]. The decrease in GF values was due to the additives of clay having the action of blocking free radicals at the formation of a three dimensional network structure.

Schmidt and Giannelis from their study on silicate dispersion and mechanical reinforcement in polysiloxane/layerd silicate nanocomposites reported that increased nanoffiller content lead to lower swelling, given that compatible nanofillers should increase the (physical) cross-link density [99]. The effectiveness of nanofillers in increasing physical cross-link density is dependent on the effective interfacial area and interfacial interaction strength.

The vulcanization kinetics of fluoroelastomer (FKM) filled with unmodified and organo-modified clay, were studied by Kader and Nah [124] with both, oscillating disc rheometer and differential scanning calorimetry under isothermal and dynamic conditions. The organo-modified clay enhanced the curing procedure through the accelerating effect of quaternary ammonium salt used in the clay modification, while the unmodified clay showed cure retardation due to the absorption of curative by the clay for in situ clay modification. The kinetic analysis showed the suitability of autocatalytic model for cure characterization. The determined kinetic parameters were in good agreement with the experimental values. The result indicated that the organoclay was efficient in reducing the energy requirement for the cure process.

4 Compatibilizing Agents, Plasticizers

4.1 Properties of Interphase Modified/Compatibilized Elastomeric Composites

The combination of different polymers is a fascinating and practical channel to get new materials with novel functions that individual components do not process. Thermoplastic elastomers not only exhibit the conventional properties of elastomeric materials at room and service temperatures but can also flow at elevated temperature and again solidify when the temperature returns, which behave like thermoplastic. However, usually due to mismatch of polarity, melt viscosity, and molecular weight the interface adhesion of rubber and thermoplastic polymer is poor, leading to an inferior mechanical property which limits the application. Therefore it is necessary to improve the compatibilization of two phases.

Compatibilization is a process of modification of the interfacial properties in an immiscible polymer blend that results in formation of the interphases and stabilization of the morphology, leading to the creation of a polymer alloy [135]. According to Utracki [136], a compatibilization process must accomplish three tasks: (i) reduce the interfacial tension, thus finer dispersion (ii) stabilize the morphology against thermal and rheological effects during the processing steps and (iii) provide interfacial adhesion in the solid state. This definition means that apart from decreases in the phase domain size and stabilization of the morphology against thermal and rheological effects, a compatibility additive must also improve the interfacial adhesion between two phases. Improved adhesion should be manifested macroscopically as improved tensile strength and modulus and perhaps as increased impact strength as well.

Elastomer blends are widely used for a variety of reasons, like improved physical properties, service life, easier processing, complex demand for the performance and reduced product costs.

In the previous years, the reactive compatibilizers have been increasingly introduced into blends. Reactive compatibilization can also provide for a degree of control over morphology development in multiphase polymer blends via manipulation of the interfacial energies within the system, which allowed the formation of a composite dispersed phase during melt blending process via encapsulation of one dispersed phase by another. With chemical reaction occurring between the functional group and the blends, the interfacial adhesion would be greatly increased and the interfacial tension would be sharply decreased.

Elastomeric composite materials consisting of cross-linked polymers and nanostructured solid fillers have become increasingly important. The interaction of the polymer molecules with the solid surface of filler particles is crucial in controlling on the performance of these elastomeric nanocomposites.

In recent years, it was reported that the anisotropic nanofillers which have a large specific surface area and high aspect ratio such CNTs and organoclay, can also be used as potential compatibilizers to improve the compatibility of mixtures of immiscible polymers. In immiscible multiphase polymer blends, one component forms the discrete phase, and it is dispersed in the host component, which act as a continuous phase. Generally, new properties are achieved through this dispersed phase, which can exist in the form of droplets, fibrils, lamellae, or even co-continuous structures after blending. The state of distribution and dispersion of filler particles between the two phases, the extent of interaction of the filler surface with either of the polymeric phases, and the resultant final phase morphology are believed to be the most important factors affecting the overall properties of the final blends. These include the affinity of the filler toward the polymeric phases, the viscosity ratio of the polymers, and the feeding routes. In the case of polymer blend nanocomposites, the nanoscale filler distribution and crystalline structure could dictate the resultant properties of the composites.

To understand the compatibilizing effect of inorganic fillers, two possible mechanisms were proposed. Mechanism I (thermodynamic compatibility): with a large specific area and high aspect ratio, anisotropic inorganic nanofillers can absorb the polymers chains on its surface to gain stabilizing energy which make the overall free-energy of mixing (ΔG_m) negative and thermodynamically driven compatibility between the immiscible components [137]. Mechanism II (kinetic compatibility): the selective localization of nanofiller in the polymer interface decreases the interfacial tension and prevents the coalescence of the domains during melt mixing which kinetically improves the compatibility between the binary phases [138].

Therefore, the addition of anisotropic nanofillers may bring about the reinforcing and compatibilization effects for an immiscible blend system simultaneously, which is a new approach to achieve high performance of polymer blend nanocomposites.

Recently rubber blends with layered silicate nanoparticles have received significant attention because of the remarkable improvements of some mechanical properties, enhanced barrier properties and reduced flammability. However, the efficiency of nanoclay with regard to improving the solid-state properties of rubber blends is extremely dependent on the degree of clay dispersion, i.e. intercalation and exfoliation as well as the distribution of the exfoliated platelets and the phase specific clay distribution, respectively. Several studies of polymer/clay systems indicated that a compatibilizing effect of clay in heterogeneous blends exists. Several studies of polymer/clay systems indicated that a compatibilizing effect of clay in heterogeneous blends exists. The droplet size of dispersed phase in NBR/styrene butadiene rubber (SBR) [139], PA6/ethylene propylene rubber (EPR) [140] was drastically reduced by addition of clay.

One of the explanations was attributed to the diminished molecular mobility, which prevents the phase separation of the components [141, 142]. Fisher et al.

[143] proposed another mechanism by which clay can form in situ grafts by absorbing large amounts of polymer, which in turn are very effective for reducing the interfacial tension and inducing compatibilization in blends from highly immiscible components.

A new method of the online measured electrical conductance (OMEC) has been successfully applied by Ali et al. [144] for the characterization of morphology development and kinetics of clay distribution in rubber blends during the mixing process. It was found that the release of the ionic surfactant from the clay galleries during the mixing was the main reason for the electrical conductance of rubber-organoclay composites. Based on the results received at mixing of pure hydrogenated acrylonitrile butadiene rubber (HNBR) with NR-clay masterbatch, the apparent clay migration from NR phase to the HNBR phase was ascribed to the different interactions with polymers and clay. Because in the clay galleries both NR and HNBR molecules can intercalate the bi-intercalated clay tactoids act as compatibilizing agent like a block copolymer. Therefore the refinement of blend morphology is caused by the compatibilizing effect of clay.

Hwang et al. [145] have successfully prepared butadiene rubber (BR)/organosilicate nanocomposites that have intercalated and partially exfoliated structures by using a two-stage melt blending method. As a result of counteracting the polarity effect and compatibility effect, the tensile properties and tear strengths of these nanocomposites increased substantially, relative to those of the neat BR, upon increasing the amounts of incorporated compatibilizer and organosilicates that contain a low weight percent of layered silicates. They attributed these results to the significantly improved compatibility and the strong interactions among the polar layered silicates, compatibilizer, and nonpolar molecular chains of the BR. The BR molecules can interdiffuse into the intergallery space of the layered silicates more readily with the aid of the compatibilizer and result in an intercalated/exfoliated structure. Additionally, these nanocomposites exhibit higher thermal stabilities and dynamic storage moduli than those of the neat BR. The relative water vapor permeabilities of the BR nanocomposites containing 10 parts of organosilicate, both in the presence and absence of the compatibilizer, were reduced to 20 % of that of the neat BR.

Being a two dimensional layered carbon nanomaterial, graphene oxide (GO) has gained much attention due to its unique structure and properties in recent years. Normally, it is oxidized from natural flake graphite and consisted of parallel pseudo two-dimensional lamellaes. The GO layers consist of randomly distributed unoxidized aromatic regions and hexahydric aliphatic regions attached with polar groups, such as hydroxyl, epoxide, ether, and carboxylate groups, as a result of oxidation. Compared to clay, GO, as one of lamellar fillers possesses much higher mechanical properties and has a larger specific surface area, which can maximize the interfacial contact between polymer and fillers and, thus, it has the potential to be effective reinforcing filler and compatibilizer.

The GO-filled natural rubber (NR)/high density polyethylene (HDPE) thermoplastic vulcanizates were successfully prepared by an ultrasonically assisted latex mixing process (ULM) [146]. GO was dispersed into NR latex by ultrasonic irradiation and followed by latex coagulation to obtain NR/GO master-batch, which was further mixed and diluted with HDPE and NR via a dynamic vulcanization process to obtain NR/HDPE/GO composites. It was found that the stacked GO platelets were successfully exfoliated and have good compatibilization efficiency for the immiscible NR and HDPE. The stacked GO platelets enhance the interfacial adhesion and phase compatibility, which results in an increase in mechanical property of NR/HDPE blends. The change of HDPE crystallinity due to the incorporation of GO is small and was not the main reason for the improvement in the mechanical properties. It was found that GO is mainly dispersed in NR phase and the interphace, which act as an obstacle to inhibit the coalescence of the NR phase.

Graphite nanosheets (GNs) have gained great interest because of their high flexibility, large surface area, and lower cost compared with other nanofillers such as nanotubes and nanowires. When natural graphite is intercalated by small molecules, such as sulfuric acid, the spacing between graphite nanolayers is increased and the resulting product is called graphite intercalated compound (GIC).

Impact and water resistant electrically conductive polyurethane (PU) foam nanocomposites with radio frequency wave absorbing characteristics were prepared by impregnation of PU foam in a designed solution of room temperature vulcanizing silicone rubber and conductive graphite nanosheets (GN) called doping solution [147]. Incorporation of bifunctional surfactant into the composition of the doping solution was found to have profound role in improving the dispersion state of GN platelets throughout the silicone rubber as well as PU matrix. Extent of GN particles and hence percolation threshold exhibited to be affected by the functionality of the compatibilizer. All PU/SR/GN nanocomposites generated from bifunctional compatibilizer exhibited higher electrical conductivity with enhanced permittivity implying enhanced formation of conductive networks by GN platelets.

Chlorinated polyethylene (CPE) was found to be an efficient compatibilizer for EPDM rubber/nylon (PA) blends, due to the interactions taking place between CPE and PA [148]. The cold milling of molten EPDM/PA/CPE blend was found to cause the formation of in situ fibril rubber compound, which display magnificently improved tear strength. It was also mentioned that small amount of PA fibril (i.e. 10 %) actually enhanced the mechanical properties of the matrix. In rubber composites reinforced with short fibres, however, that low concentrations of fibre are deleterious for the properties of the matrix rubber. This was explained by the fact that in rubber composites containing "in situ" fibrils, the polarity-matched rubber/plastic blend, the finer dispersed particles and high elongation of PA, make dispersed particles easier to stiffen or constrain the matrix rubber and permit effective load transfer across the hard-soft interface, than in the case of rubber composites containing short fibres.

Ismail and Mega [149] studied the effect of a compatibilizer, poly(propylnene-eyhylene acrylic acid) (PPEAA) and a silane coupling agent, 3-aminopropyl-triethoxy silane (3-APE), on the properties of polypropylene (PP)/natural rubber

(NR) blends reinforced with white rice husk ash (WHRA). The incorporation of WRHA in PP/NR blends increased tensile modulus, stress at yield and elongation at break of the composite system. However, the WRHA filled PP/NR blends treated with 3-APE and PPEAA showed better tensile strength, stress at yield and tensile modulus compared to similar untreated composites. SEM micrographs revealed better filler dispersion and better filler-matrix interaction as a result of the treatment with 3-APE and PPEAA.

In order to improve the compatibility between natural fibres and polypropylene (PP) and polypropylene-ethylene propylene diene terpolymer (PP-EPDM) blends, the functionalization of both matrices with maleic anhydride (MA) was attempted by López Manchado el al. [150]. The morphological observations carried out by scanning electron microscopy show that the incorporation of small amounts of functionalized polymer considerably improves the adhesion at the fibre-matrix interface. In these cases, the fibres are perfectly embedded in the matrix in relation to the composites prepared with the pure homopolymers, and a significant increase in the composite strength is also observed, particularly, after the incorporation of both modified polymers (MAPP and MAEPDM). Thus, it is possible to correlate better interfacial adhesion with the improvement of mechanical properties. It is assumed that the functionalization of the matrix reduces interfacial stress concentrations and may prevent fibre–fibre interactions, which are responsible for premature composite failure. It was also observed that both flax fibre and rubber behave as effective nucleant agents, accelerating PP crystallization. Moreover, these results are particularly relevant when the grafted matrices are added to the composite.

If rubber and plastic are blended to produce a thermoplastic elastomer (TPE), their technical properties are inferior because of the presence of a week rubbery phase and an interface dispersed in a continuous plastic matrix. This limitation can be circumvented by preferential reinforcement of the rubber phase and the interface with the plastic matrix by fillers and/or the addition of crosslinks in the rubbery phase via dynamic vulcanization without the processing characteristics being sacrificed. Pukanszky [151] proposed that the dimensions of the interface and strength of the interaction significantly influence the ultimate tensile properties of composites. The sequence of addition of the individual polymers and fillers plays an important role in determining the properties of polymer blends [152].

Ternary composites consisting of PP, an elastomer and a filler are becoming attractive materials because they possess both, high modulus and high impact strength. It has been explained that the mechanical properties of such ternary composites are strongly influenced by both the composition and the morphology. The morphology is strongly dependent on both, the interfacial interaction between the components and the mixing method. Two different morphologies are expected to form in PP/elastomer/filler ternary composites. One is the morphology with the elastomer and filler particles dispersed independently in the PP matrix, whereas the other has a core-shell type particles consisting of a filler particle core and an elastomer shell. When no functional group is introduced into the elastomer, the elastomer and the filler particles are dispersed independently in the PP matrix because there is no interaction between them. On the other hand, when the functionalized elastomer is reinforced with filler, the aforementioned core-shell structure should be formed.

Denac et al. studied the crystallization process in blends of isotactic polypropylene (iPP), with styrene rubber block copolymer blends (SRBC), as well as in their composites reinforced with talc [153, 154]. Talc crystals, homogeneously incorporated into the iPP matrix, accommodated mostly plane-parallel to the surface of samples and strongly affected the crystallization process of iPP matrix, disturbing well-developed sperulitic morphologies, Both SEBS and SEBS-g-MA block copolymers encapsulated talc crystals, thus forming core-shell morphology at significant higher extent than poly(styrene-b-butadiene-b-styrene) triblock (SBS) and poly(styrene-b-ethylene-co-propylene) diblock copolymer (SEP) in iPP/talc composites. The observed differences between the effect of SEBS and SEBS-g-MA on mechanical properties could be explained by the higher molecular weight and different miscibility with the iPP chains of the SEBS than SEBS-g-MA elastomer, as well as by the higher extent of encapsulation and disorientation of talc crystals by SEBS-g-MA than by SEBS copolymer.

5 Future Prospects

Nowadays rubber technology has become an important part in industry. Because the essential modulus and strength of neat rubber is low, it is necessary to introduce an additional reinforcing phase to achieve optimal properties required in their applications.

The expansion of the polymer industry and the continuous demand for new composites with improved properties and lower particle content are some of the new challenges in this area.

The use of nanomaterials for the reinforcement of rubber composites is a new technology that gives high expectations because it has shown enhancement in the composites properties, such as modulus of elasticity, gas-barrier property, electrical conductivity, magnetic susceptibility, hardness, abrasion resistance, scratch resistance and ultraviolet protection.

Carbon nanotubes or exfoliated graphite/graphene offers substantial opportunities in the electrical/electronics/optoelectronics areas as well as potential in specific engineering technologies. Conductive elastomer nanocomposites can be applied as artificial muscles, electromechanical actuators, touch control switches, and shape-memory polymers, etc.

Graphene oxide is another interested nanomaterial that can act as both the effective reinforcing nanofiller and the compatibilizer in the immiscible rubber-based blends in order to produce high performance, ultralight vulcanizate.

Recently, one area that has come under intense inspection is the role of nanoparticles as potential compatibilizer in immiscible polymers. The combination of two different fillers in preparation of rubber nanocomposites is an interesting tool that can present synergistic effect and gives the potential for preparing new engineering materials with specific and improved properties. There are reports in the synergistic effect of different fillers to make balance between properties and cost. A good example is the combination of CNTs with graphene.

Currently, the demand for reinforcing fillers from renewable resources, such as plant-based natural fibre, for the production of biosustainable "green-composite" materials is increasing in research area and manufacturing because of their ease of processing, low cost, low density, biodegradability and good mechanical properties. Natural fibres provide with interesting properties the final composite such as their capacity to be recyclable, renewable raw material, and less abrasive and harmful behavior.

6 Conclusions

Elastomers represent a class of natural and synthetic materials of profound scientific, technological and commercial interest. Due to the high versatility in their composition and structure, rubber products are capable of satisfying almost any requirement and standard of use. In addition, the incorporation of additives that produce reinforcing effect to the elastomeric matrix and further improve the properties, such as physicomechanical, electrical, endurance etc., greatly increases the spectrum of applications.

The reinforcing fillers commonly used in elastomers are carbon black, silica, clay particles and glass fibres. However, advanced technology reinforcements, such as aramids, might be recommended for special applications.

Hybrid composites provide new possibilities of extending the area of rubber applications, as they combine selected filler properties taking advantage of their synergistic effect.

The main reason for reinforcing rubbers is to improve the mechanical and thermal properties as well as to reduce cost, and sometimes weight of the compounds. The recent advances of nanoparticles has attracted much attention in manufacturing of rubber nanocompositres because of the small size of the filler and the corresponding increase in the surface area allowing to achieve the required mechanical properties at low filler loading. Carbon nanotubes and graphene nanoporticles are promising materials as rubber reinforcements offering good electrical properties.

Since the filler/matrix interfacial characteristics (including wetting, adhesive bonding etc.) are of prime importance for an efficient reinforcing effect, surface treatment of the filler particles with the appropriate coupling agents is necessary in order to promote proper dispersion and adequate filler/matrix interactions.

In order to improve dispersion of reinforcement in rubber matrices the use of the respective functionalized polymer, otherwise called compatibilizers, might be necessary. Among the different modifying agents, maleic anhydride is the most commonly used and seems to ensure best results at relatively low cost.

The newly developed polyblends based on mixtures of rubbers with polyolefins, require the suitable compatibilization in order to reveal their unique properties. Nanoparticles, being a very efficient reinforcing agent even at low concentrations, were also found to play the role of compatibilizer for these mixtures of immiscible polymers.

The incorporation of reinforcing fillers along with some other additives into the elastomeric matrices, often leads to very complex systems and, therefore, the influence on the crosslinking parameters must be carefully studied in each specific case, since the network characteristics are critical for the overall performance of final products.

References

1. Hull, D.: An introduction to composite materials. Cambridge University Press, New York (1987)
2. Bokobza, L.: J. Appl. Polym. Sci. **93**, 2095–2104 (2004)
3. Xanthos, M.: Polym. Eng. Sci. **28**, 1392–1400 (1988)
4. Dagli, S.S., Xanthos, M., Biesenberger, J.A.: Emerging technologies in plastics recycling. In: Andrews G.D., Subramanian, P.M. (eds.) ACS Symposium Series, vol. 513 Chapter 19 (1992)
5. Faisant, J.B., Ait-Kadi, A., Bousmina, M., Deschênes, L.: Polymer. J. Appl. Polym. Sci. **39**, 533–545 (1998)
6. Willis, J.M., Caldas, V., Favis, B.D.: J. Mater. Sci. **26**, 4742–4750 (1991)
7. Bhowmick, A.K., Stephens, H.L.: Handbook of Elastomers, 2nd edn. Marcel Dekker, New York (2001)
8. Pramanik, M., Srivastava, S. K., Samantaray, B. K., Bhowmick, A. K.: J. Polym. Sci. B. **40**, 2065–2072 (2002)
9. Mai, Y.-W., Yu, Z.Z. (eds.) Polymer Nanocomposies. Woodhead Publishing Ltd, Cambridge (2006)
10. Mathew, T., Datta, R.N., Dierkes, W.K., Talma, A.G., van Ooij, W.J., Noordermeer, J.W.M.: Macromolecular Materials and Engineering, vol. 296, pp. 42–52 (2011)
11. Lopez, J.F., Perez, L.D., Lopez, B.L.: J. Appl. Polym. Sci. **122**, 2130–2138 (2011)
12. Hui, S., Chaki, T. K., Chattopadhyay, S.:J. Appl. Polym. Sci. **110**, 825–836 (2008)
13. Bertona, A., Castellano, M., Marsano, E., Alessi, M., Conzatti, L., Stagnaro, P., Colucci, G., Priola, A., Turturro, A.: Macromol. Mater. Eng. **296**, 455–464 (2011)
14. Stöckelhuber, K.W., Svistkov, A.S., Pelevin, A.G., Heinrich, G.: Macromolecules **44**, 4366–4381 (2011)
15. Bandyopadhyay, A., Maiti, M., Bhowmick, A.K.: Mater. Sci. Technol. **22**, 818–828 (2006)
16. Hikasa, S., Nagata, K., Miyahara, K., Izumi, T., Suda, T., Toyohara, A., Kato, A., Nakamura, Y.: J. Appl. Polym. Sci. **114**, 919–927 (2009)
17. Song, X., Yoshino, H., Shibata, H., Nagatani, A., Ueda, Y.: J. Appl. Polym. Sci. **113**, 3661–3670 (2009)
18. Fekete, E., Móczó, J., Pukánszky, B.: J. Colloid Interf. Sci. **269**, 143–152 (2004)
19. Kolarik, J., Lednicky, F., Jancar, J., Pukanszky, B.: Polym. Commun. **31**, 201–204 (1990)
20. Kolarik, J., Jancar, J.: Polymer **33**, 4961–4967 (1992)
21. Mnif, F., Massardier, V., Kallel, T., Elleuch, B.: Polym. Adv. Technol. **21**, 896–903 (2010)
22. Chang, Y.W., Shin, J.Y., Ryu, S.H.: Polym. Int. **53**, 1047–1051 (2004)
23. Dasari, A., Yu, Z.-Z., Mai, Y.-W.: Polymer **46**, 5986–5991 (2005)
24. Lai, S.-M., Chen, W.-C., Chen, C.-M.: Eur. Polym. J. **44**, 3535–3547 (2008)

25. Martín, Z., Jiménez, I., Gómez, M.A., Ade, H., Kilcoyne, D.A.: Macromolecules **43**, 448–453 (2010)
26. Cheng, Q., Lü, Z., Byrne, H.: J. Appl. Polym. Sci. **114**, 1820–1827 (2009)
27. Mousa, A., Kocsis, J.-K.: Macromol. Mater. Eng. **286**, 260–266 (2001)
28. Zhang, Z., Zhang, L., Li, Y., Xu, H.: Polymer **46**, 129–136 (2005)
29. Sadhu, S., Bhowmick, A.K.: Rubber Chem. Technol. **76**, 860–875 (2003)
30. Magaraphan, R., Thaijaroen, W., Ochakun, R.L.: Rubber Chem. Technol. **76**, 406–418 (2003)
31. Varghese, S., Gatos, K.G., Apostolov, A.A., Kocsis, J.K.: J. Appl. Polym. Sci. **92**, 543–551 (2004)
32. Conzatti, L., Stagnaro, P., Colucci, G., Bongiovanni, R., Priola, A., Lostritto, A., Galimberti, M.: Appl. Clay Sci. **61**, 14–21 (2012)
33. Rajasekar, R., Nayak, G. C., Malas, A., Das, C. K.: Mater. Des. **35**, 878–885 (2012)
34. Zheng, H., Zhang, Y., Peng, Z., Zhang, Y.: J. Appl. Polym. Sci. **92**, 638–646 (2004)
35. Gatos, K.G., Thomann, R., Kocsis, J.K.: Polym. Int. **53**, 1191–1197 (2004)
36. Mohammadpour, Y., Katbab, A.A.: J. Appl. Polym. Sci.: **106**, 4209–4218 (2007)
37. Bandyopadhyay, A., Maiti, M., Bhowmick, A.K.: Mater. Sci. Technol. **22**, 818–828 (2006)
38. Kim, J.-T., Lee, D.-Y., Oh, T.-S., Lee, D.-H.: J. Appl. Polym. Sci. **89**, 2633–2640 (2003)
39. Kim, J.-T., Oh, T.-S., Lee, D.-H.: Polym. Int. **52**, 1058–1063 (2003)
40. Kim, J.-T., Lee, D.-Y., Oh, T.-S., Lee, D.-H.: Polym. Int. **52**, 1203–1208 (2003)
41. Kwon, J., Kim, H.: J. Polym. Sci. A. **43**, 3973–3985 (2005)
42. Ramanathan, T., Liu, H., Brinson, L.C.: J. Polym. Sci. B. **43**, 2269–2279 (2005)
43. Wu, H.-L., Yang, Y.-T., Ma, C.-C.M.: J. Polym. Sci. A. **43**, 6084–6094 (2005)
44. Sui, G., Zhong, W. H., Yang, X. P., Yu, Y. H.: Mater. Sci. Eng. A. **485**, 524–531 (2008)
45. Pedroni, L.G., Soto-Oviedo, M.A., Rosolen, J.M., Felisberti, M. I., Nogueira, A.F.: J. Appl. Polym. Sci. **112**, 3241–3248 (2009)
46. Cataldo, F., Ursini, O., Angelini, G.: Fuller. Nanotub. Car. N. **17**, 55–66 (2009)
47. Bokobza, L.: Polymer **48**, 4907–4920 (2007)
48. Singh, V., Joung, D., Zhai, L., Das, S., Khondaker, S.I., Seal, S.: Prog. Mater. Sci. **56**, 1178–1271 (2011)
49. Zaman, I., Kuan, H.-C., Meng, Q., Michelmore, A., Kawashima, N., Pitt, T., Zhang, L., Gouda, S., Luong, L., Ma, J.: Adv. Func. Mater. **22**, 2735–2743 (2012)
50. Bai, X., Wan, C., Zhang, Y., Zhai, Y.: Carbon **49**, 1608–1613 (2011)
51. Lian, H., Li, S., Liu, K., Xu, L., Wang, K., Guo, W.: Polym. Eng. Sci. **51**, 2254–2260 (2012)
52. Chen, B., Ma, N., Bai, X., Zhang, H., Zhang, Y.: RSC Adv. **2**, 4683–4689 (2012)
53. Ozbas, B., O'Neill, C.D., Register, R.A., Aksay, I.A., Prud'homme, R.K., Adamson, D.H.: J. Polym. Sci.B. **50**, 910–916 (2012)
54. Potts, J.R., Shankar, O., Du, L., Ruoff, R.S.: Macromolecules **45**, 6045–6055 (2012)
55. Chantaratcharoen, A., Sirisinha, C., Amornsakchai, T., Limcharoen, S.B., Meesiri, W.: J. Appl. Polym. Sci. **74**, 2414–2422 (1999)
56. Mori, M., Uyama, Y., Ikada, Y.: Polymer **34**, 5336–5341 (1994)
57. Tarantili, P.A., Andreopoulos, A. G.:J. Appl. Polym. Sci.: **65**, 267–276 (1997)
58. Bousoulas, J., Tarantili, P.A., Andreopoulos, A.G.: Adv. Compos. Lett. **10**, 239–245 (2001)
59. Ahmad, I., Chin, T.S., Cheong, C.K., Jalar, A., Abdullah, I.: American J. Appl. Sci. **Special Issue**(14–20) (2005)
60. Andreopoulos, A.G., Konstantinidou, A.V., Petsalas, H. J.: J. Appl. Polym. Sci. **38**, 2073–2078 (1989)
61. Amornsakchai, T., Sinpatanapan, B., Limcharoen, S.B., Meesiri, W.: Polymer **40**, 2993–2999 (1999)
62. Ahmad, I., Wong, P.Y., Abdullah, I.: Polym. Compos. **27**, 395–401 (2006)
63. Shibulal, G.S., Naskar, K.: J. Polym. Res. **18**, 2295–2306 (2011)
64. Amornsakchai, T., Sinpatanapan, B., Limcharoen, S.B.: Polymer **40**, 2993–2999 (1999)
65. Georgopoulos, STh, Tarantili, P.A., Avgerinos, E., Andreopoulos, A.G., Koukios, E.G.: Polym. Degrad. Stab. **90**, 303–312 (2005)

66. Baroulaki, I., Karakasi, O., Pappa, G., Tarantili, P.A., Economides, D., Magoulas, K.: Compos. A Appl. Sci. Manuf. **37**, 1613–1625 (2005)
67. Katsoulotos, G., Tarantili, P.A.: J. Appl. Polym. Sci. **107**, 2385–2393 (2008)
68. George, J., Sreekala, M.S., Thomas, S.: Polym. Eng. Sci. **41**, 1471–1485 (2001)
69. Abdelmouleh, M., Boufi, S., Belgacem, M.N., Dufresne, A.: Compos. Sci. Technol. **67**, 1627–1639 (2007)
70. Varghese, S., Kuriakose, B., Thomas, S.: J. Appl. Polym. Sci. **53**, 1051–1060 (1994)
71. Martins, M.A., Joekes, I.: J. Appl. Polym. Sci. **89**, 2507–2515 (2003)
72. Zhang, W., Zhang, X., Liang, M., Lu, C.: Compos. Sci. Technol. **68**, 2479–2484 (2008)
73. Ly, B., Thielemans, W., Dufresne, A., Chaussy, D., Belgacem, M.N.: Compos. Sci. Technol. **68**, 3193–3201 (2008)
74. Yakubu, M.K., Gumel, M.S., Umar, A., Metelerkamp, R.: J. Reinf. Plast. Compos. **29**, 2855–2868 (2010)
75. Wongsorat, W., Suppakarn, N., Jarukumjorn, K.: Adv. Mater. Res. **410**, 63–66 (2011)
76. Su, J., Chen, S., Zhang, J.: Polym. Test. **30**, 195–203 (2011)
77. Jia, X., Li, G., Yu, Y., Sui, G., Liu, H., Li, Y., Li, P., Yang, X.: J. Appl. Polym. Sci. **113**, 283–289 (2009)
78. Manjhi, S., Sarkhel, G.: J. Appl. Polym. Sci. **119**, 2268–2274 (2011)
79. Ismail, H., Mathialagan, M.: J. Appl. Polym. Sci., in Press (2012)
80. Pasbakhsh, P., Ismail, H., Fauzi, M.N.A., Bakar, A.A.: Polym. Test. **28**, 548–559 (2009)
81. Burnside, S.D., Giannelis, E.P.: Chem. Mater. **7**, 1597–1600 (1995)
82. Wang, S., Long, C., Wang, X., Li, Q., Qi, Z.: J. Appl. Polym. Sci. **69**, 1557–1561 (1998)
83. Wang, J., Chen, Y., Jin, Q.: Macromol. Chem. Phys. **206**, 2512–2520 (2005)
84. Wang, J., Chen, Y., Jin, Q.: J. Adhes. Sci. Technol. **20**, 261–276 (2006)
85. Kim, E.S., Shim, J.H., Jung, S.H., Joo, J.H., Yoon, J.-S., Lee, S.H.: Polym.Int. **59**, 479–485 (2010)
86. Labruyère, C., Gorrasi, G., Alexandre, M., Dubois, Ph: Polymer **50**, 3626–3637 (2009)
87. Labruyère, C., Monteverde, F., Alexandre, M., Dubois, Ph: J. Nanosci. Nanotechnol. **9**, 2731–2738 (2009)
88. Takeuchi, H., Cohen, C.: Macromolecules **32**, 6792–6799 (1999)
89. Ma, J., Xu, J., Ren, J.-H., Yu, Z.-Z., Mai, Y.-W. : Polymer **44**, 4619–4624 (2003)
90. Ma, J., Yu, Z–.Z., Kuan, H.-C., Dasari, A., Mai, Y.-W.: Macromol. Rapid Commun. **26**, 830–833 (2005)
91. Wang, J., Chen, Y.: J. Appl. Polym. Sci. **107**, 2059–2066 (2008)
92. Schmidt, D. F., Clément, F., Giannelis, E. P.: Adv. Funct. Mater. **16**, 417–425 (2006)
93. Kaneko, M.L.Q.A., Romero, R.B., Goncalves, M.C.: Eur. Polym. J. **46**, 881–890 (2010)
94. Yang, L., Hu, Y., Lu, H., Song, L.: J. Appl. Polym. Sci. **99**, 3275–3280 (2006)
95. Liu, Q., De Kee, D.: Can. J. Chem. Eng. **85**, 36–44 (2007)
96. Simon, M.W., Stafford, K.T., Ou, D.L.: J. Inorg. Organomet. Polym. **18**, 364–373 (2008)
97. Wang, J., Chen, Y., Wang, J.: J. Appl. Polym. Sci. **111**, 658–667 (2009)
98. Kim, E.S., Kim, H.S., Jung, S.H., Yoon, J.S.: J. Appl. Polym. Sci. **103**, 2782–2787 (2007)
99. Schmidt, D.F., Giannelis, E.: Chem. Mater. **22**, 167–174 (2010)
100. Vasilakos, S.P., Tarantili, P.A.: J. Appl. Polym. Sci. **125**(1), E548–E560 (2012)
101. Wang, M.-J.: Rubber Chem. Technol. **71**, 520–589 (1998)
102. Murphy, L.J., Wang, M.-J., Mahmud, K.: Rubber Chem. Technol. **71**, 998–1014 (1998)
103. Bokobza, L., Leroy, E., Lalanne, V.: Eur. Polym. J. **45**, 996–1001 (2009)
104. Bokobza, L., Rahmani, M., Belin, C., Bruneel, J.-L., El Bounia, N.-E.: J. Polym. Sci. B **46**, 1939–1951 (2008)
105. Ismail, H., Ramly, A. F., Othman, N.: J. Appl. Polym. Sci. doi: 10.1002/APP.38298 (2012)
106. Kotov, N. A.: Nature **442**, 254–255 (2006)
107. Park, S., Ruoff, R.S.: Nat. Nanotechnol. **4**, 217–224 (2009)
108. Hu, H., Zhao, L., Liu, J., Liu, Y., Cheng, J., Luo, J., Liang, Y., Tao, Y., Wang, X., Zhao, J.: Polymer **53**, 3378–3385 (2012)

109. Tarawneh, M.A., Ahmad, S.H., Noum, S.Y.E., Ahmad, K.Z.K.: J. Reinf. Plast. Compos. **30**, 1745–1752 (2011)
110. Tang, Z., Wu, X., Guo, B., Zhang, L., Jia, D.: J. Mater. Chem. **22**, 7492–7501 (2012)
111. Huda, M.S., Drzal, L.T., Mohanty, A.K., Misra, M.: Compos. B **38**, 367–379 (2007)
112. Anuar, H., Wan Busu, W.N., Ahmad, S.H., Rasid, R.: J. Compos. Mater. **42**, 1075–1087 (2008)
113. Lopattananon, N., Jitkalong, D., Seadan, M.: J. Appl. Polym. Sci. **120**, 3242–3254 (2011)
114. Shojaei, A., Arjmand, M., Saffar, A.: J. Mater. Sci. **46**, 1890–1901 (2011)
115. Mathew, G., Rhee, J.M., Lee, Y.-S., Park, D.H., Nah, C.: J. Indust. Eng. Chem. **14**, 60–65 (2008)
116. López-Manchado, M.A., Herrero, B., Arroyo, M.: Polym. Int. **52**, 1070–1077 (2003)
117. López-Manchado, M.A., Arroyo, M., Herrero, B., Biagiotti, J.: J. Appl. Polym. Sci. **89**, 1–15 (2003)
118. López-Manchado, M.A., Valentín, J.L., Carretero, J., Barroso, F., Arroyo, M.: Eur. Polym. J. **43**, 4143–4150 (2007)
119. Carretero-González, J., Retsos, H., Verdejo, R., Toki, S., Hsiao, B., Giannelis, E.P., López-Manchado, M.A.: Macromolecules **41**, 6763–6772 (2008)
120. Cataldo, F.: Macromol. Symp. **247**, 67–77 (2007)
121. Lopez, L.M., Cosgrove, A.B., Hernandez-Ortiz, J.P., Osswald, T.A.: Polym. Eng. Sci. **47**, 675–683 (2007)
122. Hong, I.-K., Lee, S.: J. Ind. Eng. Chem., (in Press) (2012)
123. Mathew, G., Rhee, J., Lee, Y., Park, D., Nah, C.: J. Ind. Eng. Chem. **14**, 60–65 (2008)
124. Kader, M., Nah, C.: Polymer **45**, 2237–2247 (2004)
125. Montserrat, S., Málek, J.: Thermochim. Acta. **228**, 47–60 (1993)
126. Ŝesták, C., Berggren, G.: Thermochim. Acta. **3**, 1–12 (1971)
127. Kissinger, H.E.: Anal. Chem. **29**, 1702–1706 (1957)
128. Ozawa, T.: Bull. Chem. Soc. Jpn., **38**, 1881–1886 (1965)
129. Vasilakos, S.P., Tarantili, P.A.: In-situ monitoring and modeling of the curing reaction of vinyl terminated polysiloxane/layered silicate nanocomposites, submitted for publication
130. Labruyère, C., Monteverde, F., Alexandre, M., Dubois, P.: J. Nanosci. Nanotechnol. **9**, 2731–2738 (2009)
131. Zhang, Y., Pang, M., Xu, Q., Lu, H., Zhang, J., Feng, S.: Polym. Eng. Sci. **51**, 170–178 (2011)
132. Takeuchi, H., Cohen, C.: Macromolecules **32**, 6792–6799 (1999)
133. Wang, J., Chen, Y., Jin, Q.: High. Perform. Polym. **18**, 325–340 (2006)
134. Kong, Q., Hu, Y., Song, L., Wang, Y., Chen, Z., Fan, W.: Polym. Advan. Technol. **17**, 463–467 (2006)
135. Work, W.J., Horie, K., Hess, M., Stepto, R.F.T.: Pure Appl. Chem. **76**, 1985–2007 (2004)
136. Utracki, L.A. (ed.): Polymer Blends, Handbook, Springer, Netherlands (2002)
137. Lipatov, Y.S.: Polymer Reinforcement. ChemTec Publishing, Toronto (1995)
138. Wu, D.F., Zhang, Y.S., Zhang, M., Yu, W.: Biomacromolecules **10**, 417–424 (2009)
139. Essawy, H., El-Nashar, D.: Polym. Test **23**, 803–807 (2004)
140. Khatua, B.B., Lee, D.J., Kim, H.Y., Kim, J.K.: Macromolecules **37**, 2454–2459 (2004)
141. Nesterov, A.E., Lipatov, Y.S.: Polymer **40**, 1347–1349 (1999)
142. Gritsenko, O.T., Nesterov, A.E.: Eur. Polym. J. **27**, 455–459 (1991)
143. Fisher, R., Zhang, M.W., Hu, X., Lin, M., Gersoppe, D., Sokalov, J., Rafailovich, M., Rubenstein, M., Winesett, A.H.: 7th International Conference on Atomically Controlled Surfaces, Nara, Japan, 2003
144. Ali, Z., Le, H.H., Ilisch, S., Thurn-Albrecht, T., Radusch, H.-J.: Polymer **51**, 4580–4588 (2010)
145. Hwang, W.-G., Wei, K.-H., Wu, C.-M.: Polym. Eng. Sci. **46**, 80–88 (2006)
146. Yan, N., Xia, H., Wu, J., Zhan, Y., Fei, G., Chen, C.: J. Appl. Polym. Sci. (2012). doi:10.1002/APP.37861

147. Shafieizadegan Esfahani, A.R., Katbab, A.A., Dehkhoda, P., Karami, H.R., Barikani, M., Sadeghi, S.H.H.: Compos. Sci. Technol. **72**, 382–389 (2012)
148. Ma, J., Feng, Y.X., Xu, J., Xiong, M.L., Zhu, Y.J., Zhang, L. Q.: Polymer **43**, 937–945 (2002)
149. Ismail, H., Mega, L.: Polym. Plast. Technol. Eng. **40**, 463–478 (2001)
150. López Manchado, M. A., Arroyo, M., Biagiotti, J., Kenny, J. M.: J. Appl. Polym. Sci. **90**, 2170–2178 (2003)
151. Pukanszky, B.: Composites **21**, 255–262 (1990)
152. Hui, S., Chaki, T.K., Chattopadhyay, S.: J. Appl. Polym. Sci. **110**, 825–836 (2008)
153. Denac, M., Šmit, I., Musil, V.: Compos. A **36**, 1094–1101 (2005)
154. Denac, M., Musil, V., Šmit, I.: Compos. A **36**, 1282–1290 (2005)

Fully Green Elastomer Composites

Daniel Pasquini

Abstract In this chapter we will discuss the preparation of fully green elastomer composites. First we will define the term green and when we can classify a material as a green. Then we will discuss the alternatives for the replacement of major components of composite materials, viz. filler and matrix, which are usually derived from non-renewable and synthetic materials, with materials from renewable sources. Finally we will describe some work in which fully green elastomer composites were studied.

1 The Green Concept

The concept of "green", actually employed to describe the concept of sustainability and environmental friendliness, is not new. What is new is the use of this word, because the word green is a modern term used to describe the environmentalism movement that began in the nineteenth century and that was characterized as using, preserving and renewing the available resources. Briefly, a definition of environmentalism is the sustainable management of resources and the stewardship of the environment.

In our days, the concept of what makes a product green is still evolving. There is no exact definition of the term green; on the contrary, there is some consideration of the benefit that the product brings to the environment. If a product improves the environment it can be considered a product with a degree of greenness. More specifically, a product is green if the impact on the environment is direct, measurable and intrinsic to the product. We can list some common criteria for classifying products as green:

D. Pasquini (✉)
Instituto de Química, Universidade Federal de Uberlândia, Av. João Naves de Ávila, 2121, Campus Santa Mônica, Bloco-1D, Uberlândia MG, CEP: 38400-902, Brazil
e-mail: pasquini@iqufu.ufu.br; danielpasquini2005@yahoo.com.br

P. M. Visakh et al. (eds.), *Advances in Elastomers II*,
Advanced Structured Materials 12, DOI: 10.1007/978-3-642-20928-4_5,

1. Products that have no toxic emissions;
2. Products that save energy;
3. Products made from recycled materials;
4. Products made from industrial and agroforestry residues;
5. Products that are biodegradable;
6. Products made from renewable resources.

Despite the fact that many products that have no toxic emissions are not properly green, these products can usually be considered as green products because they generally replace a hazardous material. In the same way, products that save energy are also considered green. Although they are not intrinsically green, they have a positive impact on the environment.

Nowadays, everyone knows what recycled products are and what their benefits are. Recycled materials have two direct impacts on the environment: (i) they reduce the emissions of residues and (ii) they also reduce the resource demand. As a consequence of these benefits, they can be classified as green materials.

A product made from industrial and agroforestry residues reduces the resource demand. These residues include sugar cane bagasse, sugar cane straw, wood chips, sawdust, cereal shells, corn husks, rice hulls and similar materials. They have many different applications and their use reduces the resource demand.

For a product to be considered biodegradable, it needs to be degraded in a short period of time. However, as yet there is no standard that defines the exact amount of time that qualifies a product as biodegradable. In general, the products considered as green are those that are biodegradable in a short period of time, and this time can be interpreted in different ways.

Finally, products made from renewable resources are intrinsically green. Here we can classify them in two groups of renewable resources: the first group, the rapidly renewable resources, consists of those that can typically be regenerated in a growing season in a year and the second group is the renewable resources that take more than a season to grow and to regenerate.

The use of green materials is expanding and undergoing constant growth at same time that the use of materials that are not bio based, biodegradable or recyclable is being restricted by new legislation. In this chapter the use of renewable resources will focus on the production of fully green elastomer composites. If there is a commitment of society to preserving the environment, focusing on the rational use of renewable resources, the development of new materials, particularly elastomers, has its place in our industry.

2 Composite Materials

The basic definition of composite materials is materials made from two or more constituents with significantly different mechanical properties, which remain separate and distinct within the finished structure [1]. The two main constituents of a composite are: matrix and filler (or reinforcement). The matrix surrounds and supports the filler by maintaining its relative position. The filler is responsible for the improvement of the physical and mechanical properties of the matrix. Due to the wide variety of matrices and fillers available, the association of different fillers and matrices can produce a large amount of composite materials with variable properties and with potential application in many areas.

The most common commercially available composites, polymers, are employed as matrices. The fillers are often fibres but also commonly ground minerals. Strong fibres such as fibreglass, quartz, Kevlar or carbon fibres give the composite its tensile strength, while the matrix binds the fibres together, transferring the load from broken fibres to unbroken ones and between fibres that are not oriented along the tension lines. The fillers can be divided into two main categories, macro- and microscopic fillers. In the last case, we refer to nanoparticles.

Among the options for the polymers and fillers used to form a composite, we can see that these components do not always come from renewable sources. The focus of this chapter is the production of green elastomeric composites, which have as their main components materials from renewable resources. Thus, we discuss below the options for matrices and fillers from renewable resources that could serve as substitutes for the resources that are not renewable.

2.1 Fillers from Renewable Resources

Regarding fillers from renewable sources, the subject obviously leads us to the use of plant and animal resources. Plants are the main source of potential substitutes for fillers from non-renewable resources. The use of plant components, for example sawdust, fibres and nanoparticles, as reinforcement in polymer matrices is well known and widely studied [2–8]. There are many advantages of using fillers from plant sources as reinforcement in polymer matrices, e.g. low density, low cost and low energy consumption, high specific strength, renewability and biodegradability, abundant availability in a variety of forms throughout the world, flexibility, non-abrasive nature to processing equipment, non-toxicity, ease of handling, reactive surface, organic nature and the economic development opportunity for non-food farm products in rural areas [9].

The properties of natural fibres are strongly influenced by many factors, particularly chemical composition, internal fibre structure, microfibrillar angle, cell dimensions and defects, which differ between different parts of a plant as well as between different plants. The mechanical properties of plant fibres are in general much lower when compared with those of the most widely used reinforcing glass fibres (Table 1). However, because of their low density, the specific properties, which are property-to-density ratio dependent, viz. strength and stiffness, are comparable with those of glass fibres.

Carbon fibres can be produced from vegetal macromolecules as an alternative to the conventional non-renewable polymer used for this purpose. An example is

Table 1 Physical properties of various natural fibres and their comparison with some synthetic organic and inorganic fibres [9–11]

Fibre	Density ($g \cdot cm^{-3}$)	Young's modulus (GPa)	Tensile strength (MPa)	Elongation (%)
Abaca	1.5	12	400	3–10
Alfa	0.89	22	350	5.8
Bagasse	1.25	17	290	–
Bamboo	0.6–1.1	11–17	140–230	–
Banana	1.35	12	500	5.9
Coir	1.2	4–6	175	30
Cotton	1.5–1.6	5.5–12.6	287–597	7–8
Curaua	1.4	11.8	500–1,150	3.7–4.3
Date palm	1–1.2	2.5–5.4	97–196	2–4.5
Flax	1.5	27.6	345–1,035	2.7–3.2
Hemp	1.48	70	690	1.6
Henequen	1.2	10.1–16.3	430–570	3.7–4.9
Isora	1.2–1.3	–	500–600	5–6
Jute	1.3	26.5	393–773	1.5–1.8
Kenaf	–	53	930	1.6
Nettle	–	38	650	1.7
Oil palm	0.7–1.55	3.2	248	25
Piassava	1.4	1.07–4.59	134–143	7.8–21.9
Pineapple	0.8–1.6	400–627	1.44	14.5
Ramie	1.5	24.5	560	2.5
Sisal	1.5	9.4–22	511–635	2.0–2.5
Viscose	–	11	593	11.4
Soft wood kraft	1.5	40	1,000	–
E-glass	2.5	70	2,000–3,500	2.5
S-glass	2.5	86	4,570	2.8
Aramide	1.4	63–67	3,000–3,150	3.3–3.7
Carbon	1.4	230–240	4,000	1.4–1.8

the production of carbon fibres from lignin, which has been extensively studied. Lignin is one of the most abundant natural polymers. It is produced in the chemical pulping industry as a by-product, and is used for energy generation and the recovery of pulping chemicals. In previous studies, lignin preparations were converted into fibrous materials by simple thermal treatment and/or polymer blending methods followed by thermal spinning [12, 13]. These lignin fibres were transformed into carbon fibres using a thermal conversion process similar to that found in industry. The mechanical properties of these carbon fibres were comparable with industrial general performance grade carbon fibres prepared from conventional polymers.

A special class of filler from renewable resources is nanofillers. In this group we can include whiskers [6, 9] (obtained from vegetal cellulose fibres and from animal sources) and starch nanocrystals [14]. Cellulose nanocrystals, known as whiskers or nanowhiskers, occur in the form of rod-like nanoparticles with lengths ranging from 200 to 2,000 nm and diameters ranging from 4 to 15 nm depending on the source [9]. Starch nanocrystals occur in the form of platelet-like nanoparticles with lengths ranging between 20 and 40 nm, widths between 15 and 30 nm and thicknesses between 5 and 7 nm for waxy maize starch [14]. Their nanoscale dimensions and impressive mechanical properties make nanoparticles, particularly when occurring as a high-aspect ratio, ideal candidates for improving the mechanical properties of the host material when compared with macroscopic fillers. The physical incorporation of these nanocrystals into a polymeric matrix leads to nanocomposites with outstanding properties.

However, despite their attractive properties, vegetal fibres are used only to a limited extent in industrial practice due to the difficulties associated with surface interactions. The primary drawback of agro-based fibres is associated with their inherent polar and hydrophilic nature and the non-polar characteristics of most thermoplastics. These result in difficulties in compounding the filler and the matrix, and therefore in achieving acceptable dispersion levels, which yield composites with low performances.

This hydrogen bonding phenomenon is best exemplified in paper, where these secondary interactions provide the basis for its mechanical strength. It also results in high moisture absorption and swelling of the fibres. A poor fibre–matrix interface induces a decrease in mechanical properties.

Moreover, the processing temperature of composites is restricted to about 200 °C, because of the possible degradation and/or the emissions of volatile compounds that could affect the composite properties. Higher temperatures could be used if the processing time is short. This limits the type of thermoplastics that can be used with polysaccharide fillers to commodity plastics such as polyethylene, polypropylene, polyvinyl chloride and polystyrene. However, it is worth noting that these cheap plastics constitute about 70 % of the total thermoplastic-based products consumed by the plastics industry.

Several strategies have been suggested in the literature to improve the compatibly of lignocellulosic fibres with thermoplastic polymers, by surface modifications that include (i) physical treatments, such as solvent extraction; (ii) physico-chemical treatments, like the use of corona and plasma discharges or laser, γ-ray and UV bombardment; and (iii) chemical modifications, both by direct condensation of the coupling agents onto the cellulose surface and by grafting by free-radical or ionic polymerizations [15, 16].

The different modifications of lignocellulosic materials are generally located on the surface materials. The main purpose of the modifications is to improve the compatibility of these materials with the polymer matrix. The strength of interaction depends on the type of modification performed, since only physical changes will result in weaker interactions than those from chemical modifications. Chemical modifications are more efficient because they act by changing the values of surface energy and leading to greater adhesion between the filler and the polymer matrix. Besides the reduction of surface energy values of lignocellulosic materials, another important factor is the reduction in the values of the surface polar character, which is observed by the reduction in the values of the polar contribution of surface energy. As a general rule, materials with similar values of surface energy tend to have a better affinity. Additionally, compounds with similar chemical characteristics, i.e. polar and non-polar similar characters, also tend to have a higher affinity. As a consequence, there will be better adhesion between these materials. Table 2 exhibits some values of surface energy for conventional polymers and for some lignocellulosic materials before and after surface chemical modifications.

Many papers describe chemical modifications related to the modification of natural polysaccharides, namely their hydrophobization [16, 21, 22]. A weak point in chemical modification is the fact that changes must be performed under mild conditions to prevent the degradation of the lignocellulosic materials, otherwise it would result in a loss of mechanical properties of these materials and the resulting composite, due to the lignocellulosic materials' degradation [23].

Table 2 Surface energy (γ_s) and its polar (γ_s^p) and dispersive (γ_s^d) contributions, for conventional polymers and some lignocellulosic materials before and after surface chemical modifications

Sample	Chemical modification	Surface energy (mJ·m^{-2})			References
		γ_s	γ_s^p	γ_s^d	
Avicell cellulose (AV)	–	55.7	23.9	31.8	[17]
Whatman paper cellulose (WP)	–	52.3	20.2	32.1	
TFPS-treated AV	Grafting with trifluoropropyl trimethoxysilane (TFPS)	28.9	1.1	27.8	
TFPS-treated WP		26.7	0.6	26.1	
PFOS-treated AV	Grafting with 1H, 1H, 2H, 2H-perfluorooctyl trimethosysilane (PFOS)	20.8	0.5	20.3	
PFOS-treated WP		19.9	0.1	19.8	
Bleached kraft cellulose (KC)	–	51.9	34.2	17.8	[18]
Bleached organosolv cellulose (ORC)	–	51.6	35.7	15.9	
DC-treated KC	Grafting with dodecanoyl chloride (DC)	31.3	7.0	24.3	
DC-treated ORC		33.1	4.4	28.7	
OC-treated KC	Grafting with octadecanoyl chloride (OC)	30.2	6.5	23.6	
OC-treated ORC		33.1	1.8	31.3	
Polyethylene-linear (PE)	–	35.7	0	35.7	[19, 20]
Polypropylene-isotactic (PP)	–	30.1	0	30.1	
Polystyrene (PS)	–	40.7	6.1	34.5	
Polymethylmethacrylate (PMMA)	–	41.1	11.5	29.6	
Polyethyleneterephthalate (PET)	–	44.6	9	35.6	
Polycarbonate (PC)	–	34.2	6.5	27.7	
Polyisobutylene (PIB)	–	33.6	0	33.6	

Although there are many works that aim at the surface modification of lignocellulosic materials, these are still mostly limited to materials with macroscopic dimensions, i.e. fibres, sawdust, starch, etc. Few works are found in the literature where changes are made to starch, chitin and cellulose nanoparticles [24–27]. The incorporation of nanocrystals of cellulose and starch into polymer matrices is of great interest, but their modification for compatibility with polar polymers has been little explored, and it is a challenge for the coming year.

In short, we can verify that there are many alternative types of filler from renewable resources that can be used to replace fillers that are not from renewable resources in polymeric matrices. The use of fillers from renewable sources produces materials with varying physical and mechanical properties, and choosing the correct type of filler obtains composites with similar properties to the composites using fillers that are not from renewable resources.

2.2 Elastomer Polymers

An elastomer is a polymer with the property of viscoelasticity, generally having a notably low Young's modulus (E′) and high yield strain compared with other materials. The predominant property of elastomers is their elastic behaviour after deformation by compression or tension. Elastomers are macromolecular materials that quickly recover their initial shape and dimensions after ceasing the application of pressure. Not all amorphous polymers are elastomers. Some are thermoplastics, depending on the classification of the glass transition temperature, Tg, defined as the temperature above which a polymer becomes soft and ductile and below which it becomes hard and brittle, like glass. We can say as a general rule applicable only to amorphous polymers that an amorphous polymer with a Tg below ambient temperature is an elastomer, while an amorphous polymer with a Tg higher than room temperature is a thermoplastic [28]. Table 3 displays a list of the major elastomers that are currently available and their abbreviations according to ISO standard 1629 (1987), as well as their Tg values.

The term, which is derived from elastic polymer, is often used interchangeably with the term rubber, although the latter is preferred when referring to vulcanizates. Elastomers are usually thermosets (requiring vulcanization) but may also be thermoplastic. At ambient temperatures rubbers are thus relatively soft (E′ $\sim$ 3 MPa) and deformable.

The main difference between thermoset elastomers and thermoplastic elastomers (TPE) is the type of crosslinking bond in their structures. In fact, crosslinking is a critical structural factor that contributes to imparting high elastic properties. The crosslink in thermoset polymers is a covalent bond created during the vulcanization process. On the other hand the crosslink in thermoplastic elastomer polymers is a weaker dipole or hydrogen bond or takes place in one of the phases of the material.

Table 3 Tg values of the major elastomers currently available

Elastomer		Abbreviation	Tg (°C)
Polyacrylate rubber		ACM	−22 to −40
Polyesterurethane		AU	−35
Bromobutyl rubber		BIIR	−66
Polybutadiene		BR	−112
Chlorobutyl		CIIR	−66
Chlorinated polyethylene		CM	−25
Epichlorohydrin (homopolymer)		CO	−26
Polychloroprene		CR	−45
Chlorosulfonated polyethylene		CSM	−25
Ethylene and methyl acrylate copolymer		EAM	−40
Epichlorohydrin and ethylene oxide copolymer		ECO	−45
Terpolymer ethylene (propylene) sulfur cured		EPDM, S	−55
Terpolymer ethylene (propylene) peroxide cured		EPDM, P	−55
Polyetherurethane		EU	−55
Ethylene and vinyl acetate copolymer		EVM or EVA	−30
Fluorocarbon elastomer (co or terpolymers)		FKM or FPM	−18 to −50
Fluorinated silicone		FVMQ	−70
Butadiene and acrylonitrile hydrogenated copolymer		H-NBR	−30
Butyl rubber (isoprene and isobutylene copolymer)		IIR	−66
Butadiene and acrylonitrile copolymer	Low content of acrylonitrile	NBR	−45
	Medium content of acrylonitrile		−34
	High content of acrylonitrile		−20
Natural rubber (or polyisoprene)		NR (IR)	−72
Polyglycol ether		OT	−50
Polydichlorophosphazene fluorinated		PNF	−66
Styrene and butadiene copolymer		SBR	−50
Vinyl-silicone		VMO	−120
Carboxylated nitrile rubber		XNBR	−30

In most cases the polymer's elasticity is due to covalent crosslinks, but it can be due to thermodynamic reasons. The individual segment of the chain must be flexible to take Brownian motion at room temperature. Thus, the molecules assume any statistically ordered conformation when subjected to tensile stresses. Once this tension is removed, they return to their random conformation (state of maximum entropy); the deformation process can be described thermodynamically, considering that under ideal conditions the internal energy is unchanged.

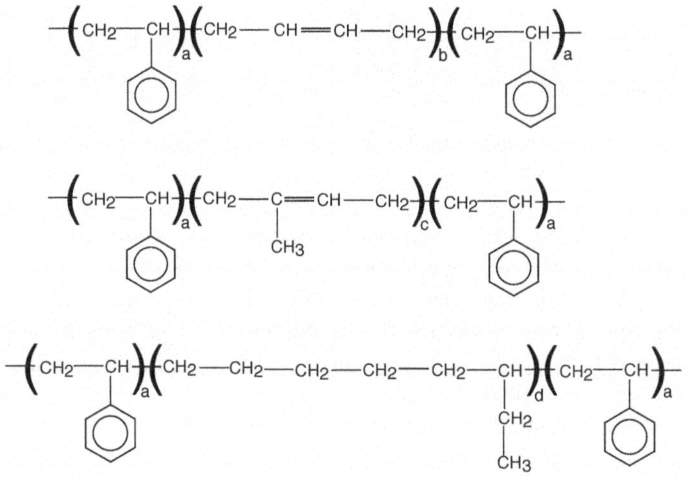

Fig. 1 Chemical structures of three common styrenic block copolymers: **a** styrene (50–80 units), **b** polybutadiene (20–100 units), **c** polyisoprene (20–100 units) and **d** polyethylenebutylene (20–100 units) segments

The TPEs may be classified on the basis of their chemistry and morphology. All of them have two polymeric phases: a soft rubbery one and a hard thermoplastic one. The TPEs' compositions can be formed by (i) block copolymers with alternating soft and hard segments, (ii) blends of a rubbery polymer and thermoplastic or (iii) rubbery polymer–thermoplastic compositions in which the former is highly vulcanized and finely dispersed in the latter.

There are six generic classes of TPEs generally considered to exist commercially. They are styrenic block copolymers, polyolefin blends, elastomeric alloys, thermoplastic polyurethanes, thermoplastic copolyesters and thermoplastic polyamides [29].

Styrenic block copolymers, commonly referred to as styrenics, are copolymers with the S-D-S structure, where S is a hard segment of polymerized styrene or styrene derivative and D is a soft central segment of polymerized diene or hydrogenated diene units. Figure 1 shows the chemical structures of three common styrenic block copolymers with the most commonly used soft diene segments: polybutadiene, polyisoprene and polyethylenebutylene.

The characteristics of these TPEs depend on the relative proportions of the polymerized styrene and diene units as well as the chemical nature of the monomers. At low styrene levels, the TPEs will be soft and rubbery with relatively low tensile properties. With increasing styrene content, the TPE progressively becomes like a harder rubber.

Copolyesters are polymers with the –A-B-A-B- structure, where A and B are alternating hard and soft polymeric segments connected by ester linkages, such as the structure of a commercial copolyester shown in Fig. 2. These block

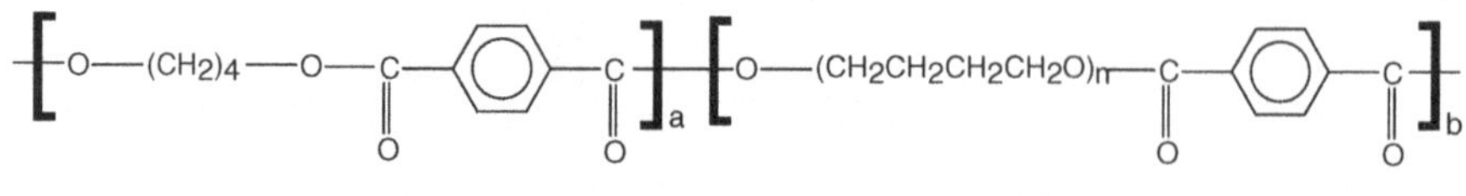

Fig. 2 Structure of a commercial copolyester: **a** hard segment (16–40 units), **b** soft segment (16–40 segments), (n) 10–50

copolymers differ from simple polyester thermoplastics, which are typically hard and crystalline polymers of organic dibasic acids and diols.

The copolyester polymers perform as TPEs if the structures of A and B are chosen to give rubbery properties to the copolymer over a useful temperature range. The structure of the soft segment provides the flexibility and mobility needed for elastomeric performance.

Thermoplastic polyurethanes are TPEs that have the same block copolymer morphology as the styrenic and copolyester polymers. Their general structure is –A-B-A-B-, where A represents a hard crystalline block derived by chain extension of a diisocyanate with a glycol. The soft block is represented by B and can be derived from either a polyester or a polyether. Figure 3 shows the typical thermoplastic polyurethane structures.

The urethane linkages in the blocks are capable of a high degree of inter- and intramolecular hydrogen bonding, and consequently increasing of the crystallinity of the hard phase, and can influence the mechanical properties of the thermoplastic polyurethanes, i.e. hardness, modulus and tear strength. As with other block copolymers, the nature of the soft segments determines the elastic behaviour and low-temperature performance of the resulting material.

The elastomeric polyamides are a class of TPEs in which amide linkages connect the hard and soft segments and the soft segment may consist of polyester, polyether or polyetherester chains. As in other block copolymer TPEs, the structure of the hard and soft blocks also contributes to the performance characteristics of the elastomeric polyamides. The nature of the hard segments determines the melting point of the elastomeric polyamides and their performance at elevated temperatures.

Thermoplastic elastomeric olefins are simple blends of a rubbery polymer, such as natural rubber (NR) or terpolymer ethylene (propylene) (EPDM), with a

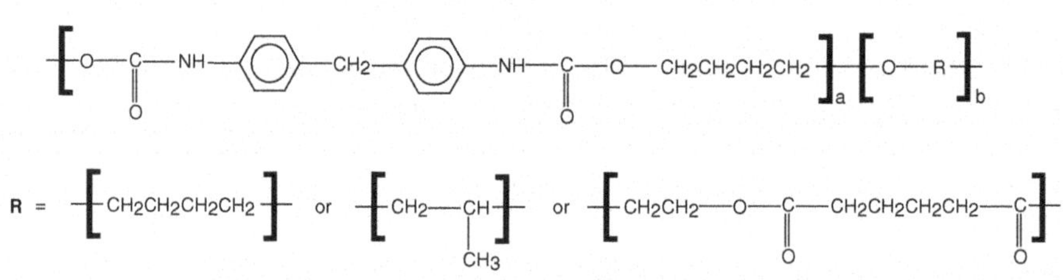

Fig. 3 Typical thermoplastic polyurethane structures: **a** hard block (30–120 units), **b** soft block (8–50 units)

thermoplastic polymer, such as polypropylene (PP) or polyvinyl chloride (PVC). Each polymer will have its own phase, and the rubber phase will have little or no crosslinking. The polymer present in a larger amount will usually be the continuous phase, with the thermoplastic being favoured due to its lower viscosity. The discontinuous phase should have a small particle size for the best properties of the thermoplastic elastomeric olefin. Reasonable compatibility of the two polymers requires their solubility parameters and polarities to be similar.

Thermoplastic vulcanizates elastomers (or elastomeric alloys) differ from thermoplastic elastomeric olefin in that the rubber phase is highly vulcanized (crosslinked). As a result, the properties and performance of a TPV are much closer to those of a conventional thermoset rubber. This difference will reflect in the final properties of the thermoplastic vulcanizates elastomers, when compared with the thermoplastic elastomeric olefin, which for a given elastomer–thermoplastic system are (i) resistance to plastic deformation, (ii) great resistance to fluids, (iii) good retention of properties at an elevated temperature relative to those at an ambient temperature and (iv) low creep and stress relaxation.

Until now we have provided a description of the common elastomers, their chemical compositions and their properties. However, within this class of polymers only natural rubber fits the criteria for classification as "green", as discussed earlier in this book chapter, because it is from a renewable source. Composite materials made from natural rubber, which serves as the polymer matrix, and reinforcing agents from plant sources have been studied and will be discussed later in this and other chapter of this book.

2.3 Polymers from Renewable Resources

The objective of including the discussion on elastomeric polymers was to provide an opportunity to compare the polymers produced from renewable sources that could serve as possible substitutes for polymers that compose the major known elastomers and that are not renewable. In nature, more specifically plant and animal sources, we can find many chemical compounds whose molecules can act as monomers or macromonomers in polymerization processes, which may be a possible replacement for fossil monomers.

Among the molecules from renewable sources that have the potential to produce new polymers, and several studies are underway in this direction, we can highlight cellulose, hemicellulose, lignin, chitin, proteins, vegetable oils, polyhydroxyalkanoate, terpenes, tannins, starch and polylactic acid. A recently published book by Belgacem and Gandini [30] describes the various recent studies concerning the use of these molecules and others, from renewable sources, in the production of polymers, monomers and composite materials. However, in this current book chapter we will emphasize the work that resulted in elastomeric polymers or blends.

Most of the polymers produced from the molecules described above are thermoplastic polymers and do not show the characteristic properties of elastomers. In the case of the thermoplastic materials obtained, the solution to using them in elastomeric composite matrices would employ them in the production of blends in the same manner as the thermoplastic elastomeric olefins described above. Another alternative is to use these polymers, which are mostly polyether, polyester or polyols, as substitutes for the corresponding polymers used in the preparation of TPEs, for example, as described above, by copolymerization.

Several works were carried out employing vegetable oils as precursors for the synthesis of vegetable oil-based polymers, in which the resulting polymers had elastomeric or rubbery properties [31–39]. In these works the vegetable oils such as linseed, tung, soybean, corn, castor oils and others, or the modified vegetable oils, were crosslinked by vinyl monomers, such as styrene, α-methyl styrene, divinyl benzene, cyclopentadiene and others. The resulting copolymers were characterized by dynamic and thermal mechanical analysis (DMA and TMA), thermogravimetry (TGA), differential scanning calorimetry (DSC) and scanning electron microscopy (SEM). The gelation time in these very different systems varied from a few minutes to a few days. The Tg values and the moduli of the prepared copolymer varied from −60 to 105 °C and from 0.1 to 2,000 MPa, respectively. These values are comparable with those of commercial rubbers synthesized from petroleum-based monomers.

The use of starch in the production of new polymers has been studied extensively. Blending is one of the most promising alternatives to make starch useful as a polymer in the replacement of other plastics, and the fast progress occurring in this field is attested by several reviews published recently [40–43]. Most polymers produced by blending starch and other polymers are thermoplastic polymers, but some of them are elastomeric or rubbery polymers. Arvanitoyannis et al. [44] studied blends of starch with 1,4-transpolyisoprene (gutta percha) with and without plasticizers by melt processing and observed that they were biodegradable because of the presence of starch. The plasticizers increased the percentage elongation considerably whereas they decreased the tensile strength, glass transition and melting points of the composite matrix. The Tg values varied from −56 to 12 °C depending on the amount of starch and 1,4-transpolyisoprene. For the mixture of starch and 1,4-transpolyisoprene 1:1 (w:w), varying the content of plasticizer glycerol, the elongation and tensile modulus varied from 6 to 40 % and from 2 to 12 MPa, respectively. Rouilly et al. [45] also prepared blends of starch and unmodified and modified natural rubber by casting mixtures of aqueous starch with glycerol and latex. The natural rubber was modified by grafting with dimethylaminoethyl methacrylate (DMAEMA) to form a latex with cationic water-soluble polymeric "hairs" of polyDMAEMA, which should form hydrogen bonds with starch. The unmodified latex acted only as a filler in the starch films, but with modified natural rubber, the mechanical properties of the films were significantly altered. The elastic modulus was greatly decreased and the strain at break greatly increased. The glass transition temperature increased from −48 to −32 °C, suggesting significant compatibilization. Carvalho et al. [46] blended native starch granules and a natural rubber latex by melt processing calling upon water as a plasticizer for starch. The blends were prepared in an intensive batch mixer at 150 °C, with natural rubber content varying from 2.5 to 20 %. The blends were characterized by mechanical analysis (stress–strain) and by scanning electron microscopy. The stable dispersion and the good adhesion between the two natural polymers were attributed in part to the natural non-rubber constituents present in the latex. The results revealed a reduction in the modulus and in tensile strength, the blends becoming less brittle than thermoplastic starch alone. Phase separation was observed in some compositions and was dependent on rubber and on the plasticizer content (glycerol). Increasing the plasticizer content made possible the addition of higher amounts of rubber. The addition of rubber was, however, limited by phase separation, the appearance of which depended on the glycerol content. Scanning electron microscopy showed good dispersion of the natural rubber in the continuous phase of thermoplastic starch matrix.

Polylactic acid (PLA) and polyhydroxyalkanoates (PHAs) are polymers that have their origin in bacteria. In the case of PLA, the precursor of this polymer is lactic acid (LA). Lactic acid is mainly prepared in large quantities by the bacterial fermentation of carbohydrates. PLA is produced by the condensation polymerization of the LA monomer, but it is difficult to obtain high molecular weights. In order to obtain high molecular weights it is necessary to use coupling or esterification-promoting agents [47, 48]. The molecular weights of PLA before and after the increase in chain length are Mw = 1,000–5,000 and Mw > 100,000. In the case of PHAs, they are produced directly by various bacteria in many grades, differing in composition, molecular weight and other parameters. The formation of a particular material, either homo- or copolymer, depends on the type of bacteria, but even more important are the conditions of polymer formation, mainly the substrate used for feeding the bacteria and the conditions of their growth. The two main PHAs, with commercial interest, are polyhydroxybutyrate (PHB) and its copolymer with valerate, polyhydroxybutyrate-valerate (PHBV) [49]. It is important to highlight that PHAs can also be produced by chemical synthesis from the relevant substituted propiolactones [50]. The typical PLA glass transition temperature (Tg) ranges from 50 to 80 °C, whereas its melting temperature (Tm) ranges from 130 to 180 °C. For instance, enantiomerically pure PLA is a semi-crystalline polymer with a Tg of 55 °C and a Tm of 180 °C. The mechanical properties of PLA-related polymers were recently reviewed by Sodergard and Stolt [48]. The mechanical properties of PLA can vary to a large extent, ranging from soft and elastic materials to stiff and high strength materials, according to different parameters, such as crystallinity, polymer structure and molecular weight, material formulation (plasticizers, blend, composites, etc.) and processing (e.g. orientation). For instance, commercial PLA has a modulus of 2.1 GPa and an elongation at break of 9 %. After plasticization, its Young's modulus decreases to 0.7 MPa and the elongation at break rises to 200 %, with a corresponding Tg shift from 58 to 18 °C [51]. With regard to the basic polyhydroxyalkanoates (PHAs), PHB and PHBV, their Tg values are 4 and 5 °C, their moduli are 3.5 and 900 MPa and their elongations at break are 5 and 15 %, respectively. However, their properties can change when copolymerized with other

PHAs. Some examples are the copolymers PHB-co-20 mol% hydroxyvalerate and PHB-co-3 mol% 3-hydroxydecano-ate/3 mol% 3-hydroxydodecanoate, the final Tg values of which, in comparison with PHB, decrease to −1 and −8 °C, the moduli decrease to 0.8 and 0.2 MPa and the elongations at break rise to 50 and 680 %, respectively [52]. Interesting materials have been developed based on copolymers of PHAs and caprolactone. It was verified that with rising caprolactone, a decrease in Tg is observed from 2 to −42 °C, for caprolactone contents of 0 and 72 %, respectively [53].

Proteins are natural macromolecules consisting of different amino acid residues arranged in a three-dimensional structure. Amino acids have two functional groups, namely an amino group (–NH2) and a carboxyl group (–COOH). The main proteins with potential applications in materials are soy protein, zein, wheat protein and casein, the first three being from plant sources and the last one from animal sources [54, 55]. In the same way as the other macromolecules described above, the properties of proteins can be modified by blending or by the addition of plasticizers [56–58]. For example, there are two glass transitions (Tg1 and Tg2) in the soy protein plasticized with glycerol, corresponding to glycerol-rich and protein-rich domains, respectively. The Tg1 of the plasticized protein decreases from −28.5 to −65.2 °C with an increase in glycerol content from 25 to 50 wt%, whereas the Tg2 is almost invariable at about 44 °C [59]. Blends of soy protein and biodegradable polyesters have been prepared using glycerol as a compatibilizing agent. Good miscibility was obtained only when the soy protein was initially combined with glycerol under high shear at elevated temperatures in an extruder. Under these conditions, partial denaturing of the soy protein led to specific interactions between the functional groups of the protein and the glycerol. Screws with large kneading blocks that produced high shear mixing were preferred and led to thermoplastic blends with high elongation and tensile strength [60].

The main constituents of plants are cellulose, hemicelluloses and lignin. These molecules have been extensively studied in many applications, using them in natural form or as derivatives. As mentioned previously, the polymers derived from these molecules and also for chitin and chitosan can be used in the production of blends with elastomeric polymers, particularly with polyurethanes and NBR polymers, in order to obtain as a result an elastomeric polymer [61–67]. These elastomeric materials can be used as a polymeric matrix in elastomer composites. Alternatively, in another approach, these molecules can be used in copolymers, in which these molecules are copolymerized with elastomeric polymers. In the latter case, we highlight the production of polyols from these and other molecules separately or lignocellulosic material containing a mixture of these molecules, by oxypropylation [68–77]. This process is employed in the polyurethane industry for the preparation of low-cost polyols from abundant and renewable biomass resources. Simple sugars and other polyols such as glycerol are commonly oxypropylated for that purpose. Following this trend, several types of polymeric substrates arising from renewable resources have been subjected to oxypropylation and the resulting macromonomer (polyol) can be used in polyurethane, polyether or polyester formulations. The oxypropylation of natural polymers and other

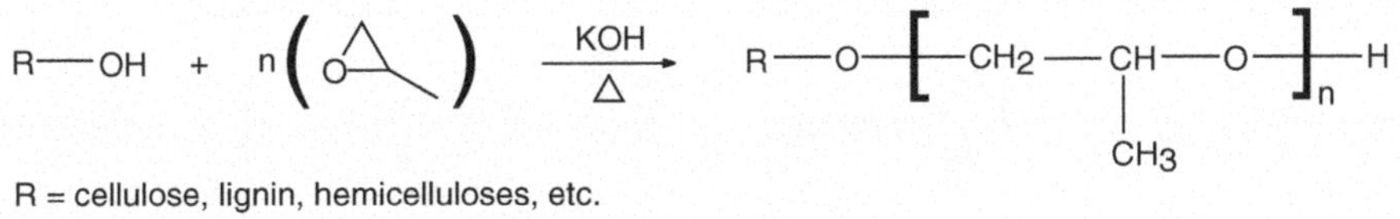

Fig. 4 Schematic view of the oxypropylation reaction

complex substrates from biomass is a method whose objective is to increase the OH functionality of the starting material by moving the hydroxyl groups to the chain end by growing a polypropylene oxide chain on the original molecule, making them much more accessible for further reaction. A scheme of reaction can be seen in Fig. 4.

The oxypropylation of natural polymeric substrates has been attracting considerable attention recently, mainly in response to the economic and environmental aspects, because this process does not require the removal of any solvent or other component, nor does it require any separation or purification procedure for the recovery of the final polyol mixture [68]. The oxypropylation can be total or partial, and in this last case a thermoplastic sleeve is formed on the material surface, and the resulting material is a composite. This process will be discussed later in this book chapter.

Finally, a promising alternative for the production of elastomeric polymers is the use of furanic molecules such as furfural and hydroxymethylfurfural. Furfural and hydroxymethylfurfural are produced from the pentose and hexose sugars, respectively. These sugars are derived from hemicelluloses, cellulose and starch. From these two compounds, furfural and hydroxymethylfurfural, many monomers can be produced through their functionalization. So, these monomers can be polymerized and the resulting polymers may have thermoplastic properties and even rubbery properties. Figures 5 and 6 show the structure of furfural and hydroxymethylfurfural as well as some monomers produced from them and their synthetic homologues monomers, respectively.

Furfural is a typical precursor to furan monomers bearing a moiety that can be polymerized by chain reaction mechanisms, and hydroxymethylfurfural, on the other hand, is ideally suited as a precursor to bifunctional furan monomers to be used in step growth reactions. All the monomers derived from furfural and hydroxymethylfurfural, shown in Figs. 5 and 6, and others not shown here, were synthesized and polymerized. These studies are described in many reviews in the literature where the synthesis route and the resulting polymer properties are described [78–82]. These monomers were employed in the preparation of polyesters, polyamides, polyurethanes and others, and their resulting properties, in many cases, were not studied. Although these properties have not been fully studied, it is believed that these properties are similar to the polymers produced with their homologous synthetic monomers. So, these monomers can be used in copolymerization with other monomers to give elastomeric polymers, or they can be used to produce polymers and for blending with elastomeric polymers. From

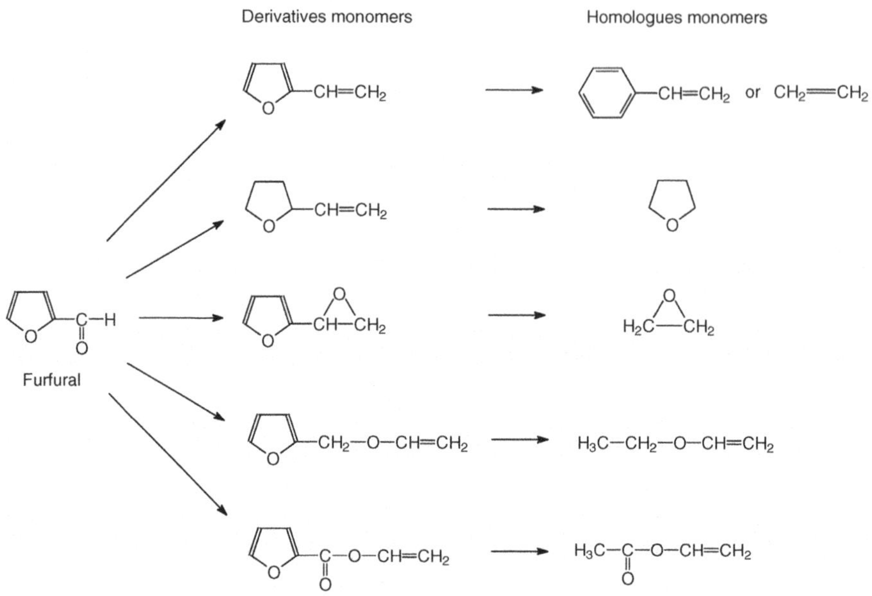

Fig. 5 Structure of furfural, some monomers produced from it and their synthetic homologues monomers

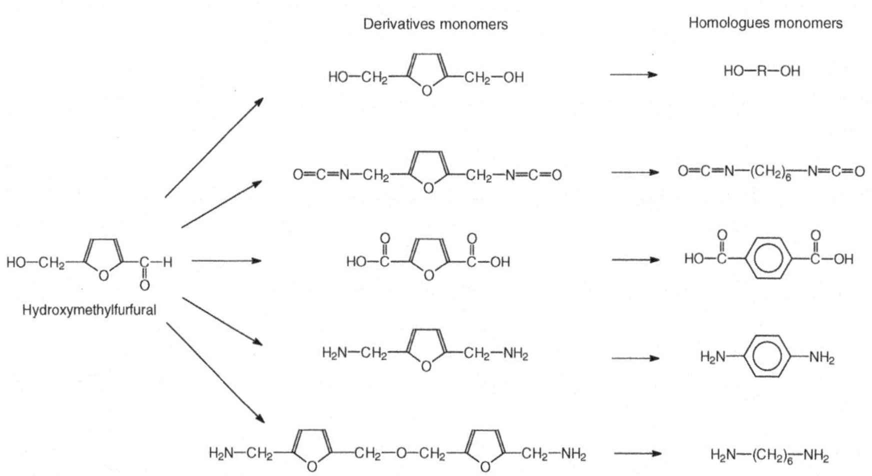

Fig. 6 Structure of hydroxymethylfurfural, some monomers produced from it and their synthetic homologues monomers

studies conducted with monomers derived from furfural and hydroxymethylfurfural, we highlight the production of polyesters [83] and polyurethanes [84, 85] by copolymerization of these furanic monomers with synthetic monomers, in which the resulting copolymer showed characteristics of elastomers. In the case of polyesters, they were semi-crystalline materials with glass transition temperatures of −51 to −62 °C and were stable up to 225 °C. In the case of the polyurethanes, thermoplastic elastomers based on furanic-aliphatic, furanic-aromatic, fully furanic moieties and other monomers such as 1,4-butanediol or 2,5-bis(hydroxymethyl)-furan as diols, methylene diphenyl diisocianate as diisocyanate and polytetramethylene oxide glycol, poly(caprolactone) or poly(butadiene)glycol as flexible macrodiols were prepared, and the resulting Tg values of the soft segments were around −60 °C and the melting point of the rigid counterparts ranged from 150 to 190 °C.

As a partial conclusion of this topic, we can verify that, except for natural rubber, all other molecules from renewable sources by themselves or in a polymerized form do not exhibit the characteristics of elastomers. The polymers produced from these molecules are mostly thermoplastic, and to provide elastomeric characteristics they need to be modified by grafting, copolymerization or blending. For the production of fully green elastomer composites, the biggest challenge is the production of elastomeric polymers from molecules from renewable sources, which will be used as a matrix in these composites. This is an area in which there are still many possibilities to explore and the production of new elastomeric polymers will be the source of many research works.

3 Green Elastomer Composites

Few works are found in the literature describing the preparation and study of fully green elastomer composites. This is mainly due to the fact that, with the exception of natural rubber, there are a few polymers derived from renewable sources that show the characteristics of elastomers. In this section we will present some studies found in the literature in which the resulting composite can be classified as a fully green elastomer composite, and we will discuss initially the composites where the polymer matrix is natural rubber and the charges were macroscopic vegetable fibres and in some cases nanoparticles. Finally, we will describe other works in which the polymeric matrix was an elastomeric polymer other than natural rubber.

The reinforcement with many kinds of macroscopic vegetal fibres in natural rubber has been extensively studied. Among the types of fibres used in these works, we can mention: coir [86, 87], sisal [88, 89], grass [90, 91], pineapple [92, 93], jute [94], isora [95], oil palm [89, 96], waste paper [97], woven sisal fabric [98], a combination of sisal and oil palm fibres [99, 100] and a combination of carbon fibre and kenaf [101]. The effects observed in these studies due to the fibre reinforcement on the mechanical properties of natural rubber composites include increased modulus, increased strength with good bonding at high fibre concentrations, decreased elongation at failure, greatly improved creep resistance over particulate-filled rubber, increased hardness and a substantial improvement in cut, tear and puncture resistance.

The uses of nanoparticles as a reinforcement agent in natural rubber-based nanocomposites have also been extensively studied. In this case, these nanoparticles were extracted from different sources. The nanoparticles from cellulose sources and chitin are called whiskers, and those from starch are called starch nanocrystals. The difference between them was discussed earlier in this chapter. So, the studies found in the literature were carried out employing nanoparticles as a reinforcing agent in natural rubber-based nanocomposites, from different sources such as cassava bagasse [102], sugar cane bagasse [103], rachis of palm tree [104–106], Crab shell chitin [107–109] and *Syngonanthus nitens* (Capim Dourado) [110]. The effects of nanoparticles' loading on tensile properties, thermal properties, moisture sorption, water vapour permeation and soil biodegradation were studied. Significant improvements in Young's modulus and the tensile strength were observed as a result of the addition of nanoparticles to the rubber matrix, especially at high nanoparticle loadings. Dynamic mechanical thermal analysis (DMA) and differential scanning calorimetry (DSC) results showed that in most of the cases there was no significant change in the glass transition temperature (Tg) of the rubber matrix upon the addition of nanoparticles but at the softening of rubber, nanoparticles have a reinforcing effect on the rubber. The presence of nanoparticles resulted in an increase in moisture sorption of rubber films with a low content while at higher nanoparticle loadings the moisture sorption tended to decrease. The barrier properties to water vapour decreased with a low content of nanoparticles, but increased with a further increase in the nanoparticle loading.

The use of regenerated cellulose as a filler in natural rubber vulcanizates composites was also studied [111]. These composites showed an unexpected reinforcement effect, with maximum tensile properties at 15 phr of cellulose. The swelling behaviour of these compounds was studied and the density of crosslinks was determined in heptane and benzene. The crosslink density is higher in vulcanizates containing regenerated cellulose than in vulcanizates containing carbon black. It is also higher when the swelling is carried out in heptane than when it is carried out in benzene. The crosslink density increases when the amount of filler increases for the experimental conditions.

The use of lignin acting as a reinforcing filler in natural rubber-based composites was also studied [112]. This new type of composite elastomer material was prepared by the addition of lignin biopolymer to styrene-butadiene rubber. The effect of lignin on the processing and mechanical properties of the prepared composites was examined. It was verified that the addition of lignin in an amount up to 30 wt% to rubber blends improved the mechanical properties of vulcanizates.

The production, characterization and properties of the natural rubber-based composites and nanocomposites will be explored more deeply in other chapters of this book.

Elastomer composites employing proteins and other carbohydrates were studied. The aqueous dispersion of defatted soy flour containing soy protein, soy carbohydrate and soy whey was blended with a styrene–butadiene latex to form elastomer composites [113]. The inclusion of soy carbohydrate increased the tensile stress in the small strain region, but reduced the elongation at break.

The inclusion of soy carbohydrate and soy whey also improved the recovery behaviour in the non-linear region. At small strain, the shear elastic modulus of 30 % filled composites at 140 °C is about 500 times higher than that of the unfilled elastomer, indicating a significant reinforcement effect generated by defatted soy flour. Compared with soy protein isolate, the stress softening effect and recovery behaviour under dynamic strain indicated that the addition of soy carbohydrate and soy whey had enhanced the filler–rubber interactions.

A novel series of protein composites has been prepared from 30 to 50 wt% polyurethane prepolymer with soy dreg (a by-product from the isolation process of soy protein), soy whole flour and soy protein isolate, by a compression-moulding process at 120 °C, without the addition of any plasticizers [114]. The toughness, thermal stability and water resistivity of these composites was significantly improved. By increasing the polyurethane prepolymer content, elastomeric materials could be obtained. With an increase in the cellulose content in the system, the tensile strength and water resistivity of the ensuing composites increased. The tensile strength, elongation at break and water resistivity were 6.9 MPa, 100 % and 0.55 for soy dreg with 50 wt% polyurethane prepolymer, and 4.8 MPa, 140 % and 0.50 for soy protein isolate with 50 wt% polyurethane prepolymer, respectively.

Composites with an elastomeric character were prepared from wheat gluten, which acts as a matrix, and methylcellulose microfibres, which act as a filler [115]. These environmentally friendly green composites were prepared by the conventional blending of wheat gluten, methylcellulose microfibres and glycerol as a plasticizer, followed by compression moulding of the mixture at 127 °C to crosslink the matrix. Their thermal and mechanical properties were evaluated in relation to their methylcellulose microfibre content. It was found that the addition of methylcellulose microfibres can significantly improve the Young's modulus and tensile strength of the composite, which is accompanied by rises in the glass transition temperatures of the wheat gluten matrix. The influences of the methylcellulose microfibre content on the thermal decomposition and gluten solubility in water are also discussed. The presence of methylcellulose microfibres can accelerate the moisture desorption of the composites during TGA testing. However, the addition of methylcellulose microfibres with content above 33.3 wt% can delay the thermal decomposition of the composites. Because both methylcellulose microfibres and glycerol are water soluble, the composites might be used as a kind of biodegradable material in low relative humidity conditions.

The combination of elastomeric polymers and filler from renewable resources to produce a fully green elastomer composite has seldom been explored when the elastomeric polymer matrix is a material other than natural rubber. This is an area to be explored, which is still deficient in publications. As previously mentioned, the difficulty is the production of elastomeric polymers from renewable sources.

4 Self-Reinforced Composites Obtained by Partial Oxypropylation

A new class of composites was prepared by the partial oxypropylation of different kinds of cellulose samples viz. regenerated cellulose Rayon, filter paper, microcrystalline cellulose Avicell, Eucalyptus bleached Kraft pulp and starch [116–119]. The oxypropylation reaction scheme can be seen in Fig. 4. The difference between total and partial oxypropylation is that, in the last case, the oxypropylation is only on the surface of the material. As a result we have a new biphasic material consisting of a low glass transition temperature component surrounding the inner core (cellulose samples or starch granules), which could be hot pressed to form films of the filler dispersed into a thermoplastic matrix. Figure 7 displays SEM micrographs of the filter paper cellulose fibres before and after partial oxypropylation.

The SEM micrographs of the filter paper fibres (Fig. 7) show the presence of a thermoplastic sleeve around the modified fibres. It was observed that, after hot pressing, these materials showed a high interfacial adhesion giving rise to a continuous morphology, unachievable with the unmodified material. The DSC tracings of the oxypropylated products showed systematically a glass transition at ca. −50 °C (albeit with different ΔCp, depending on the extent of oxypropylation), absent from the corresponding pristine substrates, which was assigned to the modified portion of the cellulose samples or starch granules, in tune with the Tg of commercial hydroxypropylcelluloses, which ranges from −60 to −70 °C. Polysaccharides grafted with oligoether chains only display the glass transition of the grafts, which explains the constant value of Tg in the present context.

Self-reinforced composites in which the oxypropylated outer sleeve of substrate generates the matrix upon hot pressing, leaving their inner core to act as reinforcing elements, were shown to be readily prepared using a variety of substrates. The interest of this process resides not only in the novel exploitation of renewable resources, but also in its green connotation, since it requires neither the use of solvents nor the application of any processing operations, given the fact that the

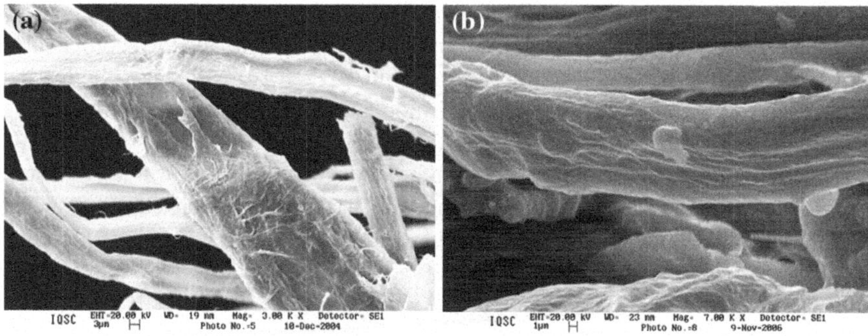

Fig. 7 SEM micrographs of the filter paper cellulose fibres **a** before (magnification 3,000X) and **b** after (magnification 7,000X) partial oxypropylation

final product is ready for further exploitation as it is removed from the reactor. There is no information about the mechanical properties of these novel materials, but they can be blended in association with other elastomeric polymers, in order to produce as a resulting material an elastomer composite.

5 Conclusions

We can conclude that the production and study of fully green elastomer composites has been largely unexplored. This is because of the difficulty in preparing an elastomeric polymer employing molecules from renewable sources. We found that the fillers from renewable sources are largely available and varied, and range from macroscopic particles to nanosized particles, and these fillers have varied reinforcement properties and can be readily employed as a substitute for fillers that are not renewable. The production of elastomeric polymers from renewable sources is a challenge and at the same time a big problem to be solved. Research efforts must be concentrated on solving this problem and thus opening new opportunities in the preparation of fully green elastomer composites and generating new applications for them as replacements for conventional elastomer composites.

References

1. Schwartz, M.M.: Composite Materials Handbook, p. 764. McGraw-Hill, New York (1991)
2. Bledzki, A.K., Gassan, J.: Composites reinforced with cellulose based fibres. Prog. Polym. Sci. **24**, 221–274 (1999)
3. Eichhorn, S.J., Baillie, C.A., Zafeiropoulos, N., Mwaikambo, L.Y., Ansell, M.P., Dufresne, A., Entwistle, K.M., Herrera-Franco, P.J., Escamilla, G.C., Groom, L., Hugues, M., Hill, C., Rials, T.G., Wild, P.M.: Review: current international research into cellulosic fibres and composites. J. Mat. Sci. **36**, 2107–2131 (2001)
4. Woodhams, R.T., Thomas, G., Rodges, D.K.: Wood fibers as reinforcing fillers for polyolefins. Polym. Eng. Sci. **24**, 1166–1171 (1984)
5. Kokta, B.V., Raj, R.G., Daneault, C.: Use of wood flour as filler in polypropylene: studies on mechanical properties. Polym. Plast. Technol. Eng. **28**, 247–259 (1989)
6. Azizi Samir, M.A.S., Alloin, F., Dufresne, A.: Review of recent research into cellulosic whiskers, their properties and their application in nanocomposite field. Biomacromolecules. **6**, 612–626 (2005)
7. Dufresne, A.: Comparing the mechanical properties of high performances polymer nanocomposites from biological sources. J. Nanosci. Nanotechnol. **6**, 322–330 (2006)
8. John, M.J., Thomas, S.: Biofibres and biocomposites. Carbohydr. Polym. **71**, 343–364 (2008)
9. Dufresne, A.: Cellulose-based composites and nanocomposites. In: Belgacem, M.N., Gandini, A. (eds.) Monomers, Polymers and Composites from Renewable Resources, pp. 401–418. Elsevier, Amsterdam (2008)
10. Neagu, R.C., Gamstedt, E.K., Berthold, F.: Stiffness contribution of various wood fibers to composite materials. J. Compos. Mat. **40**, 663–669 (2006)
11. Mohanty, A.K., Misra, M., Drzal, L.T.: Sustainable bio-composites from renewable resources: opportunities and challenges in the green materials world. J. Polym. Environ. **10**, 19–26 (2002)

12. Kubo, S., Yoshida, T., Kadla, J.F.: Surface porosity of lignin/PP blend carbon fibers. J. Wood Chem. Technol. **27**, 257–271 (2007)
13. Kadla, J.F., Kubo, S., Venditti, R.A., Gilbert, R.D., Compere, A.L., Griffith, W.: Lignin-based carbon fibers for composite fiber applications. Carbon **40**, 2913–2920 (2002)
14. Angellier, H., Choisnard, L., Molina-Boisseau, S., Ozil, P., Dufresne, A.: Optimization of the preparation of aqueous suspensions of waxy maize starch nanocrystals using a response surface methodology. Biomacromolecules **5**, 1545–1551 (2004)
15. Hill, C.A.S.: Wood Modification: Chemical, Thermal and Other Processes, p. 239. Wiley, Chichester (2006)
16. Belgacem, M.N., Gandini, A.: Chemical modification of wood. In: Belgacem, M.N., Gandini, A. (eds.) Monomers, Polymers and Composites from Renewable Resources, pp. 419–431. Elsevier, Amsterdam (2008)
17. Ly, B., Belgacem, M.N., Bras, J., Salon, M.C.B.: Grafting of cellulose by fluorine-bearing silane coupling agents. Mat. Sci. Eng. **30**, 343–347 (2009)
18. Pasquini, D., Belgacem, M.N., Gandini, A., Curvelo, A.A.S.: Surface esterification of cellulose fibers: characterization by DRIFT and contact angle measurements. J. Colloid Interface Sci. **295**, 79–83 (2006)
19. Shafrin, E.G.: Critical surface tensions of polymers. In: Brandrup, J., Immergut, E.H. (eds.) Polymer Handbook, 2nd edn, p. III:221–228. Wiley-Interscience, New York (1975)
20. Falsafi, A., Mangipudi, S., Owen, M.J.: Surface and interfacial properties. In: Mark, J.E (ed.) Physical Properties of Polymers Handbook, 2nd edn, p. 1011–1020. Springer, New York (1900)
21. Cunha, A.G., Gandini, A.: Turning polysaccharides into hydrophobic materials: a critical review. Part 1–cellulose. Cellulose **17**, 875–889 (2010)
22. Cunha, A.G., Gandini, A.: Turning polysaccharides into hydrophobic materials: a critical review. Part 2–hemicelluloses, chitin/chitosan, starch, pectin and alginates. Cellulose **17**, 1045–1065 (2010)
23. Pasquini, D., Teixeira, E.M., Curvelo, A.A.S., Belgacem, M.N., Dufresne, A.: Surface esterification of cellulose fibres: processing and characterisation of low-density polyethylene/cellulose fibres composites. Compos. Sci. Technol. **68**, 193–201 (2008)
24. Namazi, H., Dadkhah, A.: Convenient method for preparation of hydrophobically modified starch nanocrystals with using fatty acids. Carbohydr. Polym. **79**, 731–737 (2010)
25. Thielemans, W., Belgacem, M.N., Dufresne, A.: Starch nanocrystals with large chain surface modifications. Langmuir **22**, 4804–4810 (2006)
26. Feng, L., Zhou, Z., Dufresne, A., Huang, J., Wei, M., An, L.: Structure and properties of new thermoforming bionanocomposites based on chitin whisker-graft-polycaprolactone. J. Appl. Polym. Sci. **112**, 2830–2837 (2009)
27. Berlioz, S., Molina-Boisseau, S., Nishiyama, Y., Heux, L.: Gas-phase surface esterification of cellulose microfibrils and whiskers. Biomacromolecules **10**, 2144–2151 (2009)
28. Hofmann, W.: Rubber Technology Handbook, p. p. 651. Hanser, New York (1989)
29. Harper, C.A.: Handbook of Plastics, Elastomers, and Composites. McGraw-Hill, New York (1996)
30. Belgacem, M.N., Gandini, A. (eds.): Monomers, Polymers and Composites from Renewable Resources. Elsevier, Amsterdam (2008). p. 553
31. Li, F.K., Larock, R.C.: Synthesis, structure and properties of new tung oil-styrene-divinylbenzene copolymers prepared by thermal polymerization. Biomacromolecules **4**, 1018–1025 (2003)
32. Kundu, P.P., Larock, R.C.: Novel conjugated linseed oil-styrene-divinylbenzene copolymers prepared by thermal polymerization. 1. Effect of monomer concentration on the structure and properties. Biomacromolecules **6**, 797–806 (2005)
33. Andjelkovic, D.D., Larock, R.C.: Novel rubbers from cationic copolymerization of soybean oils and dicyclopentadiene. 1. Synthesis and characterization. Biomacromolecules **7**, 927–936 (2006)

34. Badrinarayanan, P., Lu, Y.S., Larock, R.C., Kessler, M.R.: Cure characterization of soybean oil-styrene-divinylbenzene thermosetting copolymers. J. Appl. Polym. Sci. **113**, 1042–1049 (2009)
35. Li, F.K., Hasjim, J., Larock, R.C.: Synthesis, structure, and thermophysical and mechanical properties of new polymers prepared by the cationic copolymerization of corn oil, styrene, and divinylbenzene. J. Appl. Polym. Sci. **90**, 1830–1838 (2003)
36. Marks, D.W., Li, F.K., Pacha, C.M., Larock, R.C.: Synthesis of thermoset plastics by Lewis acid initiated copolymerization of fish oil ethyl esters and alkenes. J. Appl. Polym. Sci. **81**, 2001–2012 (2001)
37. Andjelkovic, D.D., Valverde, M., Henna, P., Li, F.K., Larock, R.C.: Novel thermosets prepared by cationic copolymerization of various vegetable oils—synthesis and their structure-property relationships. Polymer **46**, 9674–9685 (2005)
38. Xia, Y., Larock, R.C.: Castor oil-based thermosets with varied crosslink densities prepared by ring-opening metathesis polymerization (ROMP). Polymer **51**, 2508–2514 (2010)
39. Li, F., Hanson, M.V., Larock, R.C.: Soybean oil-divinylbenzene thermosetting polymers: synthesis, structure, properties and their relationships. Polymer **42**, 1567–1579 (2001)
40. Avérous, L.: Biodegradable multiphase systems based on plasticized starch: a review. J. Macromol. Sci. Part C Polym. Rev. **C44**, 231–274 (2004)
41. Wang, X.L., Yang, K.K., Wang, Y.Z.: Properties of starch blends with biodegradable polymers. J. Macromol. Sci. Part C Polym. Rev. **C43**, 385–409 (2003)
42. Amass, W., Amass, A., Tighe, B.: A review of biodegradable polymers: uses, current development in the synthesis and characterization of biodegradable polyesters, blends of biodegradable polyesters and recent advances in biodegradable studies. Polym. Int. **47**, 89–144 (1998)
43. Yu, L., Dean, K., Li, L.: Polymer blends and composites from renewable resources. Prog. Polym. Sci. **31**, 502–576 (2006)
44. Arvanitoyannis, I., Kolokuris, I., Nakayama, A., Aiba, S.: Preparation and study of novel biodegradable blends based on gelatinized starch and 1,4-trans-polyisoprene (gutta percha) for food packaging or biomedical applications. Carbohydr. Polym. **34**, 291–302 (1997)
45. Rouilly, A., Rigal, L., Gilbert, R.G.: Synthesis and properties of composites of starch and chemically modified natural rubber. Polymer **45**, 7813–7820 (2004)
46. Carvalho, A.J.F., Job, A.E., Alves, N., Curvelo, A.A.S., Gandini, A.: Thermoplastic starch/natural rubber blends. Carbohydr. Polym. **53**, 95–99 (2003)
47. Garlotta, D.: A literature review of poly(lactic acid). J. Polym. Environ. **9**, 63–84 (2002)
48. Sodergard, A., Stolt, M.: Properties of lactic acid based polymers and their correlation with composition. Prog. Polym. Sci. **27**, 1123–1163 (2002)
49. Lenz, R., Marchessault, R.H.: Bacterial polyesters: biosynthesis, biodegradable plastics and biotechnology. Biomacromolecules **6**, 1–8 (2005)
50. Muller, H.M., Seebach, D.: Poly(hydroxyalkanoates)—a 5th class of physiologically important organic biopolymers. Angew. Chem. **32**, 477–502 (1993)
51. Martin, O., Avérous, L.: Poly(lactic acid): plasticization and properties of biodegradable multiphase systems. Polymer **42**, 6237–6247 (2001)
52. Sudesh, K., Abe, H., Doi, Y.: Synthesis, structure and properties of polyhydroxyalkanoates: biological polyesters. Prog. Polym. Sci. **25**, 1503–1555 (2000)
53. Impallomeni, G., Giuffrida, M., Barbuzzi, T., Musumarra, G., Ballistreri, A.: Acid catalyzed transesterification as a route to poly(3-hydroxybutyrate-*co*-ε-caprolactone) copolymers from their homopolymers. Biomacromolecules **3**, 835–840 (2002)
54. Osborne, T.B.: The Vegetable Proteins, p. 154. Longmans Green and Company, London (1924)
55. Kumar, R., Liu, D., Zhang, L.: Advances in proteinous biomaterials. J. Biobased Mat. Bioenergy **2**, 1–24 (2008)
56. Lieberman, E.R., Gilbert, S.G.: Gas permeation of collagen films as affected by crosslinkage, moisture, and plasticizer content. J. Polym. Sci. Part C Polym. Symp. **41**, 33–43 (1973)

57. Cuq, B., Gontard, N., Cuq, J.L., Guilbert, S.: Selected functional properties of fish myofibrillar protein-based films as affected by hydrophilic plasticizers. J. Agric. Food Chem. **45**, 622–626 (1997)
58. Liu, D., Zhang, L.: Structure and properties of soy protein plastics plasticized with acetamide. Macromol. Mat. Eng. **291**, 820–828 (2006)
59. Chen, P., Zhang, L.: New evidences of glass transitions and microstructures of soy protein plasticized with glycerol. Macromol. Biosci. **5**, 237–245 (2005)
60. Graiver, D., Waikul, L.H., Berger, C., Narayan, R.: Biodegradable soy protein-polyester blends by reactive extrusion process. J. Appl. Polym. Sci. **92**, 3231–3239 (2004)
61. Zhou, Q., Zhang, L., Zhang, M., Wang, B., Wang, S.: Miscibility, free volume behavior and properties of blends from cellulose acetate and castor oil-based polyurethane. Polymer **44**, 1733–1739 (2003)
62. Yoshioka, M., Hagiwara, N., Shiraishi, N.: Thermoplasticization of cellulose acetates by grafting of cyclic esters. Cellulose **6**, 193–212 (1999)
63. Zia, K.M., Barikani, M., Zuber, M., Bhatti, I.A., Sheikh, M.A.: Molecular engineering of chitin based polyurethane elastomers. Carbohydr. Polym. **74**, 149–158 (2008)
64. Barikani, M., Honarkar, H., Barikani, M.: Synthesis and characterization of chitosan-based polyurethane elastomer dispersions. Monatsh. Chem/Chem. Monthly **141**, 653–659 (2010)
65. Rao, V., Johns, J.: Thermal behavior of chitosan/natural rubber latex blends: TG and DSC analysis. J. Therm. Anal. Calorim. **92**, 801–806 (2008)
66. Barikani, M., Honarkar, H., Barikani, M.: Synthesis and characterization of polyurethane elastomers based on chitosan and poly(ε-caprolactone). J. Appl. Polym. Sci. **112**, 3157–3165 (2009)
67. Ciobanu, C., Ungureanu, M., Ignat, L., Ungureanu, D., Popa, V.I.: Properties of lignin–polyurethane films prepared by casting method. Ind. Crops Prod. **20**, 231–241 (2004)
68. Gandini, A., Belgacem, M.N.: Partial or total oxypropilation of natural polymers and the use of the ensuing materials as composites or polyol macromonomers. In: Belgacem, M.N., Gandini, A. (eds.) Monomers, Polymers and Composites from Renewable Resources, pp. 273–288. Elsevier, Amsterdam (2008)
69. Velazquez-Morales, P., Gandini, A., Le Nest, J.P.: Polymer electrolytes derived from chitosan/polyether networks. Electrochim. Acta **43**, 1275–1279 (1998)
70. Fernandes, S., Freire, C.S.R., Pascoal-Neto, C., Gandini, A.: The bulk oxypropylation of chitin and chitosan and the characterization of the ensuing polyols. Green Chem. **10**, 93–97 (2008)
71. Evtouguina, M., Gandini, A., Barros, A.M., Cruz-Pinto, J.J., Pascoal-Neto, C., Belgacem, M.N.: The oxypropylation of cork residues: preliminary results. Bioresour. Technol. **73**, 187–189 (2000)
72. Evtiouguina, M., Barros-Timmons, A.M., Cruz-Pinto, J.J.C., Pascoal Neto, C., Belgacem, M.N., Gandini, A.: Oxypropylation of cork and use of the ensuing polyols in the polyurethane formulation. Biomacromolecules **3**, 57–62 (2002)
73. Pavier, C., Gandini, A.: Oxyproylation of sugar beet pulp. 1. Optimization of the reaction. Ind. Crops Prod. **12**, 1–8 (2000)
74. Pavier, C., Gandini, A.: Urethanes and polyurethanes from oxypropylated sugar beet pulp. I. Kinetic study in solution. Eur. Polymer J. **36**, 1653–1658 (2000)
75. Serrano, L., Alriols, M.G., Briones, R., Mondragón, I., Labidi, J.: Oxypropylation of rapeseed cake residue generated in the biodiesel production process. Ind. Eng. Chem. Res. **49**, 1526–1529 (2010)
76. Cateto, C.A., Barreiro, M.F., Rodrigues, A.E., Belgacem, M.N.: Optimization study of lignin oxypropylation in view of the preparation of polyurethane rigid foams. Ind. Eng. Chem. Res. **48**, 2583–2589 (2009)
77. Matos, M., Barreiro, M.F., Gandini, A.: Olive stone as a renewable source of biopolyols. Ind. Crops Prod. **32**, 7–12 (2010)

78. Gandini, A., Belgacem, M.N.: Furan derivatives and furan chemistry at the service of macromolecular materials. In: Belgacem, M.N., Gandini, A. (eds.) Monomers, Polymers and Composites from Renewable Resources, pp. 115–152. Elsevier, Amsterdam (2008)
79. Gandini, A., Coelho, D., Gomes, M., Reis, B., Silvestre, A.: Materials from renewable resources based on furan monomers and furan chemistry: work in progress. J. Mat. Chem. **19**, 8656–8664 (2009)
80. Gandini, A.: Furans as offspring of sugars and polysaccharides and progenitors of a family of remarkable polymers: a review of recent progress. Polym. Chem. **1**, 245–251 (2010)
81. Gandini, A., Belgacem, M.N.: Furfural and furanic polymers. Actual. Chim. **11–12**, 56–61 (2002)
82. Gandini, A.: Polymers from renewable resources: a challenge for the future of macromolecular materials. Macromolecules **41**, 9491–9504 (2008)
83. Lasseuguette, E., Gandini, A., Belgacem, M.N., Timpe, H.J.: Synthesis, characterization and photocross-linking of copolymers of furan and aliphatic hydroxyethylesters prepared by transesterification. Polymer **46**, 5476–5483 (2005)
84. Gandini, A., Belgacem, M.N.: Furans in polymer chemistry. Prog. Polym. Sci. **22**, 1203–1379 (1997)
85. Moreau, C., Gandini, A., Belgacem, M.N.: Recent catalytic advances in the chemistry of substituted furans from carbohydrates and in the ensuing polymers. Top. Catal. **27**, 11–30 (2004)
86. Geethamma, V.G., Joseph, R., Thomas, S.: Short coir fiber reinforced natural rubber composites: effects of fiber length, orientation, and alkali treatment. J. Appl. Polym. Sci. **55**, 583–594 (1995)
87. Geethamma, V.G., Kalaprasad, G., Groeninckx, G., Thomas, S.: Dynamic mechanical behavior of short coir fiber reinforced natural rubber composites. Compos. Part A Appl. Sci. Manuf. **36**, 1499–1506 (2005)
88. Varghese, S., Kuriakose, B., Thomas, S., Koshy, A.T.: Mechanical and viscoelastic properties of short fiber reinforced natural rubber composites: effects of interfacial adhesion, fiber loading, and orientation. J. Adhes. Sci. Technol. **8**, 235–248 (1994)
89. Jacob, M., Thomas, S., Varughese, K.T.: Biodegradability and aging studies of hybrid biofiber reinforced natural rubber biocomposites. J. Biobased Mat. Bioenergy **1**, 118–126 (2007)
90. Nassar, M.M., Ashour, E.A., Washid, S.S.: Thermal characteristics of bagasse. J. Appl. Polym. Sci. **61**, 885–890 (1996)
91. De, D., De, D., Adhikari, B.: Curing characteristics and mechanical properties of alkali-treated grass-fiber-filled natural rubber composites and effects of bonding agent. J. Appl. Polym. Sci. **101**, 3151–3160 (2006)
92. Bhattacharya, T.B., Biswas, A.K., Chaterjee, J., Pramanick, D.: Short pineapple leaf fibre reinforced rubber composites. Plast. Rubber Process. Appl. **6**, 119–125 (1986)
93. Lopattananon, N., Panawarangkul, K., Sahakaro, K., Ellis, B.: Performance of pineapple leaf fiber-natural rubber composites: the effect of fiber surface treatments. J. Appl. Polym. Sci. **102**, 1974–1984 (2006)
94. Arumugam, N., Tamareselvy, K., Venkata Rao, K., Rajalingam, P.: Coconut-fiber-reinforced rubber composites. J. Appl. Polym. Sci. **37**, 2645–2659 (1989)
95. Mathew, L., Joseph, K.U., Joseph, R.: Isora fibres and their composites with natural rubber. Prog. Rubber Plast. Recycl. Technol. **20**, 337–349 (2004)
96. Joseph, S., Appukuttan, S.P., Kenny, J.M., Puglia, D., Thomas, S., Joseph, K.: Dynamic mechanical properties of oil palm microfibril-reinforced natural rubber composites. J. Appl. Polym. Sci. **117**, 1298–1308 (2010)
97. Madani, M., Basta, A.H., Abdo, A.E.-S., El-Saied, H.: Utilization of waste paper in the manufacture of natural rubber composite for radiation shielding. Prog. Rubbers Plast. Recycl. Technol. **20**, 210–287 (2004)
98. Jacob, M., Thomas, S., Varughese, K.T.: Novel woven sisal fabric reinforced natural rubber composites: tensile and swelling characteristics. J. Compos. Mat. **40**, 1471–1485 (2006)

99. Jacob, M., Varughese, K.T., Thomas, S.: Mechanical properties of sisal/oil palm hybrid fiber reinforced natural rubber composites. Compos. Sci. Technol. **64**, 955–965 (2004)
100. Jacob, M., Varughese, K.T., Thomas, S.: Natural rubber composites reinforced with sisal/oil palm hybrid fiber: tensile and cure characteristics. J. Appl. Polym. Sci. **93**, 2305–2312 (2004)
101. Anuar, H., Ahmad, S.H., Rasid, R., Ahmad, A., Busu, W.N.W.: Mechanical properties and dynamic mechanical analysis of thermoplastic-natural-rubber-reinforced short carbon fiber and kenaf fiber hybrid composites. J. Appl. Polym. Sci. **107**, 4043–4052 (2008)
102. Pasquini, D., Teixeira, E.D., Curvelo, A.A.D., Belgacem, M.N., Dufresne, A.: Extraction of cellulose whiskers from cassava bagasse and their applications as reinforcing agent in natural rubber. Ind. Crops Prod. **32**, 486–490 (2010)
103. Bras, J., Hassan, M.L., Bruzesse, C., Hassan, E.A., El-Wakil, N.A., Dufresne, A.: Mechanical, barrier, and biodegradability properties of bagasse cellulose whiskers reinforced natural rubber nanocomposites. Ind. Crops Prod. **32**, 627–633 (2010)
104. Bendahou, A., Kaddami, H., Dufresne, A.: Investigation on the effect of cellulosic nanoparticles' morphology on the properties of natural rubber based nanocomposites. Eur. Polymer J. **46**, 609–620 (2010)
105. Bendahou, A., Kaddami, H., Raihane, M., Habibi, Y., Dufresne, A.: Nanocomposite materials based on date palm tree cellulose whiskers. Rev. Roum. Chim. **54**, 571–575 (2009)
106. Bendahou, A., Habibi, Y., Kaddami, H., Dufresne, A.: Physico-chemical characterization of palm from phoenix Dactylifera-L, preparation of cellulose whiskers and natural rubber-based nanocomposites. J. Biobased Mat. Bioenergy **3**, 81–90 (2009)
107. Nair, K.G., Dufresne, A., Gandini, A., Belgacem, M.N.: Crab shell chitin whiskers reinforced natural rubber nanocomposites. 3. Effect of chemical modification of chitin whiskers. Biomacromolecules **4**, 1835–1842 (2003)
108. Nair, K.G., Dufresne, A.: Crab shell chitin whisker reinforced natural rubber nanocomposites. 2. Mechanical behavior. Biomacromolecules **4**, 666–674 (2003)
109. Nair, K.G., Dufresne, A.: Crab shell chitin whisker reinforced natural rubber nanocomposites. 1. Processing and swelling behavior. Biomacromolecules **4**, 657–665 (2003)
110. Siqueira, G., Abdillahi, H., Brass, J., Dufresne, A.: High reinforcing capability cellulose nanocrystals extracted from *Syngonanthus nitens* (Capim Dourado). Cellulose **17**, 289–298 (2010)
111. Mano, E.B., Nunes, R.C.R.: Regenerated cellulose in elastomer compounds. Eur. Polymer J. **19**, 919–921 (1983)
112. Kosikova, B., Alexy, P., Gregorova, A.: Use of lignin products derived from wood pulping as environmentally desirable component of composite rubber materials. Wood Res. **48**, 62–67 (2003)
113. Jong, L.: Characterization of defatted soy flour and elastomer composites. J. Appl. Polym. Sci. **98**, 353–361 (2005)
114. Chen, Y., Zhang, L., Du, L.: Structure and properties of composites compression-molded from polyurethane prepolymer and various soy products. Ind. Eng. Chem. Res. **42**, 6786–6794 (2003)
115. Song, Y., Zheng, Q.: Structure and properties of methylcellulose microfiber reinforced wheat gluten based green composites. Ind. Crops Prod. **29**, 446–454 (2009)
116. Gandini, A., Curvelo, A.A.S., Pasquini, D., de Menezes, A.J.: Direct transformation of cellulose fibres into self-reinforced composites by partial oxypropylation. Polymer **46**, 10611–10613 (2005)
117. de Menezes, A.J., Pasquini, D., Curvelo, A.A.S., Gandini, A.: Novel thermoplastic materials based on the outer-shell oxypropylation of corn starch granules. Biomacromolecules **8**, 2047–2050 (2007)

118. de Menezes, A.J., Pasquini, D., Curvelo, A.A.S., Gandini, A.: Self-reinforced composites obtained by the partial oxypropylation of cellulose fibers. 2. Effect of catalyst on the mechanical and dynamic mechanical properties. Cellulose **16**, 239–246 (2009)
119. de Menezes, A.J., Pasquini, D., Curvelo, A.A.S., Gandini, A.: Self-reinforced composites obtained by the partial oxypropylation of cellulose fibers. 1. Characterization of the materials obtained with different types of fibers. Carbohydr. Polym. **76**, 437–442 (2009)

Nanocomposites for Tyre Applications

Arup K. Chandra and Vivek Bhandari

Abstract Various developments in polymer nanocomposites and their understanding on application will be a major factor in future to address the emerging requirements from tyre. In this chapter, basic requirements of tyre and introduction to Nanofiller are discussed. The art of getting most of the benefit from nanofiller is lying in converting it into a suitable nanocomposites. Few popular general purpose based nanocomposites and their role in improving critical and conflicting tyre properties are included in the chapter. Polymer/Layered silicate nanocomposites have gained strong momentum in recent years because of the remarkable improvement in material properties compared to conventional micro and macro composites. In this chapter current developments in the field of rubber nanocomposites filled with layered silicates as well as nano calcium carbonate composites for various tyre applications have been addressed. An inclusive attention has been given to the structure and properties of recently developed nanosized rubber (nanoprene) and its potential application for meeting the challenges in tyre.

1 Introduction

Tyre is a very important part of a vehicle. It is the part, which comes in contact with the road. It is the most versatile and probably the first engineered product made out of rubber. Tyre has made the evolution of sophisticated personalized land transportation system possible.

A. K. Chandra (✉) · V. Bhandari
R&D Center, Apollo Tyres Ltd., Limda, Waghodia, Vadodara, Gujarat 391760, India
e-mail: arupkumar.chandra@apollotyres.com

V. Bhandari
e-mail: vivek.bhandari@apollotyres.com

P. M. Visakh et al. (eds.), *Advances in Elastomers II*,
Advanced Structured Materials 12, DOI: 10.1007/978-3-642-20928-4_6,

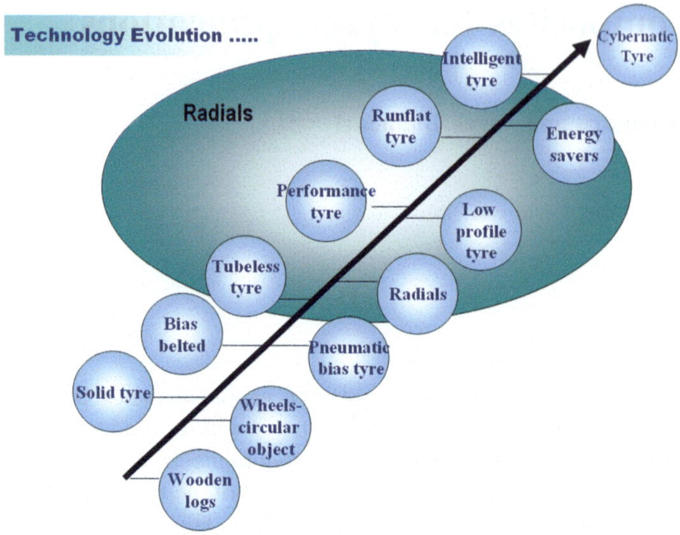

Fig. 1 Trends in tyre technology [1]

Tyre industry is the world's second largest industry among road transport industries; employing the largest man power. Without the pneumatic tyre, development of motor vehicle transport would not have been possible. The performance of the vehicle is largely dependent on the capability of tyre [1].

Over the decades tyre has undergone a series of changes and graduated to a highly technological product (Fig. 1). Increasing automobile manufacturers' requirements and ever-growing customer expectations have resulted in evolution of new product technology and an all round development in materials front.

Development in nanocomposite is a significant development of twenty-first century. The rapid growth in this field will definitely change the material demand scenario of tyre industry as well.

2 Basic Requirements of Tyre

Tyre is one of the most important components of a vehicle that provides load carrying capacity, cushioning and damping; transmits driving and breaking force; provides traction and cornering force while running.

The tyre touches the ground on an area not much larger than an average person's foot print. Roads may be covered with sharp objects, large stones, debris and many other obstacles. The tyre has to perform and overcome all these obstacles, to give the necessary traction, so that able to transmit the driving, breaking and cornering force in dry and wet; even in snow and ice conditions. Therefore, a tyre designer or compounder has to think several performance properties while designing a tyre.

The advances in the technology are continuing on many fronts to meet never ending customer demands and to accommodate new applications, thus enhancing basic functions of a tyre including lower rolling resistance, reliability, comfort and aesthetics [2, 3].

These basic requirements are found to be essentially same for all type of pneumatic tyres, though their relative importance differs with each type.

These common features are:

2.1 Cushioning the Vehicle Ride

A pneumatic tyre is the first line of protection against road shocks and its unique property of cushioning provides comfort. Cushioning is imparted by the air and the ability of the tyre membrane to deform radially.

As a result, tyre absorbs the shocks due to obstacles and cushions the vehicle from other irregularities on the road, ensuring comfort to driver and passenger as well as the long life to the vehicle. The main characteristic of a tyre is its suppleness, particularly in a vertical direction. The significant elasticity of the air contained in the tyre enables it to take the impact of any deformations caused by obstacles and uneven surfaces. The right pressure gives optimum load carrying capacity and high levels of comfort whilst retaining good steering capability.

2.2 Carrying the Load

A pneumatic tyre must carry the load when mounted on a rim and inflated to the required pressure. In other words, pneumatic tyre must form an air container of adequate strength to carry the load. Tyre support the vehicle not only when it is moving but also when it is static and must be able to transfer considerable load during acceleration and braking. A car tyre carries more than 50 times of its own weight.

2.3 Transmits Steering Response

A tyre is required to transmit driving and breaking torque efficiently to the ground to put the vehicle in motion or stand still. Good road grip coupled with adequate strength disposed in the circumferential direction underneath the tread of the tyre is required for efficient transporting.

Tyre should steer the vehicle with precision irrespective of the state of the surface and/or weather conditions. The stability of a vehicle's path depends on the tyre holding its course. In general, each vehicle has a particular inflation pressure for individual axle. Respected pressure variations between front and rear tyres ensures the ideal directional stability.

2.4 Traction

The friction between the tyre and the road surface that provides the grip is known as traction. A good all-season traction tyre provides a balance of traction in wet, dry and winter conditions.

The control of the self-propelled vehicle is determined by the tractive forces, which are acting between the tyres and the road. They can be classified into three.

1. Braking traction is the force that brings a vehicle to a stop while it is traveling.
2. Cornering traction is the force generated by a tyre that enables a vehicle to execute lateral maneuvers and round corners on a prescribed path.
3. Driving traction is the force that propels the vehicle in the intended direction.

2.5 Fatigue Life and Wear

Every revolution of the tyre leads to periodic strain radially, circumferentially and laterally. This necessitates adequate flex fatigue life of the materials and structure for economic service life of a tyre.

The wear of a tyre depends on its conditions of use e.g. load, speed, condition of the road surface, state of the vehicle, style of driving, etc. and above all, the quality of the tread materials and its contact with the ground.

2.6 Rolling Resistance

A tyre consumes energy during service as it is constantly changing its shape as the sidewalls deflect and the tread flattens into the contact patch. This consumes a small, but definite amount of energy by virtue of hysteresis loss that results in more fuel consumption and higher rolling resistance [4–6].

The rolling resistance of a tyre is defined as the mechanical energy that is converted into heat when the tyre moves for a unit distance. Generally, the component responsible for higher percentage of rolling resistance of a tyre is tread. It is demonstrated that around ~ 50 % of tyre rolling resistance is dependent on tyre tread and it's type (i.e. TBR/PCR etc.).

From last few decades, tyre technologists have been putting continuous effort for the reduction of rolling resistance as well as simultaneous improvement in

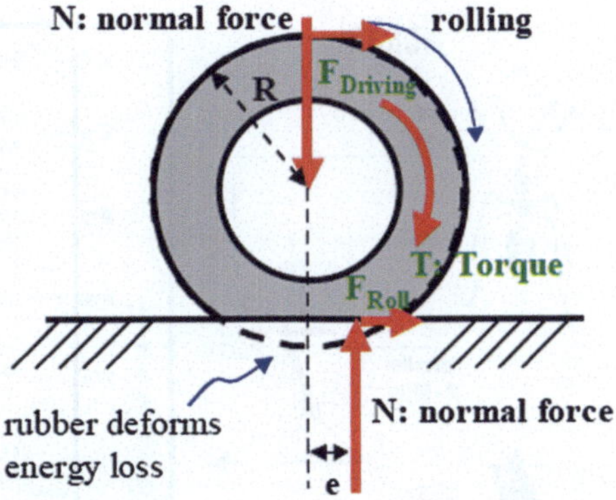

Fig. 2 Basic mechanism of rolling resistance [4]

other technical properties related to a tyre tread. A tread with higher hysteresis loss results in higher rolling resistance, but improved traction. By the proper choice of a tread compound, the rolling resistance can be improved without affecting the safety (traction) as well as other technical properties of a tyre. Rolling resistance plays a crucial role regarding the fuel consumption.

An improvement in rolling resistance around 10 % can save the fuel consumption ranging in between 0.5 and 1.5 % for passenger cars and light trucks, while the same improvement leads to a fuel saving of 1.5–3 % in case of heavy trucks [5]. In other words 5–15 % of fuel is consumed due to rolling resistance in case of passenger cars or light trucks and 15–30 % for heavy trucks. Figure 2 shows a basic mechanism of rolling resistance [4].

3 Introduction to Nano-Filler

Fillers are used in polymers for a variety of reasons; cost reduction, improved processing, density control, optical effects, thermal conductivity, control of thermal expansion, electrical properties, magnetic properties, flame retardancy and improved mechanical properties, such as stress–strain properties, hardness, tear resistance and etc. Each filler has different properties and these in turn are influenced by the particle size, shape, particle morphology and surface chemistry. Amorphous silica, carbon black and clay are the most widely used particulate fillers in polymer industry to achieve the performance requirements in various engineering applications. Degree of improvement of these performance requirements by these particulate fillers in polymer-filler composites mainly depend on the size of disperse fillers, polymer-filler interaction and morphology of the fillers.

Nanofiller is a class of new-generation fillers, which have at least one dimension in the order of nanometer and within 100 nm (Fig. 3). This new class of material has varying shapes ranging from isotropic needle-like, particulate sphere or sheet like elements. Uniform dispersion of these nanosized particles can lead to

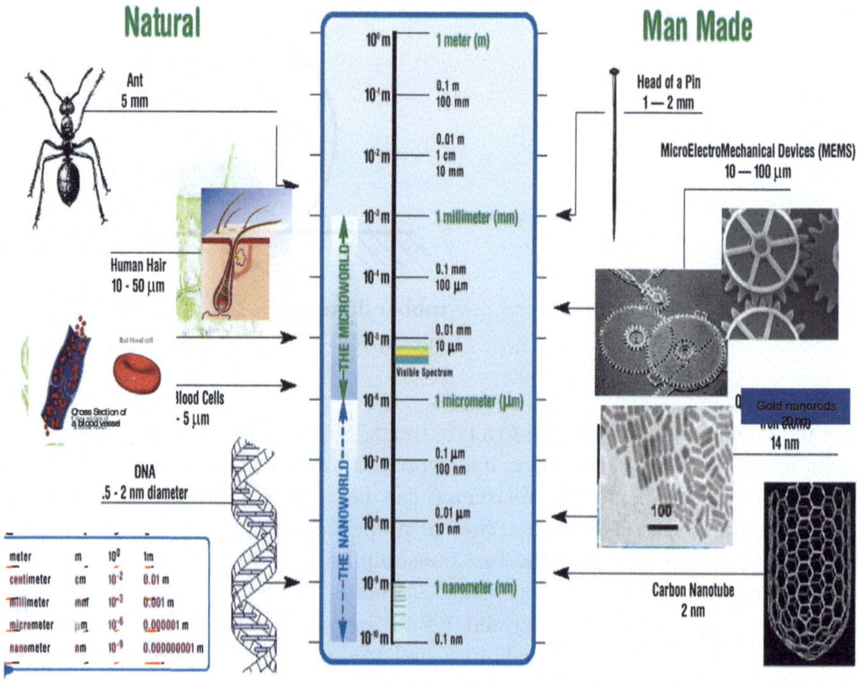

Fig. 3 An example of nanoscale materials

ultra-large interfacial area between a polymer and the fillers. This large interfacial area between the filler and a polymer and the nanoscopic dimension differentiate Polymer Nanocomposites (PNC) from traditional composites [7]. The addition of nanofillers increases the tensile, thermal and flame retardancy exponentially at lower filler dosage.

Nanofillers are necessarily nanoscopic and have a high specific surface area available for interaction with polymer. Nano-structure black is a family of new carbon blacks characterised by rough surface and enhanced filler—polymer interaction compared to standard ASTM—Black. It hinders the slippage of polymer molecule along the rough nano-structure surface and reduces the hysteresis significantly. This type of black ideally meet tyre requirement as it provides improved tread wear in addition to low hysteresis (Fig. 4). The specific surface area is one of the reasons why the nature of reinforcement is different in nanocomposites and is manifested even at very low filler loadings (<10 wt%). PNCs represent a radical alternative to conventional filled polymers or polymer blends—a staple of the modern polymer industry.

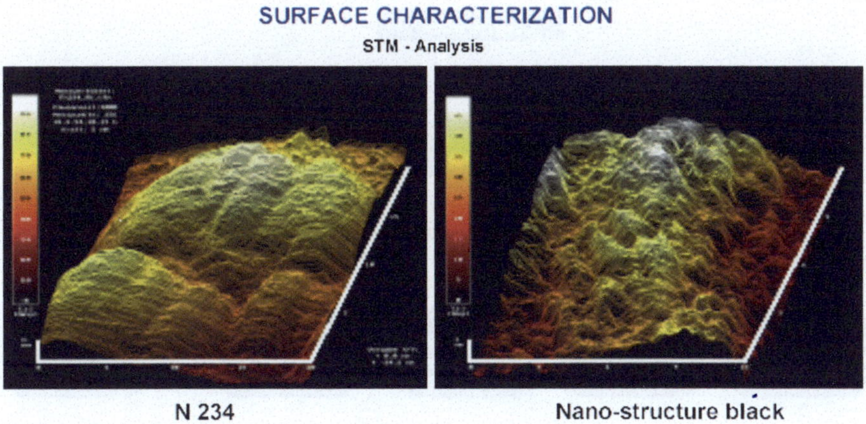

Fig. 4 New-generation nano structure black [1]

3.1 *Different Types of Nanofillers*

The nanofillers used in polymer nanocomposite are usually having different shapes and on that basis can broadly be categorized into the following classes (Fig. 5):

- **Spherical/Cubical** e.g., nanosilica, polyhedral oligomeric silsesquioxanes (POSS), nano $CaCO_3$, metal oxides etc.
- **Rod/fibre** e.g., synthetic whiskers, carbon nanotubes, carbon nanofibers, boehmite, sepiolite, nano $CaCO_3$, etc.
- **Sheet/platelet** e.g., layered silicates such as smectite group clays, synthetic mica etc.

Out of the above fillers, few types of nanofiller like Nanosilica, Carbon nanotube and silicates e.g. Montmorillonite (MMT) have high potential for future tyre industry. Some of these nanofillers are discussed subsequently;

3.1.1 Nanosilica

Silica is one of the most abundant oxide materials in the earth's crust. It can exist in an amorphous form (vitreous silica) or in a variety of crystalline forms. Silica occurs commonly in nature as sandstone, silica sand or quartzite. It is the starting material for the production of silicate glasses and ceramics.

Nanosilica can be produced through sol–gel technique. Nanosilica can be obtained in powder form. Surface treatment of nanosilica can make it organophilic and thus dispersion in elastomers can be enhanced.

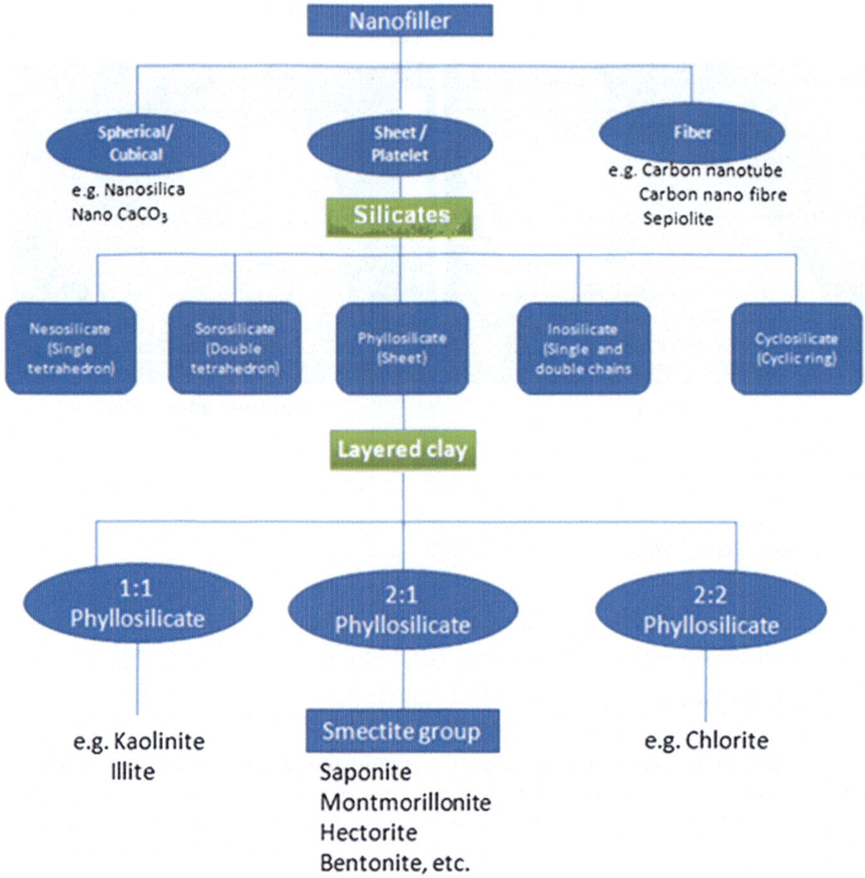

Fig. 5 Various categories of nano-fillers [7]

3.1.2 Carbon Nanotube

Carbon nanotubes are promising to revolutionize several fields in material science and are a major component of nanotechnology. Nanotube has a wide range of unexplored potential applications in various technological areas such as aerospace, energy, automobile, templates etc.

Carbon nano-tubes have a much lower percolation threshold (lower amount is required for equivalent reinforcement). Therefore, it might be possible to formulate a compound, which has all the advantages of the 'green' tyre by the incorporation of relatively small amounts of carbon nano-tubes into the silica filled compound [8].

The word nanotube is derived from their size, because the diameter of a nanotube is on the order of a few nanometers (approximately 50,000 times smaller than the width of a human hair) and can be up to several micrometers in length.

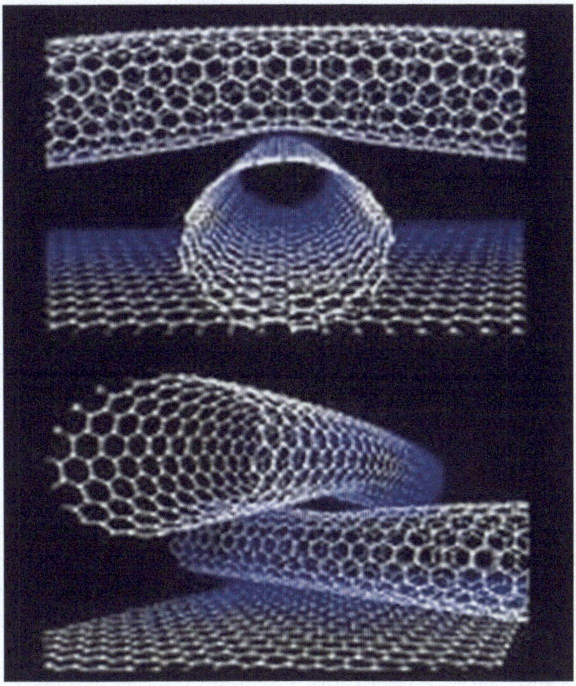

Fig. 6 Carbon nanotubes [1]

Carbon nanotubes (CNTs) are cylindrical carbon molecules (Fig. 6) with novel properties (outstanding chemical, electrical, thermal and mechanical properties: 100 times stronger than steel) which make them potentially useful in a wide variety of applications.

A nanotube is a member of the fullerene structural family. Nanotubes are composed of SP^2 bonds, similar to those observed in graphite and they naturally align themselves into ropes held together by van der Waals forces.

3.1.3 Silicates

The silicates are the largest, the most interesting and the most complicated class of minerals so far. The basic chemical unit of silicates is the $(SiO_4)^{4-}$ tetrahedron. They can form as single units, double units, chains, sheets, rings and framework structures.

In recent year's polymer/layered silicate (PLS) nanocomposites have attracted great interest because they often exhibit remarkable improvement in materials properties when compared with virgin polymer or conventional micro and macro-composites. These improvements can include high moduli, increased strength and heat resistance, decreased gas permeability. On the other hand, there has been considerable interest in theory and simulations addressing the preparation and properties of these materials. They are also considered to be unique model systems to study the structure and dynamics of polymers in confined environments. The field of PLS nanocomposites has gained momentum strongly during the last decade.

The silicates are divided into the following sub-classes, not by their chemistry, but by their structure:-

- Nesosilicates (*single tetrahedrons*)
- Sorosilicates (*double tetrahedrons*)
- Inosilicates (*single and double chains*)
- Cyclosilicates (*rings*)
- Phyllosilicates (*sheets*)
- Tectosilicates (*frameworks*)

The layered silicates mostly belong to phyllosilicate subclass which is commonly used in tyre industries.

3.1.4 Layered Clays

Layered clays are minerals that assemble regularly with the unit crystalline layer, usually at the nanoscale. A layer of clay mineral is about 1 nm in thickness and consists of platelets of around 100 nm in width, representing filler with a significantly large aspect ratio. Layered clay can be classified according to the unit and ratio of crystal types they are possessing. These are:

(i) 1:1 type—Its unit crystal is composed of one crystal sheet of silica tetrahedron combined with one crystal sheet of alumina tetrahedron. e.g. kaolinite, illite.
(ii) 2:1 type—Its unit crystal is composed of two crystal sheets of silica tetrahedron with one crystal sheet of alumina tetrahedron in between. e.g. montmorillonite (MMT), saponite, bentonite, hectorite.
(iii) 2:2 type—Its unit lamellar crystal is composed of four crystal sheets, in which crystal sheets of silica tetrahedron and alumina or magnesium octahedron are alternately arranged. e.g. chlorite.

However, the smectite group (2:1) clays are the most used nanofillers in polymers. Out of these, MMT has grabbed the utmost attention because of its

Table 1 Common 2:1 Phyllosilicates and their features

2:1 phyllosilicates	Chemical formula	CEC (mequiv/100 g)	Particle length (nm)
Montmorillonite	$M_x(Al_4\text{-}xMg_x)Si_8O_{20}(OH)_4$	110	100–150
Hectorite	$M_x(Mg_6\text{-}xLi_x)Si_8O_{20}(OH)_4$	120	200–300
Saponite	$M_xMg_6\ (Si_8\text{-}xAl_x)Si_8O_{20}(OH)_4$	86.6	50–60

natural abundance and ease of organic modification to promote dispersion in polymer matrices. Table 1 depicts common 2:1 phyllosilicates and their features.

Structure and Properties of Montmorillonite

MMT belongs to the 2:1 type of phyllosilicate group. The general formula of MMT is $[(Na,Ca)_{0.33}(Al, Mg)_2(Si_4O_{10})(OH)_2\text{-}nH_2O]$. One layer of MMT consists of two fused silica tetrahedral sheets sandwiching an edge-shared octahedral sheet of alumina, as shown in Fig. 7. Isomorphic substitution occurs when some atoms in the crystal structure are replaced with other atoms with different valence electrons. Montmorillonite can absorb water between the charged layers because of the weak binding and the large spacing between the layered sheets and it is therefore a member of a group of water-expandable clay minerals known as smectites or smectite clays.

From Fig. 7 it is clear that the layers of MMT are connected by weak van der Waals force of attraction and are arranged themselves in stacks with some gap in-between, often termed as "interlayer" or "gallery spacing". As discussed earlier, the isomorphic substitution of metal ions generates an excess of negative charge and the quantity of this governs the characterization of the clay type and in general termed as cation exchange capacity of the clay (CEC). In natural MMT, this charge is balanced by the group I and II metal cations (Na+ , Ca+ etc.) in their hydrated form and that is why it is compatible with hydrophilic polymer. But dispersion of such clay into hydrophobic polymeric matrix is a difficult task. To alleviate this problem, MMT can be organically modified by exchanging the inorganic cations such as Ca^{++}, Na^{+}, and K^{+} (those exist in the galleries of silicate

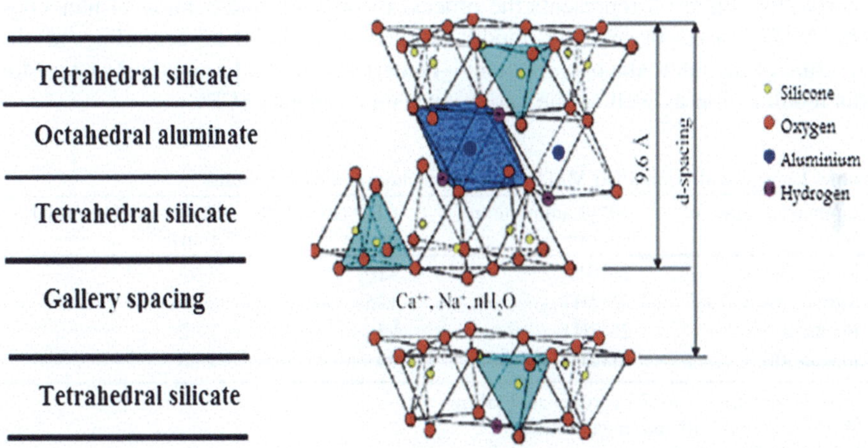

Fig. 7 Layered silicate structure of montmorllionite (MMT) [12]

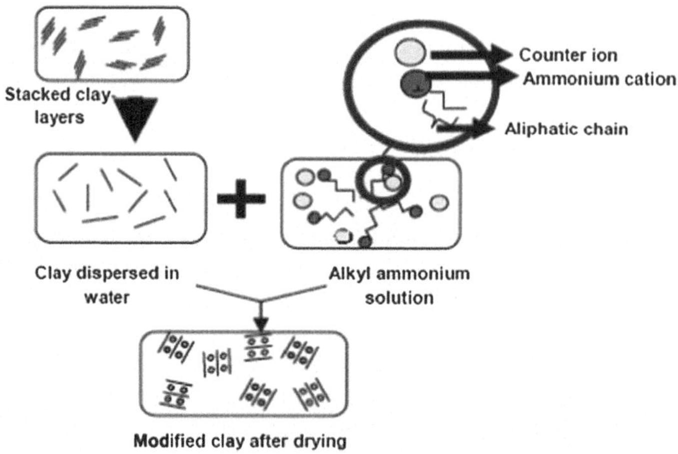

Fig. 8 Organic modification of nanoclay [13]

layer) with quaternary alkyl ammonium ions resulting in better compatibility of the silicates with the polymer matrix [9–11].

Organically Modified MMT Clays

Hydrophilic nature of clays hinders their good interaction with organic polymers. Therefore, surface modifications of clays are necessary to render them hydrophobic in order to achieve a better interaction of clay surface with the polymer matrix [13]. Figure 8 represents the process of organic modification of nanoclay. The MMT based organically modified clays are classified according to the structure of the intercalants (surfactant—quaternary alkyl ammonium ion) used for this modification as well as the cation exchange capacity (CEC).

Table 2 Organically modified MMT based Nanoclays and their features

Commercial name of ONC	Alkyl ammonium Groups	CEC (meq/100 gm)	Basal spacing (d_{001}) (nm)
Cloisite-10A	2MBHT	110	1.92
Cloisite-15A	2M2HT	125	3.15
Cloisite-20A	2M2HT	95	2.42
Cloisite-30B	MT2ETOH	90	1.85

2MBHT-Dimethyl benzyl hydrogenated tallow
2M2HT-Dimethyl dihydrogenated tallow
MT2ETOH-Bis (ethyl alcohol) methyl tallow

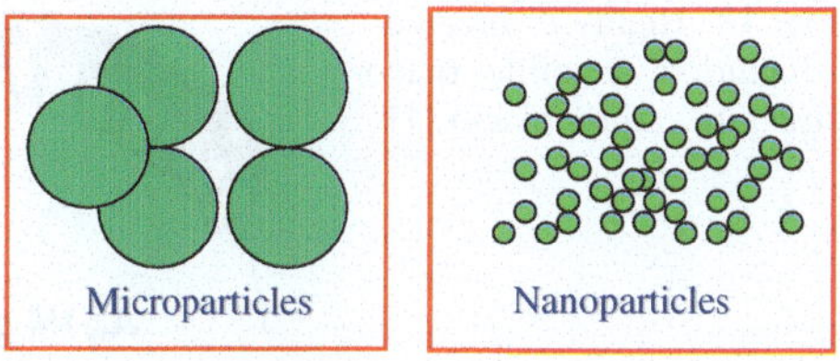

Fig. 9 Dispersion of microparticles and nanoparticles in polymer matrix [1]

Some of the commonly used organoclays are Cloisite-10A, 15A, 20A, 30B which can probably find their application in tyre industry. Table 2 depicts some of the important characteristics of these organo nanoclays (ONC).

4 Nanocomposites

A nanocomposite is defined as the composite of two materials, one material known as nanomaterial having the dimension of nanometric level at least in one dimension. These nanomaterials are very tiny in nature to the level of "Nano Meter" (nm) in size, which is even smaller than the wavelength of visible light. A typical nano particle size could be below 100 nm where radically different phenomena are manifested by the nanoparticles. The basic usage of a nanocomposite is due to many folds improvement on the physical/mechanical or on the processing properties of the composites upon addition of very minute quantity of this material.

Polymer nanocomposites are organic–inorganic hybrid materials where the inorganic phase likes nano silica, nano ZnO, nano clay are distributed in nanoscale within the organic polymer matrix (Fig. 9). Increased surface area of nano-fillers produces better interfacial polymer-nanofiller interaction. This results in enhancement of mechanical and other properties which has wide potential in the field of electronic, magnetic, optical and chemical field as well. In Polymer Nanocomposites or PNC, the polymer could be thermoplastics or thermosetting in origin. These materials provide improvement over other known composites in thermal, mechanical, electrical and even air barrier properties.

4.1 Advantages of Nanocomposites

Improvement of mechanical properties have resulted in major interest in nanocomposites materials in numerous automotive and general/industrial applications.

The main advantages offered by nanocomposites are [14–17]:-

(1) Lighter weight due to low filler loading
(2) Low cost due to lesser amount of filler used

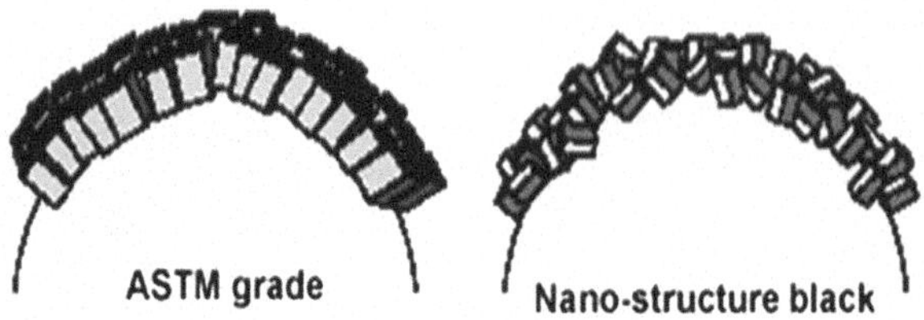

Fig. 10 Highly disordered structure of nanofiller (nano-structure carbon black) [1]

(3) Improved properties compared to conventional filler-based composites at very low loading of filler
(4) Achievement of conflicting properties
(5) High thermal stability
(6) Improved impermeability to gases, vapour and liquid
(7) Good optical clarity
(8) Improved flame retardancy with reduced smoke emission
(9) Tailored made electrical conductivity
(10) Better wear properties

Besides their improved properties, these nanocomposite materials are also easily extrudable or moldable to near final shape. Since high degrees of stiffness and strength are realized with little amount of high-density inorganic materials, they are much lighter compared to conventional polymer composites. This weight advantage could have significant impact on environmental concerns among many other potential benefits. By surveying the nanocomposites prepared so far, wear resistance polymer nanocomposites can be regarded as a successful example that brings the so-called "nano-effect" into full play. At a filler loading of less than 1 %, the wear rate of the matrix was lowered by over thousands of times. Similar enhancement due to the addition of small amounts of fillers is impossible to perceive in micro composites. The developments in this aspect have broadened the application possibility of particulate composites and solved the dilemma arising from the contradiction between performance improvement and processability deterioration, as often observed in micro particles filled composites. The main reason for improved properties is due to smaller and highly disordered structure of nanofillers that results in more active sites with a higher surface energy and ultimately leads to stronger mechanical/physiochemical interaction with polymers (Fig. 10) [1].

4.2 Classification of Nano-Composites

Depending on the nature of filler, type of dispersion and method of preparation, the nanocomposites can be divided into following subclasses.

(1) Clay based nanocomposites

(2) Silica based nanocomposites
(3) Carbon nanotube filled nanocomposites
(4) Polyhedral oligomeric silsesquioxane (POSS)-based nanocomposites and
(5) Nanocomposite based on other nanofillers like metal oxides, hydroxides and carbonates

5 Application of Nanocomposites in Tyre Industry

A tyre must deliver high traction on dry as well as wet roads, commonly called as wet and dry grip. This high traction force between tread and road is necessary for avoiding slippage while running on the road. Again, a tyre must exhibit low wear and good durability; the abrasion resistance of the tread compound should be as high as possible to achieve better mileage. With regard to the fuel consumption, rolling resistance is another important parameter, which is also related mainly to the tread compound. All these three parameters (traction, abrasion resistance, and rolling resistance) are inter-related under a common head popularly known as "Magic Triangle" (Fig. 11). The name is so because all the three properties are conflicting and interdependent and always a better balance among the three is required.

From the beginning of tyre manufacturing, researches have been keen to get the best balance of magic triangle. Filler is one of the key ingredients to balance these properties. Reinforcing fillers such as carbon black and silica individually or in combination with special chemicals are incorporated into the tyre to enhance the mechanical properties as well as to improve the magic triangle (Fig. 11). To achieve a better solution, the concept of nano technology is being introduced slowly in the tyre technology.

Many leading tyre manufacturers are now developing engineered composites to further extend the life of tyre. Pirelli has announced in 2004, the first application of nanomaterials in their winter segment [1]. Carbon black with small particles (including nanoparticles) has long been mixed with rubber to improve the wear and strength of the tyre. Cabot, one of the world's leading carbon black producers, successfully tested "PureNano" silica carbide nanoparticles designed by

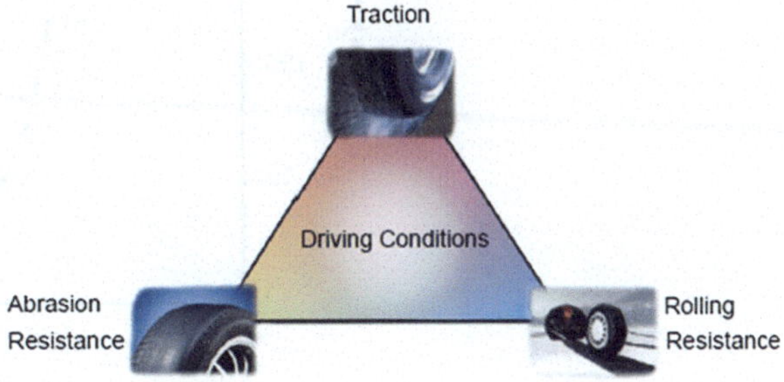

Fig. 11 Magic triangle of tyre performance

Nanoproducts Co. of Longmount, Colorado [18]. Added to tyre, the "PureNano" particles reduce abrasion by almost 50 %—a simple improvement that if widely adopted should help tyre to last twice long and thereby significantly reduce the need for new tyre/rubber.

Aeromet Technologies, Inc. of Sandy, Utah [19] crafts a cleaner process for bonding steel tyre cords to rubber during tyre manufacturing. Their new bonding material based on nanoparticles will eliminate the usage of hazardous chemicals such as cobalt and cyanide. Strong and lightweight rubber—carbon nanotube composites are coming strongly in market applications [20].

In addition to these, different types of nano-composites find their use in tyre industries. Few such types of nano-composite are discussed here.

5.1 Epoxidised Natural Rubber/Organo-Clay Nanocomposite

Epoxidised natural rubber (ENR) is such a rubber which has combine properties of natural rubber and a polar rubber. The epoxidised natural rubber (50 % epoxidation) with different grades of nano clay (10A, 15A, 20A) was mixed to prepare nanocomposites.

Nano clay shows a significant reduction of scorch time and considerable increase in maximum torque compared to conventional ENR compound. ENR/Clay nano composites have higher storage modulus which indicates the higher polymer—filler interaction of the composite and consequently better physical properties [21] (Fig. 12).

Tan delta values of the composites were found significantly reduced at high temperature which signifies lower rolling and better fuel economy for the ENR/Clay nano composites. Thus ENR/Organo clay nano composite can be used as tread compound for the future tyre.

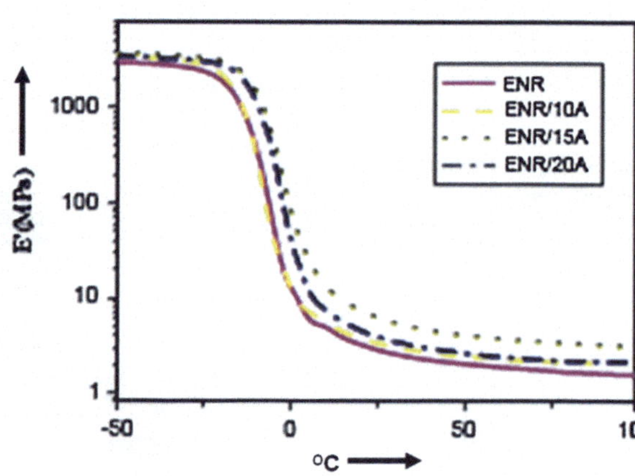

Fig. 12 Variation of storage modulus as a function of temperature [21]

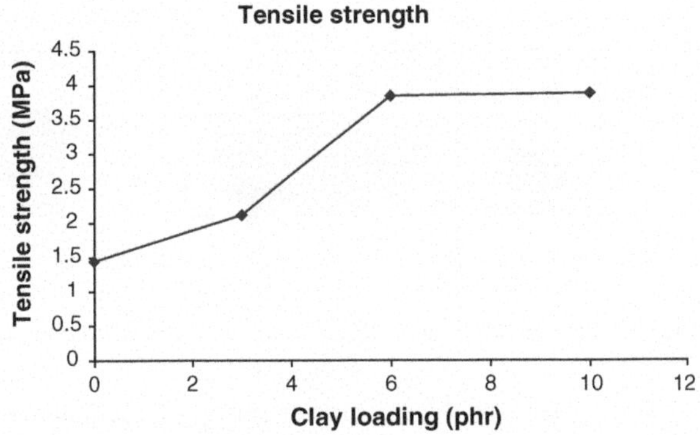

Fig. 13 Tensile strength of SBR-clay-carbon nano composite [22]

5.2 SBR/Clay Nano Composite

SBR-clay nanocomposites were made to get significant dispersion of nano clay in polymer matrix that can be characterized by X-ray diffraction. It is observed that addition of up to 10 parts of nano-fillers in SBR improves tensile strength more than 40 %. It also improves the dynamic properties of the compound significantly. But the effect of nano clay on SBR is greatly dependent on the concentration of it and it is observed that up to 6 phr of nano clay provides significant improvement of mechanical and dynamical properties [22] (Fig. 13).

The effect of nano clay in SBR can be enhanced significantly by adding the carbon black. It is observed that addition of carbon black in SBR-clay nanocomposite is giving tremendous improvement in the mechanical properties [22, 23]. Nano clay in combination of carbon black in SBR-BR based passenger car tread shows a considerable reduction in abrasion loss [24].

5.3 Butyl/Clay Nano Composite

The excellent air retention property of butyl (IIR)/halo butyl (XIIR) is very well known in tyre industries. These rubbers are extensively used in inner tube (IIR) of tube type tyre and in inner liner (XIIR) in tube less tyre. There are also some other elastomers, which are not used in tyre industries, but have very good and even better air retention properties, like, Epi Chloro-Hydrin Rubber (ECO), Poly Sulphide Rubber (TM) and Poly Urethane Rubber (PU). There are also some elastomers, like Nitrile Rubber (NBR), Ethylene/Acrylate co-polymers (EAM), which provide moderate air retention properties.

Using butyl as elastomer and clay as filler with high aspect ratio, the air retention properties could be improved to the level of 50 times better than that of normal butyl compound. Using such filler, therefore, the inner liner gauge or gauge of inner tube could be reduced for identical condition of services [25].

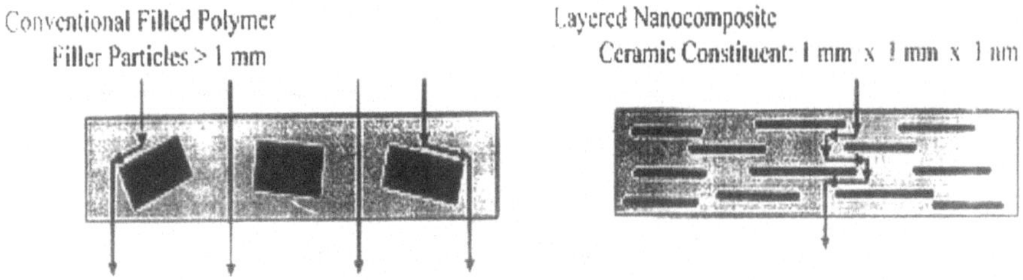

Fig. 14 Passage of air in C-Black compound and nano composite

The reason for the air barrier properties is that with carbon black as filler, the passage of air is faster because of spherical nature of the filler. With needle shaped filler in nano composite, the same air will take longer time to travel due to flat shape of the filler (Fig. 14).

5.4 Nanorubber

Nanorubber was developed jointly by professors Werner Obrecht and Lanxess, with a series of patents on the production process and on the application. It is defined as nanorubber because it is characterised by a chemical nanostructure whereby the rubber particles have a dimension of 40–65 nm. The nanorubber has been designed with a highly cross-linked particle core and consequently the particle morphology is not destroyed by high shear (Fig. 15). Additionally, due to a high degree of core crosslinking, the compounding ingredients such as oil, sulfur and accelerators are not absorbed by the core. Moreover, owing to chemical composition and crosslink density, the particle core exhibits a specific Tg, which influences temperature-dependent damping characteristics of vulcanisates (rolling resistance, wet traction). More precisely, when the nanorubber particles are dispersed in a conventional polymer matrix, they form a separate phase in the continuous phase of the polymer matrix and hence behave exactly like filler. Moreover, the particle size dimension of the nanorubber is completely comparable to that of the N550/N660 carbon blacks.

As disclosed in numerous patents on nanorubber, the development of this new material started from the observation that high levels of microgel in old emulsion SBR (E-SBR) and old emulsion polymerised BR were giving to the resulting vulcanisates high-rebound resilience at 70 °C (158 °F), which corresponds to low rolling resistance in tyre treads and low-rebound resilience at 23 °C (73.4 °F), ensuring high wet skid. While the vulcanisates with microgel had good dynamic properties, the drawback was adverse effects in mechanical properties, such as tensile, tear and wear [26].

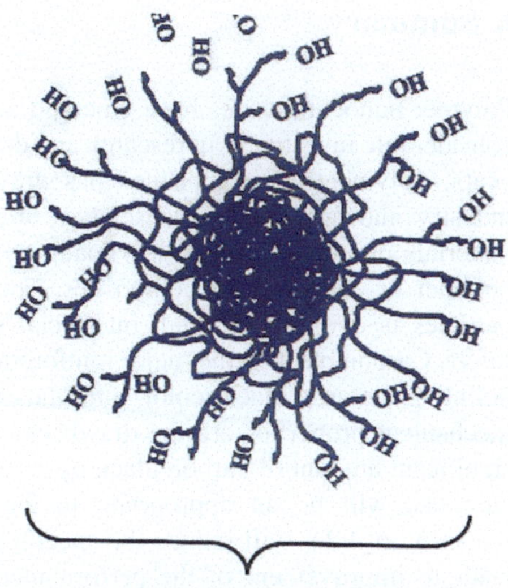

Fig. 15 General modal of nanorubber [26]

To keep the good dynamic properties offered by the polymer gel, but simultaneously compensate the low ultimate mechanical properties, the nanorubber was developed with a precise and tailor-made chemical structure as shown in Fig. 15.

The nanorubber should be added to the common polymer matrix on top of the 100 phr of the rubber content—really just as filler. Additionally, the nanorubber, as filler, has been designed to be reactive with polar fillers. For this reason the extremities of the polymer chains of the nanorubber are end-capped with hydroxyl groups to permit the linkage between the surface of the rubber nanoparticles and the silica surface in presence of Silane [26].

The nanorubber can be used as filler in a common carbon black-filled compound without the usage of silica or other polar fillers and without the use of silane coupling agents.

5.5 NR/BR/SBR–Nano Calcium Carbonate Composite

The application of an active nano-calcium carbonate in tyre was experimentally investigated. The physical and processing performance of rubber compounds is improved by adding 5–8 phr active nano-calcium carbonate in sidewall and carcass ply of tyre, together with proper loading adjustment of carbon black and oil in compound formulation. In tyre tread, with the addition of 4 phr active nano-calcium carbonate in the compound, the tear strength was improved while maintaining same abrasion resistance and other physical properties with better extrudability. The addition of active nano-calcium carbonate in tyre can reduce the production cost [27].

6 Summary

Polymer nanocomposites have emerged as a new class of materials and attracted considerable investment in research and development in last two decades. In recent years, polymer nanocomposites has drawn tremendous interest form both the industry and academia because they often exhibit remarkable improvement in materials properties at very low loading of nanofiller when compared to pristine polymer or conventional composites. The high degree of reinforcement by tiny particles is due to their high interfacial surface area compared to conventional fillers. Carbon black is the major reinforcing filler used by the tyre industry. But its polluting nature, black color and the need of high amount to enhance the mechanical properties are the drawbacks of carbon black. The replacement of a significant amount of carbon black by another reinforcing agent, which is in lesser amounts, will be an appropriate in the rubber especially tyre industry. The reduction of filler will reduce the specific gravity and heat generation, which can leads to improvement of the performance of different tyre compounds and ultimately results in better fuel efficient and long lasting tyre. The application of nanocomposite will help tyre industry to a great extent in future.

Acknowledgments Authors would like to express their sincere thanks to the management of Apollo Tyres Ltd. for kind permission to publish this chapter. Authors would also like to express their sincere thanks to their colleagues Mr. Nitya Kunti (Presently working for Phillips Carbon Black Ltd.), Mr. Sanjay Patel and Mr. Tapas Ranjan Mohanty for their valuable support.

References

1. Chandra, A.K.: Tyre technology—recent advances and future trends. Paper presented at 170th fall ACS, rubber division meeting, Cincinnati, 10–12 Oct 2006
2. Bhandari, V., Chandra, A.K.: New development in tyre technology. Paper presented at India rubber expo-2007, Chennai, 17–20 Jan 2007
3. Chandra, A.K.: Tyre technology—developments and future trends. Paper presented as workshop leader at 2nd middle east auto and aftermarket industry conference, Dubai, 28–29th May 2007
4. Schuring, D.J.: Rubber Chem. Technol. **53**, 600 (1980)
5. Indian Rubber Institute: Rubber Engineering. McGraw-Hill, New York (1999)
6. Engehausen, R., Rawlinson, A., Trimbach, J.: Tyre Technol. Int. Ann. Rev. 36–38 (2001)
7. Mohanty, T.R.: A comprehensive study of rubber based carbon black—organoclay nanocomposites in terms of optimized tyre performance properties, M. Tech. Thesis submitted to Rubber Technology Centre, IIT, Kharagpur (2011)
8. Chandra, A K.: Recent advances and future trends in tyre technology. In: Bhowmick, A.K. (ed.) Current Topics in Elastomers Research. CRC-Press/Taylor & Francis, Boca Raton, (Ref. Sec. VI; Chap-32, PP-919–934) (2008)
9. Le Pluart, L., Duchet, J., Sautereau, H., Halley, P., Gerard, J.F.: Appl. Clay Sci. **25**(3–4), 207–219 (2003)
10. Becker, O., Varley, R., Simon, G.: Polymer **43**, 4365–4373 (2002)

11. Kim, W., Kang, B., Cho, S., Ha, C., Bae, Jong-woo.: Composite Interfaces, 14, 409–425, (2007)
12. Alexandre, M., Dubois, P.: Mater. Sci. Eng. **28**, 1–63 (2000)
13. Maiti, M., Bhattacharya, M., Bhowmick, A.K.: Rubber Chem. Technol. **81**, 384–469 (2008)
14. Okada, A., Usuki, A.: Mater Sci. Eng. **C3**, 109 (1995)
15. Bandyopadhyay, A., Bhowmick, A.K., De Sarkar, M.: J. Appl. Polym. Sci. **93**, 2579 (2004)
16. Gilman, J.W.: Appl. Clay Sci. **15**, 31 (1999)
17. Messersmith, P.B., Giannelis, E.P.: J. Polym Sci. **33**, 1047 (1995)
18. www.azonano.com/Details.asp';[Article ID=1351
19. http://www.voyle.net/Nano%20biz/NanoBiz-2004
20. Strano, M.S.: Nat. Mat. **5**, 433 (2006)
21. Wang, X., Jia, D., Chen, M.: Structure and properties of epoxidised natural rubber/organoclay nanocomposite. Paper presented in 2nd international nano-electronics conference (IEEE), Shanghai, 24–27 March 2008
22. Jeenish, A.G., Patel, S.K., Chandra, A.K., Tripathy, D.K.: J. Polym. Res. **18**, 1625–1634 (2011)
23. Jeenish, A.G., Patel, S.K., Tripathy, D.K., Chandra, A K.: SBR-clay-carbon black nanocomposites for lower hysteresis, high abrasion tread. Paper presented at the international tire exhibition and conference (ITEC), Akron, 16–18 Sept'08 (Paper No. 21B) (2008)
24. Sumankumar, V.V.S.: Effect of fillers on the tread compound, M. Tech. Thesis submitted to Rubber Technology Centre, IIT, Kharagpur (2007)
25. Nah, C., Abdul Kadir, M.: Barrier Properties of Rubber Nanocomposite. In: Thomas, S., Stephen, R. (eds.) Rubber Nanocomposites: Preparation, Properties and Applications. Wiley, USA (Ref. Chap-18; Page 499)
26. Cataldo, F.: Fill the bill. Tire Technol. Int. 58–62, July (2011)
27. Liu, D., Zhang, J., Lin, X., Guan, Q., Jie, J.: Application of active nano-calcium carbonate in tire. Internal literature of Qingdao Double Star Tire Industry Co., Ltd, Jiaonan 266400, China

Elastomer-Based Bio-Nanocomposites

Sudipta Chatterjee, Arup K. Chandra and Santanu Chattopadhyay

Abstract This chapter encompasses an overview of recent advances in the field of natural, synthetic and semi-synthetic elastomer based bio-nanocomposites. Several naturally occurring polymers are considerably elastomeric and biocompatible. Elastin and its various derivatives are essential component of mammalian bio-systems. Various soft to semi-soft tissues of animals containing collagens, elastin proteins or other extra-cellular materials are classic examples of natural elastomeric bionanocomposites. There have been needs for developing synthetic or semi-synthetic elastomeric bio-nanocomposites for replacement or regeneration of such soft or semi-soft tissues or organs. Most of the elastomeric biopolymers lack in mechanical properties. Various modified cellulose components derived from plants like, cellulose whiskers, micro-fibrillated cellulose in their nano-scale size have proven promising in improving thermal, mechanical and moisture absorption properties of the elastomeric biopolymers. Similarly, biocompatible and bio-inert particulate nano filler systems including hydroxyapatite, bio-glass, silicates or other minerals can be potential reinforcing units for synthetic or semi-synthetic bionanocomposites. Non-toxicity of these precursors is utmost important for their actual bio-medical applications. Several synthetic methodologies have been adopted for the preparation of other novel and or bio-mimetic elastomeric bio-nanocomposites. Biodegradable polyurethanes and polycaprolactones, when modified with chitin or chitosan based chain extenders form novel nanocomposites, having wide applications in the area of biomaterials. Various precise and sophisticated characterization techniques namely, NMR, TEM, SEM, CD-spectroscopy have been routinely employed to evaluate the structure-property correlation of these novel bio-nanocomposites. Research is progress throughout

S. Chatterjee · S. Chattopadhyay (✉)
Rubber Technology Centre, Indian Institute of Technology, Kharagpur, India
e-mail: santanuchat71@yahoo.com

A. K. Chandra
R&D Centre, Apollo Tyres Ltd, Limda, Waghodia, Vadodara, Gujarat, India

P. M. Visakh et al. (eds.), *Advances in Elastomers II*,
Advanced Structured Materials 12, DOI: 10.1007/978-3-642-20928-4_7,

the globe for further improvement in properties of such materials and subsequently to figure out novel practical applications of these elastomeric bio-nanocomposites.

1 Introduction

Bio-nanocomposites have attracted a focal research interest in the recent past. The assembly of molecular or polymeric species of biological origin and other components through interactions, on the nano-metric scale constitutes the basis for preparation of bio-nanohybrid materials. This is entirely an interdisciplinary area encompassing biological sciences, material sciences and nanotechnology. Elastomer based bionanocomposites, which is a sub class of bio-nanohybrid materials, have several advantages of long range elasticity and viscoelasticity in addition to their inherent eco-friendly, biodegradable and renewable characteristics [1]. This special class includes composites of polysaccharides such as starch, cellulose, proteins such as collagen, elastin and synthetic bio-polymers like poly-lactic acid reinforced with particulated solids. They are also well known as green-nanocomposites [2, 3].

The modern definition of bio-elastomers can be given as: it is a material or composite having favorable bio-compatibility with adjacent tissues, with a low glass transition temperature than that of body temperature, exhibiting hyper-elasticity.

This chapter elucidates the basic idea for natural, synthetic as well as semi-synthetic varieties of elastomeric bio-nanocomposites. Traditional elastomers mainly consist of hydrocarbon chains flexible enough to allow C–C bond rotation within a specified temperature to execute long range elasticity. To make elastomeric bionanocomposites one major question is: how can we incorporate the whole range of attributes of elastomers as well as those of biomaterials in elastomeric bio-nanocomposites? If one has to do this, several parameters have to be taken into consideration. Intermolecular interaction especially inter-chains interaction and entropic elasticity are the utmost controlling factor for a material to behave as an elastomer. In true sense, most of the classical elastomeric molecules where intermolecular force is negligible are neither biological in origin nor are they biocompatible. These molecules also lack bio-relevant characteristics including biocompatibility, bio-functionality, and bio-degradability like in cellulose fibers, chitin, elastin etc.

Bio-mimetic (soft tissue) preparation of elastomeric bio-nanocomposites mainly starts with collagen or elastin based matrix mimicking cartilage type soft tissues. Amongst the biodegradable synthetic polymers, few block copolymers behave as thermoplastic elastomers with limited range of elasticity. Thus to coin the biological characteristics with long range elasticity one has a very limited choice. Thus initial developments in elastomeric bio-nanocomposites have been initiated with above mentioned class of bio-polymers following bio-mimetic preparation route.

Several synthetic methodologies have been developed in the recent past in pursuit of more versatile materials of this class. These materials exhibit excellent structural properties owing to special arrangements at the nanometric level of their assembled components. Recently, elastomers based bionanocomposites are prepared, mimicking the natural materials. Most of the works aims to develop new bio-hybrid materials having both biocompatibility and functionality. It is interesting that in the cell wall of prokaryotes, nano-structured silica nano-spheres could be assembled by the participation of cationic polypeptides [4].

As per a statistical data, the fascinating growth of research in the area of 'biomaterials including in bionanocomposites can be realized for the recent past as shown in Fig. 1 and it can be easily realized from the bar chart that the future of research in the allied area is very bright [2, 5].

Elastomers assembled with biological species are of diverse nature having different compositions, structures and textures [6]. They ultimately determine the properties of such bio-nanohybrid materials. The affinity among the components determines the stability of the morphology of resulting bio-nanocomposites guided by intermolecular interaction [7]. The coating of micro or nano-particulate solids with biopolymers occurs through hydrogen bonding or via formation of metal complexes. Magnetic ferrite nanoparticles embedded in biopolymers have extensive applications in MRI, hyperthermia and drug delivery systems [8]. Development of novel bio-hybrid materials e.g. hydro-gels displays a homogeneous dispersion resulting in multiple functionality [9]. Bio-composites are vividly useful in biomedical or eco-friendly applications. They are based on the biogenic components with well known biocompatibility, such as silica, nanoclay, iron-oxide, and hydroxyapatite to name a few.

Recent studies have shown that the great interest of interfacing a biological entity with non-biogenic metal oxides has promising futures. Such assemblies have been made with the aim of improvement of biosystems. Among protein based

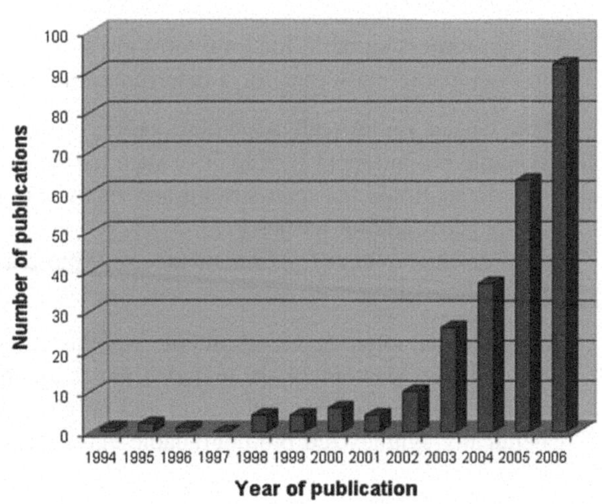

Fig. 1 Progressive development in bio-material/bio-medical related research publications in recent past. [Web of Science Databases-ISI web of knowledge, 2, 5]

polymers, gelatin is widely used in pharmaceuticals. Combination of bioactive and bio-resorbable properties are utmost important. The successful combination of bio-polymer with suitable bio-active or bio-inert hard fillers constitutes the basis of design of such materials. Elastomeric bio-nanocomposites exhibits substantial viscoelastic properties and can be easily fabricated into complex structures [10]. These materials are important for soft to semi soft-tissue replacements and making scaffolds.

2 Classification on Elastomeric Bio-Nanocomposites

Several classifications of the biomedical relevant polymers are possible. For example, some authors have distinguished between synthetic polymers like polylactic acid (PLA) and poly lactic-co-glycolic acid (PLGA) or their copolymers with polycaprolactone (PCL), and polymers of biological origin like polysaccharides (starch, alginate, chitin/chitosan, gelatin, cellulose, hyaluronic acid derivatives), proteins (say, collagen, elastin, fibrin, silk), and a variety of bio-fibers, such as ligno-cellulose containing elastomeric natural fibers [10].

Other authors have differentiated between resorbable or biodegradable polymers e.g. poly (α-hydroxyesters) polysaccharides and proteins and non-resorbable (e.g. cellulose) polymers. As synthetic polymers can be produced under the controlled conditions, they in general exhibit predictable and reproducible mechanical and physical properties such as strength, modulus, elongation at break and rate of degradation. Control of impurities is a further advantage of synthetic polymers. Essentially, the central idea has been to mimic elastic biological tissues by using a suitable polymeric matrix. The mechanical reinforcement and bioactive character of the composite are utmost important.

Based on physiological activities, classifications of bio-degradable elastomers are as follows [11]:

1. The elastomers suitable for long term physiological contact.
2. Bio-degradable elastomer for a determined time of contact.

Many tissues in the body have elastomeric properties like long range elasticity and dynamic properties [12]. Repair of such tissues projects to the development of elastomeric scaffolds that can sustain and recover in deformed state without disturbing the surrounding tissues [13].

Based on the source of precursor material, bio-elastomers can be further classified into three categories as:

1. Naturally extracted or bio-synthetic bio-elastomers; majority of protein and peptide based elastomers are included under this head.
2. Bio-synthetic degradable polyesters and polyhydroxyalkanoates.
3. Chemically synthesized bio-degradable elastomers.

Both linear and branched polyethylene (PE) can be used as a matrix and the resultant semi-flexible bio-composites have been found to give rise to a higher modulus [14]. Generally, if a reactive filler system is embedded, the polymeric matrix might be affected by the filler through reduction of molecular weight during composite processing. Formation of a shell of polymer around the particles (trans-crystallization, surface induced crystallization, or epitaxial growth) and changes in conformation of the polymer due to particle surfaces and variation in inter-particle spacing can markedly influence the characteristics of the matrix [15].

Creating nano-fibrous scaffolds with material dimensions at the nanometer scale allows for the efficient replication of the physical structure of natural bio-nano-composites. Biomaterials fabricated as nano-fibers can positively influence the physical and mechanical performance of the biomaterial scaffolds. Scaffolds fabricated from nano-fibers have been considered for use in the engineering of cartilage and wound healing applications. Some advantages of employing nano-fibrous biomaterials in tissue engineering, includes formation of polymeric nano-fibrous matrix for cellular adhesion, formation of neo-tissue and growth into a mature tissue in a bioreactor [16]. Tissue engineering process employs a nano-structured polymeric scaffold material. Advantages of nano- and bio-composites include physical mimicking of natural polymers, ease of surface functionalization, and improving mechanical properties of scaffolds of such materials can serve as a medium through which diffusion of metabolites can occur freely. Synthetic bio-polymers and their composites have been playing a pivotal role for making synthetic and semi-synthetic bio-nanocomposites. The advantages they offer includes that they can be tailored to provide a wide range of properties e.g., reduced immunogenicity, simplicity of processing and ease of sterilization. Classical example is PLGA possessing thermoplastic elastomeric characteristics and it can also easily be degraded by simple hydrolysis. PLA is an α-polyester having two enantiomeric form: D-lactide and L-lactide. PLA has certain advantages including tensile strength up to 60 MPa which is very useful for reinforcement. The high mechanical integrity possessed by these classes of polymers is often used as such for elastomeric body components and cartilage tissue engineering. The degradation and resorption kinetics need to be designed and controlled in such a way that the nano-structured scaffold retains its structural and functional integrity. Composite scaffolds consist of two or more materials. These materials together produce scaffolds that ideally draw from the properties of the individual properties of components [17]. Most of the natural polymers like chitosan and collagen display significantly lower values of tensile strength and elastic modulus, even upon cross-linking, as compared to the synthetic class of biopolymers. This in turn unveils scopes for integrating properties of synthetic and natural elastomeric composites. Their combination can overcome several drawbacks of individual components.

Chitosan and collagen based bio-nanocomposites alone could not provide higher tensile strength and enhanced mechanical properties. Nano-structured three-dimensional scaffold materials are promising options which can mimic natural soft and semi-soft tissues and facilitate growth and vasculature resulting in new tissue regeneration. For example, hydroxylapatite (HA)/collagen composites are usually prepared by self-assembly processes that can promote such tissue regeneration. Spinning of chitosan nano-fibers is a challenging task as chitosan is extensively intermolecular hydrogen bonded. As the polymer concentration is enhanced, the number of direct interacting chains increases. Thus chitosan forms a three dimensional crosslinked network and viscous gel. Addition of PEO can reduce the viscosity of chitosan and makes it spinable. Chitosan has extensively been used in wound dressing systems and tissue engineering applications. Another natural biopolymers namely, gelatin, and a synthetic biodegradable polymer, PCL, can form a favorable composition. The gelatin provides hydrophilic and cellular affinity enabling continuous release of the protein from the scaffold to create a favorable condition for cell attachment and proliferation. It gives superior mechanical properties and permits cell penetration.

Polyether and polycaprolactones are very useful precursors for preparation of elastomeric matrix of synthetic bio-nanocomposites.

3 Advancement on Elastin Based Elastomeric Bio-Nanocomposites

Elastin is an extra-cellular matrix protein which provides elasticity of tissues and organs such as lungs, muscle, and blood vessel. Elastin is mainly derived from tissues like endothelial cells, fibroblasts, chondrocytes etc. The precursor of elastin is known as tropoelastin which can self-assemble depending on physiological condition. The well known phenomenon associated with elastin is coacervation, a second order phase transition which gives proper alignment of elastin tissues [10]. The intermolecular crosslinked structure in elastin can be typically encoded by single copy gene. Coacervation process can be affected by various parameters like temperature, protein concentration, salt concentration, pH etc. Different living beings have similarities in the structure of their respective elastin proteins.

There exist elastin-laminin receptors in various types of biological cells. The receptors contain two subunits of molecular weight 61 and 55 kDa, respectively. The subunit of 67 kDa is a binding site for elastin and laminin [18]. The binding for both tropoelastin and the bound membrane at 55 kDa receptor site is significantly lower helping in release of tropoelastin. The structural features of elastin can be represented as shown in Fig. 2.

Elastin contains uncommon amino acid sequence, making it highly insoluble and resistant to fracture and rupture. Glycine and proline residues have major contribution in the sequence. Elastin has two natural verities, named as α-Elastin and κ-Elastin. The entire protein consists of a repeating polypeptide chains, used as a model of protein structures. Elasticity, glass transition temperature and coacervation are the most important parameters those determine their properties. For ideal elastin chain, energy is consumed by the backbone and recovered by the

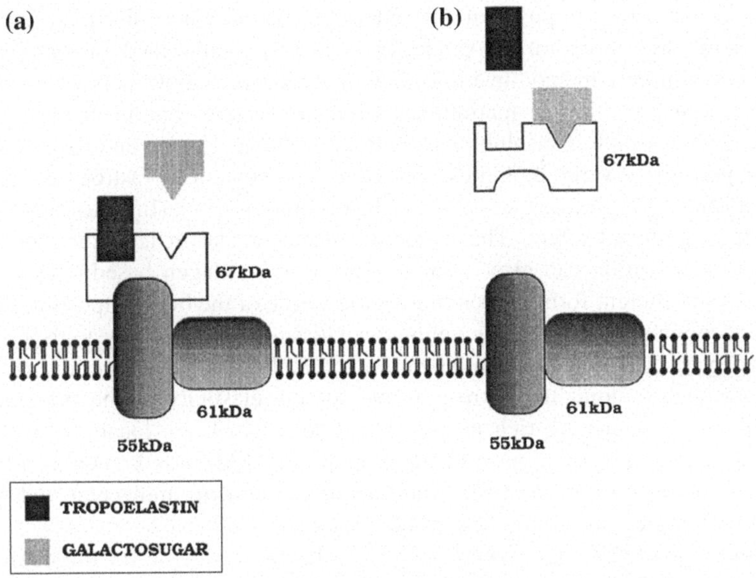

Fig. 2 Pictorial depiction of a model for the elastin-laminin receptor consisting of three subunits. Tropoelastin usually binds to the complex via elastin binding protein [19]. [Reproduced from B. Vrhovski, A.S. Weiss, Biochemistry of tropoelastin, Eur. J. Biochem, 1998, 258, 1–18]

relaxation process. It is surprising that its glass transition temperature (T_g) depends on the elastin content. In dehydrated state, its T_g is 200 °C and at 30 % dehydrated state, T_g is approximately 30 °C. Upon heat treatment, elastin becomes insoluble. Polypeptides in elastin chains lacks in hydrophobic domain. In its structure it contains type II ß-turns. By using centrifugal action, phase separation can be observed. During the turn, segmental motion increases the entropy of the whole system, which is responsible for the elasticity of the elastin itself [20]. During stretching, the damping of this system is also significant. The decrease in entropy can be compensated by release of water, which gives self-assembled structure to align tropoelastin molecule. An elastic ligament is majorly made by elastin having rod-like structure. Molecular dynamics predicts that ß- spiral structure are highly unstable. From both NMR data and MD (molecular dynamics) simulations it has been proven that the β-turns are characterized by an extensive sliding; that is, they are inter-converting to each other along the chain. This phenomenon is proposed as one of the possible sources of entropy in the relaxed state of elastin [21].

C. Hillery and S. Greenwald have carried out dynamic experiments on purified elastin samples. The elastin tissues were cycled at 20 Hz, under a load which was equivalent to arterial pressures. The fracture of elastin tissues occurred after 5–10 $\times$ 10^6 cycles. It is evident that at 1 Hz, a significant tolerance of dynamic strain has been observed as the number of cycles of application of dynamic strain is increased [22].

Elastin fibers are composed of amorphous elastin and micro-fibrils; those adopt honey-comb like structure. Tropoelastin is mainly synthesized in endoplasmic reticulum with certain modifications. This component mainly acts as a scaffold material, assisting fiber formation. Crosslinked structure has been observed in mature elastin, which is insoluble and extremely stable. Elastin and its derivatives mainly perform several biological functions in chemotaxis, astrocytes, glioma cells, arterial cell, protease activity etc. Incorporation of elastin into bio-systems generates significant effect. The major drawback of elastin is calcinations phenomena. Neucleation can also occur at elastin and collagen based sites. Elastin may exist in different forms depending on the nature of the bio-composites. Elastin has good compatibility with de-cellularized tissues, but these systems lacks in higher degree of purity. De-cellularization is mainly done by extraction methods. The hydrophobic domains are responsible for the elasticity of the protein. The hydrophobic domains are rich in glycine residues which are localized in the N-terminal and C-terminal regions of the protein [10]. Elastin is known as a fractal protein even with a comparatively short sequence showing molecular and supramolecular features resembling the whole protein.

Far-ultraviolet (UV) CD spectroscopy is a handy technique for the conformational studies of bio-molecules like peptides and proteins. This technique is based on the observation that different secondary structures show different characteristic CD spectra. Conformational analyses of α-helices as well as β-sheets are easily identified by CD spectroscopy. The spectral characteristics of conformation of elastin structures are solely dependent on solvent and temperature. The CD spectra are assumed to be a linear combination of all the different secondary structures of elastin. NMR spectroscopy is also extensively used, but due to the timescale of NMR experimentation, it is unable to predict the individual contribution of each conformers of elastin. Thus, the NMR spectroscopy provides the average preferred secondary structure of the polypeptide sequence of elastin. The polypeptide chains oscillate between extended conformations is the key player for the elasticity of elastin. A rapid flipping is suggested between extended and folded conformations [23]. The temperature induced phase transition i.e. coacervation process of elastin is utmost important. This process is completely reversible, leads to formation of two phases: one is rich in protein content, another is rich in aqueous solvent. Soluble peptides released from elastin via proteolytic decomposition, leading to slow aggregation due to the mutated microenvironment. The human tropoelastin gene controls the supramolecular structure of the entire protein which can be reproduced by the single domain of elastin.

E. Wang, S. H. Lee, S.W. Lee synthesized biomimetic matrices using elastin like polypeptides (ELP) and hydroxyapatite (HA) based bionanocomposites. Enhancements in properties of such nanocomposites with improved injectibility and mechanical properties have been observed. ELP exhibits specific interaction to bind and disperse HAP nano-particles. When these bio-composites are incorporated into calcium phosphate cements improvement in adhesion properties has been observed [24].

Mimicking of elastin based materials is an interesting topic for soft tissue engineering [25]. Various modification of elastin structure is underway of research interest. In near future, elastin based nanocomposites will play a pivotal role in the field of semi-synthetic elastomeric bio-nanocomposites.

4 Advancement on Cellulose Based Elastomeric Bio-Nanocomposites

Cellulose is the most abundant natural polymer of this planet. Cellulose and its different derivatives are extensively used during the development of Science and Technology. Chemically cellulose is a carbohydrate, which is obtained by the hydrolysis of mono-saccharides. In its structure, two glucose units are joined by a ß-1, 4 glycosidic linkages. Extensive hydrogen bonding makes a network structure of polymeric cellulose highly stable. This hierarchical structure is built up by smaller and mechanically stronger entities consisting of native cellulose fibrils. These fibrils interact strongly and aggregate to form the natural or native cellulose fibers. The lateral dimension of these fibrils depends on the source of the cellulose but it is typically of the order of a few nanometres. The chemical structure of cellulose is represented as shown in Fig. 3.

Structure of cellulose and its own crystalline organization determine the mechanical properties. Researchers have already proposed that wood contains around 10000 glucopyranose units and cotton contains 15000 repeating units [26]. The basic structural component of cellulose is cellulose microfibril that is formed during the biosynthesis. Actually, the chains of β-1,4-D-glucosyl residues aggregate to form a fibril, which is a long thread-like bundle of molecules laterally stabilized by intermolecular hydrogen bonds. The repeating units have 20 nm lengths. FTIR and XRD studies clearly reveal that the main portion of cellulose is constituted by crystallites interspersed with amorphous regions having low degree of order with parallel chains. When the cellulose is precipitated out of alkaline

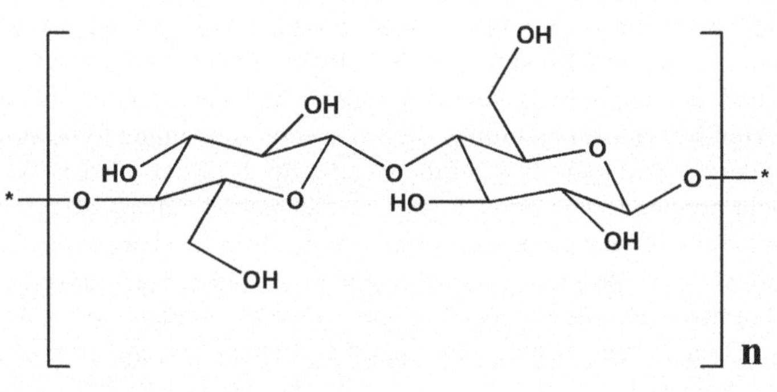

Fig. 3 Chemical structure of cellulose molecule

Table 1 Mechanical properties various cellulose based nano-fibers [28, 29]

Fiber type	Young's modulus (GPa)	Specific Young's modulus ($GPa·Mg^{-1}m^3$)	Breaking strength (GPa)	Breaking strain (%)
Flax	27.0	18.0	0.81	3.0
Jute	25.8	17.2	0.47	1.8
Hemp	32.6	21.7	0.71	2.2
Ramie	21.9	14.6	0.89	3.7

[Reproduced from **a** current international research into cellulose nanofibres and nanocomposites-S.J. Eichorn, A. Dufresne, M. Aranguren et al., J. Mater. Sci. 45 (2010) 1–33, and **b** W.E Morton, J.W.S Harle, Physical properties of textile fibers, William Heinnmann Limited, London, 1993]

solution it is known as regenerated cellulose. Inter-conversion between two forms of cellulose is performed by strong alkaline hydrolysis [27]. The material used for cellulose nano-filler contains mainly native cellulose (Cellulose I) which is extracted by traditional bleaching treatment of lignocellulosic fibers. Cellulose I is responsible for mechanical properties due to its inherent high modulus and crystalline nature. Table 1 displays the mechanical properties of various cellulose based nano-fibers. Incorporation of such nano-fibers into suitable bio-elastomeric matrices gives promising improvement of mechanical properties of the resultant nanocomposites.

Several methodologies regarding preparation of cellulose fibers have been adopted. The fibers are first milled and are made to undergo alkali treatment using NaOH followed by bleaching using sodium hypochlorite ($NaClO_2$)/acetate buffer solution. The cellulose moieties remain intact. The bleached fibers are lead to acid hydrolysis and disintegrated by mechanical shearing. Then the cellulosic nano-whiskers as highly pure single-crystal are extracted. These nano-fibers don't show surface regularity, as it contains both amorphous and crystalline domains. The surface morphology of typical Baggase whiskers has been evaluated by TEM as represented in Fig. 4. The amorphous regions are found to be randomly oriented compared to the crystalline regions.

The equatorial positions of the glucopyranose residues play the crucial role as it stabilizes the structure of cellulose. Extensive intra and intermolecular hydrogen bonding also causes insolubility in water and accordingly increase its mechanical rigidity. The mechanism of cellulose nano-crystal production has been proposed by De Souza Lima and Borsali [30]. The hydroxonium ion penetrates into the cellulose and the amorphous region promotes the cleavage of the glycosidic linkage giving individual crystallites. Reaction time is assumed to be an important parameter during acid hydrolysis of the wood-pulp. It has also been observed that sulfuric acid prepared nano-crystals present a negatively charged surface, due to the esterification of surface hydroxyl groups leading to charged sulfate groups [31]. Nano-whiskers promote reinforcement in all bio-nanocomposites. Its size, shape and dimensions solely depend on the source of cellulose. It is customary to write that cellulose nano-whiskers are devoid of chain folding, so they contain a few crystal defects.

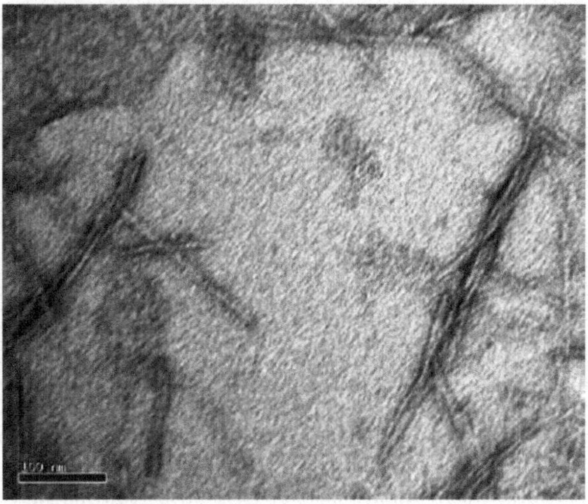

Fig. 4 TEM photomicrograph of Baggase cellulose whiskers [33], [J. Bras, M.L. Hassan, C. Bruzesse, E.A. Hassan, N.A. El-Wakil, A. Dufrense, Industrial Crops and Products, 32(2010) 627–633]

Micro-fibrillated cellulose (MFC) belongs to a special classes of cellulose derivative, obtained from wood cell by means of mechanical degradation. MFC belongs to the class of long and flexible nanoparticles which is composed of individual cellulose micro fibrils. It exhibits a typical web-like structure [32]. Homogenization treatment also gives a diluted dispersion of MFC having a sol–gel structure. Micro fluidizer is also used to destroy the fibril structure of cellulose pulps producing MFC. During production it consumes a large amount of energy, and it has been an obvious drawback of this process. An optimization process for micro-crystalline cellulose was already established depending on various parameters like concentration, temperature etc.

Another facile route for the preparation of MFC is enzymatic hydrolysis of the cellulose and its derivatives, e.g. cellulose can undergo hydrolysis by pure C-type endoglucanase. Cellulose based bio-nanocomposites has immense application as high performance materials. Improvement of mechanical properties of a model elastomeric bio-nanocomposite by the incorporation of MFC matrix can be statistically proven from literature [Table 2].

A large volume of interfacial materials are associated with nanocomposites in their micro-structure. Thus bulk properties differ significantly from the bulk elastomeric materials. It is worthy to note that prediction of the properties of cellulose nano-fiber based nanocomposites has certain limitations also. However cellulose nanoparticles like whiskers and MFC imparts significant reinforcement in composites. They have lots of advantages like—low weight, biodegradability, minimum energy consumption, lower abrasion along with some drawbacks including moisture absorption and incompatibility [35]. Preparation of cellulose based nanocomposites follows two methods namely solvent casting and mechanical extrusion. During nanocomposite preparation two methods are generally employed to modify surface of cellulose e.g., coating of the surface using

Table 2 Table of tensile modulus (E_i, MPa) and shrinkage (r_i, %) calculated for cellulose nanoparticles/NR nanocomposites during successive tensile tests [34]

Sample	E1	E2	E3	E4	E5	E6	E7
NR	0.64	0.58	0.36	0.27	0.20	0.17	0.16
NR-W1	1.58	0.75	0.38	0.27	–	–	–
NR-MF1	1.50	0.79	0.32	0.22	–	–	–
Sample	r_1	r_2	r_3	r_4	r_5	r_6	
NR	64	133	194	252	303	–	–
NR-W1	50	105	156	–	–	–	–
NR-MF1	35	84	117	–	–	–	–

[Reproduced from A. Bendahou, H. Kaddami, A. Dufresne, European Polymer Journal, 46(2010) 609-620]

specific surfactants or grafting of the hydrophobic chains onto the fibrils. Use of cellulose nanoparticles at the dry state is a challenging work as extensive hydrogen bonding forms agglomerated structures. Formation of rigid whiskers is also possible by solution cast method. Cellulose nanocomposite films are prepared in aqueous solution due to its inherent stability in water soluble polymers. Polymers in the form of latex can also be used for emulsion systems which also include natural rubber (NR), Poly-vinyl chloride (PVC) and Poly-vinyl acetate (PVAc). In non-aqueous media mainly suspension systems are used. Tunicin based whiskers are mainly used in this methods. Acid hydrolysis of MCC produces whiskers with controlled porosity. Typical nanocomposites have been prepared using cellulose tunicin whiskers in DMF solution exhibiting better thermal and mechanical properties [36]. Another method for preparation of tunicin whiskers is acid hydrolysis using HCl.

Whiskers behave differently when modified in protic solvents as protic solvents disrupt hydrogen bonding and it promotes the dispersion as mentioned earlier. Surface chemical modification is another promising technique for stabilization and dispersion of the nano-crystals. Grafting agents also improves the compatibility of cellulose fibers in bio-polymers. MFC can be chemically surface modified using silanes and isocyanates showing improved dispersion in acetone and THF. Silane coupling agents like 3-trimethoxy aminosilane used for the surface modification of cellulose nano-crystals causes a change in the hydration behavior of nano-crystals [37]. Polysaccharide nano-crystals can be transforms into a co-continuous material, improving bondability between the filler and polymer. Polycaprolactone is also a promising material with a huge scope for grafting purpose. “Grafting from” method is also a versatile one for surface modification. This results in change of crystalline features. It is evident that aspect ratio plays a vital role for reinforcement which varies significantly depending on the sources of materials [38]. Dynamic mechanical analysis (DMA) reveals important clues on complicated relaxation processes involved in such materials. But still, origin of the mechanical reinforcing effect is poorly understood for such systems. Characterization of the prepared nanocomposites in a precise and repeated manner is very essential for structure- property correlation. Transmission Election Microscopy is an essential tool for analysis of the morphology of such nanocomposites.

Cellulose with tendon produces bio-based elastomeric nanocomposites. These bio-composites give both biocompatibility and biodegradability including excellent mechanical properties.

Highly deacetylated chitosan and wood cellulose whiskers have been used to make elastomeric bio-nanocomposites. They show improved viscoelastic and mechanical properties. These materials are extensively used in biomedical and packaging applications [39].

Acetylated microfibrillated cellulose and PLA forms useful elastomeric bio-nanocomposites. A dramatic improvement of compatibility of fibre matrix interface has been observed. Accordingly, thermal, mechanical and viscoelastic properties of such nanocomposites are found to be significantly enhanced [40].

In near future, these forms of cellulose reinforced elastomeric bio-nanocomposites will occupy a pivotal role for bio-medical applications.

5 Advancement on Chitin Based Elastomeric Bio-Nanocomposites

Chitin and its derivatives are now proven to be a promising representative in the area of bio-polymers [41]. Recently, a successful method of interest is to develop chitin and chitosan based biomaterials. Hydroxy and acetamide derivatives of chitin are interesting candidates for bio-degradable elastomers. Chitin with pyranose ring has rigid cellulosic structure. Chitin is chemically composed of 1, 4-2–acetimido-2-deoxy-β-D-glucose [42]. It occurs in the skeletal materials of insects and in bacterial cell wall. Chemically, it is an aminopolysaccharide having acetamido group at C-2 position. Nanocomposites of chitin have several advantages being non-toxic, bio-degradable and anti-bacterial. Chitin as a material has vast unexplored commercial applications. Novel modification techniques of PU structure by incorporating starch, polycaprolactone and chitin as a chain extender is also a thrust area of research [43]. The hydroxyl and acetamido functionalities of chitin based bio-polymers allow chemical reaction with conventional diisocyanates or urethanes. Actually, it is evident that the –OH group at C-6 position of chitin molecule react with the pre-polymer giving final hybrid-PU. The chemical structure of chitin is represented in Fig. 5.

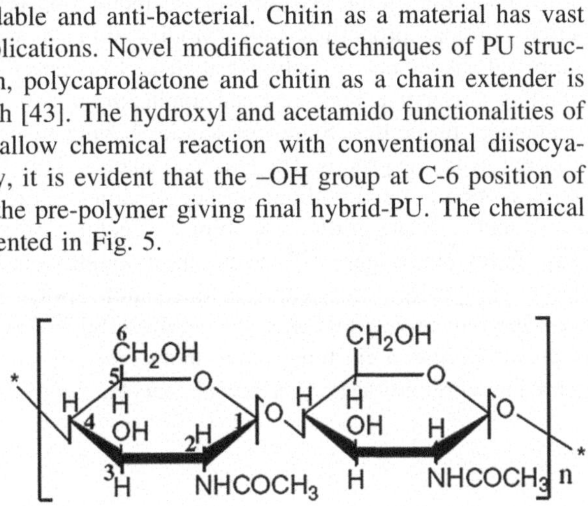

Fig. 5 Chemical structure of chitin bio-molecule [44]

It is evident that, hybrid-PU shows optimum hydrophobicity than virgin chitin based nanocomposites. Improvement in bio-compatibility of these systems has been clearly indicated in literature [44]. Structure—property relationship for elastomers based nano-structured chitin solely depends on hydrogen bonding, hydrophobicity and chitin content in the elastomeric matrix [45]. Chitin and polyurethanes gives a favorable combination to form composite structures of PU having micro-phase separated structure leading to good compatibility. Successful methods of increasing bio-degradability of such materials are developing and there have been an increasing trend of using chitin with other elastomeric nano-composites. Chitin in combination with polysaccharides is bio-degradable in bio-active environments, where bacterial population is higher. The major drawback of chitin is its inferior solubility, which mimics its application to replace the existing members of elastomeric bio-nanocomposites [46]. Chitin with pyranose ring has rigid cellulosic structure. Overall chitin based nanocomposites have sufficient scope as a potential elastomeric bio-nanohybrid material.

6 Advancement on Starch Based Elastomeric Bio-Nanocomposites

Starch is one of the most important and abundant bio-molecules since the genesis of this universe. Various types and classes of starch molecules are found in the animal and plant kingdom. Starch molecule can be modified by esterification reaction, leading to numerous applications [47]. Modified starch can also be blended with natural rubber latex, producing a typical elastomer based bio-composites. Starch-NR composites are known to have superior thermal and mechanical properties along with unique morphological characteristics [48]. The crystal structure of starch disintegrates and uniformly dispersed in the NR matrix. It can be assumed that, improved inter-phase interaction between NR and starch gives better physical properties [49]. Modification of starch reduces its size with improved dispersion in polymer matrix. Cornstarch-NR bio-composites are an effective alternative for tyre-tread applications [50]. Various factors control the mechanical properties. Starch molecules having higher sizes are bonded through adjacent hydroxyl groups. The tear resistance and elasticity usually degrades when starch is incorporated without surface modification [51]. Another approach is the use of methacrylate grafted copolymer of natural rubber and methyl methacrylate [52]. Polyvinyl acetate (PVA)/starch, poly-ethylene octane copolymer (POE)/ starch composites, poly-caprolactone/starch, polyester/starch, poly-lactic acid (PLA)/starch nano-composites are well established as elastomeric bio-nanocomposites [53]. Starch-elastomeric bio-nanocomposites has immense in application in medicine, agricultural and industrial aspects.

7 Advancement on Synthetic and Semi-Synthetic Elastomer Based Bio-Nanocomposites

To overcome several drawbacks of the available natural bio-elastomers, scientists have been engaged towards the synthesis of bio-elastomers with well optimized properties. Synthetic bio-elastomers can be associated with the certain advantages including high purity, good processibility, and optimized physical, chemical, mechanical properties. For a tissue engineered scaffold, the rate of degradation must match with the rate of tissue regeneration. Scaffolds should degrade chemically following hydrolytic or enzymatic degradation mechanism [11]. Several external factors influence the degradation pathway including types of chemical bond, crystallinity, hydrophillicity, pH and temperature. Degradation can also be assisted through the change of mechanical properties like, weight loss, surface porosity, residual monomer etc. Polyurethanes and polyphosphazenes have provoked a new era in the field of synthetic bio-elastomers. Apart from these, a handful of bio-polymers (block copolymers) have been found to exhibit thermoplastic elastomeric properties making them suitable matrices of choice for the preparation of elastomeric bio-nanocomposites.

7.1 Biodegradable Polyurethanes

Segmented polyurethanes (SPU) belong to a distinct class of synthetic biomaterial well known over the last decade. SPU provides biodegradability via hydrolysis reaction with tailored physical properties as the properties can be varied from bio-stable to rapidly bio-degradable elastomer [54]. It is customary that bio-degradable polyurethanes (PUs) are designed to undergo bio-degradation via *in vivo* mechanistic pathways [55]. SPU comprises of two segments namely, hard and soft segments. The hard segment is composed of diisocyante and the chain extender, on the contrary, the soft segment contains an amorphous macro-diol. SPU provides uniform elastomeric properties. These materials possess tensile strength (TS) ranging from 4–60 MPa with an elongation at break of 100–950 %. Average degradation time is also significantly lower for SPU [56]. The main degradation products are α-hydroxyl acids, urethanes, urea fragments, lysine etc. Biodegradable SPU contains polyether or polyester macrodiol components. The polyol are mainly made of PEO, PPO, PCL, Polylactide and polyhydroxy butyrate. It has been found that PEO based PU are weak and amorphous in nature [57]. PEO can accelerate the degradation rate but PCL enhances the crystallinity of the matrix [58]. Commercially biodegradable SPU was launched as DegraPol (Trade name) which uses mainly lysine based macrodiol. The diisocyanate component has several limitations as biomaterial because it produces toxic products during degradation leading to inflammatory response in the body. Newly invented lysine methyl ester diisocyanate (LDI) and 1, 4-diisocyanatobutane is extensively used for bio-medical application [11, 59].

Recently, degradable chain extenders have been invented which are mainly diammine based. Phosphate ester linkage containing lactic acid based chain extenders are also bio-degradable [51]. Bio-degradable SPU is designed to undergo enzymolysis/hydrolysis [49]. The degradation is controlled mainly by the degree of cross linking. Chemical links present in soft segment hydrolyze more quickly than that of the hard segment. Polyurethans can be surface modified using poly (N-vinyl pyrollidine) i.e. PolyNVP which provides well known bio-compatibility. These biomaterials have high water solubility also. These precursors have applications in coated catheters [60].

The mechanical properties and resistance to urinary encrustation of IPN network formed by polyurethane (PU) and PMMA have been studied by Jones et al [61]. Maximum elongation at break has been observed for pure PU and it decreases with increasing PMMA content. These biomaterials provide high degree of resistance to compression of the ureter in comparison with native PU [61].

Bioactive glasses are a promising reinforcing material used in the formation of several types of bio-nanocomposites. Ciobanu et al. [62] has investigated the hydroxyapatite formation on the porous surface of polyurethanes. Using solvent casting method, they have developed three dimensional porous scaffold materials. Then HA was coated on the scaffold has been precisely characterized using XRD, SEM. These elastomeric bio-nanocomposites containing uniformly coated HA can be applied to obtain uniformly coated scaffold material [61].

Zuber et al. [63] have synthesized chitin based polyurethane/clay nanocomposites. It has been observed that interaction among the clay and the polymer chain improves their dispersion in the PU matrix. Optical microscopy clearly reveals that the MMT clays have been well dispersed and intercalated layers of clays have been formed [63].

Wu et al. [64] have synthesized novel high strength elastomeric bio-nanocomposites using microcrystalline cellulose and PU. Significant improvement in tensile strength, stiffness and strain to failure properties has been observed. Due to extensive hydrogen bonding as well as covalent bonding between PU and cellulose network such improvement in properties has been observed [64].

Wang et al. [65] have embedded starch nano-crystals and cellulose whiskers in waterborne polyurethanes. It is evident that the Young's modulus and tensile strength of such elastomeric bio-nanocomposites along with thermal properties have been significantly improved than that of pristine PU. Actually, polysaccharide molecules and whiskers forms a hydrogen bonded network, giving synergistic effect. They belong to the class of eco-friendly elastomeric bio-nanocomposites [65].

Chen et al. [66] have prepared elastomeric bio-nanocomposites using starch nano-crystals and PU. Significant enhancement in tensile strength, elongation at break and Young's modulus has been observed up to an optimum loading of starch nano-crystals [66].

7.2 Bio-Degradable Polyphosphazenes

Polyphosphazenes belong to a special class of inorganic polymer used in biomedical applications. The inorganic backbone consists of alternating phosphorous and nitrogen atoms with adjacent groups. The properties of this polymer are entirely governed by the adjacent substituents [11]. During degradation, its mainly convert itself into ammonia, phosphates and alkyl moieties. The general structural representation is given as:

$-[N = PR_2]-_n$

It is evident that the structure of polyphosphazenes has a high degree of freedom of bond rotation and low glass transition temperature. Most PN derivatives show elastomeric properties. Small group containing derivatives renders lower T_g values. The movement of chains can also be restricted by inter-atomic interaction. Amine substituted polymers generally exhibits higher T_g values. The flexibility of PN backbone arises as it undergoes structural changes in the solid state [67]. Biodegradable PNs comprise hydrophilic substituents in a maintained ratio. This active group can be grafted or trapped in the polymeric matrix. Imidazole, glucose and polyether groups are incorporated to favour hydrolysis. PN also degrades via erosion mechanism depending on several factors such as lability of bonds, water permeability etc. Polyphosphazenes are classified into two categories e.g., PN substituted with amines and substituted with alcoholic moieties.

Aminated PNs are extensively used in biomedical applications [68]. Alkoxy substituted PNs, such as glyceryl substituted ones are bio-compatible in both in vitro and in vivo conditions.

Glycine containing polyphosphazenes derivatives have also been synthesized which are fully biocompatiable. They support cell adhesion and growth [69, 70]. Degradable polyphosphazenes/poly (α-hydroxy ester) blends have also been studied. The major degradation products associated with this material consist of phosphate, glucose, ammonia etc.

Liyan Qiu [71] has synthesized chitosan coated polyphosphazenes-Ca^{2+}-hydrogel i.e. a novel class of elastomeric bio-nanocomposites. Polyphosphazenes have been dropped into $CaCl_2$/chitosan gelling solution. The interaction among the polymeric components have revealed by using turbidimetric titration method. It has been observed that drug loading efficiency of such elastomeric bio-nanocomposites is significantly high [70].

Nukavarapu et al. [72] have synthesized polyphosphazene/nano-hydroxyapatite composite microsphere scaffolds. These composite have been fabricated into a three dimensional microporous scaffolds. These elastomeric bio-nanocomposites provide compressive modulus of 46–81 MPa with mean pore diameter 86–145 μm. These scaffolds have certain advantages including good cell adhesion inducing proliferation of cells [72].

Allcock et al. [68] have also studied radiation crosslinking of hydrogel based on (methoxyethoxyethoxy) polyphosphazenes. It has been observed that increasing radiation dosages enhances the degree of crosslinking. Superior mechanical properties of such elastomeric bio-nanocomposites have been observed [68].

Andrianov [73] have synthesized ionically crosslinked polyphosphazene hydrogel microsphere. The synthesized microdroplets are highly stabilized when crosslinking with Ca^{2+} ions. This microsphere can be extensively used as vaccine delivery vehicles [73].

It is already established that glycerol and sebacic acid monomers produce a novel elastomeric material, poly (glycerol sebacate) PGS a useful elastomeric material for soft tissue engineering [74]. The degradation products of PGS are non-toxic. PGS has certain advantage e.g. simple synthesis procedure, tailored mechanical and biodegradation characteristics. Extensive research on preparation and modification of novel PGS are still underway.

Elastomeric bio-nanocomposites of poly-vinyl pyrollidine, chitosan and PVA belong to a novel class of elastomeric bio-nanocomposites. The thermal and mechanical properties of such novel bio-nanocomposite are considerably high [75].

Citric acid is a non-toxic metabolic product of the body, and it can form network structure. Several advantages of poly (diol citrate) elastomers include non-toxicity, availability and lower cost. Biodegradable crosslink's can introduce elasticity in these systems. Various diols provide flexibility in the elastomer. Intermolecular hydrogen bonding can enhance mechanical properties of the elastomer. It's synthesis can be conducted under very mild condition. Citric acid reacts with various diols without using any catalyst. Improvement in tensile properties has been observed e.g., T.S increases up to 11.5 MPa with significant improvement in percentage elongation [76].

Chung et al. [77] have precisely investigated the synthesis, characterization and biological response of citric acid based elastomeric bio-nanocomposites. They have prepared a novel elastomeric bio-nanocomposites using poly (1, 8-octane-diol-co-citrate) and hydroxyapetite nanoparticles. It has been observed that mechanical properties of such elastomeric bio-nanocomposites including strength and stiffness is remarkably improved compared to the virgin polymer. The tissue response has been found to be dependent on the content of HA [77].

Webb et al. [78] have fabricated novel elastomeric bio-nanocomposites using poly (L-lactic acid) and poly-diol citrate. A significant improvement in tensile strength, elongation at break and modulus has also been reported. This system has potential to serve as novel elastomeric bio-nanocomposites for soft tissue engineering in near future [78].

Poly ether esters (PEE) belong to the group of novel elastomeric material, consisting of soft segmented polyethers and hard crystalline segment of polyesters. The ratio of ether to ester dictates wide range of mechanical properties and biodegradability. It's synthesis procedure have been widely reported and it is synthesized using PEG 1000, 1, 4 butane-diol and dimethyl terepthlate as the precursor materials [79]. The families of silk-elastin like polymer hydrogels are newly invented, consisting alternating blocks of silk-like and elastin-like blocks [80]. These types of polymers can be categorized under hydrogels as well as under bio-nanocomposites. These typical elastomeric bio-nanocomposites provide solubility, higher mechanical strength, immunogenicity and *in vivo* degradation by alternating the amino acid sequence. The properties of these typical materials can be altered by several factors including ionic strength, polymer concentration, cure time etc. It is already established that macroporus scaffolds can be fabricated from a biocompatible and bio-degradable elastomers [81]. Porous collagen matrix can be cross linked using elastomeric hyaluronic acid and fibro-nectin results improved cell attachments [82]. Elastomeric proteins have certain novel properties also e.g., long range elasticity, reversible deformation and high resilience. Synthetic polypeptides have recently applied to prepare multi-block protein based copolymers. Nanoclay based elastomeric polypeptide composites are found to be relevant in the field of bio-nanocomposites. It gives the advantages like bio-degradability and gas and water permeability. These have potential applications in living bio-systems. Immunogenic response of peptide sequences limits the choice of polypeptides for biomedical applications.

Silicone rubber-hydrogel bio-nanocomposites have been extensively investigated by P. Lopour et al. A matrix has been developed composed of polysiloxane and hydrogels. The hydrogel phase consists of crosslinked poly (2-hydroxy ethyl methacrylate). High water permeation rates have been obtained with retention of mechanical properties [83]. Elastomer based bio-nanocomposites containing silicone rubber matrix and particulated hydrogels is an emerging field of biomaterial research interest. Several factors including size, shape and aggregation of hydrogels those affect tensile properties and tear strength of such composites. It has also been observed that chemical composition of such bio-nanocomposites affect its mechanical properties [9]. In near future, silicone rubber based biomaterials are likely to serve to the entire mankind.

7.3 Bio-Inspired Preparation of Elastomeric Bio-Nanocomposites

Recent patents in the field of bio-inspired, bio-mimicked and bio-hybrid materials are an emerging facet of research. Several international patents have already been filed based on the development of such novel materials [84–88]. Biological preforms may begin to replace some classes of light duty composite materials. Collagen tissues are abundant in bio-systems which are capable of supporting tensile loads. Biomimicry is a well known method to exploit existing biologically derived materials with a nano-porous structure. Biopreform have been made from several plants like coconut, date palm etc. [89, 90]. These plants are dried under pyrolytic condition. Silicon and silicon carbide based ceramics have been made using this procedure.

Syntheses of peptide/peptide like molecule for in vitro and in vivo applications have already been patented [91]. Here, targeted protein expression can be tracked and properly monitored. Polymeric micelle for drug delivery has already been patented; a drug loaded sample contains triblock amino acids have been developed [92]. The blocks consist of one hydrophilic block, a crosslinked poly amino acid block, an uncrosslinked poly amino acid block. Surface functionalization method may be adopted for the attachment of the drug using PEG, HMPA as the novel monomer. Fluorescence resonance energy transfer method is used for their characterization. Use of amphiphilic multiblock copolymers has already been patented. A liposome is a bilayered sphere of amphiphilic molecule. Amphiphilic block copolymer contains both hydrophobic and hydrophilic block. Click chemistry is a new promising technique regarding the development of such kind of smart materials.

8 Conclusions

Elastomeric bio-nanocomposites have emerged as unique advanced material with potential biological applications. It is a challenging task for the researchers and scientists to unify characteristic properties of traditional elastomers, nano-scale reinforcements and biomaterial attributes within a single composite. A lot of these materials of this class have been invented by bio-inspired and bio-mimetic research in the allied areas. The naturally occurring bio-elastomeric matrices have been mainly protein and carbohydrate based materials having comparatively lower intermolecular force and glass transition temperature. The deficiencies of mechanical properties of the matrix are reinforced with various bio-based nano scale fillers preserving their bio-compatibility. These materials have wide range of applications including in soft tissue engineering, wound dressing, drug delivery, packaging etc. Because of the several draw backs of properties of purely bio based elastomeric bio-nanocomposites, synthetic and semi-synthetic elastomeric bio-nanocomposites are emerging on fast track. Elastomeric bio-nanocomposites have a promising future for the next decade as novel advanced materials.

References

1. http://en.wikipedia.org/wiki/composite material
2. Darder, M., Aranda, P., Hitzky, E.R.: Adv. Mater. **19**(10), 1309–1319 (2007)
3. Darder, M., Colilla, M., Hitzky, E.R.: Chem. Mater. **15**, 3774 (2003)
4. Glawe, D.D., Rodriguez, F., Stone, M.O., Naik, R.R.: Langmuir **21**, 717 (2005)
5. http://images.isiknowledge.com
6. Evans, S.L., Gregson, P.J.: Biomaterials **19**, 234 (1998)
7. de Groot, K., Geesink, R.G.T., Klein, C., Serekian, P.: J. Biomed. Mater. Res. **21**, 1375–1381 (1987)
8. Lagally, M.G.: Jpn. J. Appl. Phys. **32**, 1493 (1993)
9. Lapour, P., Plichta, Z., Volfova, Z., Hron, P., Vondracek, P.: Biomaterials, **14**(14), 1051–1055 (1993)

10. Lau, A.K.T., Hussain, F., Lafdi, K. (eds.): Nano and Biomaterials 1st edn. CRC press, Taylor and Francis group, USA (2010)
11. Shi, R., Chen, D., Li, Q., Wu, Y., Xu, X., Zhang, L., Tian, W.: Int. J. Mol. Sci. **10**, 4223 (2009)
12. Planeix, J.M., Jaunky, W., Duhoo, T., Czernuszka, J.T., Hosseini, M.W., Bres, E.F.: J. Mater. Chem. **13**, 252 (2003)
13. Suchanek, W., Yoshimura, M.: J. Mater. Res. **13**, 94 (1998)
14. Unwin, A.P., Ward, I.M., Ukleja, P., Weng, J.: J. Mater. Sci. **36**, 3165 (2001)
15. Rea, S.M., Bonfield, W.: J. Aust. Ceram. Soc. **40**, 43 (2004)
16. Zhang, Y.Z., Lim, C.T., Ramakrishna, S., Huang, Z.M.: J. Mater. Sci. **16**, 933 (2005)
17. Du, C., Cui, F.Z., Zhu, X.D., de Groot, K. : J. Biomed. Mater. Res. **44**(4) 407–415 (1999)
18. Daamen, W.F., Veerkamp, J.H.: J.C.M van Hest, T.H van Kuppevelt. Biomaterials **28**, 4378 (2007)
19. Vrhovski, B., Weiss, A.S.: Eur. J. Biochem. **258**, 1 (1998)
20. Venkatachalam, C.M., Urry, D.W.: Macromolecules **14**, 1225 (1981)
21. Debelle, L., Tamburro, A.M.: Int. J. Biochem. Cell. Biol., **31**(2), 261–272 (1999)
22. Rachev, A., Greenwald, S.E.: J. Biomechanics **36**, 661 (2003)
23. Woody, R.W.: Adv. Biophys. Chem. **2**, 37 (1992)
24. Wang, E., Lee, S.H., Lee, S.W.: Biomacromolecules **12**, 672 (2011)
25. Silva, S.S., Mano, J.F., Reis, R.L.: Crit. Rev. Biotechnol. **30**(3), 200–221 (2010)
26. Vrhovski, B., Jensen, S., Weiss, A.S.: Eur. J. Biochem. **250**(1), 92–98 (1997)
27. Anderson, M., Johansson, L.S., Tanem, B.S., Stenius, P.: Cellulose **13**, 665 (2006)
28. Eichhorn, S.J., Dufresne, A., Aranguren, M., Macrovich, N.E., Capadona, J.R., Rowan, S.J., Weder, C., Thielemans, W., Roman, M., Renneckar, S., Gindl, W., Veigel, S., Keckes, J., Yano, H., Abe, K., Nogi, M., Nakagaito, A.N., Mangalam, A., Simonsen, J., Benight, A.S., Bismarck, A., Berglund, L.A., Peijs, T.: J. Mater. Sci. **45**, 1 (2010)
29. Morton, W.E., Harle., J.W.S.: Physical Properties of Textile Fibers 3rd edn. William Heinemann Limited, London (1993)
30. de Souza, M.L., Borsali, R.: Macromole. Rapid Comm. **24**, 4168 (1991)
31. Gray, D., Raman, M., Candanedo, S.B.: Biomacromolecules **11**, 551 (2009)
32. Rowell, R.M., Han, J.S., Rowell, J.S.: Cellulose **13**, 665 (2006)
33. Bras, J., Hassan, M.L., Bruzesse, C., Hassan, E.A., El-Wakil, N.A., Dufresne, A.: Ind. Crops Prod. **32**, 6273 (2010)
34. Bendahou, A., Kaddami, H., Dufresne, A.: Europ. Polym. J. **46**, 609 (2010)
35. Henrikson, M., Henrikson, G., Berglund, L.A., Lind Storm, T.: Europ. Polym. J. **43**, 3434 (2007)
36. Azizi Samir, M.A.S., Alloin, F., Durfence, A.: Biomacromolecules **6**, 612 (2005)
37. Lu, J., Askland, P., Drazel, L.T.: Polymer **49**, 1285 (2008)
38. Favier, V., Chazy, H., Cavaille, J.Y.: Macromolecules **289**, 6365 (1995)
39. de Mesquita, J.P., Donnici, C.L., Pereira, F.V.: Biomacromolecules **11**, 473 (2010)
40. Tingaut, P., Zimmermann, T., Suevos, F.L.: Biomacromolecules **11**, 454 (2010)
41. Chandra, R., Rustgi, R., Prog, J.: Polym. Sci. **23**, 1273 (1998)
42. Muzzarelli, R.A.A. (ed.) : Chitin 1st edn. Pergamon press, USA (1977)
43. Barikani, M., Hepburn, C.: Cell. Polym. **6**, 47 (1987)
44. Zia, K.M., Barikani, M., Zuber, M., Bhatti, I.A., Seikh, M.A.: Carbohydr. Polym. **74**, 149 (2008)
45. Alfani, R., Lannancee, S., Nicholas, S.: J. App. Polym. Sci. **68**, 738 (1998)
46. Jang, M.K., Kong, B.G., Jeong, Y.I., Lee, C.H., Nah, J.W.: J. App. Polym. Sci. **42**, 3423 (2004)
47. Wang, Z.F., Peng, Z., Li, S.D., Lin, H., Zhang, K.X., She, X.D., Fu, X.: Compos. Sci. Technol. **69**, 1797 (2009)
48. Rouilly, A., Rigal, L., Gilbert, R.G.: Polymer **45**(23)7813–7820 (2004)
49. Shang, X., Fu, X., Yang, L.: J. Reinfo. Plast. Compos. **28**(3), 279–288 (2009)
50. Qi, Q., Wu, Y.P., Tian, M., Liang, G.H., Zhang, L.Q., Ma, J.: Polymer, **47**(11), 3896–3903 (2006)

51. Rosa, D.S., Lopes, D.R., Calil, M.R.: Polym. Test. **24**, 756 (2005)
52. Nakason, C., Kaesman, A., Eardrod, K.: Mater. Lett. **59**(29), 4020 (2005)
53. Yew, G.H., Mohd. Yusuf, A.M., Mohd. Ishak, Z.A., Ishiaku, U.S.: Polym. Degrad. Stabil. **90**(3), 488–500 (2005)
54. Skarja, G.A., Woodhouse, K.A.: J. Biomater. Sci. Polym. Ed. **12**, 851 (2001)
55. Santerre, J.P., Woodhouse, K., Laroche, G., Labow, R.S.: Biomaterials **26**, 7457 (2005)
56. Wang, Y., Kim, Y.M., Langer, R.: J. Biomed. Mater. Res. A **66**, 192 (2003)
57. Skarja, G.A., Woodhouse, K.A.: J. Appl. Polym. Sci. **75**, 1522 (2000)
58. Guan, J., Fujimoto, K.L., Sacks, M.S., Wagner, W.R.: Biomaterials **26**, 3961 (2005)
59. Ganta, S.R., Piesco, N.P., Long, P., Gassener, R., Motta, L.F., Papworth, G.D., Stolz, D.B., Watkins, S.C., Agarwal, S.: J. Biomed. Mater. Res. **64** A, 238–242 (2003)
60. Wetzels, G.M.R., Koole, L.H.: Biomaterials **20**, 1879 (1999)
61. Jones, D.S., Bonner, M.C., Gorman, S.P., Akay, M., Keane, P.F.: J. Mater. Sci. **8**, 713 (1997)
62. Ciobanu, C., Ignat, D., Luca, C.: Chem Bull., Politehnica. **54**, **68** (1)57–60 (2009)
63. Zuber, M., Zia, K.M., Mahboob, S., Hassain, M.: Int. J. Biological. Macromol. **47**, 196 (2010)
64. Wu, W., Henrikson, M., Liu, X., Berglund, L.A.: Biomacromolecules, **8**(12), 3687–3692 (2007)
65. Wang, Y., Tian, H., Zhang, L.: Carbohydr. Polym. **80**, 665 (2010)
66. Chen, G., Wei, M., Chen, J., Huang, J., Dufresne, A., Chang, P.R.: Polymer **49**, 1860 (2008)
67. Potin, P., Jaeger, R.D.: Eur. Polym. J. **415**, 341 (1991)
68. Allcock, H.R.: Acc. Chem. Res. **12**, 351 (1979)
69. Allcock, H.R., Kwon, S., Riding, G.H., Fitzpatrick, R.J., Bennet, J.L.: Biomaterials **9**, 509 (1988)
70. El-Amin, S.F., Kwon, M.S., Starnes, T., Allcock, H. R., Laurencin, C.T.: J. Inorg. Org. Polym. Mat. **16**(4), 387–396 (2006)
71. Qiu, L.: J. Appl. Polym. Sci. **92**, 1993 (2004)
72. Nukavarapu, S.P., Kumbar, S.G., Brown, J.L., Krogman, N.R., Weikel, A.L., Hindenlang, M.D., Nair, L.S., Allcock, H.R., Laurencin, C.T.: Biomacromolecules **9**, 1818 (2008)
73. Andrianov, A.K., Chen, J., Payne, L.G.: Biomaterials **19**, 109 (1998)
74. Wang, Y., Ameer, G.A., Sheppard, B.J., Langer, R.: Nat. Biotechnol. **20**, 602 (2002)
75. Chrissafis, K., Paraskevopolous, K.M., Papageorgiou, G.Z., Bikiaris, D.N.: J. Appl. Polym. Sci. **110**, 1739 (2008)
76. Yang, J., Webb, A.R., Pickerill, S.J., Hageman, G., Ameer, G.A.: Biomaterials **27**, 1889 (2006)
77. Chung, E.J., Qiu, H., Kodali, P., Yang, S., Sprauge, S.M., Hwong, J., Koh, J., Ameer, G.A.: J. Biomed. Mater. Res. A, 96(1), 29–37 (2011)
78. Webb, A.R., Kumar, V.A., Ameer, G.A.: J. Mater. Chem. **17**, 900–906 (2007)
79. Fairkov, S., Gogeva, T.: Macromol. Chem. **191**, 603 (1990)
80. Megeed, Z., Cappello, J., Ghandehari, H.: Pharma. Res. 19(7), 954–959 (2002)
81. Gao, J., Capro, P.M., Wang, Y.: Tissue Eng. **12**(4), 917–925(2006)
82. Doillon, C.J., Silver, F.H., Berg, R.A.: Biomaterials **8**, 195 (1987)
83. Lopour, P., Vondracek, P., Janatova, V., Sulc, J., Vaclk, J.: Biomaterials **11**, 397 (1990)
84. Layton, B.: Recent Pat. Nanotechnol. **2**, 72 (2008)
85. Pancrazio, J.J., Jr.Bey, P.P., Cuttino, D. S. et al., Portable Cell based biosensor system for toxin detection. Sens. Actuators B. Chem. **53** 179–185 (1998)
86. Marquez, G., Renn, M.J.: US20070280866A1 (2007)
87. Patel, R.A., Legum, B., Gogostri, Y., Layton, B.E: 8th biennial asme conference on engineering system design and analysis ESDA. Turin, Italy (2006)
88. Layton, B.E., Sastry, A.M., Lastoskie, C.M. et al. : Biotechniques, **37**, 564–573 (2004)
89. Foran, D.J., Chen, W.: US2003000358935 (2003)
90. Darby, P.C.: US20040126561A1 (2004)
91. Mandell, A.J., Selz, K.A., Shlesinger, M.F.: US2004000777829 (2007)
92. Breitenkamp, K., Sill, K.N., Breitenkamp, R.: US2006000396872 (2006)

Bio-Medical Applications of Elastomeric Blends, Composites

Valentine Kanyanta, Alojz Ivankovic and Neal Murphy

Abstract Elastomeric blends and composites are now extensively used for biomedical applications. Some of these applications include medical devices and utilities such as blood bags and cardiac assist pumps, and chronic medical implants such as heart valves and vascular grafts. These materials demonstrate superior biocompatibility, biostability and good mechanical properties, and as a result are now preferred over the use of metals and ceramics in most chronic medical implants applications. In addition, the chemical composition of these elastomeric blends and composites offers substantial opportunities for synthetic polymer chemists to tailor the structures to meet specific requirements. The current chapter discusses some of the recent developments in the use of elastomeric blends and composites for biomedical applications. The chapter also discusses the essential properties that materials used in these applications should possess in order to reduce the risk of severe allergic reactions in patients, implants being rejected by the host environment and premature failure of device and/or implants. An overview of the commonly used elastomer based products in biomedical applications and their fabrication/synthesis techniques is also presented.

V. Kanyanta (✉) · A. Ivankovic · N. Murphy
School of mechanical and materials engineering, University College, Dublin, Ireland
e-mail: valentine.kanyanta@e6.com

A. Ivankovic
e-mail: alojz.ivankovic@ucd.ie

N. Murphy
e-mail: neal.murphy@ucd.ie

P. M. Visakh et al. (eds.), *Advances in Elastomers II*,
Advanced Structured Materials 12, DOI: 10.1007/978-3-642-20928-4_8,

1 Introduction

Recent years have seen an increase in the use of elastomeric blends and composites for bio-medical applications. Their usage ranges from medical devices and utilities like cardiac assist pumps and blood bags, to chronic implants such as heart valves and vascular grafts [1–7]. The main compelling reasons that have promoted the use of these materials over metals and ceramics as biomaterials are due to their superior biocompatibility while at the same time offering good mechanical properties. In addition, elastomeric blends (e.g. thermoplastic elastomers (TPEs)) also have a high degree of purity (low level of extractable compounds), they are easy to recycle and are cost effective. TPE blends have also been shown to exhibit excellent barrier properties [8]. The ease with which these materials can be tailored to achieve desired material properties is yet another factor that continues to favor their use in biomedical applications.

Most importantly, a material used for medical devices and implants should possess good biocompatibility, biostability and mechanical properties. By biocompatibility it means that the material should not be harmful or cause severe allergic reactions in patients or individuals. At the same time its mechanical, physical and chemical properties should not be adversely affected by the biological environment in which it is used (i.e. biostability and biodurability). In addition the mechanical properties should be such as to withstand the predominant loading conditions to which the medical device or implant is subjected without the risk of premature failure.

Although elastomeric blends and composites are mainly used in cardiovascular and related applications, their use also covers many reconstructive medical and dental applications. These include hip-joint and knee replacements where elastomeric blends are used to compliment traditional materials such as metals and ceramics [9, 10], cosmetic surgery (e.g. breast implants and other related applications) [11–14], dental implants, skin care [15] and other medical devices like contact lenses, intraocular lenses and renal dialysers. In the year 2000, the worldwide demand for biomaterials was estimated to be in excess of 300 billion US dollars per year [16], representing between 7 and 8 % of total worldwide healthcare spending. This demand grows significantly every year, with an estimated annual growth rate of about 10 % [16]. Table 1 shows the approximate world consumption of medical devices per year, categorised according to the area of application [17]. Recent advances in the development of elastomeric blends and composites for bio-medical applications have also significantly improved consumer confidence, which has in turn increased the demand for these materials. The demand for new, improved biomaterials to replace hard and soft tissues lost due to disease, trauma, or age-related degeneration will always be significant. This chapter discusses the biomedical applications of elastomeric blends and composities. Section 2 looks at recent developments in the use of these blends and composites for medical applications. Section 3 gives an overview on common

Table 1 Approximate world annual consumption of biomaterials [17]

Application area	Approximate no. of devices used per year
Contact lens	75,000,000
Renal dialyser	25,000,000
Intraocular lens	7,000,000
Stent (cardiovascular)	>2,000,000
Hip and knee prostheses	1,000,000
Dental implants	500,000
Vascular graft	400,000
Catheter	300,000,000
Breast implants	300,000
Heart valve	200,000

elastomer based products used in biomedical applications. The preparation and fabrication of these materials is presented in Sect. 4.

2 Recent Developments on Elastomeric Composites and Blends for Medical Applications

Materials used in medical applications can be grouped into three categories: synthetic polymers, natural polymers (biomolecular materials), and inorganic materials (metals, alloys, ceramics and glasses). Elastomeric blends and composites fall into the category of synthetic polymers and are increasingly being used in a wide range of medical applications ranging from vascular grafts to hip-joint prostheses. The development of elastomeric blends and composites for medical applications is mainly driven by the need for materials with a combination of good biocompatibility, biostability and biodegradability, and mechanical properties. Natural biopolymers could be the best materials for these applications from the standpoint that these polymers are both bioactive and resorbable. However, they are subject to tedious purification techniques, processing variability, regulatory issues and the potential to induce dangerous immune responses when used as xenografts or allografts [18]. This, in part, has led to the exploration of synthetic polymers for use in medical applications. Synthetic elastomers possess a unique range of properties that makes them highly suitable for biomedical applications. Due to their superior biocompatibility, these polymers are now widely used, replacing the use of inorganic materials in several medical applications. Their good mechanical properties also ensure a minimal risk of device failure. The chemical composition of these elastomers or polymers offers substantial opportunities for synthetic polymer chemists to tailor the structures to meet specific requirements. This is easily achieved by changing the combination of their constituent monomers and processing conditions. For example the tensile strength of ultra high molecular weight polyethylene can be up to 100 times that of low-density polyethylene [19]. Elastomeric blends and composites can also be modified through secondary processes such as filling, blending, or grafting/crosslinking in a cost-effective manner to suit the application.

Synthetic polymers currently used in medical applications include thermoplastic elastomers such as polyurethanes and polyethylene, and thermoset elastomers such as Poly(glycerol-co-sebacate) and Xylitol-based elastomers. The latter are subject to several processing limitations that prohibit their use in some clinical applications. For example, their curing conditions typically employ temperatures in excess of 100 °C and vacuum environments [18]. Therefore, thermoplastic elastomers are usually favoured in most applications. Recent developments on elastomers (and their blends and composites) used in medical applications have been focused on achieving excellent biocompatibility and biostability and superior mechanical properties [2, 7, 12–15, 18]. This would ensure reduced risk of device failure, reduced allergic reactions in patients and reduced risk of device rejection by the host environment.

2.1 Biocompatible and Biostable Materials

The biocompatibility and biostability of medical devices and implants is crucial if the devices are to perform with an appropriate host response in a specific application. Over the past decade, these two properties have become key in the manufacture of medical devices and implants. The main objectives are to reduce adverse allergic reactions in patients and improve performance and service life of implanted medical devices. This is what has led to the discontinued use of latex based devices as these products were seen to cause severe allergic reactions in some patients [20]. Polyurethane and silicone elastomers have since replaced latex based elastomers in most applications, especially the ones requiring medical implants. These include vascular grafts and heart valves, among others. Polyurethane and silicone elastomers are known to offer good biocompatibility and biostability [21, 22], which has been achieved through much research performed in the last few years [2, 7, 23]. Much of this work is still on-going. Poor biostability can significantly shorten the service life of medical implants, which is very undesirable. Recent research, e.g. the work of Kang et al. [24] on polyurethane blends, has shown significant potential in achieving better material biostability and hence longer service life through elastomer blending (or design). Kang et al. [24] produced a polyurethane elastomer, which exhibited negligible degradation in its properties after 14 weeks of exposure to hydrogen peroxide/cobalt chloride solution.

2.2 Superior Mechanical Properties

Each biomaterial or medical device should possess mechanical and physical properties necessary for it to perform its physiological function. For example, a tendon material must be strong and flexible, a heart valve leaflet must be flexible and tough, while an artificial cartilage substitute must be soft and elastomeric (the ability to achieve larger deformations without rupture). At the same time, the medical device should possess good mechanical durability. For instance, good flexi-fatigue life is crucial for mechanical heart valve leaflets which must flex about 70 times per minute without tearing for the patient's lifetime (or realistically at least 10 or more years). Unfortunately, there are very few elastomeric blends that possess a combination of excellent biocompatibility and biostability, while at the same time offering good mechanical properties. The only two elastomers that seem to meet these requirements are polyurethane and silicone elastomers. Generally, polyurethane elastomers have superior mechanical properties over their silicone counterparts. Their Young's modulus is typically between 1 and 20 MPa [21]. The chemical composition of polyurethane elastomers also offers significant options for synthetic polymer chemists to tailor their structures to meet specific mechanical requirements. These elastomers also perform well under cyclic loading, a variety of temperatures and humid conditions, making them suitable for medical implants such as vascular grafts which are subjected to cyclic loading due to the pulsatile nature of blood flow under wet-37 °C conditions [21]. Polyurethane based elastomers also possess strain rate dependent viscoelastic behaviour [21], which is a characteristic of human soft tissue.

Similar to biocompatibility and biostability, mechanical properties of elastomeric blends and composites used for medical applications should be rigorously and adequately measured and characterised. This is particularly important in the case of elastomers since the mechanical properties of many of these materials are temperature, humidity and strain rate dependent. This means that the properties of a medical device once inside the host environment can significantly change from those measured in a laboratory setting under room temperature conditions. This has been shown to be the case for polyurethane elastomer medical implants [21, 23]. Therefore, when measuring these properties one should always choose test conditions that are as representative of the in vivo conditions as possible.

2.3 Intelligent Bio-Materials

One of the major achievements in recent years has been in the development/design of biomaterials used in drug delivery systems [18]. The design of biomaterials for these special applications is done at the molecular level in order to match the functionalities of molecules in a given application. Polyrotaxanes is one such example of an intelligent biomaterial that can be designed to effect dynamic molecular functions similar to those of natural tissues through the movement of cyclic compounds along the polymer's linear chain [18]. There are also a number of other elastomeric blends being used in drug delivery systems. These include new designs of integrated and combined material systems designed for drug delivery to particular sites in the body and drug coated stents, which are now widely used in vascular surgery to prevent restenosis [25].

The design and manufacture of intelligent elastomeric blends has partly been a result of the growing need for lubricious coatings and surfaces in medical devices. Smooth coatings allow physicians to easily manoeuvre medical devices through small blood vessels and delicate tissues without injuring or damaging them. As a result, surgeons today can now perform procedures that were earlier considered to be impossible. Using this coating technology, orthopaedic implants with bone growth enhancers, intensive wound care dressings, and tissue-engineered scaffoldings are now possible [18].

2.3.1 Use of Plasma Polymerised Primer Coatings to Enhance Adhesion Properties of Bio-materials

One of the major challenges faced with bio-materials is their ability to adhere to different substrates or support cell adhesion when these materials are used as substrates. This is particularly important in the manufacture of tissue-engineered scaffolds, orthopaedic implants with bone growth enhancers, and hip replacement implants where an elastomeric material is used together with a metal (Fig. 4), among others. Generally, the interface biocompatibility between synthetic prosthesis and living tissue has been seen to be a major problem in biomaterial research for several years [26]. A number of technologies have been employed to overcome these challenges, including the use of adhesion promoters, wet chemical primer application, ultraviolet light irradiation, thermal treatment, use of metallic coagents, and plasma treatment [26–28]. Most of these methods are highly complex, time-consuming and commercially unviable. Some of these processes can also lead to reduced shelf life of medical devices. For example, addition of adhesion promoters to silicone elastomer increases the reactivity of the elastomer pre-cure, and affects elastomer components' shelf life and opacity [27]. Plasma activation also tends to generate polar groups at the surface with, in some cases, a drastic reduction in adhesion on exposure to water or humidity [27, 28]. Some of these challenges can be overcome by using plasma-polymerised coatings. This is an atmospheric plasma treatment for the deposition of adhesive (primer) layers. Plasma generated gas discharge provides energy to activate gaseous or liquid monomer in order to initiate polymerization. Polymers prepared by this method are generally highly branched and highly cross-linked, and adhere very well to solid surfaces. One advantage of this method is the ability to directly attach polymers to a desired surface while the chains are growing. This significantly reduces steps necessary for other coating processes such as grafting. The use of atmospheric plasma combined with liquid deposition of precursors has the ability to plasma polymerise monomers while retaining their chemical functionality [27, 28]. The method can be used to form an interlayer (i.e. deposition of an adhesive primer layer), which due to the energetic a plasma treatment can potentially form a strong bond to the substrate, while still retaining the functional chemistry that can bond to the elastomer.

Although a relatively new technology, plasma polymerised primer coatings have been used by some researchers to enhance surface adhesion of bio-materials [26–28]. An example is the work by Nwankire et al. [27] where plasma polymerised primer coating (mixture of polyhydrogenmethyl siloxane and tetraethoxysilane) was used to improve the adhesion of silicone rubber to stainless steel. The results showed a 15-fold increase in adhesion strength or energy of the plasma coated stainless steel compared to its plasma polymerisation has also been used to deposit thin interfacial biocompatible films possessing inert properties such as biocompatibility, stability in biological environments and other specific properties depending on their function, onto medical devices and implants. This includes coatings used for knee and hip prostheses in order to increase their resistance to abrasion and improve their lubrication [26] and coating films that either contain reactive acid functional groups (by deposition of maleic anhydride (MAH) followed by hydrolysis) or are poly(ethylene glycol) (PEG)-like in nature (by using diethylene glycol vinyl ether (DEGVE) as a monomer) [29].

3 Overview of Elastomer Based Products for Biomedical Applications

This section presents an overview of elastomer blends and composites used in biomedical applications. Research in the development of these materials is ongoing and, hence, it is not possible to discuss the entire range of materials here. Therefore, only a few examples of elastomeric blends currently being used for biomedical applications are discussed.

Most of the elastomers used for biomedical applications belong to a group of synthetic polymers. These include thermoplastic and thermoset polymers. From a biological and clinical standpoint, these elastomers should exhibit biocompatibility and remodeling capacities that are comparable to natural extracellular matrix proteins [18]. From a polymer perspective, the elastomers should be able to be synthesized in large batches and exhibit tunable properties that could be realized by altering synthetic schemes and polymer processing conditions. The elastomers should also satisfy some general design principles such as; (1) allow incorporation of ester bonds to promote degradation by hydrolysis and enzymatic activity; (2) allow chemical crosslinking to achieve tunable elastomeric mechanical properties; and (3) allow the use of non-toxic monomers that can be metabolized or excreted by the host.

3.1 Thermoplastic Elastomers and Composites

Thermoplastic polymers form the largest group of elastomers and plastics used in medical applications. The general synthetic strategy used in their preparation utilizes multi-block polymers that consist of hard and soft segments that self-assemble to form a physically crosslinked network [18]. Elastomeric properties come from the physical crosslinks while biodegradation capabilities are due to the inclusion of polyester segments within the polymer blocks. The most commonly used thermoplastic elastomers for medical applications include polyurethanes, polyethylene, silicone, polyhydroxyalkanoates, and also a range of photocrosslink-able elastomers such as poly(glycerol-co-sebacate)-acrylate, poly(ε-caprolactone)-based and coumarin-functionalized polyesters.

3.1.1 Polyurethane Elastomers and Composites

Polyurethane is any polymer consisting of a chain of organic units joined by urethane (carbamate) links. Polyurethane polymers are formed through step-growth polymerization by reacting a monomer containing at least two isocyanate (diisocyanate) functional groups with another monomer containing at least two hydroxyl (alcohol) groups in the presence of a catalyst (Fig. 1). Hence, the three components used in the preparation of polyurethane elastomers are a diisocyanate, a macrodiol, and a chain extender [7, 20]. The crosslinking behaviour of a polyurethane elastomer, its color-fastness, resistance to hydrolysis and high temperature, is determined by the nature of the diisocyanate used. Macrodiol has fundamental consequences on product properties. There are two types of macrodiol used, i.e. ester diol or ether diol. Ester-based polyurethanes are primarily characterised by outstanding mechanical strength and oil resistance while ether-based polyurethanes exhibit excellent resistance to hydrolysis by hot water, acids or bases [30, 31].

Based on the chemical nature of the diisocyanate component in their formulation, they are either designated aromatic or aliphatic. Aromatic and aliphatic polyurethanes share similar properties that make them outstanding materials for

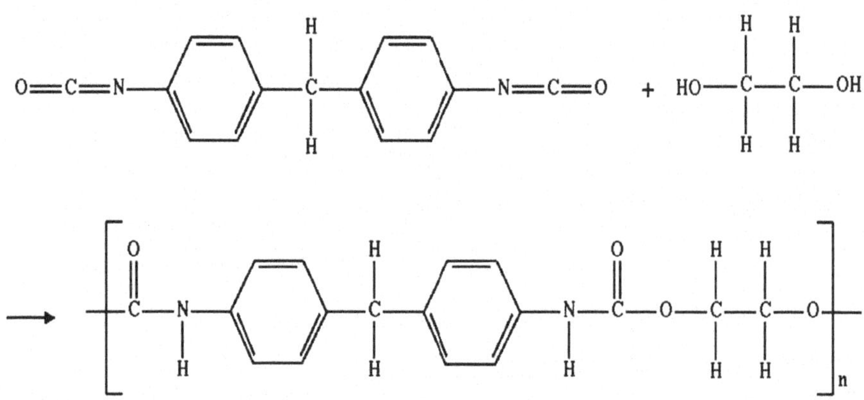

Fig. 1 Polymerisation of polyurethane (*bottom*) from a diisocyanate (*top left*) and hydroxyl group (*top right*)

use in medical devices. These properties include high tensile strength, high ultimate elongation, good biocompatibility, good hydrolytic stability, can be sterilised via ethylene oxide gas or gamma irradiation, and they are able to retain elastomeric properties at low temperatures [7, 20].

Polyurethanes are reported to have been discovered in the 1930s. However, their potential use in biomedical applications was not realised until the 1950s [20]. Today, they are among the most widely used biomaterials, with applications ranging from medical devices and utilities like cardiac-assist pumps and blood bags, to chronic implants such as heart valves and vascular grafts [2–4, 6]. They were first proposed for use as biomaterials in 1967 by Boretos and Pierce [32]. Their superior mechanical properties and blood compatibility has favoured their use and development as biomaterials, particularly as components of implanted devices [33]. One of these uses is in the manufacture of prosthetic or artificial heart valves.

Until recently, artificial heart valves were mainly of a mechanical type constructed from various material combinations such as metal and polymers (Fig. 2, bottom left). These allow unidirectional blood flow through mechanical closure of a pivoting or tilting disc valve. Recently, prosthetic heart valves are being fabricated from biochemically inert synthetic polymers such as polyurethane elastomers (Fig. 2, bottom right). The use of polymers in the manufacture of these devices helps in overcoming a variety of problems such as material fatigue, while at the same time maintaining natural hemodynamics and functional characteristics of heart valves [34, 35]. Polymeric heart valves have the ability to maintain or closely simulate natural body hemodynamics due to their flexible nature and soft texture which simulates the lubricity exhibited by natural heart valves, and are able to contract and expand freely as the blood flows through them [34, 35]. The high resistance to tear and abrasion, and fatigue [36] of polyurethane elastomers makes them well suited for use in prosthetic heart valves.

3.1.2 Polyethylene Elastomers and Composites

Polyethylene (PE) is a thermoplastic polymer consisting of long chains produced by combing the ingredient monomer ethylene (Fig. 3). It is a cross-linked polymer and can be classified into several different categories based mostly on its density and branching. Mechanical properties of polyethylene also significantly depend on variables such as the extent and type of branching, the crystal structure and the molecular weight. Ultra high molecular weight polyethylene (UHMWPE), also known as high-modulus polyethylene (HMPE) or high-performance polyethylene (HPPE), is the most widely used form of PE in biomedical applications. It has extremely long chains, with a molecular weight of between 2 and 6 million. The longer chain serves to transfer load more effectively to the polymer back-bone by strengthening intermolecular interactions, resulting in high toughness and impact strength [19]. It is also highly resistant to corrosive chemicals, with the exception of oxidizing acids and has extremely low moisture absorption, very low coefficient

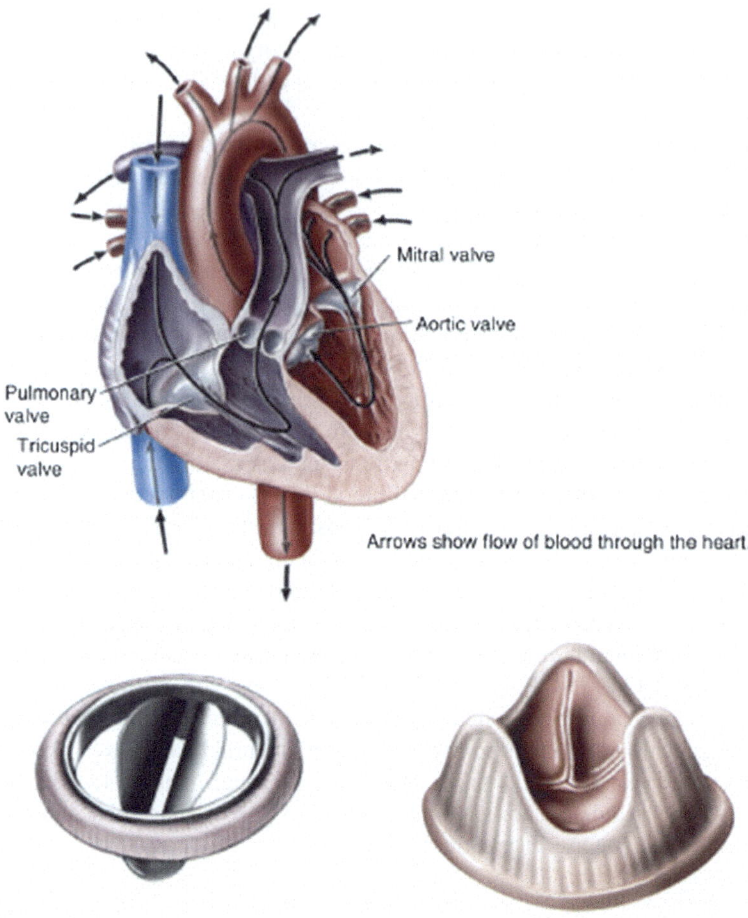

Fig. 2 Examples of artificial heart valves, *top*: heart showing heart valves positions, *bottom left*: mechanical heart valve, and *bottom right*: polymeric heart valve. Images from RelayHealth, 2009

Fig. 3 Polymerisation of polyethylene from its monomer, ethylene

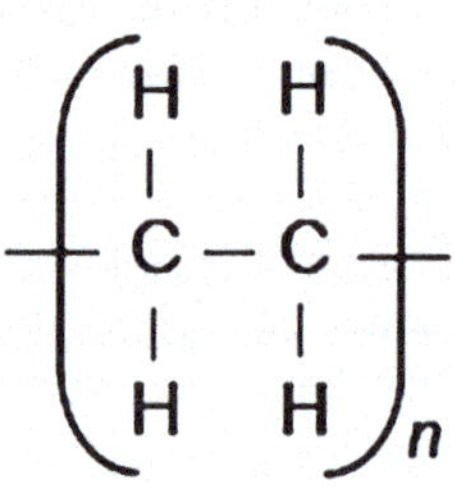

of friction, and is highly resistant to abrasion. Concerning its biocompatibility, it is nontoxic. UHMWPE is widely used in hip and knee replacements [19, 37], and most recently (since the 1980s), for spine implants [37].

Its high toughness, strength and abrasion resistance makes UHMWPE very suitable for these applications. UHMWPE was first used for a medical application in 1962 by Sir John Charnley and has since emerged as the dominant bearing material for total hip and knee replacements [19]. UHMWPE has also been used in building elastomeric composite membranes like the work of Teoh et al. [38]. Teoh et al. [38] successfully fabricated a composite membrane of UHMWPE and polyether polyurethane, combining the polyethylene's high tensile strength with polyurethane's excellent biocompatibility and biodurability properties. The end result was a membrane with nearly five times the tensile strength and 50 times the Young's modulus of polyurethane.

Figure 4 shows a schematic of an hip-joint replacement prosthesis. There are about three types of hip replacement implants used, i.e. metal and plastic, metal-on-metal and ceramic-on-ceramic. Metal-on-metal and ceramic-on-ceramic have superior wear rates although there are concerns about the wear debris generated in the case of metal-on-metal implants, which may end up being absorbed by the body (e.g. metal ions). The concern over ceramic implants is that they can break inside the body. Metal and plastic are the most commonly used types of hip replacement implants. The femoral stem and ball (Fig. 4) are all made of metal (including titanium, stainless steel, and cobalt chrome) and the acetabular cup is usually made of polyethylene. To overcome the high wear rates of polyethylene, UHMWPE is now commonly used for the manufacture of the acetabular cup.

The implant is secured to the bone either by press-fitting or cemented into place. When press-fitting, the implant is fit snugly into the bone and new bone forming around the implant secures it in position. For the other method, a special bone cement or fixation agent is used to secure the prosthesis in position as shown in Fig. 4.

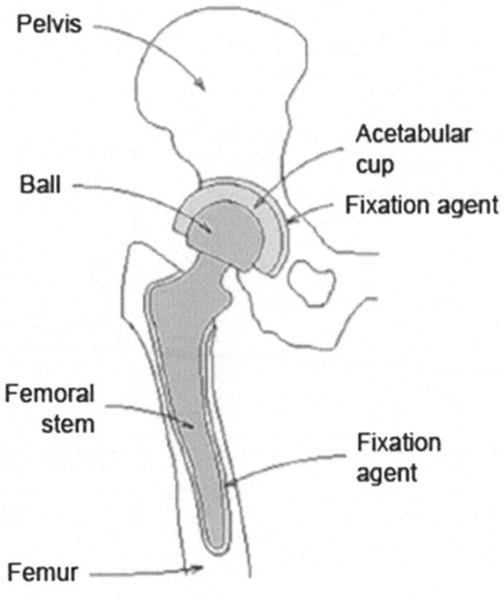

Fig. 4 Schematic of hip-joint replacement prosthesis

3.1.3 Silicone Based Elastomers

Silicones are a class of synthetic polymers whose backbone is made of repeating silicon to oxygen bonds, with silicon also bonded to organic groups (typically methyl groups) as shown in Fig. 5. The basic repeating unit is known as "siloxane". Other groups such as phenyl, vinyl and trifluoropropyl can substitute methyl groups along the silicone polymer chain. The presence of both "organic" groups attached to the "inorganic" backbone (i.e. silicon to oxygen bonds) gives silicones a combination of unique properties, making possible their use as fluids, emulsions, compounds, resins, and elastomers in a variety of applications [22]. Polydimethylsiloxanes trimethylsilyloxy terminated (Fig. 5) are the most common type of silicones. Silicones are cross-linked polymers and would usually incorporate a filler, which acts to reinforce the cross-linked matrix.

The strength of silicone elastomers without the filler is generally unsatisfactory for most applications [39]. In biomedical applications, silicone elastomers are commonly used as catheters, drains and shunts. These include medical devices fabricated with silicone extrusions and those with nonsilicone substrates that are silicone-coated to minimise host reaction [22]. An example of silicone coated devices is the Silastic Foley urology catheters, which are latex catheters whose exterior and interior is coated with silicone elastomer [22]. Silicone tubing and membranes are also widely used in extracorporeal machines, mainly due to their blood compatibility and permeability properties. Studies have also indicated that platinum-cured silicone tubing may be superior to polyvinyl chloride (PVC) in several respects [40]. The other major biomedical use of silicone elastomers is in aesthetic and reconstructive plastic surgery, including implants for breast, scrotum, chin, nose, cheek, calf, and buttocks [22]. The most prominent of these aesthetic implants is the silicone breast implant whose use dates back to 1963 [41], and still remains widely popular.

Although silicone elastomers are known to exhibit excellent biocompatibility, their mechanical properties are generally inferior to those of polyurethane

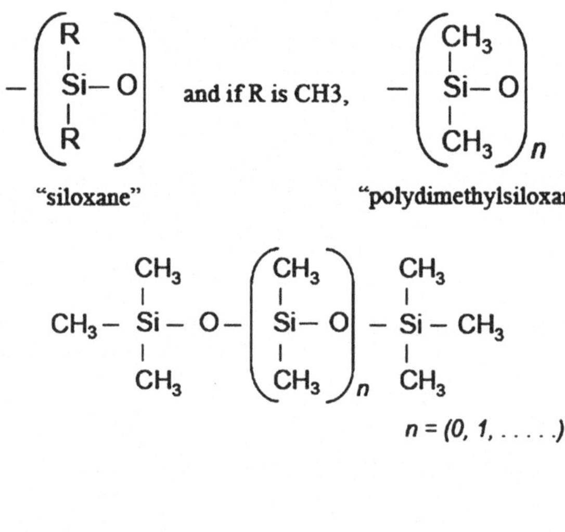

Fig. 5 Synthesis of silicone (*bottom*) from its monomer (polydimethylsiloxane)

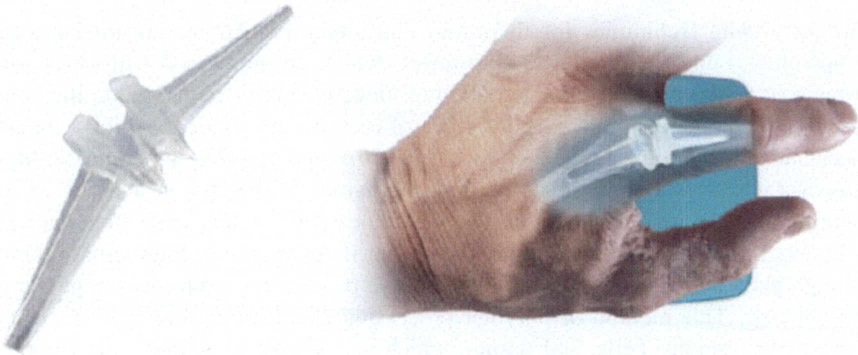

Fig. 6 Figure showing finger prosthesis implant (*left*) and when implanted (*right*). Picture from http://www.fingerreplacement.com

elastomers, making them unfavourable for applications where high mechanical loading is expected, e.g. heart valves and vascular grafts. However, silicone elastomers still find wide medical applications especially as breast implants and urinary catheters. Silicone plastics or composites are also widely used for finger prosthesis implants. Figure 6 shows an example of such as application.

3.1.4 Polyhydroxyalkanoates

Polyhydroxyalkanoates (PHA) are another class of thermoplastic polymers which are mainly used in tissue engineering and numerous regenerative medicine applications [18]. They exhibit a wide range of properties from rigid plastic to soft elastomer with tensile strengths ranging between 20 and 50 MPa and elongations at break from 5 to 850 %. PHAs are typically synthesized from bacterial cultures, and are biocompatible and generally non-toxic. The use of bacterial fermentation to produce large quantities of PHAs leads to additional purification processes such as the removal of endotoxic pyrogens [18]. PHAs degrade by surface erosion with a rate that can be tuned through material design, and exhibit melting points between 120 and 177 °C, which allows for processing by melt extrusion. In general, PHAs are a class of polymers that bridge the gap in physical properties between natural polymers and most synthetic biodegradable elastomers. These polymers are also used in the synthesis of polymer hybrids (or elastomeric blends/ composites). One of the ground-breaking contributions in this area was performed by John and Foster [42]. They produced a hybrid polymer from PHA and polyethylene glycols (PEGs), making use of the excellent biocompatibility of PHA and the high strength of polyethylene.

3.1.5 Photocrosslinkable Elastomers

The processing techniques for thermoset elastomers (which require high curing temperatures) and thermoplastic elastomers (which are processed with cytotoxic organic solvents) prohibit the integration of bioactive components within the bulk material. A good example is poly(glycerol-co-sebacate) which exhibits robust mechanical properties and exceptional ***in vitro*** and ***in vivo*** biocompatibility. However, there is a limit in its ability to polymerize directly in a tissue or to incorporate cells or temperature-sensitive molecules. In this case, free radical polymerization or photocrosslinking is used in its preparation. Photocrosslinking through photoactivation provides an alternative process for rapid low-temperature crosslinking. This method of polymer processing enables interfacing of the elastomer with proteins, cells, and tissues, which is desirable in some medical applications such as in vascular scaffolds [18]. Photocrosslinking can be employed for both aliphatic polymers as well as hydrophilic macromolecular precursors for hydrogel formation. Examples include;

- Poly(glycerol-co-sebacate)-acrylate (PGSA) whose prepolymer is first cast into its final form and then crosslinked using ultra-violet light and a photoinitiator. PGSA can also be copolymerized with poly(ethylene glycol) (PEG) to extend their properties [43].
- Poly(ε-caprolactone)-based elastomers, highly flexible elastomers whose mechanical and degradation properties can be altered predictably through varying the degree of photocrosslinking.
- Coumarin-functionalized polyesters. These multifunctional polymers are end-modified with photoactive coumarin groups, using ultra-violet light to crosslink the polymers and produce the final network structure. Coumarin groups have also been utilized as functional groups in the fabrication of shape-memory elastomers [18].

3.2 Thermoset Elastomers and Composites

Thermoset elastomers and composites are also commonly used in medical applications. The only factor limiting their use in these applications is that they present significant processing limitations such as high curing temperatures (in excess of 100 °C) and sometimes under high vacuum. These conditions prohibit the incorporation of bioactive components and ultimately limit the spectrum of downstream applications [18, 37]. Thermoset elastomers used in medical applications include poly(glycerol-co-sebacate), poly(1,3-diamino-2-propanol-co-polyol sebacate)s, and xylitol-based elastomers.

3.2.1 Poly(glycerol-co-sebacate)

Poly(glycerol-co-sebacate) (PGS) is a thermoset elastomer composed of a stoichiometric composition of sebacic acid and glycerol which is polymerized through step addition. It is a transparent, nearly colourless elastomer featuring a network of random coils with a low crosslinking density, and typically exhibits a Young's modulus between 0.3 and 1.4 MPa, an ultimate tensile strength (UTS) of between 0.4 to 0.7 MPa and a maximum elongation of up to 160 % [18, 19]. Poly(glycerol-co-sebacate) elastomers are also hydrophilic. This is due to the hydrolyzable ester bonds formed with pendant hydroxyl groups attached to their backbone. The free hydroxyl groups also participate in hydrogen bonding, which provides mechanical strength through intramolecular forces [18]. PGS exhibits good biocompatibility and its degradation is primarily through surface erosion. It belongs to a class of biodegradable elastomers, which are used in a wide variety of medical applications including drug delivery and tissue regeneration, where (cell-seeded) constructs are designed to aid or replace damaged or diseased tissue [18]. Degradation kinetics and mechanical properties of PGS can be varied modestly by altering the curing conditions.

3.2.2 Poly(1,3-diamino-2-propanol-co-polyol sebacate)s

Poly(1,3-diamino-2-hydroxypropane-co-polyol sebacate)s are a class of synthetic, biodegradable elastomeric poly(ester amide)s composed of crosslinked networks based on an amino alcohol [44]. They exhibit a low tensile Young's modulus (of the order of 1 MPa) and are capable of large deformations (reversible elongations of up to 92 %). Maximum elongations can be increased by reducing the curing time. APS polymers also possess good in vitro and in vivo biocompatibility, with degradation half-lives of up to 20 months in vivo [44].

The preparation and design criteria applied to the realization of APS is similar to that used for PGS. The only differences in preparation techniques include the engineering of the monomer feed stream to produce both amide and ester bonds, enabling sensitive tuning of the elastomer's properties. The elastomer should also contain free amines to facilitate cell attachment and allow for rapid surface modification [44]. The susceptibility of degradation of APS via hydrolysis and esterase enzymes is significantly reduced by the presence of amide bonds [18, 44]. Due to their extended biodegradation timelines and enhanced cell attachment properties, APS elastomers can be used in applications where long biomaterial half-lives are required such as scaffolds for cardiovascular grafts or peripheral nerve regeneration conduits [18, 44]. In vivo half-life of APS formulations can be tuned to up to 100 weeks. Degradation of APS formulations with high ester content is primarily by surface erosion through enzymatic activity and for formulations with higher amide content primarily by bulk erosion through hydrolysis.

3.2.3 Xylitol-Based Elastomers

Xylitol-based elastomers are also used in a number of medical applications, especially for tissue engineering scaffolds. They belong to a class of biodegradable elastomers composed of chemically multi-functional, non-toxic monomers which are endogenous to human metabolic pathways. They form a tough biodegradable crosslinked network with good mechanical and degradation properties and excellent biocompatibility properties. A wide range of crosslink densities can be achieved by altering monomer feed ratios whilst maintaining their elastomeric properties. These elastomers exhibit tensile Young's modulus values of up to 5 MPa and maximum elongations at break of up to 200 % [18]. The compressive modulus for xylitol-based hydrogels is approximately of the order of 5 kPa and compression at failure of approximately 80 %. They also exhibit little hysteresis under cyclic conditioning, which demonstrates the elastic nature of these elastomers.

Xylitol-based elastomers are prepared through polycondensation of xylitol with water-soluble citric acid, yielding a biodegradable, water soluble polymer. Subsequent acrylation of this polymer results in an elastomeric photocrosslinkable hydrogel. Xylitol-based polymers may also be processed by polycondensation with water-soluble sebacic acid. Among the advantages that xylitol-based polymers have over other synthetic polymers include rapid scalable synthesis. Their non-toxic nature also eliminates potential immune responses related to some natural polymers such as collagen. Similar to APS, the degradation rate of these elastomers can also be tuned by simple adjustments in their chemical composition.

4 Synthetic Techniques of Elastomeric Composites and Blends for Medical Applications

The use of synthetic schemes is crucial in developing elastomeric composites and blends for biomedical applications. However, it is also important to develop parallel processing capabilities in order to realize the full capacity of these materials. Structure-processing-property relationships are an essential component of biomaterials development. This section briefly discusses the application of polymer fabrication and processing techniques/methods to elastomeric composites and blends for potential medical applications.

4.1 Fabrication/Synthesis Techniques

The choice of synthetic techniques employed to produce elastomers for medical applications depends on monomer composition and its capacity for chemical or physical crosslinking. The additional requirement of crosslinking can be achieved by several methodologies such as polycondensation and photocrosslinking. However, these step-addition polymerisations require a chemical system where at least one of the components exhibits a bonding functionality of three or greater. This functionality can be incorporated by selecting the appropriate monomer.

4.1.1 Crosslinked Networks via Polycondensation

Polycondensation via step addition polymerization is commonly employed to realize a wide range of elastomeric blends and composites used in medical applications [18, 42, 45]. This polymerization mechanism involves the step by step reaction of bi-functional or multifunctional monomers to form first dimers (i.e. two identical molecules or monomers), then trimers, longer oligomers and eventually long chain polymers. Synthetic polymers produced by this method include polyesters, polyamides and polyurethanes. Due to the step-growth nature of the polymerization mechanism, a high extent of reaction is required to achieve high molecular weight. Figure 7 shows, as an example, the trimerization (formation of a trimer) of acetylene.

Apart from the need for multiple reactions to achieve high molecular weight, step-addition polycondensation has other disadvantages including sensitive reaction kinetics with small processing windows and high polydispersity indices when using two bifunctional or multifunctional monomers [18]. For example, polycondensation reactions of polyfunctional maltitol with sebacic acid can lead to polydispersity indices of as high as 4 or more.

There are, however, other synthetic routes (although much more complex) that can be used to overcome the aforementioned disadvantages. These include the creation of polymers with lower polydispersity indices to serve as prepolymer starting materials, with subsequent polymerization steps employed to realize the final crosslinked network. Atom-transfer radical polymerization (ATRP) can also be used as an alternative synthesis strategy in order to lower the polydispersity indices of network precursors [18]. The flexibility with the latter strategy lies in the decoupling of the prepolymer and crosslinking (curing) reactions.

4.1.2 Photocrosslinked Elastomers

Harsh curing conditions such as high temperatures employed in step-addition polymerizations often reduce the range of medical applications for most elastomers. For example, with high curing temperatures, the incorporation of bioactive molecules or viable cell populations is virtually impossible. High-temperature curing processes also subject polymers to rapid oxidative degradation [18]. The solution to overcoming this challenge is to employ a photopolymerisation or free radical polymerization technique. This synthesis technique can be used to produce complex three-dimensional structures using a low temperature crosslinking process, and has been successfully applied to cell-seeded hydrogel systems [18]. It involves a

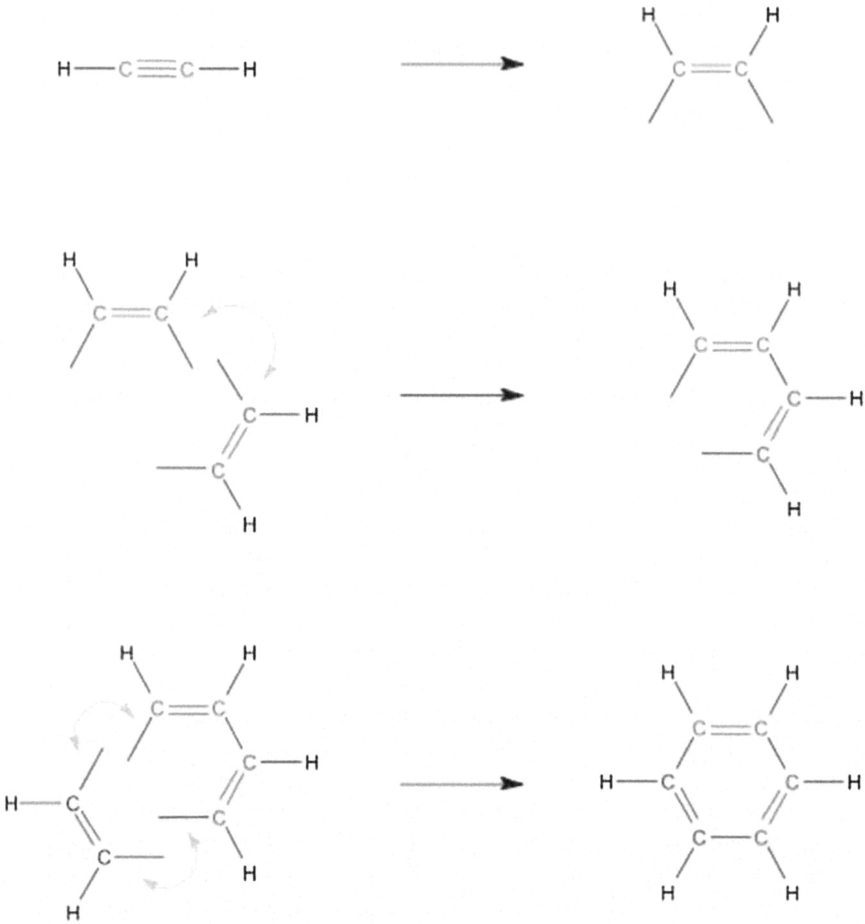

Fig. 7 Polycondensation or step-growth polymerisation of acetylene (process also known as trimerisation-formation of a trimer)

multi-step fabrication process in which microstructures are built up by employing successive deposition and photopolymerization processes. Photopolymerisation is now widely employed in the preparation of elastomers (especially biodegradable elastomers) used in medical applications. The processing capability of these elastomers relies on the presence of photocrosslinkable moieties.

Photocrosslinkable polymer chemistries typically employ photoactive diene-based modifications which may be added through terminal addition or backbone modification via esterification of free hydroxyl groups using acid chlorides with labile dienes such as acryloyl chloride or methacryloyl chloride [18]. These modifications are a straightforward procedure which can easily be utilized to extend the processing capabilities of a wide range of elastomers. Other photo-crosslinking techniques include the use of extended and end-modified (with photofunctional groups) star-based polymers [18]. Generally, the mechanical and thermal properties of these elastomers are molecular weight dependent, although their glass-transition temperature (T_g) is not.

4.1.3 Alternative Network Formation

Elastomeric networks can also be achieved by bonding macromolecular precursors with labile pendant groups, using multifunctional crosslinking additives. Although this strategy is normally employed for the preservation of natural biological materials through crosslinking with dialdehydes [18], it can also be used for synthetic systems. Alternative crosslinking chemistries such as the use crosslinking reagents (e.g. diisocyanates) can also be used to expand the functionality of crosslinked systems. For example, diisocyanates have specifically been used to prepare crosslinked elastomers using star-based precursors such as 6-armstar poly(L-lactide-co-[epsilon]-caprolactone) and star poly(glycolide-co-[epsilon]-caprolactone) [18]. This synthetic approach using crosslinking reagents can be applied to form a wide variety of monomers including polyester macromolecular network precursors.

4.2 Elastomer Processing Methods

Once elastomers are synthesised, they need to be fabricated into products/devices for use e.g. in medical applications. There are several elastomer-processing methods available. The choice of the processing method/considerations depends on the type of elastomer being processed, and is governed primarily by the intended function of the elastomer as well as the final intended application. Elastomers can be processed into thin films or complex structures. The most popular processing methods can be categorised into two groups based on the type of elastomer being processed, i.e. thermoplastic or thermoset elastomers.

4.2.1 Thermoplastic Elastomer Processing Methods

Since thermoplastic elastomers soften when heated, most processing methods involve the application of heat to soften or melt the polymer, after which the elastomer is formed into a desired shape. The formed shape is retained on cooling. The softening and forming processes are reversible allowing the materials to be easily recycled. Common processing methods for thermoplastic elastomers include extrusion, injection moulding, blow moulding, rotational moulding and thermoforming.

4.3 Extrusion

The extrusion process is commonly employed to manufacture medical tubing from a wide variety of thermoplastic elastomers. These include catheters, artery prostheses, surgical tubes, and hypodermic tubes. During the extrusion process, the elastomer or plastic is melted and then, using a die, formed into a desired shape. The raw thermoplastic material or resin, usually in the form of small beads, is gravity fed from a top mounted hopper into the barrel of the extruder as shown in Fig. 8. The material flows through the feed throat and is forced by a rotating screw into the barrel which is heated to the desired melt temperature of the molten plastic or elastomer. Depending on the elastomer, this temperature can be between 200 and 275 °C. In most processes, a heating profile is set for the barrel in which three or more independently controlled heater zones gradually increase the temperature of the barrel from the rear (where the plastic enters) to the front in order for the resin to melt gradually as it is pushed through the barrel. This also lowers the risk of overheating which may cause degradation in the polymer. Extrusion can also be used to produce thin/film materials and also in coating applications, which is necessary in some medical applications [22].

4.4 Injection Moulding

The injection moulding process can be used for manufacturing parts from both thermoplastic and thermosetting elastomers. The process is quite similar to the

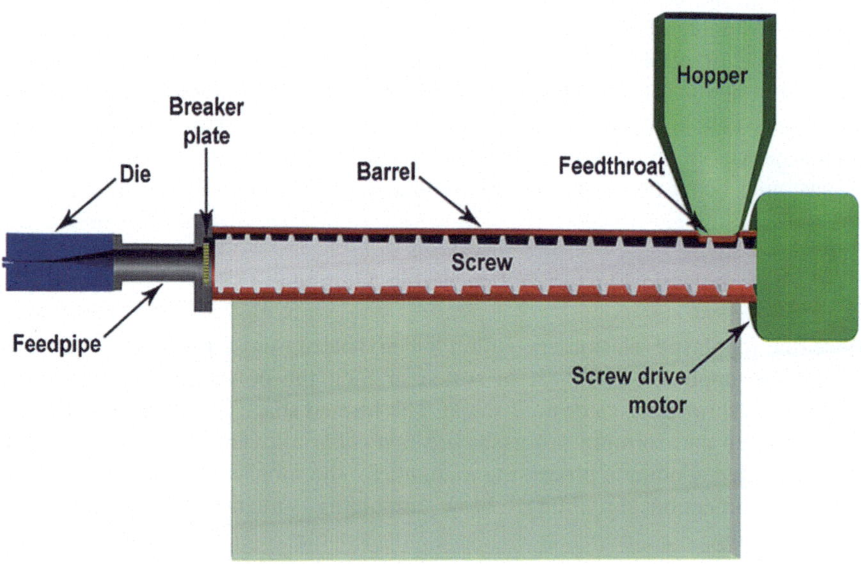

Fig. 8 Schematic drawing of the extrusion process

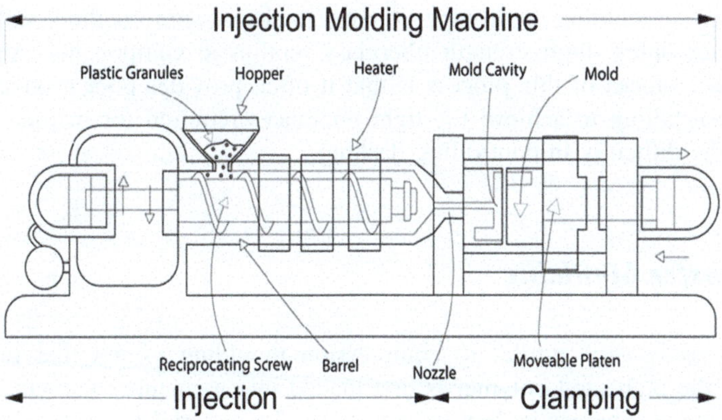

Fig. 9 Schematic of the injection moulding process (www.thefullwiki.org)

extrusion process. The material or resin is fed into a heated barrel, mixed, and forced into a mould cavity where it cools and hardens to the configuration of the mould cavity as shown in Fig. 9. Among the advantages of injection molding are high production volumes/rates, repeatable high tolerances, low labour cost, and minimal scrap losses. Several elastomer based medical implants including cardiovascular and orthopaedic implants are manufactured by this method.

4.5 Compression Moulding

During compression moulding, the generally preheated moulding material is first placed in an open, heated mould cavity and then using a plugger the material is compressed into the mould to achieve the desired shape (see Fig. 10). Heat and pressure are maintained until the moulding material has cured. The process normally employs thermosetting resins in a partially cured state, either in the form of granules or preforms but can also be used for moulding thermoplastics.

Fig. 10 Schematic of compression moulding process

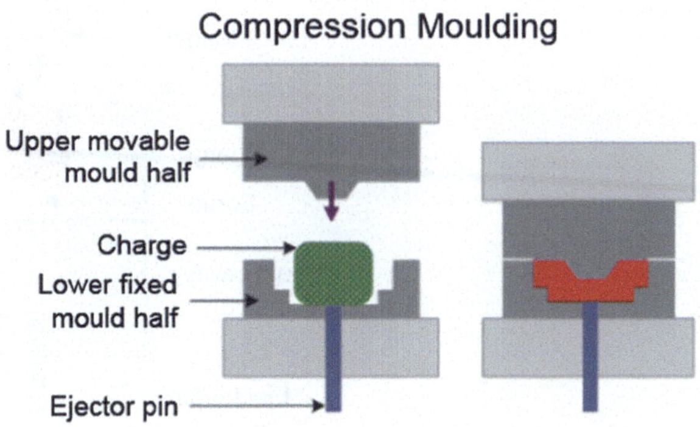

Compression moulding is a high-volume, high-pressure method suitable for moulding complex, high-strength fiberglass reinforced components. Among the main disadvantages of this process is that it often provides poor product consistency, often failing to achieve the tight tolerances required for medical devices, and also the difficulty in controlling flashing.

4.6 Transfer Moulding

This process is very similar to compression moulding except that in transfer moulding the preheated polymer is first loaded into a chamber known as the pot (see Fig. 11). A plunger is then used to force the material from the pot through channels known as a sprue and runner system into the mould cavities. The walls of the mould are heated to a temperature above the mould material melting point in order to allow a faster flow of material through the cavities. The moulds in both compression and transfer molding remain closed until the curing reaction within the material is complete. Ejector pins, usually incorporated into the mould design, are used to push the part from the mould once it has hardened.

Other common moulding processes include blow moulding, rotational moulding, dip moulding and thermoforming. These moulding processes (except dip moulding and thermoforming) work on the same principle where mould material is melted and then formed in a preform (or mould cavity) with the final product taking the shape of the preform. Blow molding is normally used for manufacturing hollow parts. The process involves forming the melted material into a parison or preform (a tube-like piece of plastic with a hole in one end through which compressed air can pass), and then using air pressure to push the plastic walls to match the mould. Once cooled, the formed (and hardened) part is removed. In rotational moulding (also mostly used for producing hollow parts) the melted material inside

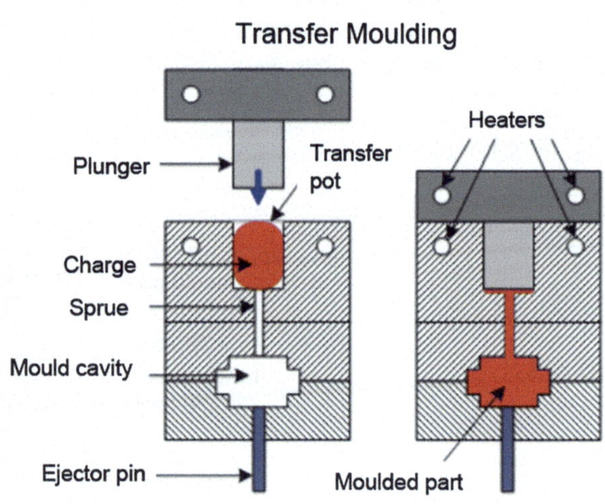

Fig. 11 Schematic of transfer moulding process

a heated hollow mould is slowly rotated causing the material to disperse and stick to the walls of the mould. The mould is rotated continuously during the heating and cooling phases in order to maintain an even thickness throughout the part and also to avoid sagging or deformation during the cooling phase. Dip moulding, as the name suggests, involves dipping a heated mold (mandrel made in the shape of the object being molded) into the resin. The mold is then extracted from the resin solution and run through a baking process to cure the resin. This method is widely used for preparation of tubular products including artery prostheses and surgical gloves [23]. Multiple dipping steps (coatings) build up the tube thickness until the required thickness (usually up to around 1mm) is achieved . During thermoforming a plastic sheet is heated to a pliable forming temperature (without melting the material as is the case for the moulding processes described above), formed into the desired shape in a mould, and trimmed to create a usable product.

4.6.1 Thermosetting Elastomer Processing Methods

Thermosets are materials that are formed whilst warm or hot but once formed they permanently set in the formed shape. This process is irreversible, i.e. the material can not be softened again by reheating. Processing of thermosetting elastomers presents a much greater challenge than thermoplastic elastomers. The most commonly used processing methods for this class of elastomers, excluding injection moulding, which is already discussed above, are compression moulding and transfer moulding.

4.6.2 Additional Fabrication/Processing Techniques

In addition to the synthesis, fabrication and processing techniques described above, special surface material preparation may also be necessary for biomaterials used for implantable devices. This is because the micro-and nano-structural cues of the underlying substrate or biomaterial can have enormous effect on the behaviour of human and animal cells [18, 46, 47]. These nanotopographic cues can lead to altered morphology, adhesion, proliferation, differentiation, and migration in many cell types. The interaction between these cells and the biomaterial (i.e. medical device) is crucial to the performance the implanted device. By employing two-dimensional fabrication (or surface preparation) of elastomers, the interaction between the cells and the biomaterial can be significantly improved [18, 46, 47]. Biomaterial surface can specifically be engineered to improve cell adhesion and proliferation, e.g. tissue engineered vascular scaffolds [18], and/or prevent/minimise protein absorption [27–29]. Plasma treatment (discussed in Sect. 2.3.1) is one of the most recent technologies being employed to achieve this.

The other well-known surface preparation/fabrication technique is that of electro spinning. This method enables the fabrication of biomaterial surfaces with tunable porosity and nano-scale topography, the latter of which can modulate cell functions such as adhesion, migration, and proliferation [18, 46]. Electro spinning has also been used to fabricate fibrous scaffolds using a variety of natural materials and synthetic polymers and can be adopted for thermoplastic elastomers as well.

5 Conclusion

Elastomeric blends and composites form the largest group of materials used in medical applications, especially for implantable devices. This group of materials also offers substantial opportunities to polymer scientists to tailor their structures to achieve desired properties, which can not easily be performed for most traditional biomaterials such as metals and ceramics. Elastomeric blends and composites also have far superior biocompatibility properties compared to metals and ceramics, while at the same time offering good mechanical properties.

With an ever-growing need for materials with enhanced biocompatibility, biostability and mechanical properties in order to; (i) reduce risk of severe allergic reactions in patients, (ii) reduce risk of device rejection by the host environment, and (ii) reduce risk of premature device failure, active research and development into optimised elastomeric blends and composites for biomaterials is on-going and must be seen as a high priority for the medical device industry.

References

1. Gogolewski, S.: In vitro and in vivo molecular stability of medical polyurethanes: A review. Trends Polym. Sci. **1**, 47–61 (1991)
2. Christenson, E.M., Wiggins, M.J., Anderson, J.M., Hiltner, A.: Surface modification of poly(ether urethane urea with modified dehydroepiandros-terone for improved in vivo biostability). J. Biomed. Mater. Res. **73A**, 108–115 (2005)
3. Lelah, M.D., Cooper, S.L.: Polyurethanes in medicine. CRC Press, Boca Raton (1986)
4. Szycher, M.: Polyurethanes in vascular grafts. Elastomer World 218, p44 (1998)
5. Szycher, M., Reed, A.: Biostable polyurethane elastomers. Med. Device Technol. **3**, 42–51 (1992)
6. Stokes, K., McVenes, R.: Polyurethane elastomer biostability. J. Biomater. Appl. **9**, 321–354 (1995)
7. Gunatillake, P.A., Martin, D.J., Meijs, G.F., et al.: Designing biostable polyurethane elastomers for biomedical implants. Aust. J. Chem. **56**, 545–557 (2003)
8. Ripple, W.S., Simons, J.: Thermoplastic elastomers in medical devices. Technical contribution for MedPlast Supplement (2007)
9. Khorasani, M.T., Zaghiyan, M., Mirzadeh, H.: Ultra high molecular weight polyethylene and polydimethylsiloxane blend as acetabular cup material. Colloids Surf. B **41**, 169–174 (2005)
10. Onatea, J.I., Cominb, M., Bracerasa, I., et al.: Wear reduction effect on ultra-high-molecular-weight polyethylene by application of hard coatings and ion implantation on cobalt chromium alloy, as measured in a knee wear simulation machine. Surf. Coat. Technol. **142–144**, 1056–1062 (2001)

11. Brandon, H.J., Young, V.L., Jerina, K.L., et al.: Variability in the properties of silicone gel breast implants. Plast. Reconstr. Surg. **108**(3), 647–655 (2001)
12. Koo, N.: The fabrication of a flexible mold for high resolution soft ultraviolet nano-imprint lithography. Nanotechnology **19**, 1–4 (2008)
13. Barr, S., Bayat A.: Current implant surface technology: An examination of their nanostructure and their influence on fibroblast alignment and biocompatibility. Eplasty **9**, e22 (2009)
14. Barr, S., Hill, E., Bayat, A.: Patterning of novel breast implant surfaces by enhancing silicone biocompatibility, using biomimetic topographies. Eplasty **10**, 246–268 (2010)
15. Shanshan L., Daniel, M.D., Yi, C., et al.: Designed biomaterials to mimic the mechanical properties of muscles. Nature **465**(7294), 69–73 (2010)
16. Lysaght, M.J., O'Loughlin, J.A.: Demographic scope and economic magnitude of contemporary organ replacement therapies. ASAIO J. **46**(5), 515–521 (2000)
17. Ratner, B.D.: An introduction to biomaterials. University of Washing-ton Engineered Biomaterials. http://www.uweb.engr.washington.edu/research/tutorials
18. Bettinger, C.J.: Biodegradable elastomers for tissue engineering and cell biomaterial interactions. Macromol. Biosci. (2011). doi:10.1002/mabi.201000397
19. Kurtz Steven, M.: UHMWPE Biomaterials Handbook-Ultra-High Molecular Weight Polyethylene in Total Joint Replacement and Medical Devices (2nd edn), Elsevier, pp. 543 (2009) SBN: 978-0-12-374721-1
20. Tilak M.S.: Dip molding of polyurethane and silicone for latex-free, nonallergic products. Medical device and diagnostic Industry (2001)
21. Kanyanta, V., Ivankovic, A.: Mechanical characterisation of polyurethane elastomer for biomedical applications. J. Mech. Behav. Biomater. **3**, 51–62 (2010)
22. Colas, A., Curtis, J.: Biomaterials science. High Molecular Weight Polyethylene in Total Joint Replacement and Medical Devices. Academic Press, Elsevier (2009)
23. Kanyanta, V.: Towards early diagnosis of atherosclerosis -accurate prediction of wall shear stress. PhD thesis, University College Dublin, Ireland (2009)
24. Kang, J., Erdodi, G., Brendel, M.C., et al.: Polyisobutylene-based polyurethanes. v. oxidative-hydrolytic stability and biocompatibility. J. Polym. Sci. Part A: Polym. Chem. **48**(10), 2194–2203 (2010)
25. Dibra, A., Kastrati, A., Mehilli, J., et al.: Paclitaxel-eluting or sirolimus-eluting stents to prevent restenosis in diabetic patients. N. Engl. J. Med. **353**, 663–670 (2005)
26. Holvoet, S., Chevallier, P., Turgeon, S., Mantovani, D.: Toward high-performance coatings for biomedical devices: Study on plasma-deposited fluorocarbon films and ageing in pbs. Materials **3**, 1515–1532 (2010)
27. Nwankire, C.E., Ardhaoui, M., Dowling, D.P.: The effect of plasma-polymerised silicon hydride-rich polyhydrogenmethylsiloxane on the adhesion of silicone elastomers. Polym. Int. **58**(9), 996–1001 (2009)
28. Nwankire, C.E., ONeill, L., Byrne, G., Dowling, D.P.: The effect of plasma polymerised si-h rich polymethylhydrogen siloxane (phms) on the adhesion of silicone elastomer. In: Proceedings of the 31st Annual Meeting of the Adhesion Society, pp. 436 (2008)
29. Ademovic, Z., Wei, J., Winther-Jensen, B., Hou, X., Kingshott, P.: Surface modification of pet films using pulsed ac plasma polymerisation aimed at preventing protein adsorption. Plasma Process. Polym. **2**, 5363 (2005)
30. Knoerr, K., HHomann, U.: Millable Polyurethane Elastomers, Hand-book of Elastomers, 2nd edn. Marcel Decker, Inc., New York (2001)
31. Recker, K.: Cast Polyurethane Elastomers, Handbook of Elastomers, 2nd edn. Marcel Decker, Inc., New York (2001)
32. Boretos, J.W., Pierce, S.W.: Segmented polyurethane: A new elastomer for biomedical applications. Science **158**, 1481–1482 (1967)
33. Lamba, N.M.K., Woodhouse, K.A., Cooper, S.L.: Polyurethanes in Biomedical Applications. CRC Press, Boca Raton (1997)

34. Lakshmi, P.D., Helene, A.S., Philippe, S., Ajit, P.Y.: Fluid mechanics of artificial heart valves. Clin. Exp. Pharmacol. Physiol. **36**(2), 225–237 (2009)
35. Bloomfield, P.: Choice of heart valve prosthesis. Heart **87**(6), 583–589 (2002)
36. Wiggins, M.J., Anderson, J.M., Hiltner, A.: Effect of strain and strain rate on fatique-accelerated biodegradation of polyurethane. Biomed. Mater. Res. **66A**, 463–475 (2003)
37. Kurtz S.M.: The UHMWPE Handbook. Academic Press, New York (2004)
38. Teoh, S.H., Tang, Z.G., Ramakrishna, S.: Development of thin elastomeric composite membranes for biomedical applications. J. Mater. Sci. Mater. Med. **10**(6), 343–352 (1999)
39. Noll, W.: Chemistry and Technology of Silicones. Academic Press, New York (1968)
40. Harmand, M.F., Briquet, F.: In-vitro comparative evaluation under static conditions of the hemocompatibility of four types of tubing of cardiopulmonary bypass. Biomaterials **20**(17), 1561 (1999)
41. Cronin, T.D., Gerow, F.J.: Augmentation mammaplasty: A new
42. John, L., Foster, R.: Biosynthesis, properties and potential of natural-synthetic hybrids of polyhydroxyalkanoates and polyethylene glycols. Appl. Microbiol. Biotechnol. **75**, 1241–1247 (2007)
43. Nijst, C.L., Bruggeman, J.P., Karp, J.M., Ferreira, L., Zumbuehl, A., et al.: Synthesis and characterization of photocurable elastomers from poly(glycerol-cosebacate). Biomacromolecules **8**, 3067–3073 (2007)
44. Bettingera, C.J., Bruggemanb, P., Borensteinc, J.T., Langerb, R.S.: Amino alcohol-based degradable poly(ester amide) elas-tomers. Biomaterials **29**(15), 2315–2325 (2008)
45. Barrett, D.G., Luo, W., Yousaf, M.N.: Aliphatic polyester elastomers derived from erythritol and α, ω-diacids. Polym. Chem. **1**, 296–302 (2010)
46. Cui, W., Zhou, Y., Chang, J.: Electrospun nanofibrous materials for tissue engineering and drug delivery. Sci. Technol. Adv. Mater. **11** (2010). doi:10.1088/1468-6996/11/1/014108
47. Liua, X., Wona, Y., Ma, P.: Porogen-induced surface modification of nano-fibrous poly(l-lactic acid) scaf-folds for tissue engineering. Biomaterials **27**(21), 3980–3987 (2006)

Other Applications: Engineering

L. A. Gracia, J. M. Bielsa, F. J. Martínez, J. M. Royo, J. L. Pelegay and B. Calvo

Abstract This chapter describes how the finite element technique can be used for the design of elastomeric components for automotive and railway applications. In the first section a description of the industrial needs regarding the design with these types of materials and the reasons why they arouse so much interest for engineering applications is given. Also, a complete literature review and explanation of fundamentals are included concerning different features these materials exhibit from the mechanical point of view: elasticity, inelasticity, fatigue, and tribology behavior. The second section includes several details about constitutive models used for the finite element (FE) modelling of elastomeric materials. Among them, some basic kinematics of finite elastic deformations are explained as well as details about constitutive behavior for rubbers and rubber-like materials such as strain energy potentials usually implemented in FE codes for modelling hyperelasticity, time and frequency domain viscoelasticity, constitutive models for modelling inelastic effects, and available approaches for modeling fatigue behavior. In the third section, a methodology for the design of elastomeric components by means of the FE method is explained, including valuable information about experimental testing for material characterization focused on the calibration of former explained constitutive models. In the fourth and last section, four examples are presented,

"Use of finite element (FE) techniques for the design of elastomeric components for automotive and railway applications".

L. A. Gracia (✉) · J. M. Bielsa · F. J. Martínez · J. M. Royo · J. L. Pelegay
Grupo de Investigación Aplicada en Simulación, Caracterización, Diseño y Desarrollo de Materiales (SICADDEMA), Instituto Tecnológico de Aragón (ITA), Zaragoza, Spain
e-mail: lgracia@ita.es

B. Calvo
Aragón Institute of Engineering Research, University of Zaragoza, Zaragoza, Spain

B. Calvo
Centro de Investigación Biomédica en Red en Bioingeniería, Biomateriales y Nanomedicina (CIBER-BNN), Zaragoza, Spain

P. M. Visakh et al. (eds.), *Advances in Elastomers II*,
Advanced Structured Materials 12, DOI: 10.1007/978-3-642-20928-4_9,

related to the application of FE techniques for the analysis and the design of components for automotive and railway applications. These examples cover the modelling of different aspects and features of elastomeric materials and demonstrate the advantages provided by FE techniques in comparison to the experimental design procedures used until the recent past in the industry.

1 Introduction

1.1 Motivation. Problem Description and Industrial Needs

The use of elastomeric materials in engineering has increased considerably during recent decades and many products are now made of this type of materials. Elastomers are mainly used in the automotive and railway industries and in numerous mechanical, civil, electronics and electrical engineering applications.

Reinforced elastomers are materials composed of a matrix of entangled rubber molecules with reinforcement particles, such as carbon black, oxides of zinc or sulphur, among others, embedded in the matrix (see Fig. 1). These additives cause an increase in the stiffness of the material [1] and at the same time significantly modify their inelastic, hysteretic [2] and stain rate dependent properties So and Chen [3].

The unique properties of these materials make them very useful for a large variety of industrial applications such as couplings between stiff structures or for avoiding or at least reducing transmission of vibrations. Examples of these components are: pipes, top mounts, bushings and hydro bushings for suspensions and shock absorbers, torsion axes, supports for stabilizing bars, compression blocks, seals and membranes (see Fig. 2).

The complex nature of the behaviour of reinforced elastomers and the huge variety of existing compounds make it quite difficult to establish general rules and design guidelines. However, in order to increase competitiveness in high-tech

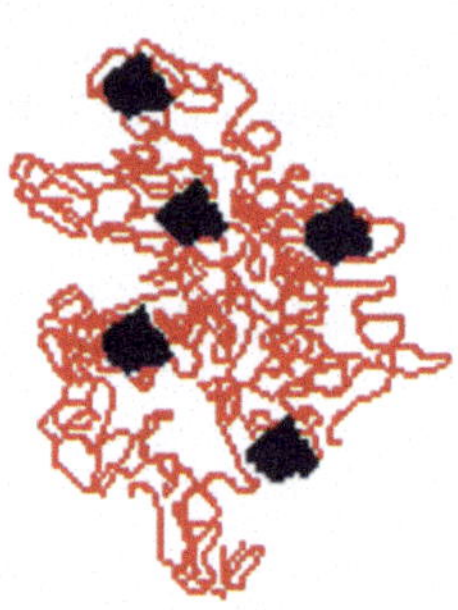

Fig. 1 Drawing of the molecular structure of an elastomeric material

Fig. 2 Example of metal-rubber components for the automotive industry

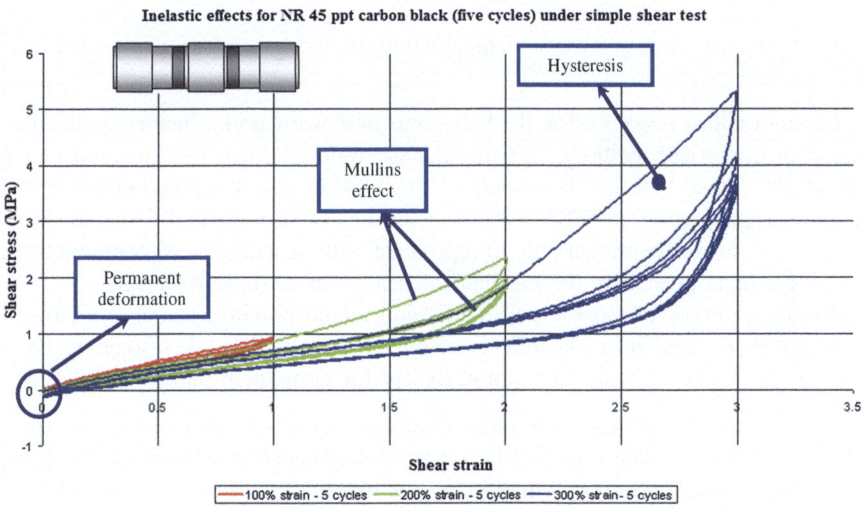

Fig. 3 Inelastic effects in the mechanical behaviour of a reinforced elastomer subjected to cyclic simple shear

applications, it is nowadays necessary to have reliable and sufficiently fast design methods, including those for the characterization of material properties.

The mechanical behaviour of reinforced elastomers is highly non-linear and strain-rate dependent. It shows hysteresis, permanent deformations and stress softening under cyclic loads. When a rubber sample is loaded in simple tension, is un-loaded and is again re-loaded, the stress for the second load is lower than the first for higher strains than those reached under the initial load. This phenomenon is known as the Mullins effect and is observed during the first cycles (see Fig. 3).

Another phenomenon, known as the Payne effect or the Fletcher-Gent effect (see Fig. 4), involves the thixotropic behaviour of the material under dynamic

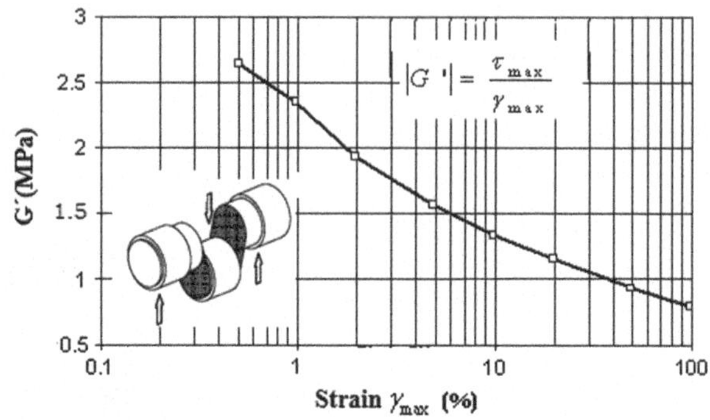

Fig. 4 Graphical representation of the Payne effect for a reinforced elastomer subjected to cyclic simple shear

loads and consists of a substantial decrease of the stiffness modulus (1) when the strain amplitude of the oscillations of the dynamic load increases.

$$|G'| = \frac{\tau_{\max}}{\gamma_{\max}} \tag{1}$$

During the last two decades, the behaviour of elastomers has been simulated by means of numerical methods, in particular the finite element (FE) method [4, 5]. Classically, these inelastic phenomena—the Mullins effect [6, 7] and strain rate dependant properties [8–12] have been studied and modelled independently, without a constitutive model able to reproduce with accuracy or to combine these inelastic effects present in the mechanical behaviour of these materials.

The mechanical response experimentally observed in reinforced elastomers can basically be divided into four effects or phenomena, which taken together characterise their general typical response, except for permanent deformation [13]:

(a) a response characterized by large elastic deformations (behaviour known as hyperelasticity—see Fig. 5). Elastomeric materials are able to reach strains up to 500 % under relatively small loads (the maximum tensile resistance of these materials is between about 10 and 15 MPa, while that of some metals reaches 3000 MPa for strains below 5 %).
(b) a superimposed response of finite viscoelasticity that governs the rate dependent effects such as relaxation (see Fig. 6) and creep (see Fig. 7).
(c) a superimposed behaviour of finite plasticity responsible for the hysteresis phenomena which are independent of the strain rate associated to the relaxed equilibrium states (see Fig. 8) and
(d) a damage response (stiffness reduction) within the first cycles, which induces in the material a considerable amount of stress softening (the phenomenon known as the Mullins effect—see Fig. 9).

Although the Mullins effect takes place with the first and more severe load events, it can influence significantly the long-term behaviour of a reinforced rubber [14]. The dependence of the stress–strain response in the pre-conditioning implies that each material point in the non-homogenous deformed component exhibits

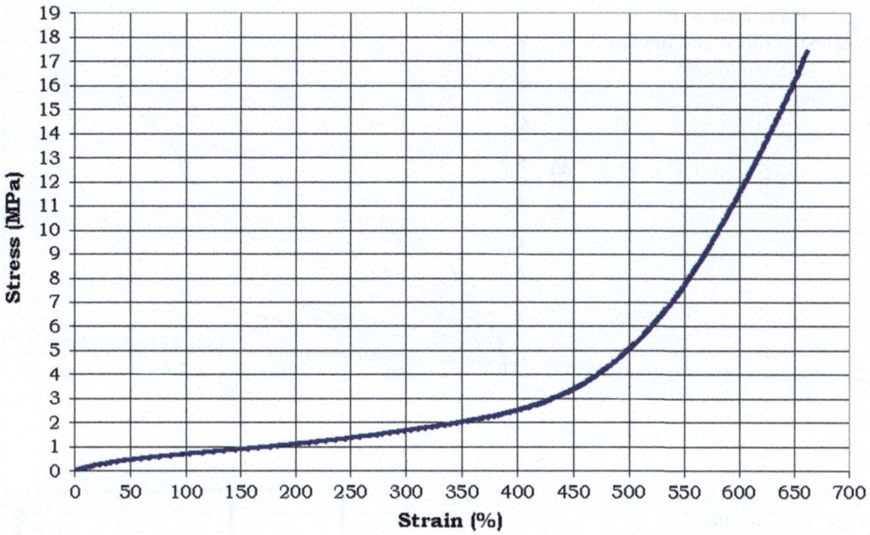

Fig. 5 Hyperelastic behaviour of an elastomer

Fig. 6 Phenomenon of relaxation for the material for different temperatures

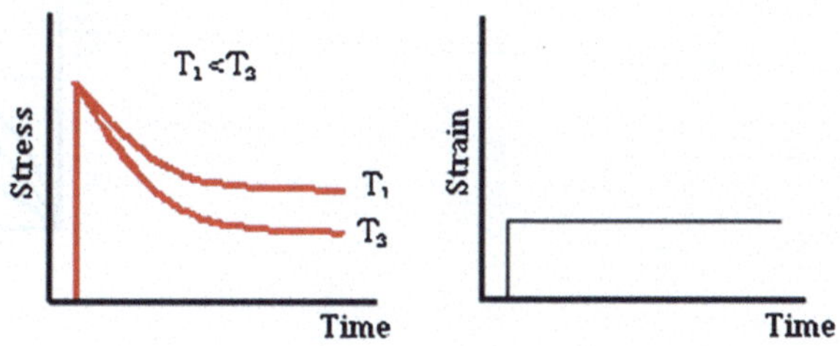

Fig. 7 Creep phenomenon for the material

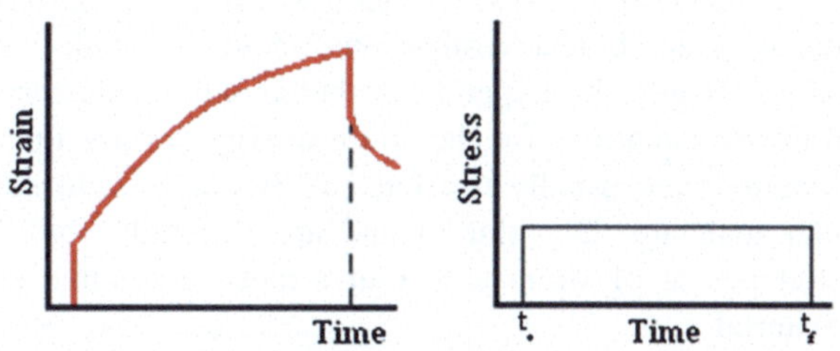

different stress–strain behaviour. Although all these material points can behave more or less according to a hyperelastic constitutive law, such a particular law varies from point to point. Because of this, a single hyperelastic constitutive law probably cannot provide realistic predictions. The inclusion of the Mullins effect allows load and unloading events to be modelled accurately [15].

The basic response under finite elastic tension governs the response of the vulcanized elastomeric polymers and has been widely investigated as much

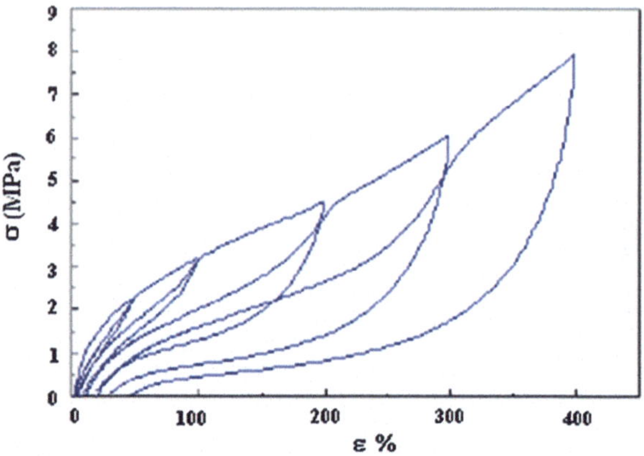

Fig. 8 Phenomenon of hysteresis in load and unload cycles of a reinforced elastomer

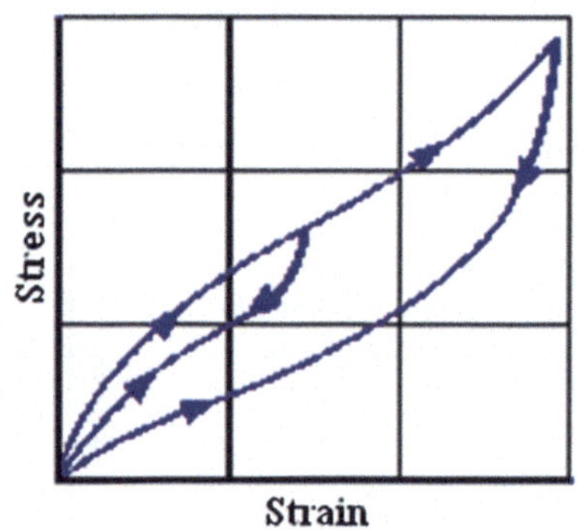

Fig. 9 Idealized behaviour of the Mullins effect

experimentally [16–19] as theoretically [20–24], and numerically [25–28]. As a result of all these investigations, there is a wide variety of constitutive models which enable the hyperelastic behaviour of elastomers to be modelled through different equations for the strain energy density function [18, 27, 29–31]. These functions are usually restricted to isochoric deformations (that is, deformations that maintain the volume) and are generally formulated in terms of principal stretches or of strain tensor invariants, assuming an isotropic behaviour of the material.

The viscoelastic response is evident in relaxation or creep tests as well as in load cyclic processes. These last show the typical frequency-dependent hysteresis in which the width of the hysteresis cycles increases with the applied deformation. This phenomenon can be explained, from a microscopical point of view, as a rearrangement of the secondary weak bonds among the polymer chains during the strain process. Experimental research works are reported in Ferry [32], Hauser and Sayir [33], Johnson et al. [34] and Lion [8] and other works of a theoretical and computational nature in Simo [35], Kaliske [36] and Lion [8] and Holzapfel et al. [37].

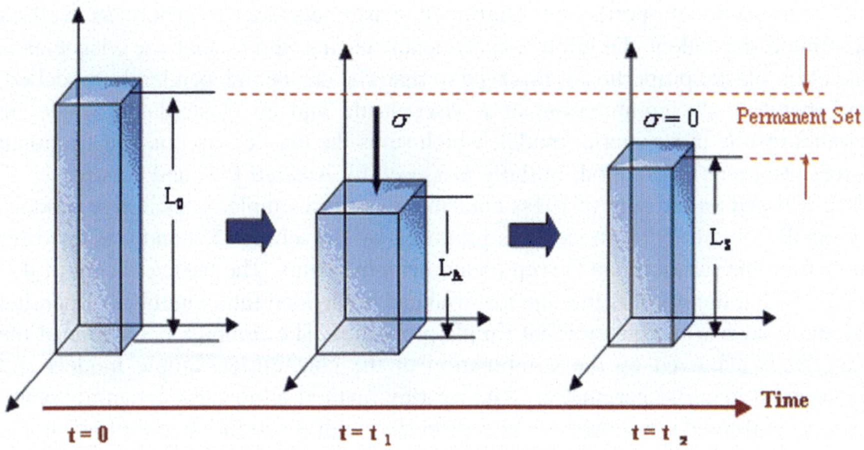

Fig. 10 Schematic representation of the permanent strain (Lo-Ls)

The strain-rate independent elastoplastic response, can be identified as hysteresis in the relaxed equilibrium response within the cyclic deformation process [38] and can be attributed, from a micromechanical point of view, to the processes of irreversible sliding occurring among the reinforced particles and also with the polymeric matrix [39, 40]. This behaviour results in the appearance of residual or permanent strains in the reinforced elastomers subjected to deformation [37]. This phenomenon is known in the industry as permanent set (see Fig. 10).

The effect of damage independent of the strain rate is identified with the stress softening that reinforced elastomers suffer during the first load cycles, known as the Mullins effect [41–45]. From a micro-mechanical point of view this can be explained as the rupture of the bonds among the polymer chains and the reinforced particles [46–48].

The Mullins effect can be interpreted as an effect of damage [49, 50], in which the evolution of the damage depends on the maximum stretch occurring in the strain history [37] or on the maximum strain energy [35, 51, 52]. The energy required to cause the damage is not recoverable and is dissipated in the form of heat.

The constitutive models available in commercial finite element (FE) codes for the modelling of hyperelastic behaviour of elastomeric materials are calibrated from experimental data coming from uniaxial deformation modes. It is highly desirable that the material be tested under uniaxial deformation modes similar to those predominant in the component in order to achieve an accurate characterization.

These hyperelastic models, well known and widely used in the industry, at least in the automotive sector, are insufficient to model the dynamic behaviour of such materials. Although there are other viscoelastic or damage models implemented in FE codes for reproducing the inelastic effects of these materials, it has been demonstrated that they are not able to model accurately the global behaviour of reinforced elastomers.

The dynamic properties of reinforced elastomers are frequency as well as amplitude dependent. Different experimental testing shows that the viscoelastic and elastoplastic properties of this type of material can be independently modelled, and therefore the combination of a viscoelastic and an elastoplastic model in parallel results in a material model, which sums the elastic, viscous and frictional forces. The overlay method, initially proposed by Austrell [53] and Austrell et al. [54], is based on the sum of stress contributions from simpler constitutive models. Originally, this fraction model was proposed by Besselling [55] and was used for modelling the plasticity and creep phenomena in metals. The basic concept of this model is that it considers that the material can be divided into a number of parallel fractions, each with conventional simple properties. The complex behaviour of the material is achieved by the combination of the constitutive simple models and appropriate material parameters. The overlay method allows the dynamic behaviour of reinforced elastomers to be reproduced with a certain degree of accuracy, including the amplitude dependency [56, 57], frequency dependency as well as the mechanical hysteresis typical of this kind of material. Olsson and Austrell [58] propose a procedure for the fitting of the constants of the overlay method from dynamic stationary shear tests. Ahmadi et al. [59] propose a methodology for calibrating the material parameters of the overlay method from quasi-static experimental data.

Anti-vibration components are usually made of reinforced elastomers and work under a preload to which a cyclic strain history is superimposed; dynamic stiffness and the loss angle are essential properties defining their behaviour. Therefore, their prediction with a degree of accuracy is quite important at their design stage.

The model of Morman et al. [60], implemented in commercial FE codes such as ABAQUS [97], MSC. MARC [61] for the analysis of low-frequency vibrations in viscoelastic solids submitted to a initial static strain, is not able to model the behavior of reinforced elastomers. These materials show a strong dependency of the dynamic stiffness with the amplitude [56, 57] but the model only includes the frequency dependency and does not reproduce the dependency either with the preload or with the amplitude.

In practice, there is not too much information available about the dynamic response of reinforced elastomers under complex deformation modes and neither there is any accepted methodology for the consideration of their properties in FE analysis for the prediction of dynamic behaviour under cyclic loads. The methodology used to date consists of the characterization of dynamic properties of the material through testing, scanning a frequency range, average deformations and amplitudes, resulting in a huge matrix with stiffness modulus and loss angle values for each of the analyzed conditions [62]. The analyst must decide which conditions to simulate, depending on the range of values and the design of the applications.

Gómez and Royo [63] developed a methodology which permits considering these dependencies by modifying the viscoelastic model implemented through calibrating the viscoelastic parameters and by defining the material behaviour in a particular way for each element of the model in function of its strain level, preload as well as amplitude. This methodology is based on the combination of the results analysis derived from the material characterization testing and the implementation of the shift factor functions in commercial FE codes, through user subroutines.

Concerning the fatigue life prediction of elastomeric materials, designers define this fatigue life through minimising the stress, strain or energy density values, which are calculated through the FE method, making exclusive use of hyperelastic material models. The inelastic effects, characteristic of this type of material, are not taken into account. The main requirement is to obtain material constants that describe the material behaviour in service conditions. These constants are frequently unknown. In the particular case of elastomers, there remain many unresolved questions, which converge in this simple requirement. The properties of elastomers are strongly dependent on several factors such as the temperature, humidity, light and load conditions to which they are submitted. All these factors are able to modify the fatigue properties of the material (crack initiation and propagation), and therefore the first challenge is to test and asses these properties accurately, quickly and efficiently. The second, more complex challenge is to predict the life through numerical models. Furthermore, being able to simplify the in-service conditions in the experimental testing is important so that the tests may be viable in cost and duration.

The phenomenon of fatigue is described by Ellul [64] as the progressive weakness of several physical variables, i.e. stiffness loss, as the result of the slow growth of cracks produced by the application of cyclic loads or strains. The microscopic process that starts fatigue in elastomers is not so well known as that of metals, and consequently the macroscopic and phenomenological approach remains the most common today. In fact, fatigue treatment in elastomers is mainly empiric. The problem is faced through durability testing in components or in simple samples, whose results are extrapolated for fatigue predictions. This practice is quite limited due to the large number of factors affecting the fatigue life of this type of material, including mechanical factors such as frequency, load sequence, load type—relaxing or nor-relaxing load, uniaxial or multiaxial loads, … -, thermal factors such as temperature, oxygen concentration, ozone, UV rays; and chemical factors such as material composition, additives and vulcanization. Up to now, the consideration of all these factors in damage models for rubber materials has been achieved by means of the calibration of empiric equations to test results. Several authors such as Lake [65, 66], Cadwell [67] and Roach [68] have studied several of the aforementioned aspects. It is important to remark that fatigue failure in elastomers occurs in a fragile way, that is, without previous plastic deformations, as was indicated by Lake [69] and Lake and Yeoh [70]. It occurs in two phases: the first, in which the cracks start nucleation around agglomerate particles in the material, and the second one in which the cracks start growing until the material fails. For some time, it was believed that the crack nucleation, growth and final failure could be modelled in terms of the mechanical behaviour of elastomer fracture [71, 72]. However, there are aspects that only occur during the nucleation phase which require more careful study. In particular, it is necessary to understand the aspects of the multiaxial fatigue in the crack nucleation phase [73].

One of the weak points relating to fatigue life prediction in elastomers is the determination of the accumulated damage and the treatment of multiaxial loads. Flamm et al. [74] demonstrate that, in most cases, the Miner's linear accumulation damage rule is inadequate. Besides, an added problem is related with the testing frequency, especially for bulk samples, in which the heat generated with the cyclic loading causes degradation of the material. Therefore, the testing temperature becomes an influential factor especially important in the fatigue life of a reinforced elastomer.

The non-linear behaviour of rubber (finite strains and quasi-incompressibility) and the fact that scalar criteria make no reference to a specific material failure plane imply that it is always possible to construct a non-proportional multiaxial history that holds the scalar equivalence criterion value constant while simultaneously varying the individual components of the history [75, 76]. Therefore, scalar equivalence criteria predict infinite life under certain kinds of non-proportional cyclic loading which in fact result in finite life. It can therefore be concluded that an analysis approach that makes specific reference to the failure plane is needed.

Due to the aforementioned limitations of scalar fatigue criteria, different multiaxial fatigue criteria have been proposed in literature to overcome these limitations. However, it is worth pointing out that their level of development has been uneven since most of them have been applied to natural rubber at specimen level only, excepting the cracking energy density proposed by Mars [72] that has been applied successfully to automotive components [77].

The influence of non-relaxing cycles hinders crack growth and therefore increases the expected fatigue life (see Fig. 36). The effect of non-relaxing loads is of great importance in real components such as shock absorbers, which are submitted to static preload (because of the assembly process, external loads or vehicle weight) superimposed on small periodic oscillations. The effects of non-zero minimum loading on fatigue crack growth have been analysed in the literature for strain crystallizing rubbers [78, 79] and for non-crystallizing rubbers [80]. The effects of non-relaxing cycles were incorporated in a fatigue life prediction model using the model proposed by Mars and Fatemi [75, 76].

Relating tribology in elastomers, it is extremely important to develop a scientific knowledge on tribological behaviour on microscale and even more at nanoscale levels in order to avoid expensive "trial and error" method, broadly used in industry. Historically, the majority of friction investigations have been carried out for metals, being the most significant those proposed by Coulomb and Amonton (see [81] and [82]). In contrast with other rigid materials, friction of rubbers is characterized by several macroscopical dependencies: contact pressure, relative sliding speed and temperature. The particular mechanical characteristics of rubbers influence their frictional behavior, as has been demonstrated by many authors [83, 84]. Thirion [85] demonstrated that rubber friction coefficient falls markedly when the contact pressure increases. The surface morphology of the countermaterial also plays a fundamental role in rubber friction. According to most recent investigations found in the literature [86, 87], dry friction on rubbers is mainly governed by the hysteretic and the adhesion phenomena, which can be modelled according to the bulk properties of the rubber material (complex modulus) and the surface roughness structure of the metallic counterpart. Regarding wear in elastomers, there is still not a clearly set up classification. The most general classification of wear types is set up along several decades by authors such as Kragelskii [88], Blau [89], Zhang [90] and Myshkin [91], including in it wear by abrasion, by erosion, by fatigue and by adhesion, although other types of wear such as corrosion, tribochemical or fretting wear, caused in the process during the first wear types, are also considered in literature (Burris [156], [92–94]). The different influences set up in the wear process in the elastomer must also be included in the wear modelling. The vast characterisation was carried out by Meng and Ludema [95], considering three main approximations about wear modelling: models based on empirical relationships, models based on contact mechanics and models based on material failure mechanisms.

1.2 Overview of Mechanical Behaviour of Elastomeric Materials: Elastic, Inelastic and Fatigue Behaviour

1.2.1 Elastic Behaviour

The most obvious as well as the most important physical feature of elastomers is their capacity of being deformed under relatively small stresses. The most relevant features of the elastic behaviour of these materials are as follows [17]:

(i) The stress–strain curve is highly non-linear and therefore it is not possible to define the elastic behaviour of these materials through the Young modulus (Hooke law) as in metals. In this type of material, the elastic modulus is dependent on the strain level [96]. At low strains the modulus is high because the connections between the reinforcement particles and the matrix are active. As the strain level increases, this property decreases because the interactions between the reinforcement particles break and for high strain levels, the modulus increases again because of the reaction caused by the finite extensibility of the polymer chains.

(ii) The material, depending on its composition, is able to reach high strains under relatively small loads.

(iii) The behaviour of these materials is practically elastic, that is, once the stress is removed, the elastomeric material recovers its original shape. This property is more or less true depending on the composition of the elastomer compound. Depending on the added reinforcement particles this property can decrease and the material can exhibit the so-called permanent set or residual strains, which reduce the ability of the material to recover itself.

The non linear elastic behaviour of rubber can be successfully modelled by means of the hyperelasticity models available in commercial FE codes [97].

In general, the compressibility of elastomers is quite low. A material is defined as incompressible when its volume does not vary as it deforms, except for when the deformation is due to thermal expansion.

Elastomers can be considered as incompressible material, because they show a big difference between the initial shear modulus (μ_0) and the initial bulk modulus (K_0). A typical reinforced elastomer for industrial purposes presents a shear modulus ranging between 0.5 and 6 Mpa and bulk modulus ranging between 2000 and 3000 MPa [98]. The high volumetric compared to shear stiffness indicates a practically incompressible behaviour.

The relative compressibility of a material can be evaluated by the relationship K_0/μ_0. This expression can also be expressed in terms of the Poisson coefficient through (2).

$$v = \frac{3^{K_0}/_{\mu_0} - 2}{6^{K_0}/_{\mu_0} + 2} \tag{2}$$

Some investigations [99] have demonstrated that natural rubbers reinforced with carbon fibres, when submitted to different load conditions (uniaxial, hydrostatic, monotonic, cyclic), suffer from volume changes. The mechanisms of volume change seem to be related with the damage evolution, chain orientation and viscoelasticity. At the microscopic scale, thanks to observations with the scanning electron microscope (SEM), the evolution of the damage with the elongation has been measured and it has been concluded that, the volume change and the damage evolution are proportional.

1.2.2 Inelastic Behaviour

The response of elastomeric materials under cyclic loading is of interest for several applications, especially those related to the absorption and reduction of vibrations (i.e. for shock absorbers, bumpers and silent-blocks) [100]. Under dynamic loading elastomers exhibit dissipative phenomena such as the Mullins effect, hysteresis and strain-rate dependency due to viscoelasticity.

The parameters usually used for describing the dynamic properties of these materials are the dynamic modulus and the loss angle. These parameters come from the linear viscoelasticity theory.

Suppose that the excitation load A and the response, F, vary in a sinusoidal way:

$$A = A_0 sin(\omega t) \qquad F = F_0 sin(\omega t + \delta) \tag{3}$$

where, A_0 and F_0 are the amplitude of the excitation (displacement) and of the response (force), respectively, ω is the oscillation frequency and δ is the delay between the application of the excitation and the system response.

With these magnitudes, the two parameters defining the dynamic behaviour of the material are:

- the dynamic stiffness, defined as the relationship between the amplitudes of the response and the excitation

$$K_{din} = \frac{F_0}{A_0} \quad (4)$$

- the damping, which is related to the delay introduced into the system by the damper element

$$\eta = \tan \delta. \quad (5)$$

Similarly, instead of relating displacements and forces, stresses and strains are related and the following expressions are obtained for the dynamic modulus

$$E_{din} = \frac{\sigma_0}{\varepsilon_0} \quad (6)$$

and the damping:

$$\eta = \frac{W_c}{\pi \sigma_0 \varepsilon_0} \quad (7)$$

where W_c is the energy dissipated in one cycle (see Fig. 11).

These expressions are quite simple to obtain when the behaviour of the model is fully linear and are useful for showing clearly the physical meaning of the two magnitudes. However, their extension to non-linear behaviour, as occurs in real components, is not so easy, due to the fact that harmonics deform the time signal making the measurement of the amplitude as well as the loss angle difficult [101]. The sources for non linearities can be geometrical effects (friction and large

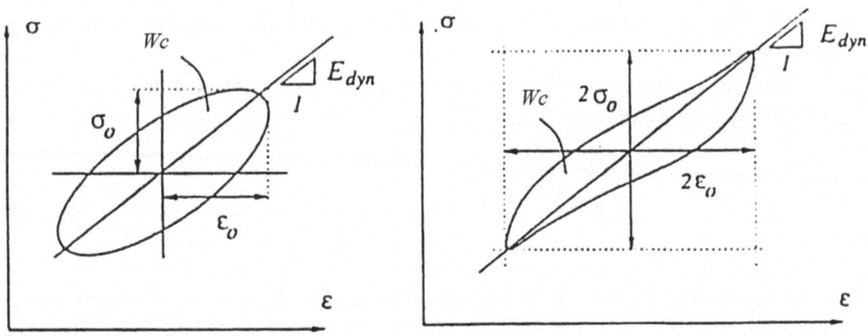

Fig. 11 Hysteresis cycles with linear behaviour (*left*) and non-linear behaviour (*right*)

displacements), type of loading or material properties, and in the case of reinforced elastomers all these factors at the same time.

Many rubber industrial components suffer from a large variety of dynamic load types, including non-regular periodic oscillations resulting from a combination of different frequencies, pulses and random noise. Together with the non-linear properties of reinforced elastomers, this means that the linear viscoelastic theory cannot be applied [102, 103].

Viscoelastic damping has a significant presence in many polymeric materials and this internal damping is a very useful feature in many industrial components. It has its origin in the molecular structure of the material. The damping comes from the relaxation and recovery of the polymeric net after deformation. An important relationship exists between the effect of the frequency and the temperature because of the direct connection existing between the material temperature and the molecular movement.

The damping grows with the active presence of reinforcement additives which result in a two-phase material with constituents of very different mechanical properties. The phenomenon derives fundamentally from two mechanisms [53]:

- Viscous, a result of resistance to the reorganization of the rubbery phase chains. This reorganization of very long chains cannot occur suddenly. It depends on the strain rate causing the viscous nature of the material.
- Hysteretic, caused by the additives that are much stiffer than the rubber matrix in which they are embedded. These compounds make connections inside the rubbery net. When the material is submitted to strain, these C–C and C-rubber joints break, a process that is rate independent (frequency non-dependent). These breakages are responsible for the non-linear behaviour with the amplitude.

During a load cycle elastomers dissipate energy causing hysteresis. This means that the material behaviour is different during the load path than during the unload path. The hysteresis is a direct consequence of the viscoelastic behaviour of the elastomers and it provokes a delay between the stress and the strain history. The reason that the stress, for a certain strain, is lower during the loading paths is that the material relaxes during the time of the cycle and dissipates energy in the form of heat. This dissipated energy in the cycle corresponds to the closed area in-between the loading and unloading paths and it is used in the industry, for instance, for the damping of vibrations.

The lost energy per cycle W_c (represented by the area enclosed in the cycle) for a certain frequency is expressed as:

$$U_c = \oint \sigma d\varepsilon = \sigma_0 \varepsilon_0 \omega \int_0^T \cos(\omega t) sin(\omega t + \delta) dt = \pi \sigma_0 \varepsilon_0 sin\delta = \pi A_0 F_0 sin\delta \quad (8)$$

where ε_0, σ_0, A_0 and F_0 are the amplitude of the strain, the stress, the displacement and the force, respectively and δ is the loss angle, which is related to the material damping by (5).

When non-linear behaviour is present, the hysteresis cycles are deformed from the ellipsoidal shape that they exhibit when non-linearities exist. Usually in these cases the equivalent damping is identified as the area enclosed by the distorted non-linear cycle (see Fig. 11 with (8)).

When reinforced elastomers are submitted to cyclic loading, from the first, until the fourth or fifth cycles depending on the compounding, they suffer from significant stress-softening. This is known as the Mullins effect [104]. When these materials are submitted to large strain cycles, they suffer from stress softening, but only for the subsequent strains lower than those reached previously [105]. After this softening and for strains higher than the maximum strain level reached previously, the material tends to retake the stress–strain behaviour of the virgin state. This phenomenon is attributed to the progressive breakage of the connections between the rubber matrix and the reinforcement particles, and to the configuration changes of matrix itself [53]. The Mullins effect can have important implications for the way in which characterization tests are carried out because the material behaviour can be considerably affected by the previous testing that it has been subjected to [96]. In order to obtain stationary behaviour in the testing of reinforced elastomers, it is necessary to pre-strain the samples before executing the tests, a practice commonly known as material preconditioning.

The anisotropy of an elastomer is strain driven [31, 41] and it is usually defined by the model proposed by Spencer [106] in which the strain energy density function (W) includes the predominant behaviour directions through one-dimensional arrays ($\mathbf{m}_i^0$ and $\mathbf{n}_i^0$): $W = W\left(\mathbf{C}, \mathbf{m}_i^0, \mathbf{n}_i^0\right)$. This formulation has been used by different authors to define the behaviour of soft biological materials reinforced with collagen fibres as blood vessels [107], ligaments [108] or cornea [109].

The use of reinforcement loads or particles increases the damping as well as the stiffness of the rubber compound and, as has been previously mentioned, the stress–strain behaviour becomes non-linear, the stiffness being greater at lower strain amplitudes. This phenomenon is known as the Fletcher-Gent effect or Payne effect [57]. The loss angle is also amplitude dependent and reaches its maximum value at a low strain percentage, while decaying considerably for higher strain amplitudes. This amplitude dependency of the dynamic properties makes the linear viscoelasticity theory non valid: the higher the percentage of the reinforcement load, the less applicable is this theory. A feature especially relevant for the isolation of vibrations at high frequencies (usually associated to small amplitude vibrations) is that the dynamic stiffness value tends asymptotically to a finite value at small amplitudes [110].

An additional difficulty induced by the use of reinforcement particles is that the dynamic properties experience a delay in reaching the stationary or equilibrium state after the application of a sinusoidal strain [56]. If the previous deformation is lower or null compared to the measurement strain, the dynamic stiffness will decay until the equilibrium state, while if the previous strain amplitude is higher, there will be a recovery towards the dynamic stiffness associated to the lower strain.

This fact suggests that a complete constitutive model for reinforced elastomers requires frictional as well as viscoelastic elements.

Medalia [111] provides a detailed review of the dynamic properties of reinforced elastomers and their dependency with the strain amplitude.

For most industrial applications of reinforced elastomers, the Payne effect, the time dependency as well as the Mullins effect are non desired but unavoidable phenomena. It is therefore desirable to be able to quantify them.

1.2.3 Fatigue

When mechanical rubber components are submitted to dynamic loading, they suffer from fatigue. This phenomenon appears in this type of material as the progressive weakness of several physical variables, i.e. stiffness loss as the result of the slow growth of cracks produced by cyclic loads or strains. There is evidence that the fracture of rubber materials occurs through the presence of defects or imperfections in the parts. From these imperfections, intrinsic in these materials, the cracks can grow under a certain load till they reach a sufficient size to cause the fracture of the material. Due to the initiation of cracks being produced from very small defects in the parts and to the complex behaviour of the elastomeric materials, there is a wide disparity in the prediction of the fatigue life in samples without predefined cracks.

Typical models for predicting fatigue life in rubber follow two overall approaches. The first one focuses on predicting crack nucleation life, given the history of quantities defined at a material point, in the sense of continuum mechanics. Stress and strain are examples of such quantities. The other approach is based on ideas from fracture mechanics, focusing on predicting the growth of a particular crack, given the initial geometry and energy release rate history of the crack.

The crack nucleation approach or ε-N approach considers that the material has a life determined by the stress and strain history at a certain point. The fatigue life according to this approach could therefore be defined as the number of cycles necessary to obtain a crack of a certain length. This approach is quite familiar and convenient for designers as it is formulated in terms of stresses and strains. It is particularly appropriate when the component under study exhibits cracks or initial defects of a size several orders of magnitude lower than its characteristics and when the multiaxial stress state can be related with some accuracy with the stress state of the fatigue material characterization tests. In order to model the effect of the multiaxial loads on the fatigue life of elastomers during the nucleation phase, it is necessary to refer to an equivalence criterion that defines the basis on which to confirm if a component is valid or not for standing fatigue loading and involves one or more parameters characterising the mechanical severity of the load history. The parameters traditionally used in the crack nucleation approach as equivalence criteria are the maximum principal strain and the strain energy density. To date, it seems that only scalar equivalence criteria have been applied for the fatigue life prediction in elastomers, which cannot account for the crack growth direction.

Critical plane theories rely on the physical process of fracture and make use of the continuum variables on the actual fracture plane. Many critical plane approaches have been successively developed in the metal fatigue field, for instance, the models of Brown–Miller [112], Fatemi–Socie [113], Smith et al. [114], Wang and Brown [115–117] and Chen-Xu-Huang [118]. However, for rubber fatigue only a few authors have published crack nucleation parameters associated with a critical plane idea [75, 76, 119, 120]. The use of strain as a life parameter has advantages since it can be obtained directly from measured displacements. When the strain energy density is used, it is often evaluated using hyperelastic material models based on the strain invariants and therefore it is also based on strains. Strain energy density has been used as a fatigue parameter in metals [121], although the correlation between experimental and predicted results is not satisfactory. Rivlin and Thomas [122] proposed a model to study the fracture of rubber under static loading based on the strain energy density, and this has been used by many researchers to correlate analysis results to experimental component life data, considering the strain energy density as a measurement of the energy release rate of the different flaws present in the material. The application to components of this approximation carried out by some authors [123, 124] shows differences in fatigue life to computed strain energy density levels. The main limitations of the strain energy density are that (a) it is unable to predict the fact that the crack surface appears in a specific orientation, and only part of the total spent energy plays a role in the crack nucleation process for multiaxial conditions; (b) it does not account for crack closure and (c) it fails to predict large life differences between simple tension and simple compression loadings. Stress has rarely been used as a fatigue life parameter in rubber [125]. This is related to the fact that fatigue testing in rubber has traditionally been carried out by displacement control, and the accurate stress determination in rubber components can be difficult. The maximum principal strain and the octahedral shear strain have also been used as fatigue parameters based on strains. The maximum principal strain criterion was introduced by Cadwell et al. [67] for unfilled vulcanized natural rubber and remains in use nowadays, particularly for uniaxial strain loadings. It also provides a good correlation for axial/torsion tests, whereas the octahedral shear strain criterion makes a prediction that is roughly similar to the principal strain criterion for rubber [75, 76]. However, for an incompressible material both strain based criteria always satisfy that their values are positive and they are therefore unable to account for compression states where the crack is closed.

The complementary focus to the ε-N approach for the fatigue life prediction of reinforced elastomers is fracture mechanics Lake [69]. This approach assumes explicitly the existence of cracks or defects (material inhomogeneities, agglomerates or contaminants introduced in the material mixing procedure) in the material. The fundamental hypothesis is that the crack growth occurs because the stored potential energy of the component turns into surface energy associated with new crack surfaces. The basis of the energetic approach is the use of the strain energy relaxation velocity or tear energy (T) as a mean of characterizing the behaviour of the crack growth in the material. Usually, this relationship is obtained experimentally through crack growth testing in samples where, for a known strain deformation mode, the tear energy is a function of the sample geometry and/or applied load. With this process together with quantifying the crack growth speed through the applied load cycle, it is possible to characterise the crack growth speed in the material. This approach was successfully applied by Lindley and Stevenson [126] and Gent and Wang [127] for predicting the fatigue behaviour of elastomeric components submitted to compression and shear loads by relating the energy release rate with the strain energy density instead of with the maximum principal strain. However, there remains much to investigate in relation to the suitable material characterization for crack growth, influential factors (initial crack growth, temperature, frequency and load type) and equivalence criteria, which become essential aspects if FE models are to be used for fatigue life prediction.

Fatigue life prediction for rubber combining the crack nucleation and growth approaches has been applied successfully by different authors [66, 128]. This analysis is based on the integration of a crack growth law relating the crack advance per cycle and the energy release rate or tearing energy. The basis of the energetic approach is the use of the strain tearing energy, as a means of characterising crack growth behaviour. The relationship between the crack growth rate dC/dN and tearing energy T is known as the crack growth characteristic of the material since T is independent of the sample geometry. Typical curves for a natural rubber (NR) compound cycled under relaxing and non relaxing conditions (R-ratio of 0.05 and 0.1) are shown in Fig. 36.

The fatigue life of a certain structure can be considered as the number of cycles necessary for a certain crack present in the material at the beginning of the fatigue process (c_0) to grow up to a critical length (c_1) that provokes the final failure of the component. Given the crack growth behaviour and the energy release rate history, the fatigue life can be computed via the integration of the crack growth law between the correct limits [64]. The fatigue life of any rubber component can be predicted by integrating (9) depending on the energy release rate and its description according to the crack growth behaviour only.

$$N = \int_{c_0}^{c_1} \frac{dc}{f(T_{\min}, T_{\max})}$$
$$N = \int_{T_i}^{T_f} \left(\frac{1}{dT/dc} \right) \cdot \left[\frac{dT}{f(T)} \right] \tag{9}$$

The Cracking Energy Density (CED) proposed by Mars [72] rationalizes fatigue life for specific failure planes across a wide range of states, relates physically to the fracture mechanical behaviour of small flaws under complex loading and is well defined for arbitrarily complex strain histories. This parameter accounts for the effect of crack closure, which occurs when the stress state causes compression on a material plane. This fact is of great importance because rubber is most commonly used in applications which experience a compressive load. Mars and Fatemi [75, 76] proposed three different critical plane criteria for its use in the computation of CED, the plane that maximizes the CED peak value, the plane of maximum CED range and the plane of minimum life.

Saintier et al. [119, 120] investigated fatigue crack initiation under multiaxial non-proportional loadings in a natural rubber and tested under multiaxial loading. The proposed fatigue crack criteria is based on the micro mechanisms of crack initiation such as cavitation, decohesion and micro-propagation, consisting of a critical plane approach under large strain conditions using a micro to macro approach. This criterion gives promising results, by predicting the fatigue life, crack orientations and location even in cases with internal crack initiation although for the moment this approach is limited to proportional loading histories.

Verron et al. [129, 130] and Andriyana et al. [131] proposed a new predictor for crack nucleation in rubber based on the configurationally stress tensor to propose a fatigue life predictor for rubber. This criterion is formulated in terms of continuum mechanics quantities in order to be combined with the standard FE method in engineering applications. It takes into account the presence of microscopic defects by considering that macroscopic crack nucleation can be seen as the result of the propagation of those microscopic defects. For elastic materials, it predicts privileged regions of rubber parts in which macroscopic fatigue cracks might appear.

Wang et al. [132] proposed a continuum damage model to investigate the fatigue damage behaviour of elastomers. The elastic strain energy of a damaged material is expressed based on the Ogden model [24], and the damage strain energy release rate is derived in the context of continuum damage mechanics. The damage evolution equation is established to develop a formula to describe the fatigue life as a function of the nominal strain amplitude under cyclic loading. The results indicate that the theoretical formula for the fatigue life as a function of the nominal strain amplitude, derived from the proposed damage model, can describe experimental data for carbon-filled natural rubbers.

1.3 Overview on Tribological Behaviour of Rubber-Like Materials: Friction and Wear

In recent decades, tribology has played a remarkable role in mechanical systems in which components are made of rubber-like materials working under sliding conditions. Such components are important in most industrial sectors, particularly in automotive and railway applications. Two of the main aspects related with the tribology of rubber-like materials are friction and wear, which are explained in detail below.

1.3.1 Friction of Rubber-Like Materials

As it is commonly known, the classical Coulomb and Amontons friction laws, which mainly establish that the friction coefficient is independent of the area of contact, have been proved to be non-valid in the case of rubber-like materials. For this material type, due to their specific mechanical properties, the friction coefficient should be expressed as a function of contact pressure, sliding speed, temperature and lubrication regime.

$$\mu = \mu(L, T, \dot{\gamma}, \upsilon, R_a \ldots) \tag{10}$$

The dependence with the contact pressure is associated to the varying ratio of real (microscopic level) to apparent (macroscopic level) area of contact when the vertical load (contact pressure) rises. The problem increases in complexity when neither the contact pressure distribution nor the ratio of real to apparent area of contact are uniform along the apparent area of contact, cylindrical contact geometry being a typical example of this situation.

In the proposed expression (10), several dependencies on external parameters such as vertical load (contact pressure), temperature, sliding speed, etc., are introduced as key variables. Rubber-like materials in general and rubbers in particular have high friction characteristics, a consequence of their low elastic modulus and their viscoelasticity. Thus, under contact pressure they deform to a large extent, resulting in high values of the real area of contact. Hence, classical models for metals are no longer valid in the case of rubber friction.

The high friction coefficient has been exploited in many applications, for example: tyres, shoe soles, bicycle brake blocks, etc. However, there are many other applications in which the frictional behaviour of the rubber is expected to have the opposite characteristics, as for example in the case of windscreen wipers and seals. In such cases the rubber must be treated to produce low frictional properties in the case of dry friction, or else the working conditions must be ensured to be in the hydrodynamic lubrication regime.

It is commonly known that friction coefficient values are difficult to find in the literature. This is because the friction coefficient can rarely be assumed to be constant and, as stated in expression (10), it depends on several factors such as contact pressure (vertical load), sliding velocity, temperature, surface roughness and lubrication regime, where applicable.

As described in the Sect. 1.2, Amontons and Coulomb established that friction force is proportional to the vertical load and independent of the geometry of the contact. Coulomb defined the friction coefficient μ as the ratio between friction and vertical load. For materials obeying this law, μ is independent of the vertical load and thus of the normal stress. Rubber does not obey Amontons' and Coulomb's laws since the friction coefficient falls markedly when increasing normal stress. For this particular behaviour, an analytical law which became widely used was defined by Thirion [85]:

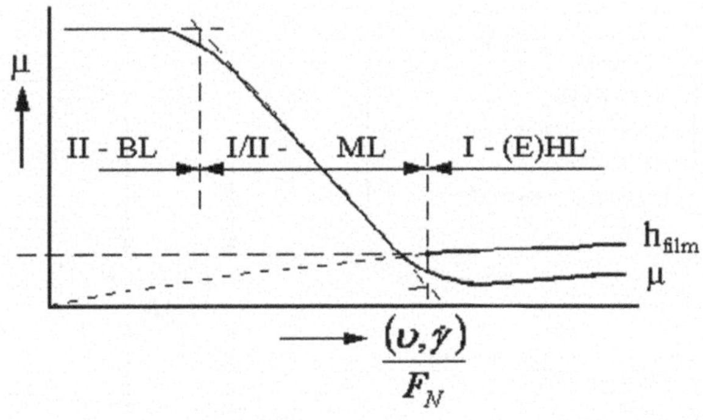

Fig. 12 Variable dependency in friction between rubber-like materials and wear

$$\frac{1}{\mu} = a + b\left(\frac{P}{E}\right) \tag{11}$$

where μ is the friction coefficient, P is the normal stress, E is the elastic modulus of the rubber and a and b are empirical constants. Schallamach [133] later showed how the behaviour described in Eq. (11) may be explained on the assumption that the friction force is proportional to the true area of contact, resulting in:

$$\mu = const\left(\frac{P}{E}\right)^{\frac{-1}{n}} \tag{12}$$

where the value of n is derived from a model which considers the deformation of the rubber on the asperities of the metallic counterpart and depends on the geometry and distribution considered for peaks and valleys. In general, n depends on the nominal normal stress, but for restricted ranges it is considered to be constant. At sufficiently high normal stresses, the real area of contact becomes equal to the apparent area of contact, so that the frictional force becomes constant and μ is inversely proportional to P, as described in (12). This particular condition is referred to as "saturation".

In addition, the lubrication regime plays a substantial role in the frictional behaviour of rubbers in lubricated (fluid) conditions. The influence of the lubrication regime can cause a drop in the friction coefficient value from one order of magnitude, depending on whether the lubrication regime is in the boundary state (direct interaction between the rubber and the micro-asperities of the metallic counterpart), the elasto-hydrodynamic state (the rubber and the asperities of the counterpart are separated by the lubricant placed in between, and the frictional force is caused by the viscous shearing of the fluid) or finally in the mixed regime (the lubricant thickness is of the order of a couple of molecular chain lengths). The influence of the lubrication regime on friction is described by the Stribeck curves (see Fig. 12, [84]), which establish a relationship between the lubrication regime and a given physical magnitude consisting of the ratio of lubricant viscosity times sliding speed to vertical load.

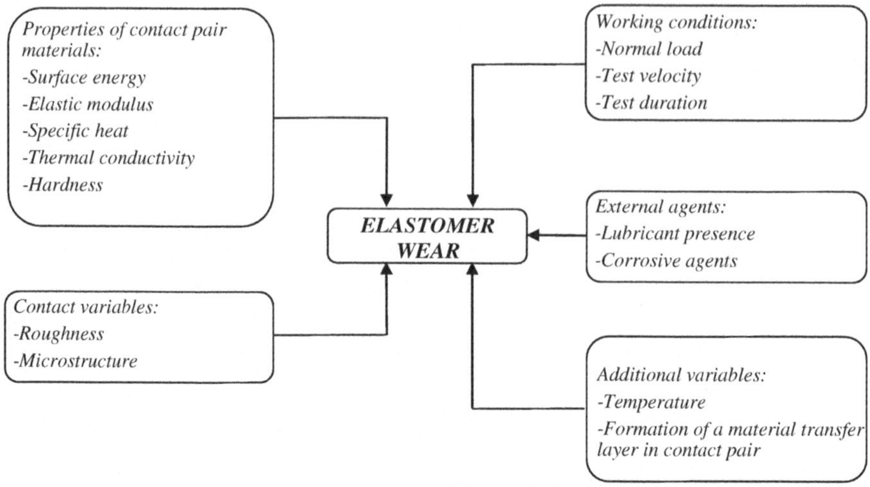

Fig. 13 Variable dependency in wear between rubber-like materials and metals

The first stage of the curve corresponds to boundary lubrication (BL), the second to mixed lubrication (ML) and the third to elasto-hydraulic lubrication (EHL). In the curve, μ is the friction coefficient, υ the kinematic viscosity of the fluid, $\dot{\gamma}$ the sliding velocity, F_N the normal load and h_{film} the height of the fluid film.

Finally, prior to the definition of a test configuration for the analysis of the friction coefficient in rubber-like materials, the micro-scale effects on their friction mechanisms briefly described above must be taken into account. Thus, within the experimental and numerical work to be carried out for the definition of the test configuration, the effect will be evaluated of the macroscopic parameters (vertical load, sliding speed and temperature) on the measured friction force.

1.3.2 Wear on Rubber-Like Materials

Wear in polymers in general and in rubber-like materials in particular is defined as the damage done to a solid surface, involving progressive material loss and caused by the relative movement between contacting surfaces. Some authors, such as Zhang [90], exclude from the category of wear the fracture or fatigue damage caused by inner cracks of the solids, the pure corrosion or aging of rubber surfaces resulting from chemical reactions and plastic deformation without loss of material. Wear depends on several factors, such as the nature of materials at the surface zone as well as in the bulk far from the contact zone, on operating parameters, on geometry at both macroscopic and microscopic level, and on environmental conditions. Surveys carried out by Rymuza [134], Viswanath and Bellow [93] and Stachowiak and Batchelor [82] show several dependencies of different variables in the dynamics of wear between rubber-like materials and metals, as illustrated in Fig. 13.

The influence of the most relevant parameters is detailed below.

Formation of a Material Transfer Layer in Contact Pair

Several authors have examined transfer layer formation in the sliding of a rubber-like material over a counter material with a harder contact surface. A notable example is the study performed by Buckley [135] with polytetraethylene (PTFE) sliding over a metallic surface, where a strong adhesion between both materials is produced, caused on the one hand by the chemical reaction between fluorine and carbon from PTFE with the metallic counter surface, and on the other hand by the ease of movement of the material molecules under load conditions. Other authors, such as Makinson and Tabor [136] and Tanaka et al. [137], have analysed the transfer mechanism of this material and other materials such as high density polyethylene (HDPE) and ultra-high molecular weight polyethylene (UHMWPE), observing a discrete sheet formation of material over the metal, producing wear increase and friction decrease.

In other research, such as that carried out by Thorpe [138], the material was detached at the asperity zones separately and stuck over the counter surface in a material transfer type known as lump transfer.

According to analyses by Jain and Bahadur [139], a similar behaviour is produced in the contact between two rubber-like materials, setting up a material transfer layer in the material with the weaker cohesion.

Influence of the Counter Material Roughness

Several authors have focussed their studies on the analysis of sliding contact pairs between rubber-like materials and metals to obtain the optimum material roughness which generates the lowest material wear. On the one hand, authors such as Birkett and Lancaster [140] suggest that the lower the counter material roughness, the lower the wear of the rubber-like material. On the other hand, Dowson et al. [141] established an optimum roughness level within the material manufacture tolerances, also depending on the sliding velocity of the application. Other surveys, such as that carried out by Barrett et al. [142] on ultra high molecular weight polyethylene (UHMWPE), established a dependency loss of the roughness with the wear rate at high sliding velocities due to the faster transfer layer formation and also to the wear rate stabilization.

In another study, Stackowiak and Batchelor [82] established the penetration depth of the metallic asperities and the sliding distance as the most influential parameters on the wear rate. According to the same authors, the wear does not remain constant over time. After a fast initial increment of the wear rate, it then decreases over time once the asperities have been covered by the rubber-like material transfer layer.

Another additional factor affecting the wear rate, apart from roughness, is the height distribution in the counter material asperities. Play [143] found important differences in the wear rate between surfaces with a height Gaussian distribution in asperities and surfaces with a non Gaussian distribution. This fact also influences the amount of detached debris, which determines the transfer layer formation modifying the wear rate as well as the friction, as shown by Blanchett and Kennedy [144] and by Barrett et al. [142].

Influence of the Temperature

Several authors, such as Tanaka and Uchiyama [145], Kar and Bahadur [146] and Stachowiak and Batchelor [82], have analysed the influence of the temperature in the wear process of rubber-like materials under sliding conditions with metal. The temperature rise of the rubber-like material in the wear process, caused by its low melting point and by its low thermal conductivity, implies a wear process modification in the contact pair, decreasing the friction and increasing the wear rate. This process is known by authors like Stackowiak and Batchelor [82] as melting wear. During this process, the melted rubber-like material is placed on the counter material surface which is not affected by the temperature rise of the rubber-like material because the melting point of the counter material is much higher.

Another influential aspect in the rubber-like material wear process is the latent heat of melting, already analysed by Mc C. Ettles [147] in specimens of polypropylene and nylon sliding over steel. This parameter imposes a limit to the temperature attained at the contact pair, so that although the friction coefficient changes with the sliding velocity or with the load when the melting point of the rubber-like material is attained, the temperature at the contact pair remains constant at the yield limit.

Besides, for soft materials such as aluminium, particles of the counter material can also be transferred to the rubber-like material surface, implying a friction coefficient increase, according to the researches of Mizutani et al. [148].

Thermal conductivity of the material is another influential factor in rubber-like material wear, as shown by Watanabe and Yamaguchi [149] in wear tests with nylon specimens on steel and glass surfaces. For high test velocities and the same test conditions, melting wear is found in nylon on glass, but not on steel, since the thermal conductivity of this material is higher than that of glass.

Finally, the combined effect of counter material roughness and the contact temperature when both parameters have high values causes the wear rate of the rubber-like material to be extremely high, even without reaching the melting point of the material [150]. This was proved by Barrett et al. [142]. In this case, severe abrasion of the rubber-like material surface takes place, beginning with a linear wear rate of shorter periods of higher wear combined with longer periods of less significant wear.

Influence of Lubricant and Oxidant Agents

In general, the inclusion of lubricants reduces friction in contact pairs, the quantitative reduction of the friction depending on the type of rubber-like material and on the lubricant. Authors such as Cohen and Tabor [151] have noticed a sharp drop in the friction coefficient in tests with nylon and glass in a water bath, obtaining lower drops in tests with organic substances such as hexane or benzene. The differences are caused by the polar nature of the polyamide, the main constituent of nylon. The formation of a transfer layer is also modified by lubricants, decreasing to very low values under wet conditions, especially under water [152].

The wear rate of the contact pair is also affected by the solubility of the rubber-like material. In some cases, if a solvent is able to penetrate through the rubber-like material surface, an accelerated wear increase with cracking in the rubber-like material surface is produced, mainly caused by the action of the solvent. According to an analysis by Evans [152], several solvents such as acetone, benzene, tetraclochlorinemethane or toluene show solubility parameters near to that of different rubber-like materials, causing an accelerated wear as contact is produced.

Regarding the effects of oxidative agents, some polymers such as nylon or polyethylene show a wear process similar to the corrosive wear of metals, leading to a reduction in friction coefficient values and to an increase in the wear rate. This effect is proportional to the surface damage level due to the corrosive agents. This feature is consistent with the resistance of UHMWPE to chemical agents, observed by authors such as Batchelor and Tan [153] and Scott and Stachowiak [154].

Influence of Rubber-Like Material Microstructure

It is worth mentioning, according to studies from Bartenevev and Lavrentev [155], the low friction values and low wear rates at temperature values near the glass transition point, showing high wear rates at different temperatures, higher or lower than that point. This behaviour is different from that of other crystalline polymers, showing a wider temperature range for low friction coefficient and wear rate values.

Types of Wear in Rubber-Like Materials

A classification of the wear types in rubber-like materials has not yet been clearly established. Over more than half a century, several researchers including Kragelskii [88] and Blau [89] have proposed several types of classifications from different points of view. However, to date there is no generally recognised methodology. Zhang [90] takes into consideration a classification including abrasive, erosive and fatigue wear, while other authors such as Myshkin et al. [91] include adhesive wear in this classification. These wear types are the most broadly accepted in the literature. In some cases, corrosive or tribo-chemical wear are also

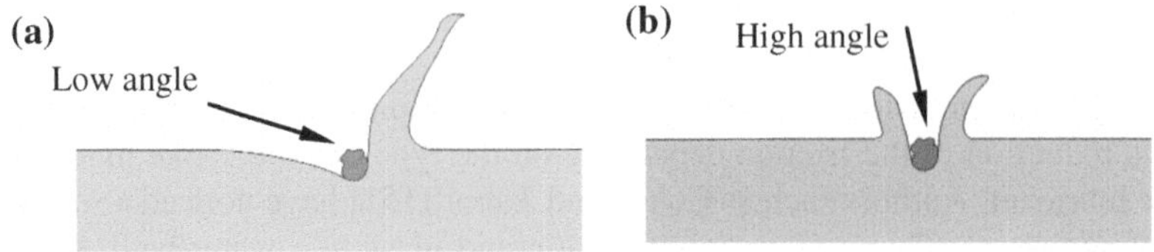

Fig. 14 Types of erosive wear. **a** Abrasive erosion. **b** Impact erosion

considered, as analysed by Burris [156] and Viswanath and Bellow [93], or fretting wear, as studied by Je et al. [92] and by Nah [94]. However, these are taken as particular cases of the most influential phenomena previously referred to. The most relevant wear types are explained in detail below.

(a) Erosive wear

This is defined as the wear resulting from the interaction between a solid surface and a fluid stream containing abrasive particles at a certain speed. The types of erosive wear can be abrasive erosion, if it is caused by a fluid stream parallel to the solid surface (see Fig. 14a), or impact erosion if it takes place under a fluid stream perpendicular to the solid surface (see Fig. 14b).

It is important to mention the surveys of Zhang [157, 158] into the wear mechanism by abrasive erosion of rubber-like materials such as natural rubber (NR), styrene-butadiene rubber (SBR), nitrile-butadiene rubber (NBR) or polyurethanes (PU). These surveys were carried out in an abrasive erosion testing machine under wet conditions with sodium hydroxide in water. The effects analysed in these tests by SEM examinations were: delamination, micro cut, micro crack initiation and propagation or mechanic-chemical degradation. Besides, this author also studied the influence of the particle wear speed, flow velocity, particle size and particle concentration for the same rubber-like materials. It is also worth noting the work of Arnold and Hutchings [159] which studied the erosive wear of non-reinforced rubber-like materials, finding two different mechanisms: the first at low impact angles and the second under conditions of perpendicular impact to the surface, eliminating the material from the surface in both cases by means of fatigue crack propagation.

(b) Adhesive wear

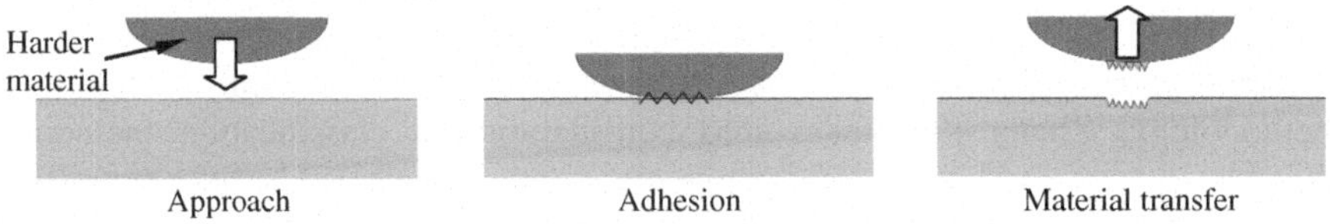

Fig. 15 Adhesive wear

This takes place as a result of the adhesive forces present in the contact pair between two solid surfaces, where part of the rubber-like material is transferred from its surface and adheres to the counter material surface (see Fig. 15). Bely et al. [160] noticed that the material transfer is the most important characteristic of adhesive wear in rubber-like materials. The processes associated with other wear types such as abrasive wear and fatigue wear can also take place together with adhesive wear.

The material transfer phenomenon caused by friction, where micro size particles are transferred from one surface to another, is a very common effect in contact pairs between rubber-like material and metal. This effect was studied by Makinson and Tabor [136] and by Tanaka et al. [137]. Usually, the soft material film, that of the rubber-like material, is transferred to the harder material surface, in this case the metal surface. If the transferred rubber-like material film is continuously being placed and eliminated, the wear rate increases. If, on the other hand, the rubber-like material film is maintained, even with changes in the friction force, the changes in the wear rate vary very slightly. In general, however, it can be stated that a high degree of adhesion is produced in a contact pair in which rubber-like material and metal are involved.

In rubber-like materials with low concentrations of additives in the components, a very slight transfer to the metallic counter surface is produced due to the low adhesion between both surfaces. This fact implies that the contact pair shows very inadequate tribological properties [161]. Under other circumstances, the harder material can be transferred over (onto?) the softer material surface; for instance, bronze may be transferred to rubber-like material. This results in harder material particles being set in the rubber-like material surface, acting as abrasive particles, as analyzed by Myshkin et al. [91].

(c) Abrasive wear

The wear type known as abrasive wear is the most common in polymers in general and in rubber-like materials in particular. It is caused by the relative movement between a solid surface and sharp particles of the same or higher hardness acting against the surface of the first body. Studies carried out by Swain [162] revealed that indirect mechanisms such as micro cutting, detachment of individual particles of the material or accelerated fatigue by repetitive deformations can be involved in the cutting process of the material (see Fig. 16).

In the wear caused by a cutting mechanism, a very sharp tool or a very rough surface cuts the softer surface, eliminating the material by debris formation.

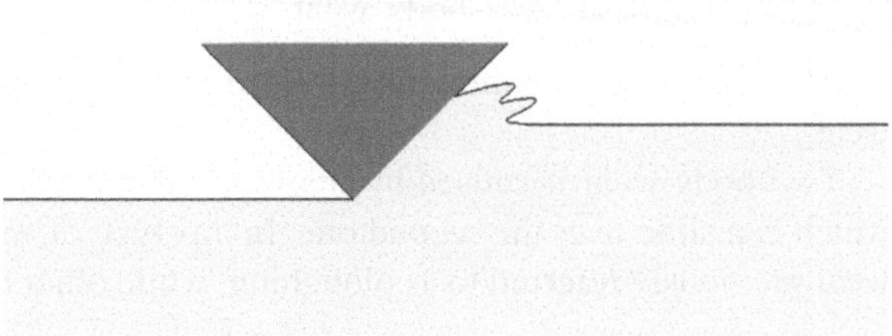

Fig. 16 Abrasive wear

Depending on the material hardness, the debris can be detached as a result of crack propagation or of repetitive deformation of the material [82].

According to Zhang's [90] surveys, rubber abrasion can be classified into two categories: pattern and intrinsic abrasion. In the first type of abrasion, parallel ridges, called abrasion patterns, appear on the surface at right angles to the sliding direction. Pattern abrasion occurs under unidirectional sliding direction conditions. On the other hand, intrinsic abrasion arises when the direction of the relative motion changes periodically. Usually, and under identical conditions, the wear rate of pattern abrasion is higher than that of intrinsic abrasion. Zhang [90] also classifies abrasion wear into wet abrasion and dry abrasion, depending on whether or not liquid exists on the frictional surface. Another classification made by the same author is based on the type of contact between both surfaces: point contact abrasion, line contact abrasion and multiple-point contact abrasion.

One of the most widespread techniques for observing the wear process in rubber-like materials is by means of electronic microscopy (SEM). Some authors, such as Bhowmick et al. [163], have investigated this process and established that a cutting mechanism at micromechanical level is present, in which the material is eliminated by means of wavy sheets. This process is known as micro-cutting. Kayaba [164] revealed another mechanism involving the formation of grooves along the sliding direction, identified by means of several observations of specimens tested in a tribometer under the pin-on-disc configuration. This mechanism, known as ploughing, is less destructive than micro-cutting and does not involve material being detached, Other authors, such as Myshkin et al. [91], have named this wear type grooving.

In harder materials, such as thermoplastic polyurethanes (TPU), two different mechanisms are present in the wear process: macro-delamination and micro-molecular fracture [165, 166]. Macro-delamination consists of the formation and growth of cracks, leading the material to tear in terms of parallel grooves, finally breaking due to tensile stresses. Micro-molecular fracture consists of the detachment of small particles due to the breaking of simple material molecules or any of its aggregates. At the same time, abrasive wear particles in these materials are also related with the presence of additives, fillets or plasticizers. Bartenevev and Lavrentev [155] noticed that plasticizers have a negative effect on the abrasive wear of several polymers due to their softening.

Zhang [90] has also presented valuable surveys in the quantitative evaluation of rubber abrasion by means of different methods such as fractal theory, computerized simulation technology and computer-generated image analysis.

(d) Two-body and three-body wear

Abrasive wear is commonly divided in two groups: two-body and three-body wear.

Two-body wear is caused by sharp protuberances present in one of the surfaces which can slide over the second one. In this type of wear, some asperities cause the wear previously referred to as ploughing, while other asperities cause micro-cutting wear, depending on two factors: the attack angle of the particle and the shear stress generated between both contact surfaces [91].

On the other hand, in three-body wear, there are particles trapped in both contact surfaces which can rotate and slide freely between them. These particles are detached from the worn surface of the contact pair or come from the lubrication of the contact pair if the test is carried out under wet conditions. They may also be particles from the environment. The effect, according to research by Singer and Wahl [167], is a decrease in the friction between both surfaces, setting up transfer layers but also inducing wear tracks with the detached particles.

For some time, it was believed that these two wear types were very similar. However, Zum Gahr [168] identified several differences between them. The wear rate in three-body wear is around one order of magnitude lower than that obtained in two-body wear, since the abrasive particles in three-body abrasion wear the contact surfaces only 10 % of the sliding time, while during the remaining 90 % of the time they merely rotate between the surfaces [169]. According to Johnson [170], another difference lies in the fact that two-body wear corresponds to a material elimination model typical of cutting or micro-cutting, while three-body wear involves slower mechanisms of material elimination. In the latter case, the mechanisms common in two-body wear, such as micro-cutting or ploughing, do not occur. There is instead a random wear mechanism due to the non-controlled presence of a third body [171].

(e) Fatigue wear

This is a type of wear similar to abrasive wear, produced against a rough surface. The difference between them, according to Zhang [90], is that the surface for fatigue wear is formed by small soft rough projections, while for abrasive wear the surface is formed by hard sharp projections. Fatigue wear as a concept was presented by Kragelskii [88], being a low intensity wear type compared with abrasive wear. The main feature of this type of wear is the irreversible damage suffered by the material under the repetitive action of compressive, tensile and shear strains in the contact pair. Along the relative sliding between both surfaces, the polymer interacts with the sharper projections of the rough counter surface, which leads to the initiation and development of cracks, also helped to propagate by the presence of internal voids in the material [169]. Several authors, including Zhang [90], have modified the term of fatigue wear to frictional wear or rolling wear if the rubber-like material shows low tearing resistance and slides over low rough surfaces with a high friction coefficient, causing the formation of rolling or spiral particles at the contact pair. These particles are continuously detached during sliding. In this type of wear, each asperity of the worn material surface suffers a sequential load from the asperities present in the contact. Subsequently, stresses arise at different scales in the surface and subsurface regions. These stresses are the cause of the material fatigue, which leads to the initiation and propagation of cracks and to the formation of worn particles. These cracks are formed at points where the maximum shear stresses occur, their position also depending on the friction coefficient between both surfaces. The higher the friction

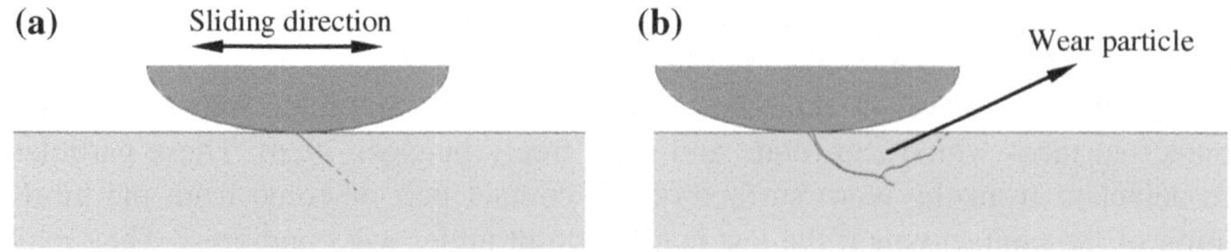

Fig. 17 Fatigue wear. **a** Crack initiation as result of fatigue process. **b** Crack growth and propagation and formation of wear particle

coefficient, the nearer the point of maximum shear stress to the surface, its depth increasing as the friction coefficient decreases. At the same time, the initiation of fatigue cracks is helped by the presence of defects in the material, such as marks, scratches in the surface, internal voids or impurities. These defects are responsible for the stress concentrations. With the repetitive action of the load and, consequently, of the material stress, the cracks grow, join each other and intersect generating the material detachment [91]. Figure 17 shows the fatigue wear process.

Other authors such as Jain and Bahadur [172] or Neale and Gee [173] consider this type of wear as abrasive wear on a small scale since the asperities of the counter surface, at micromechanical level, act as initiating particles of the rubber-like material abrasion. According to analyses by Jia and Ling [174] of fatigue wear in polyurethanes caused by the repetitive action of abrasive particles on the material, and considering that its elastic modulus is within that of a rubber-like and of a plastic material, the effects of ploughing or crack formation are not directly generated. Nevertheless, mechanical fatigue is more likely to take place. According to this study, the repetitive impacts of the abrasive particles with the material lead to tensile, compressive and shear strains and stresses in the contact layer, forming fatigue cracks due to the repetitive actions with the interactions. Other surveys carried out by Liu et al. [175] and by Marchenko [176] show the highest shear stress at a certain depth under the contact surface. On the other hand, the highest material strain is located at the surface, a propitious place for crack initiation although at the same time this is where the highest compressive stresses are also located which in some way act against crack initiation. As the distance from the contact surface increases, the compressive stresses decrease faster than the shear stresses. This means that almost all the stresses are shear stresses which, being cracks, are more easily formed at a distance from the contact surface.

Another effect taken into consideration by Jia and Ling [174] is the temperature influence, which is higher in TPU layers near to contact due to friction and material deformation hysteresis. This heat can be more easily dissipated at the surface where the temperature quickly drops due to the contact with the environment. However, its dissipation is more difficult at a certain depth, and this decreases the cohesive material energy and consequently cracks are initiated. The repetitive contact of these particles implies that cracks propagate and intersect with each other, leading to material being detached as debris. Stackowiak and Batchelor [82] also studied the temperature effect on wear behaviour. They demonstrated that with the low temperature at which polymers melt, as well as their low thermal conductivity, the high temperature reached at the contact pair is higher than the melting point of the material and it thus begins to deform in an effect known as melting prow. This effect, spread over all the contact surface of the polymer, is known by other authors such as Bartenevev and Lavrentev [155] as fatigue wave formation.

Other authors establish in their studies that cracks present in the material subsurface are exacerbated during application cycles by the plastic deformation of the material, being propagated to near cracks in a process defined as delamination by authors such as Johnson [177], Suh et al. [178] and Da Silva [179]. According to these authors, any particle generated and detached in the wear surface implies a higher dragging, thereby increasing the friction force. This in turn accelerates the delamination process.

2 Constitutive Models for F.E. Modelling of Elastomeric Materials

Hyperelasticity refers to the quality of materials which can experience large elastic strain that is recoverable. Elastomers such as rubber and many other polymer materials fall into this category. The microstructure of polymer solids consists of chain-like molecules. The chain backbone is mostly made of carbon atoms. The flexibility of polymer molecules allows different types of arrangement such as amorphous and semi crystalline polymers. As a result, the molecules possess a much less regular character than metal crystals. The behaviour of elastomers is therefore very complex. On a macroscopic scale, they usually behave as elastically isotropic initially, and anisotropic at finite strain as the molecule chains tends to realign in the loading direction. However, under essentially monotonic loading conditions a larger class of elastomers can be approximated by an isotropic assumption, and this has been historically popular in their modelling.

The constitutive behaviour of hyperelastic materials is usually derived from the strain energy potentials. Also, hyperelastic materials generally have very small compressibility, often referred to as incompressibility. The hyperelastic material models assume that the material response is isothermal. This assumption allows strain energy potentials to be expressed in terms of strain invariants or principal stretch ratios. Except as otherwise indicated, the materials are assumed to be nearly or purely incompressible. Material thermal expansion is always assumed to be isotropic.

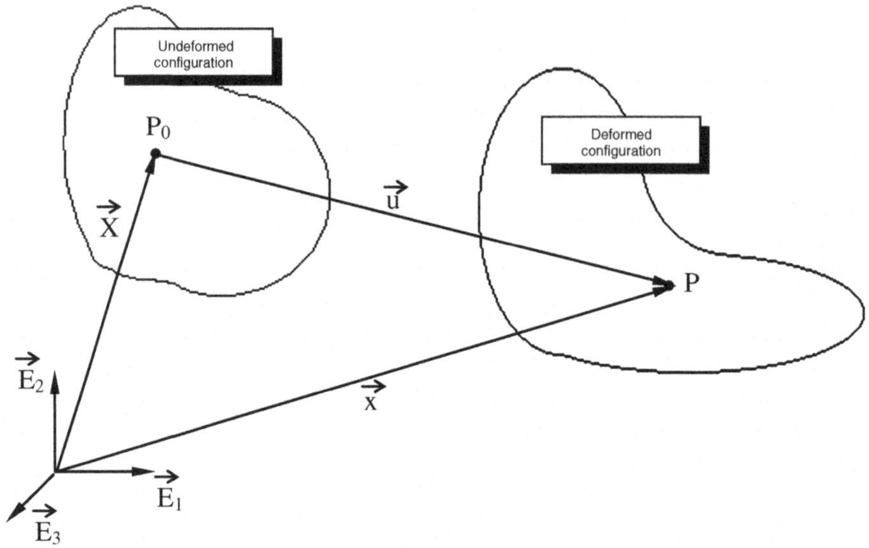

Fig. 18 Deformed and undeformed shape for the deformable body

2.1 Modelling the Elastic Behaviour of Elastomers

2.1.1 Basic Kinematics of Finite Elastic Deformations

This section gives a brief description of the finite strain theory, introducing concepts such as gradient tensors, finite strain tensors, stretch ratios as well as compatibility conditions. For a detailed description of these concepts, see Holzapfel [180].

A body whose configuration in terms of size or shape changes under external loads is defined as a deformable body. A direct result is that distances between points of the solid vary. Therefore, it is appropriate to consider two different configurations of this type of solid, the undeformed configuration and the deformed configuration after applying external loads.

Given a three-dimensional solid deformable body, as shown in Fig. 18, and a Cartesian reference system $(\vec{E}_1, \vec{E}_2, \vec{E}_3)$, the position vector $(\vec{X})$ for the point P_0 in the undeformed configuration can be written as:

$$\vec{X} = X_1\vec{E}_1 + X_2\vec{E}_2 + X_3\vec{E}_3 \tag{13}$$

After a load application, the solid body is deformed and the point P_0 is now in the position P, whose coordinates are:

$$\vec{x} = x_1\vec{E}_1 + x_2\vec{E}_2 + x_3\vec{E}_3 \tag{14}$$

The relationship between deformed and undeformed coordinates is the displacement field in the spatial configuration:

$$\vec{u} = \vec{x} - \vec{X} = u_1\vec{E}_1 + u_2\vec{E}_2 + u_3\vec{E}_3 \tag{15}$$

From a Lagrangian point of view, the coordinates for points in the deformed configuration can be expressed as a function of the coordinates of points in the undeformed configuration.

$$\vec{x} = \vec{x}(\vec{X}) \tag{16}$$

The deformation gradient tensor is defined, based on the relationship between the position vector and the deformed and undeformed configurations, as:

$$\mathbf{F} = \left[\frac{\partial x_i}{\partial X_j}\right] \tag{17}$$

In a deformable body, those properties which change along with the deformation of the body might be described either by the evolution of its value along the trajectory of a given material point, material description (also known as *Lagrangian description*), or by the change of its value at a fixed location in space occupied by (different for each time instant) particles of the body, spatial (*Eulerian*) description.

The rigid body motion can be decomposed in a displacement and a rotation. Equation (18) relates the initial and final coordinates of the deformable body:

$$\vec{x} = R \cdot \vec{X} + \vec{c} \tag{18}$$

where

$K_0 = \frac{2}{D}$ translation vector (independent of position).

R rotation tensor, due to its antisymmetry verifies:

$$\mathbf{R}^{\mathrm{T}} \cdot \mathbf{R} = \mathbf{R} \cdot \mathbf{R}^{\mathrm{T}} = \mathbf{I} \tag{19}$$

In order to separate the movement due to the deformation from the rigid body motion, the right Cauchy Green strain tensor is defined as:

$$\mathbf{C} = \mathbf{F}^{\mathrm{T}} \cdot \mathbf{F} \tag{20}$$

If there is only rigid body motion, this tensor is constant and unitary:

$$\mathbf{C} = \mathbf{F}^{\mathrm{T}} \cdot \mathbf{F} = \mathbf{R}^{\mathrm{T}} \cdot \mathbf{R} = \mathbf{I} \tag{21}$$

The left Cauchy-Green deformation tensor is obtained reversing the order of multiplication in the formula for the right Cauchy Green strain tensor:

$$\mathbf{b} = \mathbf{F} \cdot \mathbf{F}^{\mathrm{T}} \tag{22}$$

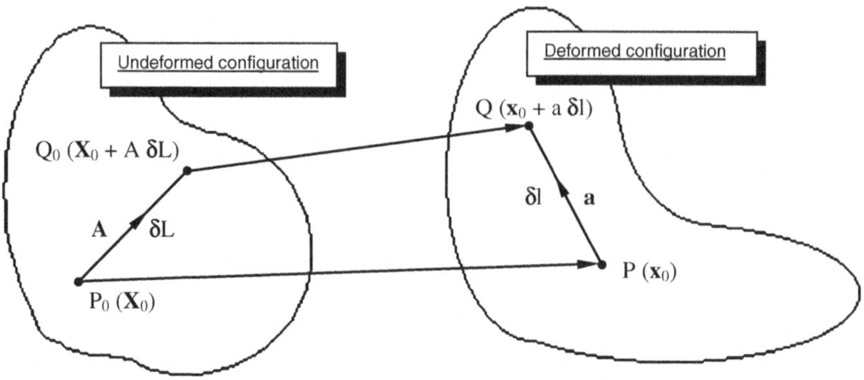

Fig. 19 Deformation for a deformable body

The deformations can also be expressed using the Green Lagrange strain tensor defined according to the following expression:

$$\mathbf{E} = \frac{1}{2}\left(\mathbf{F}^{\mathrm{T}} \cdot \mathbf{F} - \mathbf{I}\right) = \frac{1}{2}(\mathbf{C} - \mathbf{I}) \tag{23}$$

This tensor is null when there is no deformation.

When the movement of the solid body is due to a displacement and a rotation, the deformation gradient tensor could be expressed as:

$$\mathbf{F} = \mathbf{R} \cdot \mathbf{U} \tag{24}$$

where $\boldsymbol{U}$ is a symmetric tensor called the right stretch tensor, related with the right Cauchy Green strain tensor through the following expression:

$$\mathbf{C} = \mathbf{F}^{\mathrm{T}} \cdot \mathbf{F} = \mathbf{U}^2 \tag{25}$$

(a) Principal stretches

The principal stretch coefficients are now described. Consider two points, P_0 and Q_0, in the undeformed configuration of a deformable body, as shown in Fig. 19, both related by the vector δL. If the body is subjected to a displacement field defined through the deformation gradient tensor $\boldsymbol{F}$, the displaced points P and Q in the deformed configuration are related by means of the δl vector, and this verifies:

$$\mathbf{F} \cdot \mathbf{A} = \mathbf{R} \cdot \mathbf{U} \cdot \mathbf{A} = \lambda \cdot \mathbf{a} \tag{26}$$

where

$\boldsymbol{A}$	unity vector of vector δL on undeformed configuration.
$\boldsymbol{a}$	unity vector of vector δl on deformed configuration.
$\lambda = \frac{\delta l}{\delta L}$	stretch coefficient of vector δL, due to deformation.

Assuming that there are no rotations and δL only changes its modulus due to deformation, then the following conditions are satisfied:

$$\mathbf{F} = \mathbf{U} \tag{27}$$

$$\mathbf{A} = \mathbf{a} \tag{28}$$

Therefore (26) can be transformed into the following expression:

$$\mathbf{U} \cdot \mathbf{A} = \lambda \cdot \mathbf{a} \tag{29}$$

Reordering this expression, it is possible to obtain:

$$(\mathbf{U} - \lambda \mathbf{I}) \cdot \mathbf{A} = 0 \tag{30}$$

According to (30), $\boldsymbol{A}$ is an eigenvector of $\boldsymbol{U}$ associated to the eigenvalue λ. Since $\boldsymbol{U}$ is a 3×3 positive definite matrix, it has three positive and real eigenvalues called principal stretches

$$\lambda_1 \geq \lambda_2 \geq \lambda_3 \tag{31}$$

The right stretch tensor $\boldsymbol{U}$ has three principal directions associated to these eigenvalues, forming an orthogonal coordinate system named the principal axis system, where the right stretch tensor adopts the following form:

$$\mathbf{U} = \begin{bmatrix} \lambda_1 & 0 & 0 \\ 0 & \lambda_2 & 0 \\ 0 & 0 & \lambda_3 \end{bmatrix} \tag{32}$$

Observing the right stretch tensor expressed in the principal axis, and considering a displacement field defined through the deformation gradient tensor $\boldsymbol{F}$, the movement of L_O, vector joining two near points in a deformable body, could be decomposed in two parts:

First, a variation of the L_O modulus expressed in the principal axis as shown in Fig. 20 $\bar{L}_f = \bar{L}_0 + \overline{\Delta L}$.

The change in modulus for the components of L_O is defined by the principal stretches $(\lambda_1, \lambda_2, \lambda_3)$ since:

$$\lambda_i = \frac{L_i \pm \Delta L_i}{L_i} \tag{33}$$

And second, the movement is completed with a rotation of the L_f vector according to the rotation tensor $\boldsymbol{R}$.

(b) Strain invariants

The deformation of a body can be expressed in terms of the right Cauchy Green strain tensor invariants. These invariants are defined as:

$$I_1 = C_{11} + C_{22} + C_{33} \tag{34}$$

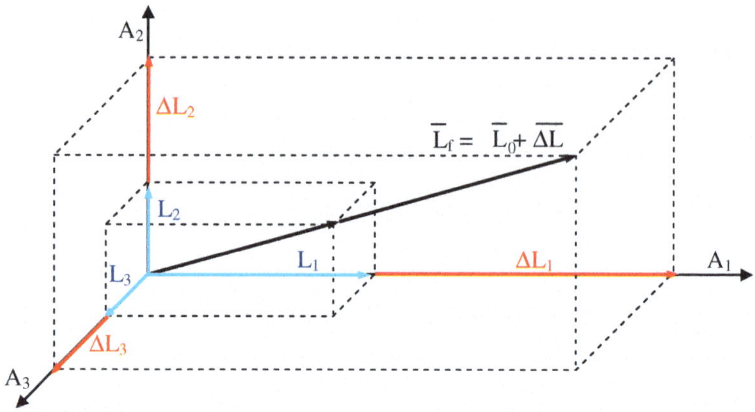

Fig. 20 Principal stretches

$$I_2 = C_{11}C_{22} + C_{22}C_{33} + C_{33}C_{11} - C_{12}^2 - C_{23}^2 - C_{31}^2 \tag{35}$$

$$I_3 = C_{11}C_{22}C_{33} + 2C_{12}C_{23}C_{31} - C_{11}C_{23}^2 - C_{22}C_{31}^2 - C_{33}C_{12}^2 \tag{36}$$

According to (25), the principal directions of $\boldsymbol{C}$ are the same as of $\boldsymbol{U}$, and the eigenvalues of $\boldsymbol{C}$ are $\lambda_1^2, \lambda_2^2, \lambda_3^2$. The right Cauchy Green strain tensor expressed in the principal axis has the following shape:

$$\mathbf{C} = \begin{bmatrix} \lambda_1^2 & 0 & 0 \\ 0 & \lambda_2^2 & 0 \\ 0 & 0 & \lambda_3^2 \end{bmatrix} \tag{37}$$

Expressing $\boldsymbol{C}$ invariants in terms of principal stretches:

$$I_1 = \lambda_1^2 + \lambda_2^2 + \lambda_3^2 \tag{38}$$

$$I_2 = \lambda_1^2\lambda_2^2 + \lambda_2^2\lambda_3^2 + \lambda_3^2\lambda_1^2 \tag{39}$$

$$I_3 = \lambda_1^2\lambda_2^2\lambda_3^2 \tag{40}$$

(c) Elasticity for incompressible materials

A material is incompressible when its volume does not change when is deformed, excepting deformations due to thermal expansion (see Fig. 21). In any other case, after applying a load, the volume in the undeformed configuration is the same as in the undeformed configuration, therefore the volumetric change coefficient J is equal to 1.

Near incompressibility is often a device by which incompressibility can more readily be enforced within the context of computational formulations. In this case,

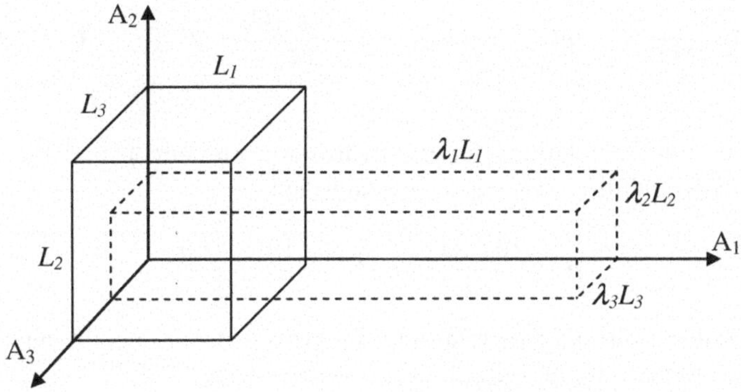

Fig. 21 Deformation for an incompressible solid

it is usual to split the deformation locally into a so-called volumetric part and an isochoric part.

$$\begin{aligned} \mathbf{F} &= J^{\frac{1}{3}}\bar{\mathbf{F}} \quad \bar{\mathbf{F}} = J^{-\frac{1}{3}}\mathbf{F} \\ \mathbf{C} &= \mathbf{F}^T\mathbf{F} \quad \bar{\mathbf{C}} = J^{-\frac{1}{3}}\mathbf{C} = \bar{\mathbf{F}}^T\bar{\mathbf{F}} \end{aligned} \tag{41}$$

$J^{\frac{1}{3}}$ is associated with volume-changing deformations, while $\overline{\mathbf{F}}$ is associated with volume-preserving deformations. We shall call $\overline{\mathbf{F}}$ and $\overline{\mathbf{C}}$ the modified deformation gradient and the modified right Cauchy-Green tensors, respectively.

2.1.2 Constitutive Behaviour

To characterize isothermal processes, we postulate the existence of a unique decoupled representation of the strain-energy density function $W = W(\mathbf{C})$ based on the kinematic assumption (41) such as

$$W(\mathbf{C}) = W_{vol}(J) + W_{dev}(\bar{\mathbf{C}}) \tag{42}$$

where W_{vol} and W_{iso} are given scalar-valued functions of J and $\mathbf{C}$ respectively that describe the volumetric (or dilational) and the isochoric (or distortional) responses of the material.

The constitutive equations for compressible hyperelastic materials in the standard form

$$\mathbf{S} = 2\frac{\partial\Psi(\mathbf{C})}{\partial\mathbf{C}} = \mathbf{S}_{vol} + \mathbf{S}_{dev} \tag{43}$$

where the second Piola-Kirchhoff stress $\mathbf{S}$ consists of a purely volumetric contribution and a purely isochoric one. This split is based on the definitions

$$\mathbf{S}_{vol} = 2\frac{\partial\Psi_{vol}(J)}{\partial\mathbf{C}} = Jp\mathbf{C}^{-1} \text{ and } \mathbf{S}_{des} = 2\frac{\partial W_{dev}(\bar{\mathbf{C}})}{\partial\mathbf{C}} = J^{-\frac{2}{3}}(\mathrm{I} - \frac{1}{3}\mathbf{C}^{-1}\otimes\mathbf{C}) : \bar{\mathbf{S}} \tag{44}$$

with the constitutive relations for the hydrostatic pressure p and the modified second Piola-Kirchhoff stress tensor $\mathbf{S}$ established as

$$p = \frac{dW_{vol}(J)}{dJ} \text{ and } \bar{\mathbf{S}} = 2\frac{\partial W_{dev}(\bar{\mathbf{C}})}{\partial\bar{\mathbf{C}}} \tag{45}$$

Compressible isotropic hyperelasticity in terms of invariants, by analogy with the decouple representation (42) is

$$W = W_{vol}(J) + W_{dev}(\bar{I}_1(\overline{\mathbf{C}}), \bar{I}_2(\overline{\mathbf{C}})) = W_{vol}(J) + W_{dev}(\bar{I}_1(\overline{\mathbf{b}}), \bar{I}_2(\overline{\mathbf{b}})) \tag{46}$$

with $\overline{\mathbf{b}} = J^{-2/3}\mathbf{b}$ the modified left Cauchy-Green tensor and $\bar{I}_1$ and $\bar{I}_2$ the first two modified strain invariants of the symmetric modified Cauchy-Green tensor ($\overline{\mathbf{C}}$ and $\overline{\mathbf{b}}$ have the same eigenvalues) defined by

$$\bar{I}_1 = tr\overline{\mathbf{C}} = tr\overline{\mathbf{b}} \tag{47}$$

$$\bar{I}_2 = \frac{1}{2}(tr(\overline{\mathbf{C}}))^2 - tr\mathbf{C}^2 = \frac{1}{2}(tr(\overline{\mathbf{b}}))^2 - tr\mathbf{b}^2 \tag{48}$$

$$\bar{I}_3 = \det(\overline{\mathbf{C}}) = \det(\overline{\mathbf{b}}) \tag{49}$$

The Cauchy stress tensor $\boldsymbol{\sigma}$ is 1/J times the push-forward of $\mathbf{S}$, that is

$$\boldsymbol{\sigma} = J^{-1}\boldsymbol{\chi}_*(\mathbf{S}), \qquad \sigma_{ab} = J^{-1}F_{aA}F_{bB}S_{AB} \tag{50}$$

Elasticity tensors in the material description. The linearized constitutive equations are required to obtain numerical solutions of nonlinear (boundary value) problems using iterative solution techniques of Newton's type. Consider the nonlinear second Piola-Kirchhoff stress tensor $\boldsymbol{S}$ at a certain point and configuration. Its variation with respect to the right Cauchy-Green tensor $\boldsymbol{C}$ may be written as

$$d\mathbf{S} = \mathbf{C} : \frac{1}{2}d\mathbf{C} \qquad \mathbf{C} = \mathbf{2}\frac{\partial\mathbf{S}(\mathbf{C})}{\partial\mathbf{C}} \tag{51}$$

with $\boldsymbol{C}$ the elasticity tensor in the material description or the referential tensor of elasticities. If we assume the existence of the strain energy density W (hyperelasticity), then using expressions (42) and (51), we obtain the well known relation

$$C = 4\frac{\partial^2 W(\mathbf{C})}{\partial\mathbf{C}\partial\mathbf{C}} \tag{52}$$

Given the structure of the decoupled stress relation, the associated decoupled elasticity tensor may be written as

$$\mathbf{C} = \mathbf{C}_{vol} + \mathbf{C}_{des} = 2\frac{\partial \mathbf{S}_{vol}}{\partial \mathbf{C}} + 2\frac{\partial \mathbf{S}_{des}}{\partial \mathbf{C}} \tag{53}$$

The elasticity tensor in the spatial description or the spatial tensor of elasticities, denoted by c, is defined as the push-forward of $\boldsymbol{C}$ times a factor of J^{-1}, that is, the Piola transformation of $\boldsymbol{C}$ on each index.

$$\mathbf{c} = J^{-1}\chi_*(\mathbf{C}), \qquad \mathbf{c}_{abcd} = J^{-1}F_{aA}F_{bB}F_{cC}F_{dD}\mathbf{C}_{ABCD} \tag{54}$$

For a detailed description of the previous expressions (50) and (54), see Holzapfel [180].

Numerous specific forms of strain-energy functions have been proposed in the literature to describe the elastic properties of incompressible as well as compressible materials, and more or less efficient new specific forms are published almost daily. In the next subsections, we present only some of the most used models of the many available for hyperelastic materials.

2.1.3 Particular Forms of the Strain Energy Potential

FE codes offer several forms of the strain energy density potential in order to model the behaviour of elastomer materials. The models may be divided into two groups, according to the formulation of the strain energy density function (based on strain invariant or principal stretches) or according to continuum mechanics or statistical theories (phenomenological or predictive models). Additionally, the strain energy density functions may be classified as phenomenological or predictive models.

Phenomenological models require a large number of characterization tests in order to fit correctly the behaviour of the material. Predictive models are based on micromechanics considerations. The main advantage of these models is their ability to fit the material behaviour from a reduced set of characterization tests. They are able to predict behaviour of the material for strain states without any experimental data to fit the model. Models based on the first strain invariants and those based on statistical theories fall within this classification.

(a) Polynomial

The form of the polynomial strain energy potential is shown as follows:

$$W = W_{dev} + W_{vol} = \sum_{i,j=1}^{N} C_{ij}(\bar{I}_1 - 3)^i(\bar{I}_2 - 3)^j + \sum_{i=1}^{N}\frac{1}{D_i}\left(J^{el} - 1\right)^{2i} \tag{55}$$

where W is the strain energy per unit of reference volume, N is a material parameter with values equal to or lower than six and C_{ij} and D_i are temperature dependent material parameters.

The initial shear modulus and bulk modulus are defined according to (56) and (57).

$$\mu_0 = 2(C_{10} + C_{01}) \tag{56}$$

$$K_0 = \frac{2}{D_1} \tag{57}$$

From the general polynomial form shown in (55), some particular forms of strain energy potential are shown below without considering the volumetric contribution:

Mooney-Rivlin (N = 2)

$$W_{dev} = C_{10}(\bar{I}_1 - 3) + C_{01}(\bar{I}_2 - 3) \tag{58}$$

Three term Mooney-Rivlin (N = 2)

$$W_{dev} = C_{10}(\bar{I}_1 - 3) + C_{01}(\bar{I}_2 - 3) + C_{11}(\bar{I}_1 - 3)(\bar{I}_2 - 3) \tag{59}$$

Signorini (N = 3)

$$W_{dev} = C_{10}(\bar{I}_1 - 3) + C_{01}(\bar{I}_2 - 3) + C_{20}(\bar{I}_1 - 3)^2 \tag{60}$$

Third Order Invariant (N = 4)

$$W_{dev} = C_{10}(\bar{I}_1 - 3) + C_{01}(\bar{I}_2 - 3) + C_{11}(\bar{I}_1 - 3)(\bar{I}_2 - 3) + C_{20}(\bar{I}_1 - 3)^2 \tag{61}$$

Third Order Deformation or James-Green-Simpson (N = 5)

$$\begin{aligned} W_{dev} = {} & C_{10}(\bar{I}_1 - 3) + C_{01}(\bar{I}_2 - 3) + C_{11}(\bar{I}_1 - 3)(\bar{I}_2 - 3) \\ & + C_{20}(\bar{I}_1 - 3)^2 + C_{30}(\bar{I}_1 - 3)^3 \end{aligned} \tag{62}$$

(b) Reduced polynomial

Reduced polynomial models are related with the polynomial strain energy potential and correspond with certain selections of material parameters C_{ij}. For this model, dependency with the second strain invariant is removed. Sensitivity to the second strain invariant of strain energy density function is much lower than for the first strain invariant. The form of the reduced polynomial strain energy potential is as follows:

$$W = W_{dev} + W_{vol} = \sum_{i=1}^{N} C_{i0}(\bar{I}_1 - 3)^i + \sum_{i=1}^{N} \frac{1}{D_i}\left(J_1^{el} - 1\right)^{2i} \tag{63}$$

where U is the strain energy per unit of reference volume, N is a material parameter with values equal to or lower than six and C_{i0} and D_i are temperature dependent material parameters.

The initial shear modulus and bulk modulus are defined according to expressions (64) and (57).

$$\mu_0 = 2C_{10} \tag{64}$$

Derived expressions from the reduced polynomial potential are the Neo-Hookean and Yeoh forms. The form of both strain energy potentials, without considering the volumetric part, are shown in (65) and (66) for the Neo-Hookean and Yeoh potentials, respectively.

$$W = W_{dev} + W_{vol} = C_{10}(\bar{I}_1 - 3) + \frac{1}{D_1}(J^{el} - 1)^2 \tag{65}$$

$$\begin{aligned} W = W_{dev} + W_{vol} &= C_{10}(\bar{I}_1 - 3) + C_{20}(\bar{I}_1 - 3)^2 + C_{30}(\bar{I}_1 - 3)^3 \\ &+ \frac{1}{D_1}(J^{el} - 1)^2 + \frac{1}{D_2}(J^{el} - 1)^4 + \frac{1}{D_3}(J^{el} - 1)^6 \end{aligned} \tag{66}$$

(c) Gent

The Gent model is a phenomenological model based on the concept of chain extensibility. The form of the Gent strain energy potential is as follows:

$$W = W_{dev} + W_{vol} = \frac{\mu J_m}{2} \ln\left(1 - \frac{\bar{I}_1 - 3}{J_m}\right)^{-1} + \frac{1}{D}\left(\frac{J_{el}^2 - 1}{2} - \ln J_{el}\right) \tag{67}$$

where U is the strain energy per unit of reference volume, M is the initial shear modulus, J_m is the limiting value of $I_1 - 3$ and D is the material incompressibility parameter. As the parameter J_m goes to infinity, the model is converted into Neo-Hookean form.

The initial bulk modulus is defined according to expression (68).

$$K_0 = \frac{2}{D} \tag{68}$$

(d) Potential functions based on principal stretches:

(d.1) Ogden

The form of the Ogden strain energy potential is:

$$W = W_{dev} + W_{vol} = \sum_{i=1}^{N} \frac{2\mu_i}{\alpha_i^2}(\bar{\lambda}_1^{\alpha_i} + \bar{\lambda}_2^{\alpha_2} + \bar{\lambda}_3^{\alpha_3} - 3) + \sum_{i=1}^{N} \frac{1}{D_i}((J^{el} - 1)^{2i} \tag{69}$$

where U is the strain energy per unit of reference volume, N is a material parameter with values equal to or lower than six and μ_i, α_i and D_i are temperature dependent material parameters.

The initial shear modulus is defined according to (70) while the initial bulk modulus is a function of the parameter D_1, as shown in (57).

$$\mu_0 = \sum_{i=1}^{N} \mu_i \qquad (70)$$

Other models are based on statistical theories including the definition of the strain potential parameters with physical interpretation. These models are derived from simplified models of polymeric chains and statistical considerations of chain length.

(d.2) Marlow

The form of the Marlow [181] strain energy potential is according to (42), where U_{dev} depends on the deviatoric part of the first invariant $W_{dev}(\bar{I}_1)$.

(e) Statistical models:

(e.1) Arruda-Boyce

The form of the Arruda-Boyce strain energy potential is as follows:

$$W = W_{dev} + W_{vol} = \mu\left\{\frac{1}{2}(\bar{I}_1 - 3) + \frac{1}{20\lambda_m^2}(\bar{I}_1^2 - 9) + \frac{11}{1050\lambda_m^4}(\bar{I}_1^3 - 27) + \frac{19}{7000\lambda_m^6}(\bar{I}_1^4 - 81) + \frac{519}{673750\lambda_m^8}(\bar{I}_1^5 - 243)\right\} + \frac{1}{D}\left(\frac{J_{el}^2 - 1}{2} - \ln J_{el}\right) \qquad (71)$$

where W is the strain energy per unit of reference volume, and μ, λ_m and D_i are temperature dependent material parameters. λ_m corresponds to the maximum stretch until the material stiffness grows significantly.

The initial shear modulus and bulk modulus are defined according to (72) and (68).

$$\mu_0 = \mu\left(1 + \frac{3}{5\lambda_m^2} + \frac{99}{175\lambda_m^4} + \frac{513}{875\lambda_m^6} + \frac{42039}{67375\lambda_m^8}\right) \qquad (72)$$

(e.2) Van der Waals

The form of the Van der Waals strain energy potential, also known as the Kilian model, is:

$$W = W_{dev} + W_{vol} = \mu\left\{-(\lambda_m^2 - 3) + [\ln(1-\eta) + \eta] - \frac{2}{3}a\left(\frac{\tilde{I} - 3}{2}\right)^{\frac{3}{2}}\right\} + \frac{1}{D}\left(\frac{J_{el}^2}{2} + \ln J_{el}\right) \qquad (73)$$

where U is the strain energy per unit of reference volume, $\tilde{I} = (1-\beta)I_1 + \beta I_2$ and $\eta = \sqrt{\frac{\tilde{I}-3}{\lambda_m - 3}}$.

μ is the initial shear modulus (μ_0), λ_m is the locking stretch, a is the global interaction parameter, b is an invariant mixture parameter whose value is comprised between 0.0 and 1.0 and D defines the volumetric behaviour.

The initial shear modulus and bulk modulus are defined according to expressions (68) and (74).

$$\mu_0 = \mu \tag{74}$$

2.2 Modelling the Inelastic Behaviour of Elastomers

Several authors have developed constitutive models, some of them phenomenological models [13, 49, 50, 182–187] and others statistical or micromechanical based models [11, 12, 188–192], to predict the typical inelastic effects of elastomer materials (hysteresis, Payne effect, Mullins effect and permanent set). These models are able to reproduce one or two inelastic effects normally, but a constitutive model including all the effects with sufficient accuracy has yet to be developed.

A brief summary of the main damage models available in scientific publications is included in this section as well as detailed descriptions of constitutive models to incorporate viscoelastic behaviour in the time and frequency domains or inelastic effects such as the Mullins or Payne effects, frequency dependent behaviour or mechanical hysteresis into FE codes (basically, descriptions of the Overlay method, and the Bergström and Boyce, Ogden and Roxburgh and Simo models).

2.2.1 Time Domain Viscoelasticity

The viscoelastic behaviour of elastomers is usually approached by means of rheological models, basically the generalized Maxwell model, which is implemented in commercial FE codes by means of a Prony series. These concepts are described in the following paragraphs.

Viscoelastic properties are modelled by means of linear viscoelastic models (rheological models), combining linear springs and dashpots in the time domain. The springs model the elastic behaviour while the dashpots are responsible for viscous dissipation. A basic description of the behaviour of both elements is shown in Fig. 22.

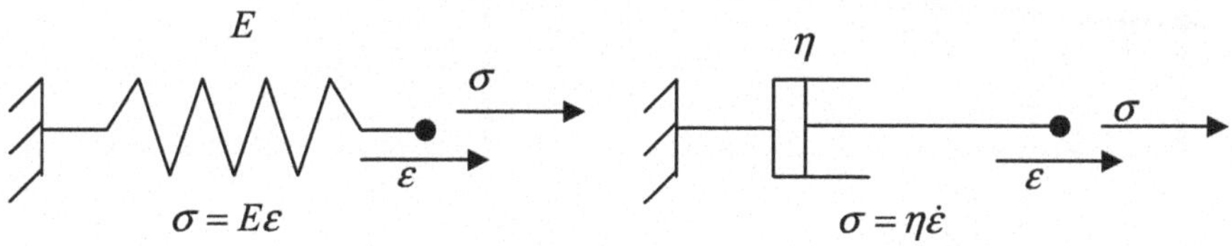

Fig. 22 Basic elements of a rheological model: linear spring and dashpot

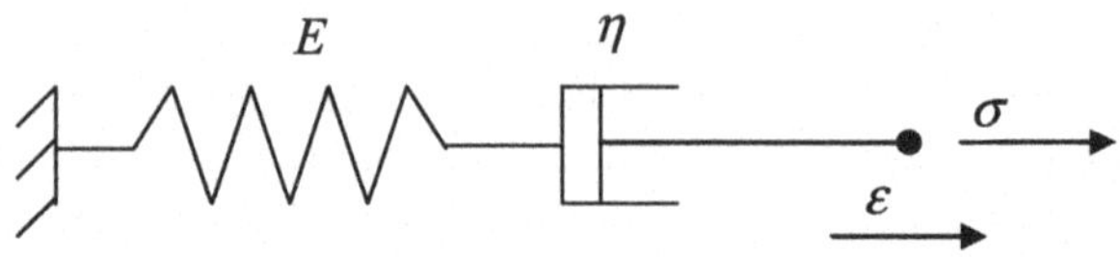

Fig. 23 Maxwell's model

Combining these elements in serial or parallel, it is possible to construct different viscoelastic models such as those of Maxwell, Kelvin, Zener, etc. Different models are now analysed in terms of quasi-static and cyclic behaviour in order to obtain the relaxation and complex moduli.

(a) Maxwell's model

Maxwell's model consists of a linear spring-dashpot series, as shown in Fig. 23.

The relaxation behaviour of this model, where the stresses are fully relaxed, suggests that this element behaves like a viscoelastic linear fluid. Relaxation, defined through $E_R(t)$, is the fundamental function defining viscoelastic behaviour. By applying a step strain to the model it is possible to obtain its output in terms of stress.

Since both components are placed in series, the whole element strain is defined as $\varepsilon = \varepsilon_{spring} + \varepsilon_{dashpot}$ and the temporal derivative is:

$$\dot{\varepsilon} = \dot{\varepsilon}_{spring} + \dot{\varepsilon}_{dashpot} \tag{75}$$

Replacing and reordering (75) with the terms $\dot{\varepsilon}_{spring} = \dot{\sigma}/E$ and $\dot{\varepsilon}_{dashpot} = \sigma/\eta$ it is possible to obtain (76), defining the stress–strain relationship for Maxwell's model (see Fig. 24).

$$\dot{\sigma} + \frac{E}{\eta}\sigma = E\dot{\varepsilon} \tag{76}$$

The relaxation modulus $E_R(t)$ could be obtained solving (76) for a step strain condition. For $t > 0$ the strain derivative is $\dot{\varepsilon} = 0$ and therefore:

$$\dot{\sigma} + \frac{E}{\eta}\sigma = 0 \quad t > 0 \tag{77}$$

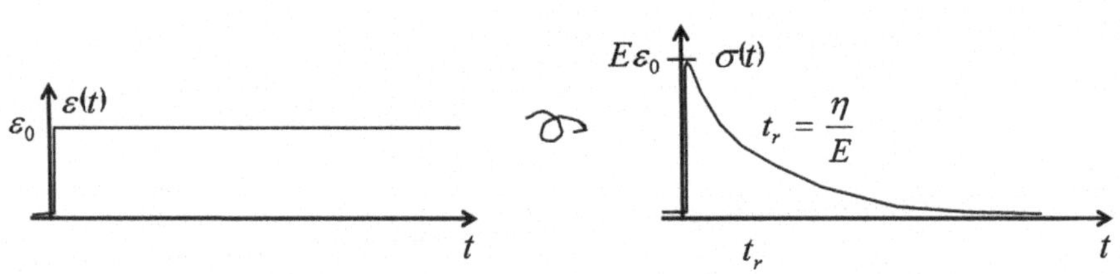

Fig. 24 Relaxation of Maxwell's model. Stress is relaxed to zero

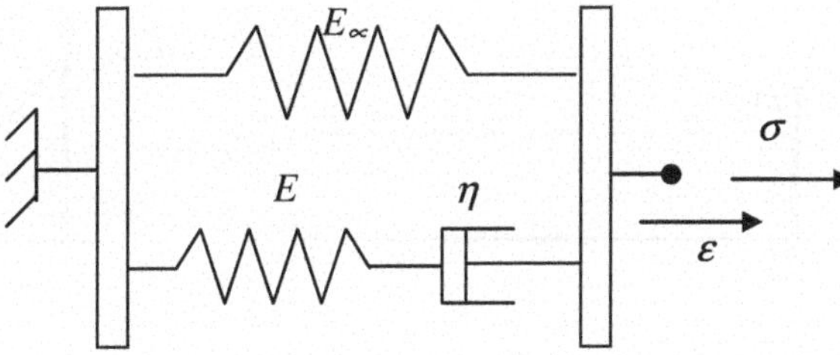

Fig. 25 Zener's model

When a step deformation is applied to the element, the dashpot behaves like a rigid element due to the infinite velocity at the instant of strain application. In this way, initial stress is defined through the spring behaviour, considering it as an initial condition to solve the differential equation $\sigma(0) = E\varepsilon_0$, that is, the instantaneous elastic response. The expression (78) is obtained solving (77):

$$\sigma(t) = E\varepsilon_0 e^{-\frac{E}{\eta}t} \tag{78}$$

This stress relaxation can be observed in Fig. 23 and the relaxation modulus of Maxwell's model is extracted as:

$$E_R(t) = Ee^{-\frac{t}{t_r}} \tag{79}$$

where $t_r = \frac{\eta}{E}$ is the relaxation time.

To obtain Maxwell's complex modulus, it is necessary to solve again Eq. (76), but for a stationary sinusoidal strain input $\varepsilon^* = \varepsilon_0 e^{i\omega t}$ instead of the step strain input used previously.

$$C = E\frac{i\omega}{i\omega + E/\eta}\varepsilon_0 = E\frac{i\omega t_r}{i\omega t_r + 1}\varepsilon_0 \tag{80}$$

The complex modulus for Maxwell's model is:

$$E^*(\omega) = E\frac{i\omega t_r}{1 + i\omega t_r} \tag{81}$$

(b) Zener's model

Zener's model consists of a set of linear springs and a dashpot placed in parallel. This model is based on Maxwell's model, adding an additional spring in parallel to the spring-dashpot set, as shown in Fig. 25.

Zener's model is the simplest viscoelastic model with solid properties with reasonable physical properties, in the sense that the relaxation values are reasonable and the creep time is finite.

If σ_∞ is the stress due to the spring and σ_M the stress in the Maxwell element, the total stress for the Zener model is defined according to (82)

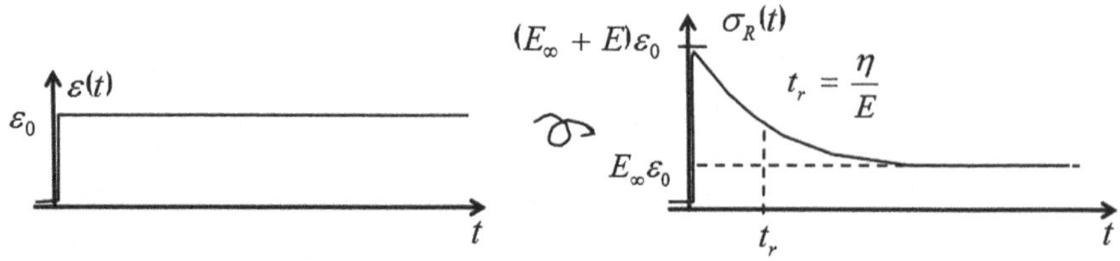

Fig. 26 Relaxation of Zener's model. Stress is relaxed to a given value for infinite time

$$\sigma = \sigma_{\infty} + \sigma_M \tag{82}$$

Combining Hooke's law to model the spring behaviour and the constitutive equation for the Maxwell element, the output of the Zener element to a step strain input is:

$$\sigma_R(t) = E_{\infty}\varepsilon_0 + Ee^{-\frac{E}{\eta}t}\varepsilon_0 \quad t > 0 \tag{83}$$

And therefore, the relaxation modulus for Zener's model is:

$$E_R(t) = E_{\infty}\left(1 + \frac{E}{E_{\infty}}e^{-t/t_r}\right) \tag{84}$$

These relationships are shown in Fig. 26.

Zener's complex modulus can be obtained solving Eq. (82) for a stationary sinusoidal strain.

$$\sigma^* = E_{\infty}\varepsilon^* + E\frac{i\omega t_r}{1 + i\omega t_r}\varepsilon^* \tag{85}$$

The complex modulus for Zener's model is:

$$E^*(\omega) = E_{\infty}\left(1 + \frac{E}{E_{\infty}}\frac{i\omega t_r}{1 + i\omega t_r}\right) \tag{86}$$

(c) Generalized Maxwell model

Zener's model can be generalised adding more Maxwell elements in parallel, as shown in Fig. 27. This model has the same qualitative properties as Zener's model but with better capacity to fit experimental data.

Following the nomenclature used in Zener's model, the total stress in the model is expressed as:

$$\sigma = \sigma_{\infty} + \sigma_{M1} + \sigma_{M2} + \cdots + \sigma_{Mn} \tag{87}$$

The relaxation modulus is obtained adding the individual terms of the different Maxwell models, obtaining in this way the so called Prony series:

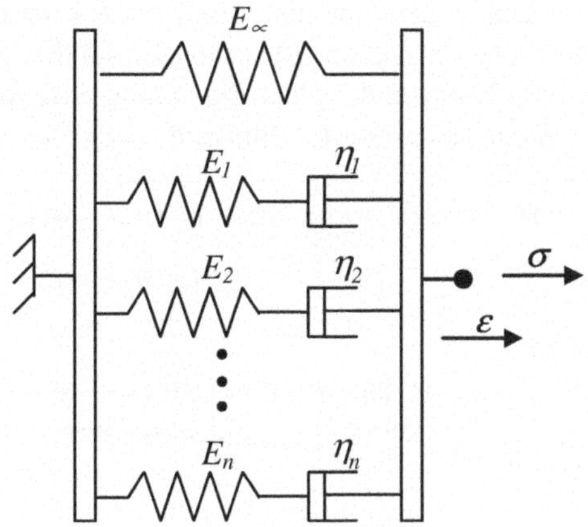

Fig. 27 Maxwell's model

$$E_R(t) = E_\infty + \sum_{j=1}^{n} E_j e^{-t/t_{rj}} \tag{88}$$

The complex modulus is derived from Maxwell's model in a similar way:

$$E^*(\omega) = E_\infty + \sum_{j=1}^{n} E_j \frac{i\omega t_{rj}}{1 + i\omega t_{rj}} \tag{89}$$

The viscoelastic behaviour of elastomers is approached by a generalized Maxwell model, which is implemented in commercial FE codes by means of a Prony series and takes into account the effect of the strain rate.

2.2.2 Frequency Domain Viscoelasticity

Several constitutive models have been developed by Lubliner [193], Johnson et al. [34] and Simo [35] to analyse large strain viscoelastic behaviour. In the particular case of simulations of small amplitude oscillations around a base state, the model proposed by Morman [60] is commonly available in commercial FE codes. This method can be applied to the analysis of small amplitude oscillations in viscoelastic solids superimposed on an initial static deformation. The superimposed oscillation can be small enough in order to linearise the constitutive equations. The basic constitutive equation was initially formulated by Lianis [194], who derived viscoelastic small strain theory superimposed on large static deformations using the finite linear viscoelasticity theory formulated by Coleman and Noll [195]. This model assumes a separation between preload and time effects. Stiffness equations defined in the FE method are identical to those that define non linear elasticity, except for considering complex values for the stiffness matrices and displacement vectors. Using this method, which is valid for viscoelastic analysis in the frequency domain, the steady state response to a small oscillation for a given frequency is calculated simply by solving a linear equation system (with complex coefficients) and it has the major advantage of an important reduction of the computational cost compared with incremental procedures used in the time domain.

The general formulation of small oscillation loads is:

$$L^*(\varepsilon, \omega) = D(\varepsilon) + i\omega\Phi^*(\varepsilon, \omega) \tag{90}$$

where

$L^*(\varepsilon, \omega)$ is the complex stiffness modulus expressed as a function of static deformation and frequency.

$D(\varepsilon)$ is the static stiffness modulus expressed as a function of the static modelling parameters.

$\Phi^*(\varepsilon, \omega)$ is the cyclic behaviour term depending on the static deformation and excitation frequency.

As can be observed in the model definition, amplitude dependence is not considered and this is the main limitation of this model for reproducing the behaviour of filled elastomers.

Functions in the tensor Φ^* are dependent on the static strain and frequency. Frequency dependent behaviour is incorporated through the Fourier transform of the viscous shear relaxation function $g^*(\omega)$ defined in real and imaginary parts as:

$$g'(\omega) = \frac{1}{2\omega} \frac{G'(\omega) - G^\infty}{G^\infty} \tag{91}$$

$$g''(\omega) = \frac{1}{2\omega} \frac{G''(\omega)}{G^\infty} \tag{92}$$

where

$G^*(\omega)$ is the dynamic shear modulus

G^∞ is the infinite (long term) shear modulus.

2.2.3 Constitutive Models for Inelastic Effects

This section provides a brief summary of different constitutive models available in the technical literature for including inelastic effects in the material modelling of elastomers.

Holzapfel et al. [37] formulated a continuous phenomenological model extending the pseudo-elastic model proposed by Ogden–Roxburgh to include the effects of permanent set. This damage model allows the Mullins effect and permanent set to be reproduced simultaneously. Mechanical and thermal hysteresis, essentially dependent on the strain rate, is not included. The model idealizes its stress response so the unloading path is the same as the loading path. It is based on a strain energy density function describing the loading path and a damage function for the unloading path, as Ogden and Roxburgh proposed. The main difference between the models is that Holzapfel-Stadler-Ogden consider three damage values to govern the anisotropic damage mechanism as a function of the principal stretches. (The Ogden–Roxburgh model only considered a single damage variable for the description of isotropic damage as a function of the strain energy density instead of principal stretches).

Holzapfel [180] propounded a strain rate independent isotropic damage model for the finite strain domain. This model can describe the Mullins effect but fails to reproduce permanent set and hysteresis. Besides, it is limited to loading histories with very low strain rates, where viscous effects can be neglected. This model is based on a pure phenomenological approach using a continuous damage model to describe macroscopic constitutive behaviour for materials with distributed micro cracks.

Lion [8] suggested a constitutive model based on 3D finite strain visco-plasticity theory to model the inelastic behaviour of reinforced elastomers. The model has the capacity to predict the Mullins effect as well as nonlinear dependency with the strain rate and weak equilibrium hysteresis. The basic structure of this model is the decomposition of the total stress in an equilibrium stress independent of the strain rate and a strain rate dependent on overstress. The uniaxial model is generalized in a three dimensional one using dual variables and their derivatives. Non linear strain rate dependency is represented using a stress-dependent relaxation time, while equilibrium hysteresis is introduced by means of arc-length. The Mullins effect is taken into account through a continuous damage model applying the effective stress concept. An important characteristic of the model is that it does not use decomposition of the deformation gradient into elastic and inelastic parts. A notable drawback, especially from the point of view of applicability, is that the model needs to fit seventeen material parameters, six for the overstress modelling, eight for equilibrium stress conditions and three for the Mullins effect.

Miehe and Keck [13] propounded a phenomenological material model to simulate superimposed elastic, viscoelastic and plastic stress responses including large strain damage. This formulation is suitable for modelling reinforced elastomer behaviour in monotonic and cyclic strain processes in isotherm conditions. The fundamental idea of this model, based on experimental observations, is to consider locally stored free energy decomposed in an additive way into three different contributions acting in parallel: a basic elastic stress response, a strain rate dependent viscoelastic overstress and a strain rate independent elastoplastic overstress. Damage is considered isotropic in these three parts. The model is able to reproduce the hysteresis of the equilibrium response for the material, the frequency dependent hysteresis that provokes an increment in the width of hysteresis cycles when the strain rate and Mullins effect are increased.

The model proposed by Besdo and Ihlemann [38, 196] is a phenomenological constitutive model that allows the material hysteresis and permanent set to be reproduced taking into account the influence of the load history influence to simulate damping effects. However, the load direction, temperature and long deformation processes are not taken into account. The final configuration of the model is fully three dimensional and the defining expressions contain Lagrangian stress and strain tensors exclusively. Strain history is represented through a simple scalar parameter, therefore the loading direction that could provoke an anisotropic behaviour is not taken into account and it is assumed that stress values are independent of strain rate and depend on deformation magnitude only. Rubber quasi-incompressibility is modelled using linear compression-pressure behaviour with a high Bulk modulus. For inelastic models, strain energy density is not a potential, therefore the material model is expressed directly in terms of stresses. Its formulation is based on an approximation of hysteresis cycles measured between two asymptotic limiting curves. Stress changes in the range compressed between those lines is described by means of a specific differential equation. Total stresses are composed of the sum of basics stresses and the solution of the differential equation. Deviatoric and volumetric contributions are treated separately, so each variable can be decomposed into its deviatoric and volumetric part.

Qi and Boyce [197] proposed a 3D micromechanical constitutive model to model the Mullins effect adopting the concept of Mullins and Tobin [105], who considered material softening as an evolution in the hard and soft domains within the elastomer while the stretch produces a quasi-irreversible rearrangement of molecular chains, due to a non-related local deformation resulting from short polymeric chains reaching the extensibility limit. This non-related deformation provokes a displacement in the molecular network joints with regard to the initial configuration, producing a rearrangement of the hard and soft domains in the elastomeric phase with the strain, increasing the effective volume of the soft domain and therefore a material softening with deformation.

Govindjee and Simo [192] proposed a model to predict the Mullins effect in reinforced elastomers from a micromechanical point of view. The model is based on a free energy function with first order precision which is derived from the compound of free energy densities from its components. Subsequently, the same authors proposed a phenomenological model based on the former one [6, 7], more efficient from a computational point of view. The main drawback of micromechanical models is that the constitutive equation for the material involves the integration of a complex expression over a domain dependent on the strain history in the polymer network space phase. The real application of the model is unworkable in most cases due to computational problems.

Marckmann et al. [198] proposed a constitutive model to reproduce simultaneously the Mullins effect, the hysteretic behaviour of elastomers and strain amplitude dependence. This model brings together three available models, those of Arruda and Boyce [28] to model hyperelastic behaviour, Bergström and Boyce [11, 12] to model material hysteresis and Simo [35] to simulate the Mullins effect. According to the authors, the three models are based on the physics of polymer chains and can be put together efficiently in a single constitutive equation. Loss of stiffness after each successive cycle due to viscoelastic effects is not taken into account. To improve this situation, especially in the first loading cycles, it would be necessary to add a time dependent term to the model. Besides, relaxation parameters would reduce the influence of the Mullins effect that is overestimated in the first loading cycles.

Four constitutive models have been selected for detailed description due to their special interest. One of them has been implemented in commercial FE codes with modifications (Ogden and Roxburg), another without modifications (Bergstöm and Boyce), while the Overlay method and Simo damage model could easily be incorporated into FE analysis of elastomers [199].

(a) Overlay method

The overlay method, proposed by Austrell et al. [54], is based on the sum of contributions obtained in simple models. Originally, this model was suggested by Besselling [55] to analyse yield and plasticity in metals. The basic idea behind this model is that the material is divided into several parallel fractions, each of them modelled by means of conventional properties. Complex behaviour of the material is obtained using a combination of simple constitutive models in parallel and suitable material parameters. The model enables the dynamic behaviour of filled elastomers to be reproduced with relative accuracy, including the Payne effect as well as frequency dependent behaviour and mechanical hysteresis. However, this model is not able to reproduce permanent set or the Mullins effect. Olsson and Austrell [58] proposed a procedure to fit the model constant using steady state dynamic experimental shear tests.

The most general form of the model is a multiaxial version of a unidimensional model proposed by Kümmlee, represented in Fig. 28.

Each fraction has elastic, plastic and viscous strains that are added with the same strain level for all the fractions considered.

The elastic contribution of the strain in each fraction *i,* results in an elastic stress in each fraction $\boldsymbol{\sigma}_i^e$, according to Hooke's law:

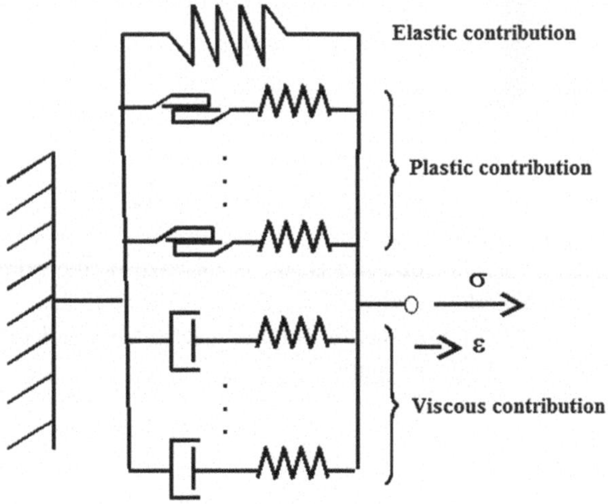

Fig. 28 Unidimensional rheological model proposed by Kümmlee

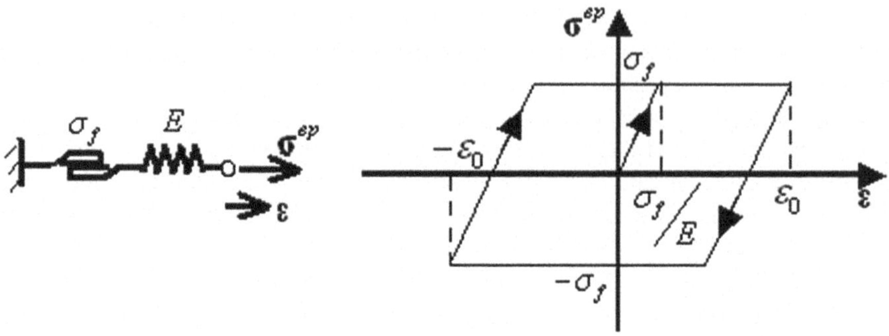

Fig. 29 Elastoplastic stress for a simple frictional model

$$\boldsymbol{\sigma}_i^e = \mathbf{C}^e \boldsymbol{\varepsilon}_i^e \tag{93}$$

where $\mathbf{C}^e$ is the elastic modulus tensor and $\boldsymbol{\varepsilon}_i^e$ is the elastic strain in fraction i.

The viscoelastic stress response in the j fraction, $\boldsymbol{\sigma}_j^v(t)$, is defined through the convolution integral (93) which describes the linear viscoelastic response for an arbitrary strain history

$$\boldsymbol{\sigma}_j^v(t) = \int_{-\infty}^{t} E_{R_j}(t - t')d\boldsymbol{\varepsilon}_j^v(t') \tag{94}$$

where E_{Rj} is the relaxation modulus for a Maxwell element (see Fig. 23) for the fraction j. t and t' are two time instants and $\boldsymbol{\eta}$ is the viscosity coefficient.

The elastoplastic stresses for a fraction l, $\boldsymbol{\sigma}_l^{ep}$, can be modelled as shown in Fig. 29, where σ_f is the stress in the frictional element and E_p is the elastic modulus of the fraction.

$$\begin{cases} \text{If } \sigma_l^{ep} < \sigma_f \text{ (elastic deformation)} \\ \text{For plastic deformation} \end{cases} \quad \boldsymbol{\sigma}_l^{ep} = \frac{E_p \boldsymbol{\varepsilon}}{\boldsymbol{\sigma}_f} \tag{95}$$

The total stress is the sum of stresses in each fraction

$$\boldsymbol{\sigma} = \sum_{k=1}^{N} \psi_k \boldsymbol{\sigma}_k \tag{96}$$

where $\boldsymbol{\sigma}$ is the total stress, ψ_k is the size of the relative volume for the fraction k and N is the number of fractions in the model.

The total strain rate deformation tensor is divided for each fraction into elastic, plastic and viscous parts:

$$\dot{\boldsymbol{\varepsilon}} = \dot{\boldsymbol{\varepsilon}}_k^e + \dot{\boldsymbol{\varepsilon}}_k^p + \dot{\boldsymbol{\varepsilon}}_k^v \tag{97}$$

For reinforced elastomers, the model proposed by Austrell [53], stress fractions are divided into viscoelastic and elastoplastic fractions with the same weights for both. Therefore, the Cauchy stress tensor, $\boldsymbol{\sigma}$, is the sum of contributions of elastic, viscoelastic and elastoplastic stresses.

$$\boldsymbol{\sigma} = \boldsymbol{\sigma}^{e} + \boldsymbol{\sigma}^{ve} + \boldsymbol{\sigma}^{ep} \tag{98}$$

where the elastic stress contribution is defined according to Eq. (87), the visco-elastic contribution as the sum of viscoelastic fractions is:

$$\boldsymbol{\sigma}^{ve} = \sum_{j=1}^{M_{ve}} \boldsymbol{\sigma}_j^{ve} \tag{99}$$

And the elastoplastic stress contribution is the sum of the elastoplastic fractions,

$$\boldsymbol{\sigma}^{ep} = \sum_{l=1}^{M_{ep}} \boldsymbol{\sigma}_l^{ep} \tag{100}$$

expressed by using the ideal plasticity model of Von Mises for simplicity, although more sophisticated elasto-plastic models could be used. M_{ve} and M_{ep} represent the number of visco-elastic and elasto-plastic fractions, respectively.

The overlay method can be used in a FE analysis constructing a model with several identical meshes in parallel, sharing nodes and each one with a different

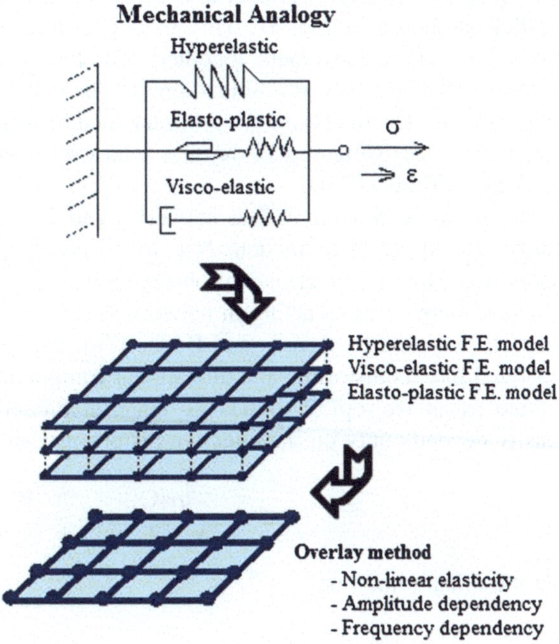

Fig. 30 Graphical representation of the overlay method for its use in FE analysis

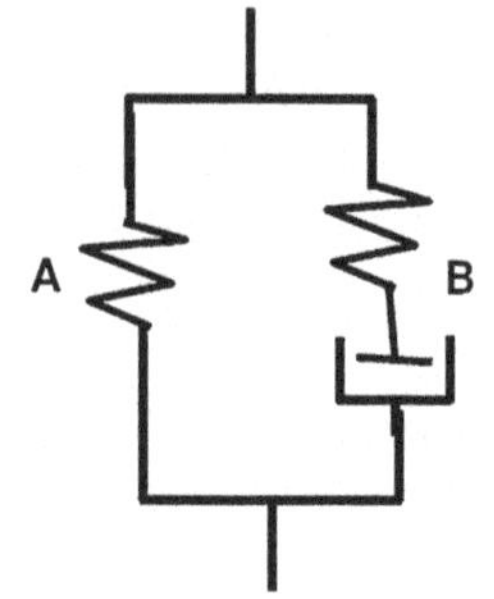

Fig. 31 Unidimensional rheological model proposed by Bergström and Boyce

material behaviour (visco-hyperelastic and elasic-plastic), as shown in Fig. 30. The overlay method has a significant advantage as it uses constitutive models available in non-linear FE codes.

(a) Bergström and Boyce model

The model proposed by Bergström and Boyce [11] is micromechanical based, mainly related with the movement of polymer chains when the material is stretched. Time dependent behaviour is governed by the reptational displacement combined with the frictional force of polymer chain sections subjected to a reconfigurational movement due to molecular slipping. The mechanical behaviour of the material can be broken down into two different parts: an equilibrium network corresponding to stress free configuration after a relaxation time and a second network which captures the deviation from the equilibrium state depending on the strain rate. In this way, the material is modelled as two polymer networks acting in parallel as shown in Fig. 31. This model is able to reproduce the mechanical stiffness of filled elastomers together with the time dependent behaviour and hysteresis. Additionally, it is able to predict the material behaviour as a function of the strain rate accurately and for arbitrary load histories. However, it is not able to model the stress softening in the first load cycles experienced by the elastomer material (Mullins effect).

The network A is an "ideal network" and can be modelled through typical constitutive hyperelastic models. Network B incorporates inelastic effects into the model according to experimental observations.

The Cauchy stresses acting in network A can be obtained from the eight chain network proposed by Arruda and Boyce [28], where the deformation gradient is separated into its distortion and dilatational components. This eight chain model is defined as an isotropic hyperelastic material model in which the strain energy density depends only on the effective distortional stretch.

$$\bar{\lambda^*} = \sqrt{\frac{tr[\mathbf{C}^*]}{3}} = \sqrt{\frac{tr[\mathbf{B}^*]}{3}} \tag{101}$$

and the dilatational one

$$J = \det(\mathbf{F}) \tag{102}$$

Therefore, the Cauchy stresses in network A are defined according to the equation:

$$\boldsymbol{\sigma}_A = \frac{\mu_A^0}{J\bar{\lambda}^*} \frac{L^{-1}\left(\frac{\bar{\lambda}^*}{\lambda_A^{lock}}\right)}{L^{-1}\left(\frac{1}{\lambda_A^{lock}}\right)} dev[\mathbf{B}^*] + \kappa[J-1]\mathbf{1} \tag{103}$$

where μ_A^0, λ_A^{lock} and $\kappa \in \mathbf{R}_+$ are material constants, $\mathbf{B}^* = J^{-2/3}\mathbf{F}\mathbf{F}^T$ and $\bar{\lambda}^* = \sqrt{\frac{tr(\mathbf{B}^*)}{3}}$ is the chain effective distortional stretch.

The deformation gradient in network B can be separated into an elastic and a viscous part using multiplicative decomposition:

$$\mathbf{F}_B = \mathbf{F}_B^e \mathbf{F}_B^v \tag{104}$$

The Cauchy stresses in this network are obtained in a similar way to those evaluated for network A:

$$\boldsymbol{\sigma}_B = \frac{\mu_B^0}{J_B^e \bar{\lambda}_B^{e*}} \frac{L^{-1}\left(\frac{\bar{\lambda}_B^{e*}}{\lambda_B^{lock}}\right)}{L^{-1}\left(\frac{1}{\lambda_B^{lock}}\right)} dev\left[\mathbf{B}_B^{e^*}\right] + \kappa\left[J_B^e - 1\right]\mathbf{1} \tag{105}$$

where

$$J_B^e = \det(\mathbf{F}_B^e) \tag{106}$$

and

$$\bar{\lambda}_B^{e*} = \sqrt{\frac{tr(\mathbf{B}^{e*})}{3}} \tag{107}$$

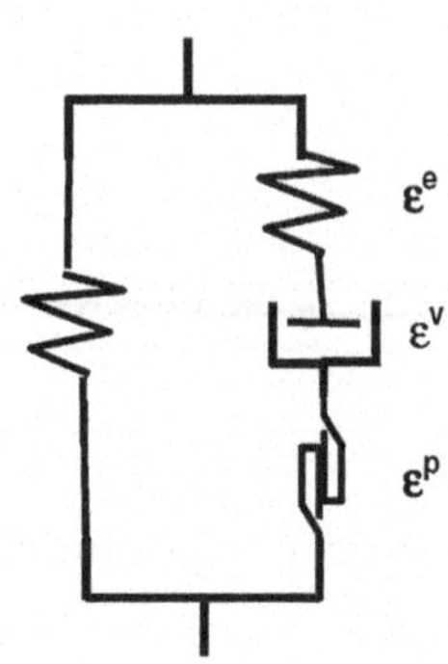

Fig. 32 Unidimensional rheological model proposed by Bergström and Boyce to improve unloading behaviour

The total stress in the element is the sum of the stress contributions of network A and B: $\boldsymbol{\sigma}_A + \boldsymbol{\sigma}_B$.

The model, defined in this way, addresses reasonably most situations excepting unloading behaviour at high strain rates. By adding a frictional element to network B in the previous model, it is possible to improve this aspect, as is shown in Fig. 32.

The motivation for this modification is based on the experimental observation of unloading stresses and their lower dependency on the strain rate, suggesting that the mechanism involved in the unloading behaviour of network B has a frictional character.

This idea can be included in a constitutive model for a uniaxial loading state using a strain rate dependent frictional element governed by the expression:

$$\dot{\boldsymbol{\varepsilon}}^p = |\dot{\boldsymbol{\varepsilon}}| \frac{\boldsymbol{\sigma}_B}{\eta} \tag{108}$$

where $\dot{\boldsymbol{\varepsilon}}$ is the strain rate tensor, $\boldsymbol{\sigma}_B$ is the stress in network B and η is the viscosity coefficient which takes different values for the loading or unloading paths. This idea can be generalised to a finite strain network using multiplicative decomposition of the deformation gradient into three different contributions (elastic, plastic and viscous):

$$\mathbf{F}_{\mathrm{B}} = \mathbf{F}_B^e \mathbf{F}_B^v \mathbf{F}_B^p \tag{109}$$

The stress in network B is calculated according to (105) and the kinematic is defined through the additive decomposition of the total strain velocity gradient in the three contributions:

$$\mathbf{L}_{\mathrm{B}} = \mathbf{L}_B^e + \mathbf{F}_B^e \mathbf{L}_B^v \mathbf{F}_B^{e-1} + \mathbf{F}_B^e \mathbf{F}_B^v \mathbf{L}_B^p \mathbf{F}_B^{v-1} \mathbf{F}_B^{e-1} = \tilde{\mathbf{L}}_B^e + \tilde{\mathbf{L}}_B^v + \tilde{\mathbf{L}}_B^p \tag{110}$$

where $\mathbf{L}_B^{v,p} = \dot{\mathbf{F}}_B^{v,p} \mathbf{F}_B^{v,p-1} = \mathbf{D}_B^{v,p} + \mathbf{W}_B^{v,p}$ and $\tilde{\mathbf{L}}_B^{v,p} = \tilde{\mathbf{D}}_B^{v,p} + \tilde{\mathbf{W}}_B^{v,p}$, $\mathbf{D} = {}^1\!/_2(\mathbf{L} + \mathbf{L}^T)$ is the symmetric part of the strain deformation gradient or strain rate tensor and $\mathbf{W} = {}^1\!/_2(\mathbf{L} - \mathbf{L}^T)$ is the anti-symmetric part of the strain deformation gradient or rotational velocity tensor.

In order to develop a single decomposition, $\tilde{\mathbf{W}}_B^v = \tilde{\mathbf{W}}_B^p = 0$ and therefore,

$$\tilde{\mathbf{D}}_B^v = \dot{\gamma}_B \frac{\boldsymbol{\sigma}'_B}{|\boldsymbol{\sigma}'_B|_F} \text{ and } \tilde{\mathbf{D}}_B^p = \frac{|\dot{\mathbf{F}}|}{\eta} \boldsymbol{\sigma}'_{\mathrm{B}} \tag{111}$$

with $\dot{\gamma}_B$ the effective strain rate defined as

$$\dot{\gamma}_B = C_1 \left(\bar{\lambda}_B^v - 1\right)^{C_2} \left(\frac{\tau}{\hat{\tau}}\right)^m \tag{112}$$

where $\tau = \|dev(\boldsymbol{\sigma}_B)\|_F$ is the effective stress, and $C_1, C_2, \hat{\tau}$ and m are material constants. For the proposed model, constants $\hat{C}_1 \equiv {}^{C_1}\!/_{\hat{\tau}^m}$ and m are positive while

C_2 is a constant that, based on molecular dynamics, has a probable value near to – 1. And finally, the η parameter is expressed as:

$$\eta = \begin{cases} \eta_{up}, & \text{If } \left(\boldsymbol{\sigma}' \cdot \mathbf{D}_B^v\right) > 0 \\ \eta_{down}, & \text{Otherwise} \end{cases} \tag{113}$$

(b) Ogden and Roxburgh model

Ogden and Roxburgh [51] suggested a phenomenological model to reproduce the Mullins effect (considering it as a quasi-static effect) and it is valid for the multiaxial general strain and stress state. The model is based on incompressible isotropic elastic theory supplemented by an additional continuous parameter, interpreted as a damage parameter, which modifies the strain energy density. This parameter controls the material properties allowing a strain energy density function to be used for the loading and modified in unloading conditions, reproducing the stress softening. Both behaviours are different from the virgin response of the material. Dissipation is measured as a damage function depending only on the damage parameter and the point on the loading path where the unloading starts. Therefore, this damage function is directly related with the dissipated energy in the first loading–unloading cycle and the damage parameter can be expressed, in general, in an implicit form in terms of strain. When the damage parameter is active, an equation to model damage evolution can modify the strain energy density function. Since the material response is governed by different strain energy density functions in the first loading path than in the unloading path, the model is known as pseudo-elastic.

The model describes the response of the material in terms of a strain energy density function with the form $W(\mathbf{F}, \eta)$. This definition is used in standard non-linear elasticity, excepting for the scalar variable η. This function is called the pseudo-energy function and is the basis of the pseudo-elasticity theory. In this model, η is considered as a continuous scalar parameter, and the material is considered incompressible. Therefore, $\det \mathbf{F} = 1$, being $\mathbf{F}$ the deformation gradient.

The second stress tensor of Piola-Kirchoff and the Cauchy stress tensor $\mathbf{S}$ and $\boldsymbol{\sigma}$ are defined as:

$$\mathbf{S} = \frac{\partial W}{\partial \mathbf{F}}(\mathbf{F}, \eta) - p\mathbf{F}^{-1} \text{ and } \boldsymbol{\sigma} = \mathbf{F}\frac{\partial W}{\partial \mathbf{F}}(\mathbf{F}, \eta) - p\mathbf{1} \tag{114}$$

where p is a Lagrange multiplier associated to the incompressibility constraint and $\mathbf{1}$ is the unity tensor.

Equilibrium equations in a body without external forces are expressed by means of the equation:

$$div\boldsymbol{\sigma} = \mathbf{0} \ in \ \Omega \tag{115}$$

And the following equation is necessary due to the inclusion of η in the model:

$$\frac{\partial W}{\partial \eta}(\mathbf{F}, \eta) = 0 \ in \ \Omega \tag{116}$$

During a deformation process, the variable η can be active, inactive or can be activated in a continuous way. When it is inactive, the material behaves as an elastic material governed by the strain energy density function $W(\mathbf{F}, \eta)$, being η a constant parameter. When η is active, its value is determined in terms of the deformation gradient tensor using (117),

$$\eta = \chi(\mathbf{F}) \tag{117}$$

The material stills remains elastic but following the behaviour imposed by the modified strain energy density function $W(\mathbf{F}, \chi(\mathbf{F}))$. The model is only applicable to isotropic materials, therefore the pseudo-strain energy function can be expressed in terms of positive principal stretches, $\lambda_1, \lambda_2, \lambda_3$, where λ_i^2 are the eigenvalues of the right Cauchy stress tensor.

$$\mathbf{C} = \mathbf{F}^T\mathbf{F} \tag{118}$$

And taking into account the incompressibility constraint,

$$\lambda_1\lambda_2\lambda_3 = 1 \tag{119}$$

the pseudo-strain energy functions are expressed as $\tilde{W}(\lambda_1, \lambda_2, \eta)$ and the Cauchy stress components are

$$\sigma_i = \lambda_i \frac{\partial \tilde{W}}{\partial \lambda_i}(\lambda_1, \lambda_2, \eta) - p \quad (i = 1, 2, 3) \tag{120}$$

Therefore, the equilibrium equation (116) is simplified to

$$\frac{\partial \tilde{W}}{\partial \eta}(\lambda_1, \lambda_2, \eta) = 0 \tag{121}$$

The first loading path for a perfectly elastic material is also the unloading path, considering $\tilde{W}(\lambda_1, \lambda_2)$ as a way of characterising the loading path. Standard strain energy density functions can be used for this function, for instance the Yeoh model (66). Using equation (120) and removing *p,* the following expression can be obtained:

$$\tilde{\sigma}_\beta - \tilde{\sigma}_3 = \lambda_\beta \tilde{W}_\beta \ with \ \beta = 1, 2 \tag{122}$$

where the swung dash indicates the primary loading path and therefore the constraint $\frac{\partial W}{\partial \eta} = 0$ is inoperative. The usual constraints are also imposed:

$$\tilde{W}(1,1) = 0;\ \tilde{W}_\beta(1,1) = \frac{\partial \tilde{W}}{\partial \lambda_\beta}(1,1) = 0 \text{ with } \beta = 1,2$$

$$\tilde{W}_{11}(1,1) = \frac{\partial^2 \tilde{W}}{\partial \lambda_1^2}(1,1) = \tilde{W}_{22}(1,1) = \frac{\partial^2 \tilde{W}}{\partial \lambda_2^2}(1,1) = 2\tilde{W}_{12}(1,1) = \frac{\partial^2 \tilde{W}}{\partial \lambda_1 \partial \lambda_2}(1,1) = 4\mu \tag{123}$$

where μ is the shear modulus of the material.

Unloading is possible from any point in the primary loading path, and the beginning of unload is the point to activate η. For the first loading and successive loading–unloading cycles, η varies according to (124), which is supposed to be solved explicitly for η,

$$\eta = \chi(\lambda_1, \lambda_2) = \chi(\lambda_2 \lambda_1) \tag{124}$$

Therefore, a strain energy density function for the unload, symmetric in terms of (λ_1, λ_2) and expressed as $w(\lambda_1, \lambda_2)$, is defined next:

$$w(\lambda_1, \lambda_2) = W(\lambda_1, \lambda_2, \chi(\lambda_1, \lambda_2)) \tag{125}$$

Considering $(\lambda_{1m}, \lambda_{2m})$ as the values of (λ_1, λ_2) at the starting point of the unload implies that $\chi(\lambda_{1m}, \lambda_{2m}) = 1$, therefore χ and w depend on the point where the unload starts.

The pseudo-strain energy function proposed by Ogden and Roxburgh is:

$$W(\lambda_1, \lambda_2, \eta) = \eta \tilde{W}(\lambda_1, \lambda_2) + \phi(\eta) \tag{126}$$

where $\phi(\eta)$ is a damage function with the following values for the first unloading cycle: $\eta = 1$ and $\phi(1) = 0$. The next expression is obtained substituting (126) into (121), which defines η in terms of strain.

$$-\phi'(\eta) = \tilde{W}(\lambda_1, \lambda_2) \tag{127}$$

Considering $\eta = 1$ at any point in the first load path where an unloading path is started, we have

$$-\phi'(1) = \tilde{W}(\lambda_{1m}, \lambda_{2m}) \equiv W_m \tag{128}$$

and deriving equation (127):

$$-\phi''(\eta) \frac{d\eta}{d\lambda} = \hat{W}_\lambda(\lambda_1, \lambda_2) \tag{129}$$

If the unloading path is associated with the reduction of the value associated to η, the next constraint based on stress softening is $\phi''(\eta) < 0$ which is satisfied by (129). If the material is unloaded completely $\lambda_1 = \lambda_2 = 1$, η reaches its minimum value η_m, and it satisfies

$$\phi'(\eta_m) = -\tilde{W}(1,1) = 0 \tag{130}$$

When the material is damaged, it is unloaded completely and the pseudo-strain energy function has a residual value

$$w(1,1) = W(1,1,\eta_m) = \phi(\eta_m) \tag{131}$$

This residual value of energy, $\phi(\eta_m)$, is unrecoverable and can be interpreted as a measure of the dissipated energy required to damage the material. For our convenience and according to (128), the function f is defined independently of W_m,

$$\phi'(\eta) + W_m = \phi'(\eta) - \phi'(1) = f(\eta) \tag{132}$$

that satisfies the same constraints as function ϕ, that is,

$$f(1) = 0, \quad f(\eta_m) = W_m \tag{133}$$

and integrating (132) with regard to η, we obtain:

$$\phi(\eta) = \int_1^{\eta} f(\eta)d\eta + (1-\eta)W_m \equiv \Phi(\eta, W_m) \tag{134}$$

where function Φ is defined to reflect the dependency of ϕ with regard to W_m, and then

$$\phi(\eta_m) = \Phi(\eta, W_m) \tag{135}$$

The dissipation ratio is obtained differentiating (134)

$$\dot{\phi}(\eta_m) = \frac{d\Phi}{dW_m}(\eta_m, W_m)\dot{W}_m = (1-\eta_m)\dot{W}_m \tag{136}$$

where the dot indicates a temporal derivative or the derivative with regard to any other strain parameter that grows with the primary loading path. Since $\eta_m \leq 1$ only happens at the beginning of the first loading cycle and the stored energy in the primary loading path should increase, that is $\dot{W}_m > 0$, the dissipation ratio is non-negative according to (136) (in agreement with the second law of thermodynamics and the Clausius-Duhem inequality).

The term $W_m - \phi(\eta_m)$ is the recoverable energy and increases with the ratio $\eta_m\dot{W}_m$. Function ϕ determines the damage parameter in terms of the strain state through (127). The selection of $\phi(\eta)$ is arbitrary, but subject to the following constraints:

$$\phi(1) = 0, -\phi'(1) = \tilde{W}(\lambda_{1m}, \lambda_{2m}) \equiv W_m, \phi''(\eta) < 0 \;\; con \;\; 0 < \eta \leq 1 \tag{137}$$

Ogden and Roxburgh selected the function $\phi(\eta)$ to be

$$-\phi'(\eta) = m \;\; erf^{-1}(r(\eta - 1)) + W_m \tag{138}$$

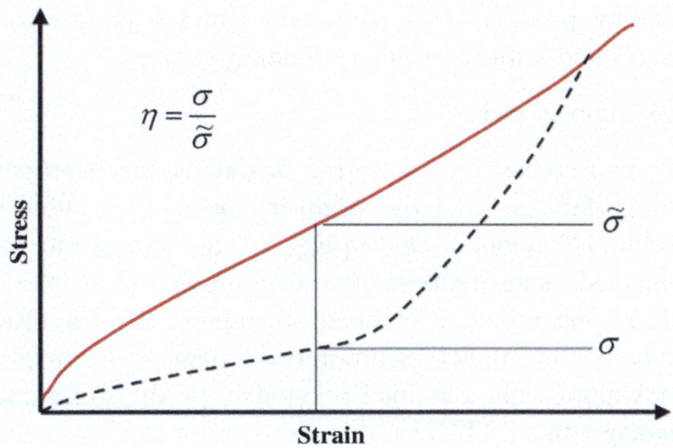

Fig. 33 Physical interpretation of damage parameter for Ogden and Roxburg model

with m and r being parameters of the material and $erf^{-1}(\bullet)$ the inverse of error function

$$erf(x) = \frac{2}{\sqrt{\pi}} \int_0^x e^{-t^2} dt \tag{139}$$

Parameter m has energy dimensions while parameter r is dimensionless. The damage parameter is expressed as

$$\eta = 1 - \frac{1}{r} erf\left(\frac{1}{m}(W_m - \tilde{W}(\lambda_1, \lambda_2))\right) \tag{140}$$

that can be interpreted as the relationship between stress with damage (σ) and stress without damage ($\tilde{\sigma}$) in the primary response, see Fig. 33.

The minimum values of η, η_m, corresponding to the undeformed configuration and the dissipation ratio, are defined as:

$$\eta_m = 1 - \frac{1}{r} erf\left(\frac{W_m}{m}\right) \tag{141}$$

$$\dot{\phi}(\eta_m) = \frac{1}{r} erf\left(\frac{W_m}{m}\right) \dot{W}_m \tag{142}$$

Parameter r is a measurement of relative damage to the virgin state. Specifically, the higher the parameter r, the further from 1 is the value of the damage parameter η, and the lower the damage in the material.

Parameter m governs the damage dependency on the strain amount. For small values of m, significant damage takes place for small deformations and the material response in the small strain region is not very affected for successive primary loads. For high values of m, there is little damage for small strains, but the material response changes markedly in the small strain region after the first loading process. This parameter can be interpreted as a strain energy density associated with the primary loading curve.

(c) Simo model

A viscoelastic constitutive model has been proposed by Simo [35], incorporating damage for large strain by means of a linear law for isotropic and anisotropic behaviours. Decoupling of the deviatoric and volumetric responses is obtained using multiplicative decomposition of the deformation gradient. Hyperelastic behaviour is obtained asymptotically for fast and slow processes. Additionally, the model incorporates a damage parameter characterized through the maximum value obtained previously by the strain energy of the material without damage. In a cyclic test, this viscoelastic damage model predicts a progressive loss of stiffness and an increasing dissipation as the maximum amplitude is increased, in accordance with the Mullins effect. It is crucial to assume a free energy potential function allowing consideration of the stress tensor divided into initial and equilibrium parts.

Simo proposed a viscoelastic model formulated in terms of three dimensional finite strain, which incorporates a damage mechanism based on irreversible thermodynamics theory to model the Mullins effect. This model is applicable to non-isotropic behaviour.

The basic characteristics of the model are:

(i) Additive decomposition of the stress tensor into initial and non-equilibrium parts, that allows for a general anisotropic response.
(ii) Uncoupled deviatoric and volumetric response over any strain range due to the multiplicative decomposition of the deformation gradient into volumetric and deviatoric parts.
(iii) Viscous response characterised through a constitutive equation with linear ratio, implying a convolution representation generalising viscoelastic models with regard to linearised kinematics (general relaxation functions).
(iv) Isotropic damage mechanism, incorporating stress softening behaviour (Mullins effect), governed through the maximum value of strain energy for a material without damage. For a cyclic test, this viscoelastic damage model predicts the stiffness reduction (Payne or Fletcher-Gent effect) and the increasing dissipation as the maximum cyclic amplitude is increased.

The constitutive model is especially well-suited for large scale computation, since the constitutive integration algorithm has the following main characteristics:

(v) It is a generalised mid-point algorithm, which ensures incremental objectivity (that is, the constitutive equations are independent of the reference during a rigid body motion).
(vi) It can be linearised exactly in a closed form and due to the structure of the inelastic rate equation, the tangent operator is also symmetric.
(vii) The incompressibility constraint is imposed by means of a mixed FE technique, based on a Hu-Washizu formulation.

The damage mechanism is modelled from a phenomenological point of view, based on continuous damage mechanics and the “equivalent stress” concept (originally used by [49]). In this context, Simo proposes an isotropic three dimensional damage model suitable for its numerical implementation. The fundamental hypothesis is based on a damage parameter characterized through the maximum value obtained previously by the strain energy of the material without damage. Using thermodynamic concepts, this concept is generalised for a three dimensional situation introducing a strain energy density for the material without damage as a scalar measure of maximum strain.

Simo proposes the following strain energy density function, W, neglecting residual stresses in order to characterise the isothermal process and considering that damage only affects the deviatoric part of the strain energy density function,

$$W(\mathbf{C}, D) = W_{vol}(J) + (1 - D)\bar{W}_0(\bar{\mathbf{C}}) \tag{143}$$

where $W_{vol}(J)$ is a convex function (with a minimum for $J = 1$) which describes the elastic volumetric response, $\bar{W}_0(\bar{\mathbf{C}})$ represents the effective deviatoric density function (convex) without damage or elastic energy stored in the material without damage, $D \in [0, 1]$ and $(1 - D)$ are reduction factors. Stresses for an isotropic compressible hyperelastic material are obtained deriving this function with regard to time. The following expression can be obtained using the chain rule,

$$\dot{W} = \frac{dW_{vol}(J)}{dJ}\dot{J} + (1 - D)\frac{\partial \bar{W}_0}{\partial \bar{\mathbf{C}}} : \dot{\bar{\mathbf{C}}} - \bar{W}_0\dot{D} \tag{144}$$

The second Piola-Kirchoff stress tensor $\mathbf{S}$ (formulated as the sum of deviatoric and volumetric responses) is:

$$\mathbf{S} = \mathbf{S}_{vol} + (1 - D)\bar{\mathbf{S}}_0 \tag{145}$$

where

$$\mathbf{S}_{vol} = 2\frac{\partial W_{vol}(J)}{\partial \mathbf{C}} = J\frac{\partial W_{vol}(J)}{\partial J}\mathbf{C}^{-1} \tag{146}$$

$$\bar{\mathbf{S}}_0 = J^{2/3} DEV\left(2\frac{\partial \bar{W}(\bar{\mathbf{C}})}{\partial \bar{\mathbf{C}}}\right) \tag{147}$$

And internal dissipation

$$D_{\text{int}} = f\dot{D} \geq 0 \text{ with } f = \bar{W}_0(\bar{\mathbf{C}}) \geq 0 \tag{148}$$

Inequality (148) shows that damage is a dissipative process. The thermodynamic force f governs the damage process and its physical interpretation is an effective deviatoric energy $\bar{W}_0$, released by the unity reference volume of the material. The thermo-dynamic force f is related with the internal variable D through the expression:

$$f = \bar{W}_0(\bar{\mathbf{C}}) = -\frac{\partial W}{\partial D} \tag{149}$$

Therefore, the damage process is controlled as a function of the conjugated variable instead of the internal variable. The evolution of the damage parameter D is characterised through the equation for irreversible evolution described as follows. Defining an equivalent strain Ξ_s

$$\Xi_s = \sqrt{2\bar{W}_0(\bar{\mathbf{C}}(s))} \tag{150}$$

where $\bar{\mathbf{C}}(s)$ is the time-modified left Cauchy-Green strain tensor, $s \in R$. If Ξ_t is the maximum value of Ξ_s along the load history up to time t, that is:

$$\Xi_t = \max_{s \in (-\infty, t]} \sqrt{2\bar{W}_0(\bar{\mathbf{C}}(s))} \tag{151}$$

The damage criterion is defined in the strain space using the constraint:

$$\varphi(\mathbf{C}(t), \Xi_t) = \sqrt{2\bar{W}_0(\bar{\mathbf{C}}(t))} - \Xi_t \leq 0 \tag{152}$$

Equation $\varphi(\mathbf{C}(t), \Xi_t) = 0$ defines the damage surface for the material in the strain space. Naming $\mathbf{N} := {}^{\partial \varphi}/_{\partial \mathbf{C}} \equiv \left({}^{1}/_{\Xi_s}\right){}^{\partial W_0}/_{\partial \mathbf{C}}$ the normal of damage surface, the following options are available:

$$\varphi < 0 \text{ or } \varphi = 0 \text{ and } \begin{cases} \mathbf{N} : \dot{\mathbf{C}} < 0 \\ \mathbf{N} : \dot{\mathbf{C}} = 0 \\ \mathbf{N} : \dot{\mathbf{C}} > 0 \end{cases} \tag{153}$$

Finally, the evolution equation specifies the evolution of the damage variable D

$$\frac{dD}{dt} = \begin{cases} \bar{h}(\Xi, D)\dot{\Xi}, \text{ if } \phi = 0 \text{ and } \mathbf{N} : \dot{\mathbf{C}} > 0 \\ 0, \text{in other cases} \end{cases} \tag{154}$$

where $\bar{h}(\Xi, D)$ characterises the damage process in the material. If $\bar{h}(\Xi, D)$ is independent of D, then the second deviatoric stress tensor of Piola-Kirchoff can be expressed in the following way, where the maximum strain in time t, Ξ_t, is defined using (151)

$$\bar{\mathbf{S}}(t) = \bar{g}(\Xi_t)\frac{\partial \bar{W}_0(\bar{\mathbf{C}}(t))}{\partial \mathbf{C}} \tag{155}$$

with

$$\bar{h}(\Xi) = -\frac{d\bar{g}(\Xi)}{d\Xi} \tag{156}$$

In order to complete the damage model, function $\bar{g}(\Xi_t)$ is specified, or equivalently, $\bar{h} = -\frac{d\bar{g}}{d\Xi}$. This function can be specified using experimental data adopting the exponential form:

$$\bar{g}(x) = \beta + (1-\beta)\frac{1-e^{-x/\alpha}}{x/\alpha}, \quad \beta \in [0,1], \quad \alpha \in [0,\infty) \tag{157}$$

where β and α are considered material parameters.

The starting hypothesis of this phenomenological damage model is that damage only affects the deviatoric stresses, therefore the evolution of the stress tensor for the material is defined according to (155). The following expression is obtained using the chain rule:

$$\bar{\mathbf{S}} = \begin{cases} \left[\bar{g}\bar{\mathrm{C}}_0 - \bar{g}'\bar{\mathbf{S}}_0 \otimes \bar{\mathbf{S}}_0\right] : \frac{\dot{\mathrm{C}}}{2} & si\ \varphi = 0\ y\ \mathbf{N} : \dot{\mathbf{C}} > 0 \\ \bar{g}\bar{\mathrm{C}}_0 : \frac{\dot{\mathrm{C}}}{2} & \text{otherwise} \end{cases} \tag{158}$$

with $\bar{g}' = {}^{d\bar{g}}/_{d\mathrm{E}} = -\bar{h}$. Therefore, the deviatoric contribution to the elasticity tensor is expressed as:

$$\bar{\mathrm{C}}(\mathrm{t}) = \begin{cases} \bar{g}\bar{\mathrm{C}}_0 - \bar{g}'\bar{\mathbf{S}}_0 \otimes \bar{\mathbf{S}}_0 \text{ in loading} \\ \bar{g}\bar{\mathrm{C}}_0 \ \ \text{otherwise} \end{cases} \tag{159}$$

According to Lion [8], the damage model proposed by Simo considers the Mullins effect as the only source of inelasticity in the behaviour of reinforced elastomers, without taking into account other sources such as hysteresis or permanent set.

2.3 *Fatigue Behaviour*

Typical models for predicting fatigue life in rubber follow two overall approaches focussing (a) on predicting crack nucleation life and (b) on predicting the growth of a particular crack.

The crack nucleation approach [67, 200–202] considers that fatigue life is determined for a given stress and strain history at a certain point within a material. Therefore, the fatigue life according to this approach could be defined as the number of cycles necessary to obtain a crack of a certain length. The most usual parameters used to predict the nucleation life are based on strain or energy. Figure 34 summarizes the typical approach to predicting fatigue life following a crack nucleation scheme, using strain-controlled fatigue tests to obtain the uniaxial fatigue curve for the rubber material and FE analysis of the component to detect the critical zones as a function of strain.

The use of strain as a life parameter has advantages since it can be obtained directly from measured displacements. When the strain energy density is used, it is

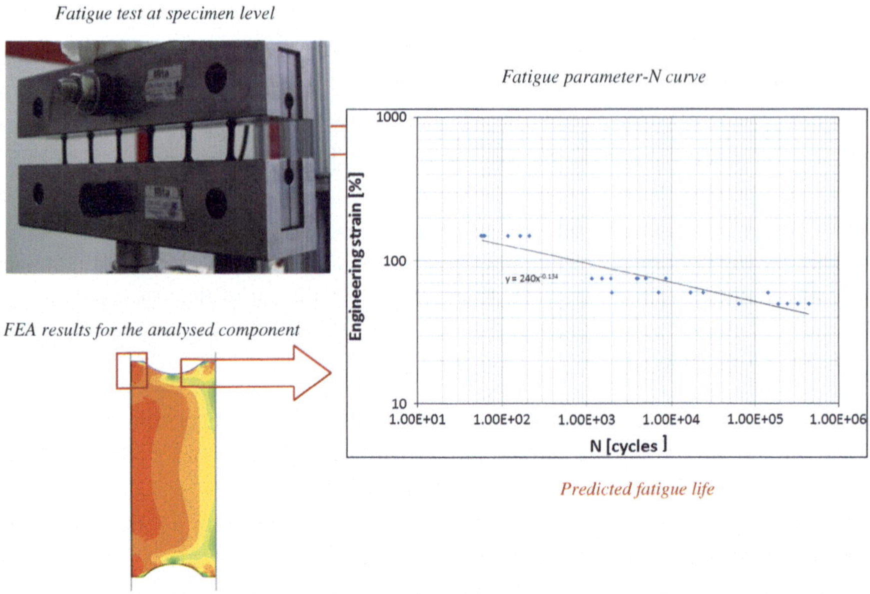

Fig. 34 Typical fatigue life prediction scheme based on a crack nucleation approach

often evaluated using hyperelastic material models based on the strain invariants and therefore it is also based on strains. Strain energy density has been used as a fatigue parameter in metals [121] although the correlation between experimental and predicted results is not satisfactory.

Rivlin and Thomas [122] proposed a model to study the fracture of rubber under static loading based on the strain energy density. This has been used by many investigators to correlate analysis results to experimental component life data, considering the strain energy density as a measurement of the energy release rate of the different flaws present in the material. The application to components of this approximation carried out by some authors [123, 124, 203–205] shows differences between fatigue life and computed strain energy density levels. The main limitations of the strain energy density are that it is unable to predict the fact that the crack surface appears in a specific orientation, only part of the total spent energy plays a role in the crack nucleation process for multiaxial conditions, it does not account for crack closure and it fails to predict large life differences between simple tension and simple compression loadings.

Stress has rarely been used as a fatigue life parameter in rubber (Andre N. [125]). This is related to the fact that fatigue testing in rubber has traditionally been carried out under displacement control, and the accurate stress determination in rubber components can be difficult. The maximum principal strain and the octahedral shear strain have also been used as strain-based fatigue parameters. The maximum principal strain criterion was introduced by Cadwell [67] in unfilled vulcanized NR and remains in use nowadays, particularly for uniaxial strain loadings. It also gave a good correlation for axial/torsion tests [121]. The octahedral shear strain criterion makes a prediction that is roughly similar to the principal strain criterion for rubber [75, 76], but for an incompressible material both strain based criteria always satisfy that their values are positive and therefore are unable to account for compression states where the crack is closed.

Finally, fatigue life prediction on rubber combining the crack nucleation and growth approaches, also known as the energetic approach to rubber fatigue, has been applied successfully by Mars [77] to an automotive component. This approach is described in detail in its principal aspects in the following paragraphs.

When a rubber part is subjected to repeated stresses a certain time passes before visible fatigue cracks appear. The distinction between crack initiation and propagation is not clear. What matters in practice is how fast a crack of a certain size will propagate under certain loading conditions. The focus in fracture mechanics is thus on fatigue crack propagation. An initial crack size is chosen that reflects the typical size or the maximum crack size detectable present in a given material. Typical industrial rubber exhibits inhomogeneities on a microscopic scale containing gel particles, voids, and roughness at cut or moulded edges. These inhomogeneities form highly localized stress concentrations that initiate the fatigue failure. For natural rubber the effective size of initial flaws are estimated to be about 25 μm.

Fracture mechanics seeks to obtain fracture parameters that represent failure characteristics of the material, independent of geometry and loading. An energy criterion for crack growth was proposed by Griffith [206]. He contended that every body contains a distribution of flaws and the failure starts at the largest of these. Griffith proposed that an initial crack C in a body of thickness l and a fixed displacement t will grow if the decrease in the total elastic energy W of the body per unit increase in the crack C is at least equal to the surface energy T required to form new crack surfaces. Expressed in mathematical terms, this is:

$$-\left(\frac{1}{t}\right)\cdot\left(\frac{\partial W}{\partial C}\right)_{l}\geq T \tag{160}$$

Under conditions where the geometry thickness is not constant, the Griffith criterion is modified to include the work of the external forces. This concept was first applied successfully to the tearing of rubber by Rivlin and Thomas [122]. In this case, T in (160) is no longer equal in magnitude to the surface energy because rubber is not perfectly brittle. In fact, stresses induced at the tip of a flaw cause large local deformation and result in much more energy being dissipated. The strain energy release rate criterion is still valid for rubber, however, provided the energy dissipation is confined to a small zone at the crack tip.

The tearing energy concept is valid for mechanical fatigue cracking, since the crack propagation rate dC/dN is a function only of tearing energy T. To apply the Griffith criterion to a rubber component in an engineering application, it is necessary to calculate the tearing energy for a particular geometry. Fracture mechanics may also be employed to predict in a given structure the crack size

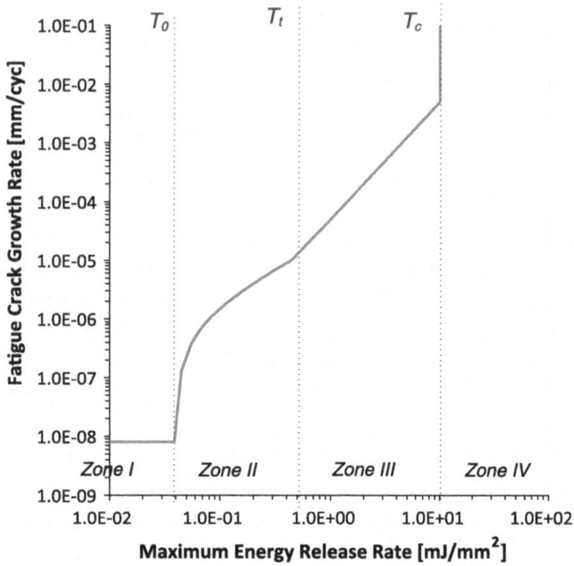

Fig. 35 Typical crack growth curve in a rubber material and identification of the different crack growth regimes

range that will propagate slowly under specified loading until it reaches the critical size from which it will propagate rapidly to catastrophic failure. The fatigue life is the number of cycles required to break a specimen into two pieces at a particular load state.

The approach to rubber fatigue known as the energetic approach is based on the integration of a crack growth law relating the crack advance per cycle and the energy release rate or tearing energy. The basis of the energetic approach is the use of the strain tearing energy defined according to (161), as a means of characterising crack growth behaviour.

$$T = -\frac{\partial W}{\partial A} \tag{161}$$

where W is the total elastic strain energy stored in a part containing a crack and A is the area of the fracture surface associated with the crack. The partial derivative indicates that the test piece is considered to be held at constant extended length, so that the external forces do not work.

The relationship between the crack growth rate dC/dN and tearing energy T is known as the crack growth characteristic of the material since T is independent of the sample geometry. Typical curves for a natural rubber (NR) compound cycled under relaxing conditions (minimum tearing energy equal to zero) are shown in Fig. 35.

The curve can be described easily in terms of four regions. In region I the tearing energy T is lower than the threshold tearing energy T_0, hence no mechanical crack growth occurs. T_0 is the mechanical fatigue limit, and below T_0 crack growth is caused by ozone attack only [207]. In region II, crack growth is dependent on ozone and mechanical factors in an approximately additive way and following a linear evolution. In region III a power law dependency between crack growth rate and tearing energy has been found for many rubbers as well as for non rubbery materials. Finally, region IV corresponds to uncontrollable crack growth and catastrophic failure.

The crack growth characteristic can be described using a crack growth law. For example, Lake and Lindley [71] identified the four aforementioned crack growth regimes and proposed the following set of equations, shown in (162), in order to describe the crack growth behaviour of a rubber material.

$$\frac{dc}{dN} = \begin{cases} r_z & T \leq T_0 \\ A \cdot (T - T_0) + r_z & T_0 < T \leq T_t \\ r_c \cdot \left(\frac{T}{T_c}\right)^F & T_t < T \leq T_c \\ \infty & T_c < T \end{cases} \tag{162}$$

Although this description, dividing the crack growth characteristics into four distinct regimes governed by a set of empirical equations is useful to describe the behaviour of crack growth, in practice it is oversimplified since many materials follow more complex empirical relationships. The different parameters used in (162) are the rate of crack growth under the influence of atmospheric ozone (r_z) when the energy release rate is above the mechanical threshold value of the energy release rate (T_0). The constant A describes the behaviour of the crack growth rate in the transition regime, when the energy release rate falls between the mechanical threshold energy release rate (T_0) and the transition value of the energy release rate (T_t). The power law regime is comprised between the transition tearing energy and the critical value of the energy release rate at which spontaneous fracture ensues (T_c). It is described by means of the power law exponent (F) and the crack growth rate where the power law intersects the vertical asymptote at T_c (r_c). For a natural rubber, F is about 2, whereas in general F lies between 2 and 6 for most rubber vulcanizates depending mainly on the elastomer used and on secondary factors such as rubber formulation.

The effects of non-zero minimum loading on fatigue crack growth have been analysed in the literature for strain crystallizing rubbers [78, 79] and for non-crystallizing rubbers [80]. An important difference between the crack growth behaviour of crystallizing and non-crystallizing rubbers is that under a static load, non-crystallizing rubbers can exhibit steady crack growth, while crystallizing rubbers typically exhibit no crack growth. For non-crystallizing rubbers, the crack growth rate under static load is a function of the energy release rate. Under cyclic load, the crack growth rate in non-crystallizing rubbers can be computed as the sum of steady and cyclic contributions to the crack growth. The fatigue crack growth rate *dc/dN* in strain crystallizing rubbers, for instance natural rubber-NR, depends on the maximum energy release rate T_{max} experienced during a load cycle, but also on the minimum T_{min}. Even when *dc/dN* is expressed as a function of $\Delta T = T_{max} - T_{min}$, a dependence on T_{min} remains. The fatigue crack growth rate in strain crystallizing rubbers depends not only on the maximum energy

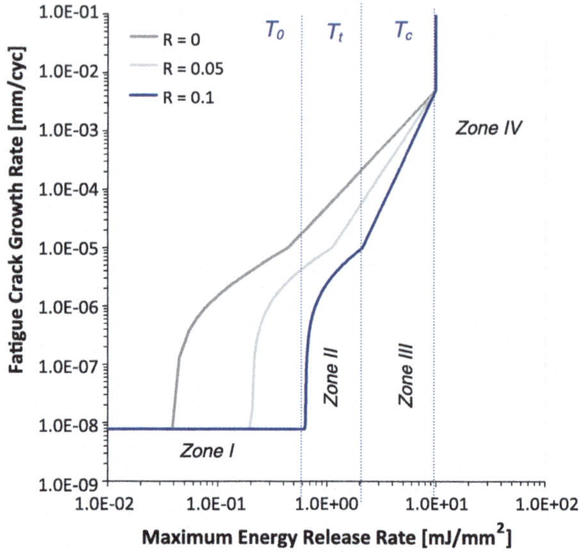

Fig. 36 Typical crack growth curve in a rubber material and identification of the different crack growth regimes showing the influence of non-relaxing conditions

release rate experienced during a loading cycle, but also on the minimum, or equivalently, the *R* ratio. The *R* ratio quantifies the minimum load attained during a cycle as a fraction of the maximum load.

Typical curves for a natural rubber (NR) compound cycled under relaxing and non relaxing conditions (R-ratio of 0.05 and 0.1) are shown in Fig. 36, where the different crack growth regimes are clearly identified. As is also shown, the influence of non-relaxing cycles hinders the crack growth and therefore increases the expected fatigue life.

The effects of non-relaxing cycles were modelled by Mars and Fatemi [208], modifying the crack growth law as a function of the R-ratio:

$$F(R) = F \cdot e^{C \cdot R} \tag{163}$$

where F is the power law exponent of the crack growth law proposed by Lake and Lindley [71], and C is a parameter of the material.

Finally, the fatigue life of a certain structure can be considered as the number of cycles necessary for a certain crack present in the material at the beginning of the fatigue process to grow up to a critical length that provokes the final failure of the component. Given the crack growth behaviour and the energy release rate history, the fatigue life can be computed via the integration of the crack growth law between the right limits and according to Ellul [64], and the fatigue life of any rubber component can be predicted by integrating the following equation depending on the energy release rate and its description according to the crack growth behaviour only:

$$N = \int_{c_0}^{c_1} f(T)\,dc \tag{164}$$

The integration limits indicated in the expression (164), parameters c_0 and c_1, are the initial and final crack size. The initial crack size is related to the presence of flaws, inhomogeneities and defects in the material and can be obtained by means of transmission electronic microscopy (TEM) on a sample of the material, while the final crack size is usually neglected since most of the life of a rubber component is spent to nucleate a crack. Alternatively, the final crack size is chosen for each specific application.

Equation (164) allows the fatigue life of a rubber component to be predicted in a very general way, since the question of how to obtain or describe the evolution of the tearing energy along a load cycle for a piece of material remains unsolved. Common methods available to obtain this parameter are based on analytical approximations through scalar parameters, by means of the combined use of FEA and numerical analysis or using multiaxial fatigue criteria. These methods are described briefly below.

The energy release rate can be derived from scalar parameters such as the strain energy density for simple geometries and loading conditions using empirical equations that are available in the bibliography. For example, the expression for a tensile specimen with an edge crack of length c is shown in (165).

$$T = 2kWc \tag{165}$$

where W is the strain energy density at the point where the energy is being evaluated and the parameter k is a function of the engineering strain:

$$k = \frac{2.95 - 0.08\varepsilon}{(1+\varepsilon)^{1/2}} \tag{166}$$

These relationships can be applied to the catastrophic tearing of rubber. The tearing energy concept is valid for mechanical fatigue cracking, since the crack propagation rate *dC/dN* is a function only of the tearing energy *T*. For more complex geometries or loading conditions, the tearing energy can be obtained using a FE simulation of the studied structure with an embedded crack using the J-integral approach to evaluate the energy release rate for a given crack length.

The non-linear behaviour of rubber (finite strains and quasi-incompressibility) and the fact that these criteria make no reference to a specific material failure plane imply that it is always possible to construct a non-proportional multiaxial history that holds the scalar equivalence criterion value constant while simultaneously varying the individual components of the history Mars [77]. Therefore, scalar equivalence criteria predict infinite life under certain kinds of non-proportional cyclic loading which actually result in finite life and it can be concluded that an analysis approach that makes specific reference to the failure plane is needed.

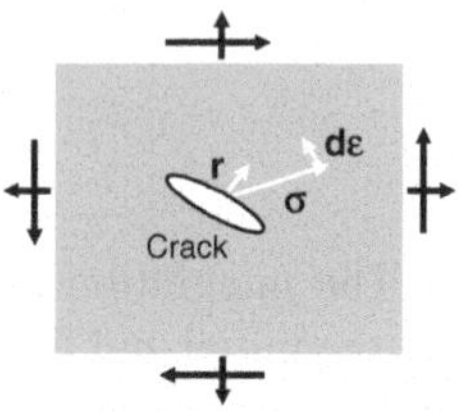

Fig. 37 Cracking energy density definition

Due to the aforementioned limitations of scalar fatigue criteria, different multiaxial fatigue criteria have been proposed in the literature to overcome such limitations, but it is worth pointing out that their level of development is uneven since most of them have been applied for natural rubber at specimen level only, excepting the Cracking Energy Density (CED) proposed by Mars [77]. The CED rationalizes fatigue life for specific failure planes across a wide range of states, relates physically to the fracture mechanical behaviour of small flaws under complex loading and is well defined for arbitrarily complex strain histories. This parameter accounts for the effect of crack closure, which occurs when the stress state causes compression on a material plane. This fact is of great importance because rubber is most commonly used in applications which experience a compressive load. Mars and Fatemi have proposed three different critical plane criteria [75, 76] for use in the computation of CED: the plane that maximizes the CED peak value, the plane of maximum CED range and the plane of minimum life.

The cracking energy density W_c represents the strain energy density available to be released due to the growth of small cracks in specific planes, and is defined according to the expression:

$$dW_c = \boldsymbol{\sigma} : d\boldsymbol{\varepsilon} \tag{167}$$

where $\boldsymbol{\sigma}$ and d are defined according to Fig. 37 and the following expressions:

$$\boldsymbol{\sigma} = \mathbf{r}^T \sigma \tag{168}$$

$$d\varepsilon = d\varepsilon\mathbf{r} \tag{169}$$

With $\boldsymbol{r}$ the unity vector normal to the selected plane (Fig. 37) and d

$$d\varepsilon = \mathbf{F}^T d\mathbf{E}\mathbf{F}^{-1} \tag{170}$$

where $\boldsymbol{F}$ and $\boldsymbol{E}$ are the gradient deformation tensor and the Cauchy-Lagrange strain tensor respectively.

The relationship between the cracking energy density and the tearing energy, as proposed by Mars [77], is:

$$T = KW_c c \tag{171}$$

where K has a value of 2π, and c is the radius of the analysed crack.

A brief summary of this methodology is shown in Fig. 38 as a typical fatigue analysis scheme for rubber components following the energetic approach.

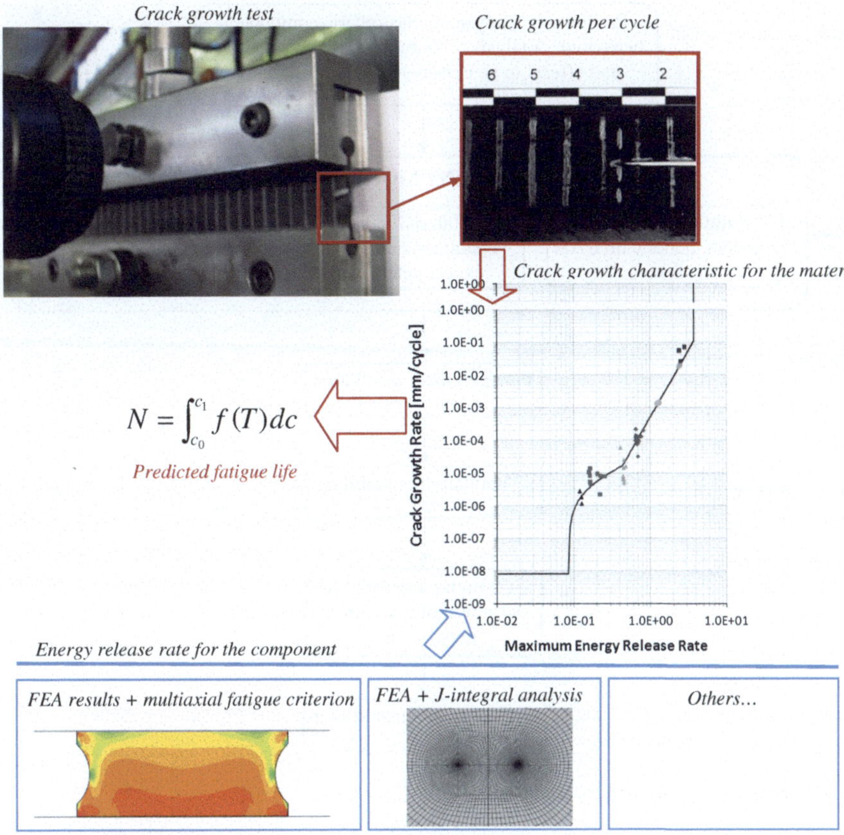

Fig. 38 Typical fatigue life prediction scheme based on the energetic approach

3 Methodology for Design of Elastomeric Components

3.1 Introduction

In the design of elastomeric components, a broadly used technique is FE method, computer assisted technique within CAE (Computer Aided Engineering) to obtain approximated numerical solutions of the response of physical systems working under external loads. This technique is based on principle of virtual work or weak formulation, modelling the whole structure in discrete parts, elements, which are set up and connected between them by means of nodes. Computers are used, once restrictions and external loads are applied to the model, to solve the equations resulting from it, obtaining the unknown variables of the system, displacements and rotations in nodes, and all the variables derived from them, such as, for instance, stresses, strains, contact pressures or reaction forces.

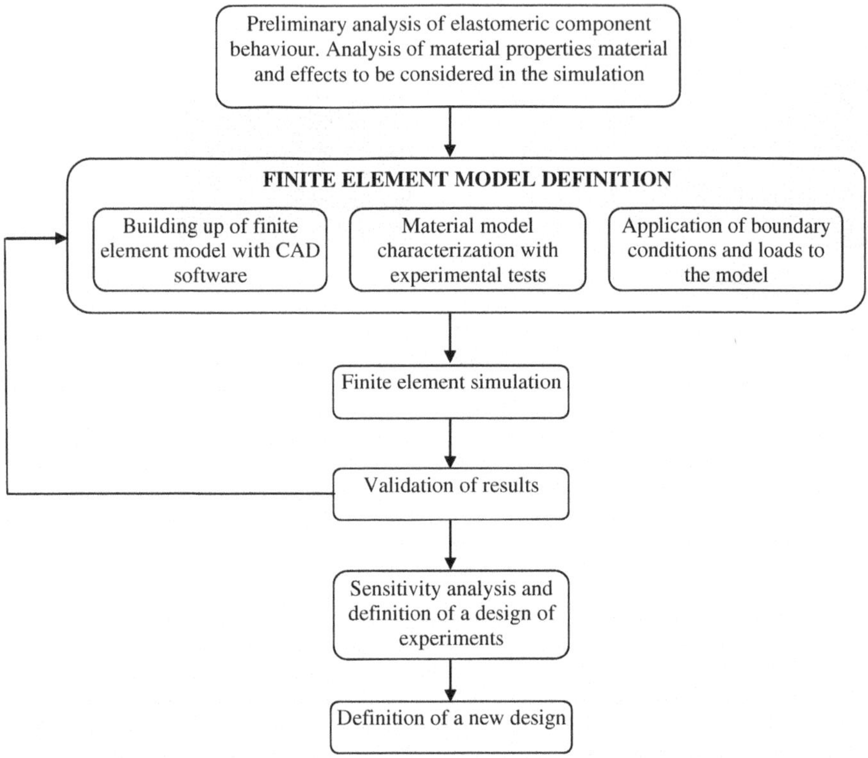

Fig. 39 Stages set up in a FE calculation

Benefits obtained from the use of FE method in design of elastomeric components, preferably beginning from design phase, include: improvement of characteristics and quality of the final product, lower development time of the product or time-to-market, optimum use of the material, weight saving, verification of the whole structure before prototyping phase and reduction of development and production costs. Besides, capacity to manage reliable predictions can help to reduce rejection rate of the product during production phase, assuring an increase of its competitiveness.

The use of codes based on FE method allows carrying out different types of analyses that reproduce the real behaviour of a component of or a system. These analyses can be classified into: mechanical or structural, modal, buckling, thermal, electrical, magnetic, fatigue or mass diffusion. Additionally, some of these types of analyses can be combined between them resulting in a more complex analysis. Anyway, the different stages to be followed in a FE analysis are always common. A scheme of them is shown in Fig. 39.

According to this Figure, the first task consists in analysing which is the behaviour of the elastomeric component and which are the material effects to be considered in the FE simulation. Thus, by means of this preliminary analysis, it must be set up if the material should be characterised only with hyperelastic material properties or also with other effects such as inelastic, relaxation or fatigue properties, in case they are determinant in the real component behaviour. Additionally, other effects such as friction or wear may be also important and in this case, they must be characterised, too.

Once the preliminary analysis is carried out, the next step consists in the FE model definition, including its building up by means of pre-processing software, the material characterisation with results of those experimental tests required from the first analysis, and the application of boundary conditions and loads according to the real working conditions of the component.

Whenever is possible, the results obtained from the FE simulation should be validated with experimental tests in order to check, on the one hand, if the FE model being built is suitable for simulating the real component behaviour and, on the other hand, if the considered material effects included in the simulation are valid or enough, at least. If any of both suppositions is not valid, another loop in the FE model definition must be done, rebuilding the FE model or carrying out new experimental tests to include those material effects not included in the first iteration of the process. If the validation is accurate enough, the next step consists in carrying out an analysis of the most influential parameters on material behaviour or which are the most critical on the design improvement. The aim of this analysis is to optimise the component design in order to improve its efficiency and its competitiveness, proposing a new design if it is necessary.

3.2 Experimental Material Characterisation Tests

Different tests can be planned to characterise behaviour of elastomeric components. On one hand, properties of hyperelastic behaviour can be characterised by means of quasi-static tests on universal test machines under different deformation modes; usually the deformation modes at which the component works. In these tests, the sensitivity of the material to the strain rate and to the temperature can be taken into consideration. On the other hand, inelastic material properties, such as viscoelastic properties, material hysteresis, permanent set, Mullins effect or Payne effect, can be set up by means of tests on universal test machines (UTM) or on dynamic-thermo-mechanical analyzers (DMTA) tests in dynamic analyses, obtaining the time-dependent response at different frequencies and temperatures. The importance of the material characterisation is pointed out in Sect. 4, examples of industrial applications, for instance, in the simulation of a silent block in automotive industry or in the analysis of rubber block on wheel of railway coach.

Fatigue material properties are also obtained by means of tests on universal test machines. Finally, by means of tribometer tests, the tribological characterisation of a contact pair between the elastomeric component and another part in contact with the first one which usually works as counter material can be set up.

The tribological characterisation includes the measurement of the friction force, of the friction coefficient and also the quantification of wear involved in the contact pair. In this characterisation, different parameters such as contact pressure, sliding velocity, temperature, lubricant viscosity or coating on the elastomeric surface can be also analysed. Next, a more detailed description of the required tests to characterise this material properties is carried out.

3.2.1 Obtaining of the Hyperelastic Properties of the Material

Tests on universal testing machines are planned to obtain the hyperelastic properties of the elastomeric material under different deformation modes, those at which the component usually works. With the results of these tests, the material models available in FE codes to reproduce hyperelastic behavior can be calibrated. The tests usually performed on universal test machines are uniaxial tensile tests, uniaxial compression tests, shear tests, equibiaxial tests or volumetric compression tests.

Uniaxial Tensile Tests on Universal Test Machine

These tests should be done if tensile deformation mode is the predominant one of the elastomeric component. The tests and dimensions of the specimens are detailed in ISO 37:1994 standard. The specimens used in these tests correspond to dumbbell specimens, according to dimensions defined in Fig. 40, which values are enclosed in Table 1.

Dumbbell specimen of Type 1 is the most suitable one to carry out uniaxial tensile tests. If those dimensions cannot be fulfilled, specimen of Type 2 is that one recommended and finally, specimen of Type 3 should be tested if Type 1 and Type 2 specimens are not possible to be built up. Three different repetitions under the same test conditions are recommended to be carried out at least. For each test, initial preconditioning can be taken into consideration depending on the estimated strain level attained in the real application. Therefore, the preconditioning level will be set up as the maximum tensile strain expected in the elastomeric component. The velocity to be chosen in the test will depend on the strain rate of the

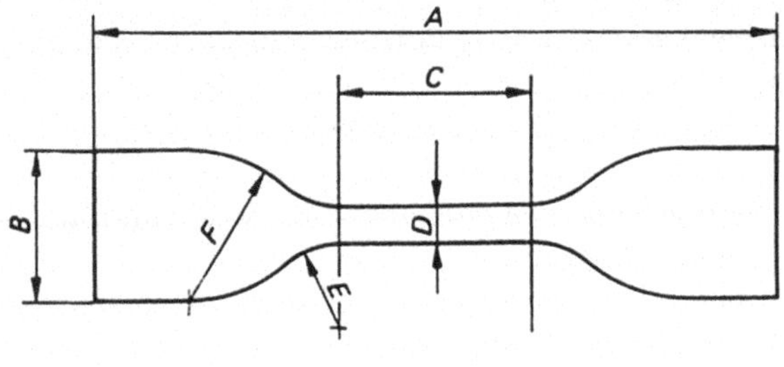

Fig. 40 Dumbbell specimens used for uniaxial tensile tests

Table 1 Dimensions of dumbbell specimens

Dimensions	Type 1 (mm)		Type 2 (mm)	Type 3 (mm)
A. Total length (min)	115		75	35
B. Width of ends	25 ± 1		12.5 ± 1	6.0 ± 1
C. Length of straight thin part	33 ± 2		25 ± 1	12 ± 0.5
D. Width of straight thin part	6.0	+2 −0	4.0 ± 0.1	2.0 ± 0.1
E. Minor radius	14 ± 1		8.0 ± 0.5	3.0 ± 1
F. Mayor radius	25 ± 2		12.5 ± 0.1	3.0 ± 0.1
Thickness	2.0 ± 0.2		2.0 ± 0.2	1.0 ± 0.1

component, but a usual test velocity to consider quasi-static conditions is 10 mm/min.

Tests will be carried out in equipments with devices able to measure load, grips displacement and marks distance in the test specimen at all times. This equipment must have a 1 % accuracy within working range with a constant test velocity application. To measure deformation, contact or laser extensometer located at the straight thin part can be used. With contact extensometers, the specimen torsion must be avoided.

Equipment grips must hold the specimen in the wide ends and must be fixed when stress is progressively increased along test. To measure specimen dimensions, calibres or micrometers with enough precision must be used, as well as a durometer to measure elastomeric material hardness. An example of this test is shown in Fig. 41.

At the same time, other parameters such as the humidity of the test temperature must be controlled at the beginning of the test.

As result of these tests, engineering stress and stretch values can be expressed according to (177):

$$\sigma = \frac{F}{S_f} = \frac{F}{S_o} \cdot \lambda \tag{172}$$

where F is the axial force applied to the specimen and S_o and S_f are the initial and final cross sections of the specimen and lambda the strain to be calculated according to (173).

The strain is calculated as the displacement between specimen marks, expressed in percentage with regard to the initial distance between them:

$$\lambda = \frac{H_t - H_0}{H_0} \cdot 100 \tag{173}$$

where H_t is the distance between marks in the time t and H_0 is the initial distance between marks.

Once the tests are carried out, the curves obtained from them must be corrected to be used in the numerical characterisation by means of the material models

Fig. 41 Uniaxial tensile test equipment

available in FE codes. The evolution of strain–stress curves obtained in each actuation cycle is shown in Fig. 42. The curve chosen to characterise the material model is that one corresponding to the last actuation cycle. Previously to carry out the characterisation, it is necessary to correct the permanent strain of the curve, in this case $\varepsilon_{0(4)}$, in order to begin the curve without any permanent strain.

To impose that the curve obtained in last cycle passes through the origin in order to use it in the material model fitting, the next formulas must be applied:

$$\varepsilon = \frac{(\varepsilon' - \varepsilon_P)}{(1 + \varepsilon_P)} \tag{174}$$

$$\sigma = \sigma' \cdot (1 + \varepsilon_P) \tag{175}$$

where ε' and σ' are strain and stress values to be corrected, ε and σ are the values already corrected and ε_P the permanent strain ($\varepsilon_0(4)$ in Fig. 42).

Uniaxial Compression Tests

These tests must be done if compression deformation mode is the predominant one in the elastomeric component. The tests and dimensions of the specimens are detailed in ISO 7743:1989 standard. The specimens used in these tests, shown in Fig. 43, correspond to cylindrical specimens with an aspect ratio (height/diameter)

Fig. 42 Strain-stress curve with different preconditioning cycles

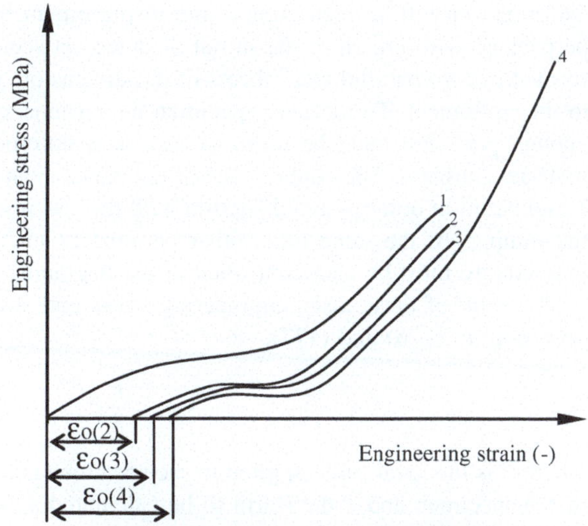

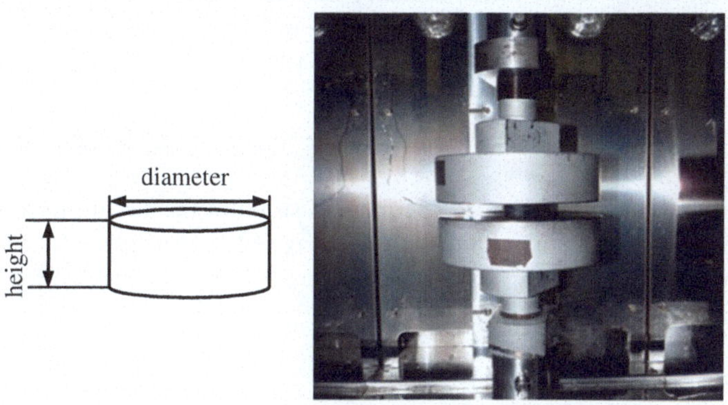

Fig. 43 Cylindrical specimens and equipment used for uniaxial compression tests

of 0.4. Specimen is compressed by flat plates placed top and bottom of the specimen. Three different repetitions under the same test conditions are recommended to be carried out at least.

As in case of uniaxial tensile tests, for each test, initial preconditioning can be taken into consideration depending on the estimated strain level attained in the real application. Therefore, the preconditioning level will be set up as the maximum tensile strain expected in the elastomeric component. The velocity to be chosen in the test will depend on the strain rate of the component, but a usual test velocity to consider quasi-static conditions is 10 mm/min.

Tests will be carried out in equipments with devices able to measure load and plates displacement in the test specimen at all times. This equipment must have a 1 % accuracy within working range with a constant test velocity application.

Deformation will be measured as the displacement between plates, expressed in percentage with regard to the initial distance between them. Both plates must be formed by two parallel steel sheets of flat dimension, perfectly polished and fitted to the equipment. To measure specimen dimensions, calibres or micrometers with enough precision must be used, as well as a durometer to measure elastomeric material hardness. The contact face of the plates must be covered by a thin film of Vaseline oil in order to avoid friction with the specimen and so avoid barrelling in the sample. At the same time, other parameters such as the humidity of the test temperature must be also controlled at the beginning of the test.

As result of these tests, engineering stress and strain values can be expressed according to (176) and (177).

$$\sigma = \frac{F}{S} \tag{176}$$

where F is the axial force applied to the specimen and S is the initial cross section of the specimen and λ the starin to be calculated according to (173).

The strain is calculated as the displacement between plates, expressed in percentage with regard to the initial distance between them:

$$\varepsilon = \frac{L_t - L_0}{L_0} \cdot 100 \tag{177}$$

where L_t is the distance between plates in the time t and L_0 is the initial distance between plates.

Regarding the correction of the permanent strain of the strain–stress curve, the criteria set up for the uniaxial tensile test in Fig. 42 is also set up for the uniaxial compression test.

Shear Tests

These tests must be carried out when shear deformation mode is the predominant one of the elastomeric component. The tests and dimensions of the specimens are detailed in ISO 1827:1991 standard. The specimens used in these tests, shown in Fig. 44, correspond to identical blocks of 4 mm ± 1 mm thickness, 20 mm ± 5 mm width and 25 mm ± 5 mm length, glued in each face of maximum surface to four rigid plates of same width and enough length to avoid bending during test. As in the case of uniaxial tests, three different repetitions under the same test conditions are recommended to be carried out at least.

Initial preconditioning can be taken into consideration depending on the estimated strain level attained in the real application. Therefore, the preconditioning level will be set up as the maximum tensile strain expected in the elastomeric component. The velocity to be chosen in the test will depend on the strain rate of the component, but a usual test velocity to consider quasi-static conditions is 10 mm/min.

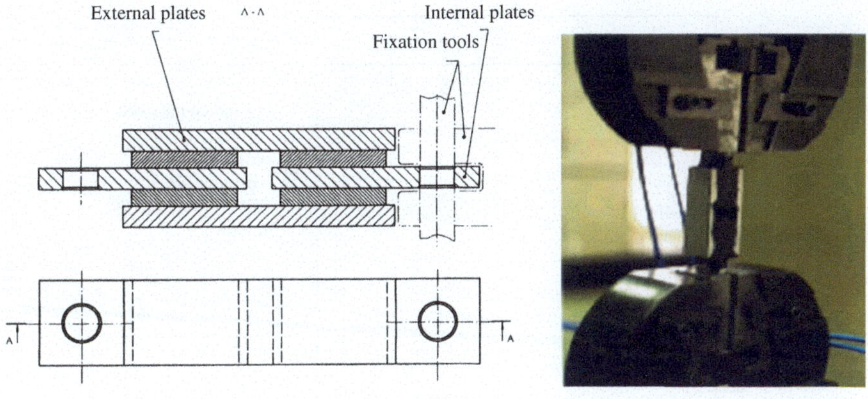

Fig. 44 Specimens and equipment used for shear tests

Tests will be carried out in equipments with devices able to measure load and plates displacement in the test specimen at all times. This equipment must have 1 % accuracy within working range with a constant test velocity application.

As result of these tests, engineering stress values are calculated by only one rubber blocks, with an applied force of *F/2*. Stress formula is expressed as:

$$\sigma = \frac{F}{2 \cdot S} \tag{178}$$

where F is the axial force applied to the grip and S is the initial shear section of one of the rubber blocks.

For calculating strain, shear angle γ is considered, expressed as (see Fig. 45):

$$\gamma = \arctan \frac{d}{2e} \approx \frac{d}{2e} \tag{179}$$

where d is the strain of one of the blocks and e corresponds to its thickness.

Fig. 45 Calculation of shear strain

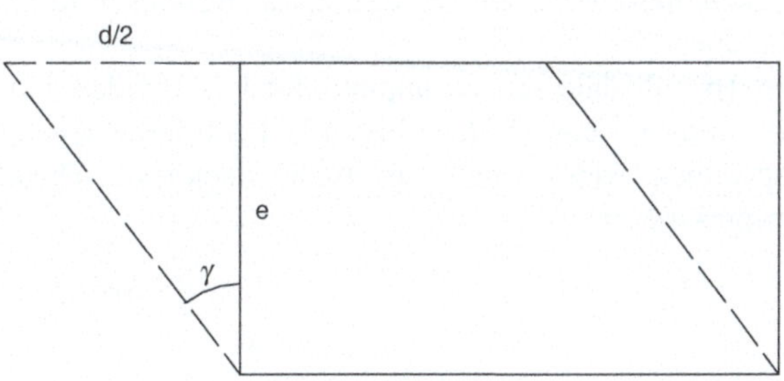

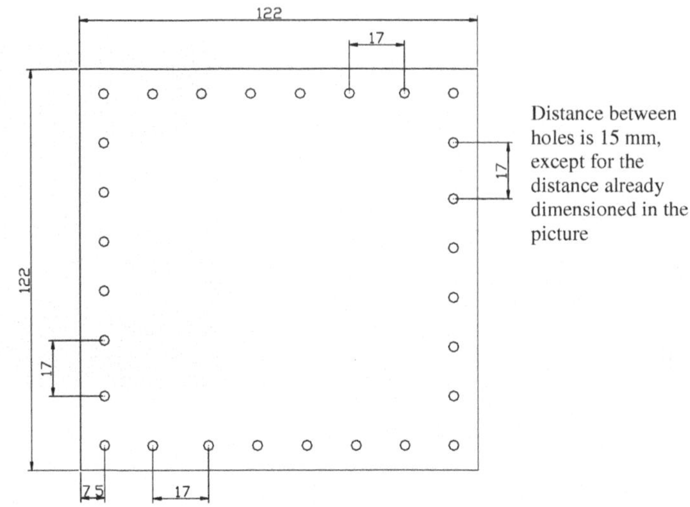

Fig. 46 Specimens used for biaxial tensile tests

Biaxial Tensile Tests

These tests must be carried out when tensile deformation mode is the predominant of the elastomeric component in two perpendicular directions at the same plane. The specimens used in these tests are square specimens of dimensions 122 mm by edge, according to Fig. 46. Diameter of holes is the same as the diameter of screws used to hold it to grips. Three different repetitions under the same test conditions are also to be carried out at least.

Tests must be carried out in equipment with load capacity of 10 kN, with devices able to measure load and plates displacement in the test specimen at all times. This equipment must have 1 % accuracy within working range with a constant test velocity application. The velocity to be chosen in the test will depend on the strain rate of the component, but a usual test velocity to consider quasi-static conditions is 10 mm/min. A tool to stretch the specimen in two perpendicular directions is required in the test, being the specimen fixed to this tool by means of screws.

To measure specimen dimensions, calibres or micrometers with enough precision must be used, as well as a durometer to measure elastomeric material hardness.

To calculate stress, applied load is divided into two perpendicular forces, resulting $F \cdot \cos 45^\circ$ (see Fig. 47). Each force acts against a section equals to the specimen width times specimen thickness. Therefore, engineering stress is expressed as:

$$\sigma = \frac{F}{S} = \frac{F \cdot \cos 45^\circ}{A \cdot e} \tag{180}$$

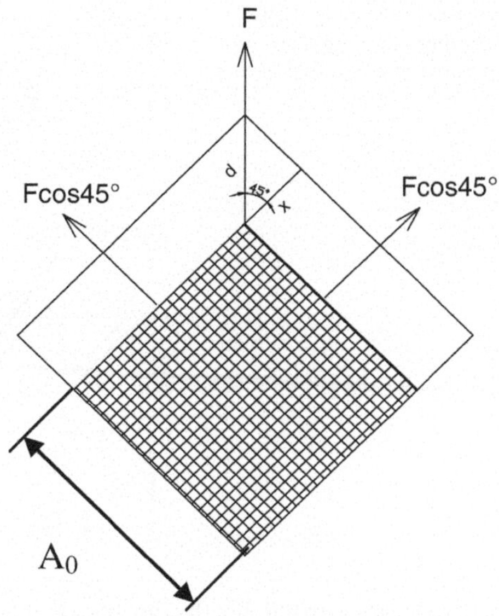

Fig. 47 Specimens used for biaxial tensile tests

To calculate strain values, specimen is deformed according to Fig. 47 so that the displacement registered by the equipment is d, while real value produced by force F · cos 45° is x = d · cos 45°. Therefore, strain equals to:

$$\varepsilon = \frac{\Delta L}{L} = \frac{x}{A} = \frac{d \cdot \cos 45^{\circ}}{A} \tag{181}$$

Volumetric Compression Tests

These tests aim to obtain the compressibility of the elastomeric material. This property is used in the definition of numerical models of elastomeric material. Specimens of approximated aspect ratio of 1.45 (height Ho/diameter Do, see Fig. 48) are used. To manage this aspect ratio, several specimens could be piled.

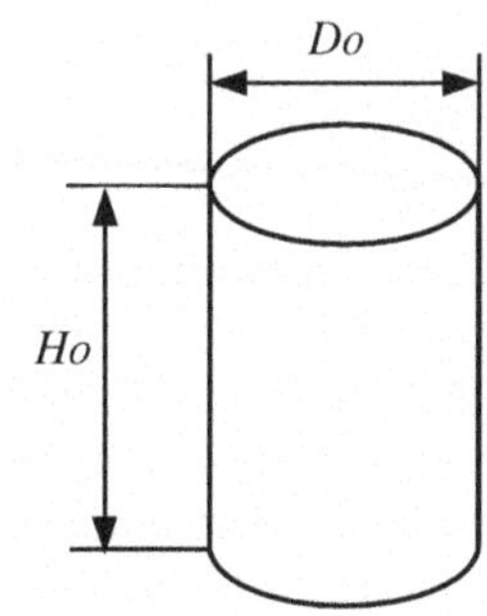

Fig. 48 Cylindrical specimens used for volumetric compression tests

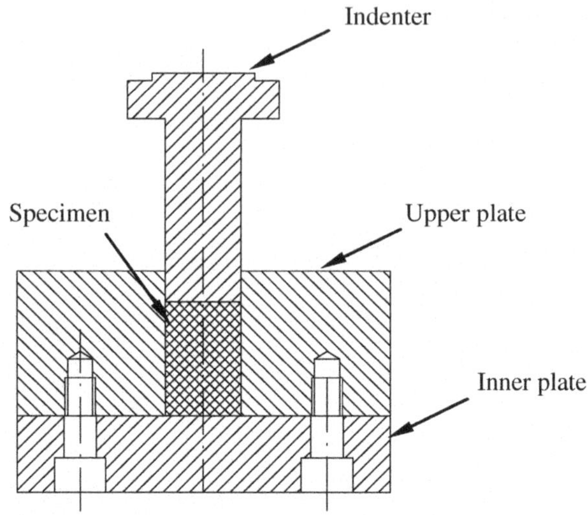

Fig. 49 Assembly of the specimen in equipment for volumetric compression test

Tests are carried out by means of two steel plates of flat face, joined to an indenter and a steel block in which the specimen to be tested is introduced (see Fig. 49). Tests must be carried out in equipments with load capacity of 10 kN, with devices able to measure load and plates displacement in the test specimen at all times. This equipment must have 1 % accuracy within working range with a constant test velocity application. The velocity to be chosen in the test will depend on the strain rate of the component, but a usual test velocity to consider quasi-static conditions is 10 mm/min. Force and displacement between plates must be measured along test. Three different repetitions must be done.

To calculate engineering stress, next formula is applied (being F the force to compress the specimen and S its initial section):

$$\sigma = \frac{F}{S} \tag{182}$$

As strain value, the displacement between plates is considered, expressed as percentage of the specimen height after preconditioning, so that (being Ht the displacement between plates in a time t and Ho the initial height of the specimen):

$$\varepsilon = \frac{Ht}{Ho} \cdot 100 \tag{183}$$

3.2.2 Obtaining of Inelastic Properties of the Material

Inelastic material properties can be characterized by means of tests on universal test machines or on DMTA tests in dynamic analyses. Next, the way of obtaining some of these properties, such as viscoelastic, material hysteresis, permanent set, Mullins or Payne effects, are briefly described.

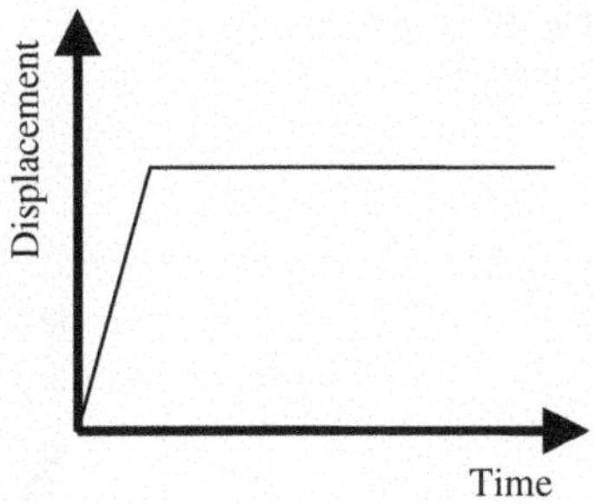

Fig. 50 Test control in relaxation tests

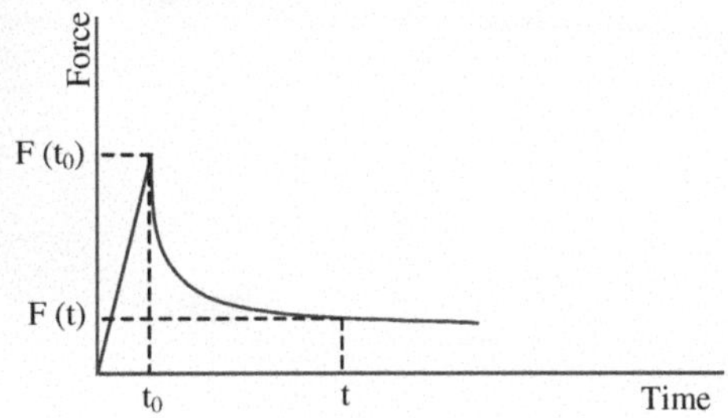

Fig. 51 Data obtained from relaxation tests

Viscoelastic Properties

Viscoelastic properties can be characterised by means of relaxation tests on universal test machines or by means of DMTA tests in dynamic analyses.

Relaxation Tests on Universal Test Machines

To carry out relaxation tests, and depending on the predominant deformation mode of the elastomeric component, the type of test detailed in Sect. 3.1 must be followed.

In this case, the test control is carried out by strain up to a reference value and once that strain is attained, the tests is controlled by position, maintaining fixed the specimen during an established time and registering force data along time. Figure 50 shows the test control carried out in relaxation tests and Fig. 51 shows the data obtained from relaxation tests.

From Fig. 51, relaxation rate can be set up according to the next equation:

$$\text{Relaxation rate (t)} = \frac{F(t_0) - F(t)}{F(t_0)} \cdot 100 \tag{184}$$

Tests on DMTA

Elastomeric materials present viscoelastic nature with a strong dependence of their mechanical properties on temperature, time and frequency. This kind of materials

Fig. 52 DMTA test equipment

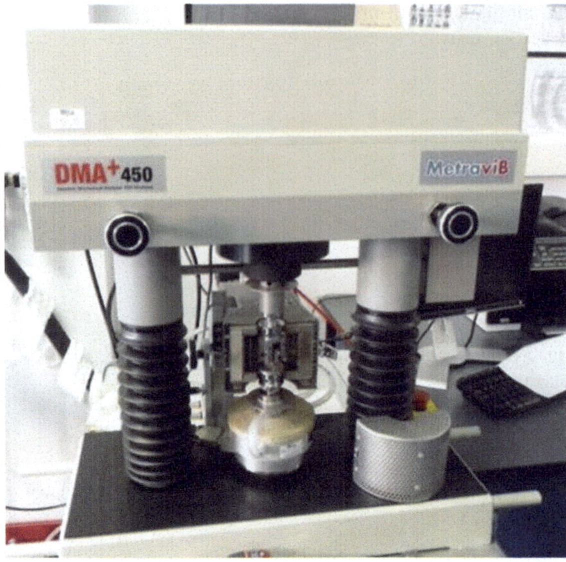

are usually used in industrial applications involving extreme conditions of pressure, temperature, load cycles and frequency, which implies significant changes in the material properties with regard to the original state. In perfectly elastic materials the energy involved in the system is completely stored; on the other hand, in purely viscous liquids the energy is totally dissipated. But in polymers, only part of the energy is dissipated [209]. Nowadays, mechano-dynamic technique have evolved to the study of the sinusoidal movement at which the visco-elastic material is undergone, so that in the majority of the devices the specimen is tested under an oscillatory stress, measuring as result the generated strain or vice versa, and additionally the offset between excitement and response. Figure 52 shows an example of DMTA test equipment.

Generally, the study of this offset is carried out by means of the mathematical basis of the simple harmonic movement, which allows separating the elasticity complex modulus (or shear modulus), $E*$ (or $G*$), in two orthogonal components, storage modulus, E' (or G') and loss modulus, E'' (or G'').

$$E^*(G^*) = E'(G') + iE''(G'') \tag{185}$$

where

$$E'(G') = \left(\sigma_0/_{\varepsilon_0}\right)\cos\delta \tag{186}$$

$$E''(G'') = \left(\sigma_0/_{\varepsilon_0}\right)\sin\delta \tag{187}$$

and σ_0, ε_0 and δ are, respectively, stress amplitude, strain amplitude and lost angle between both of them, according to Fig. 53.

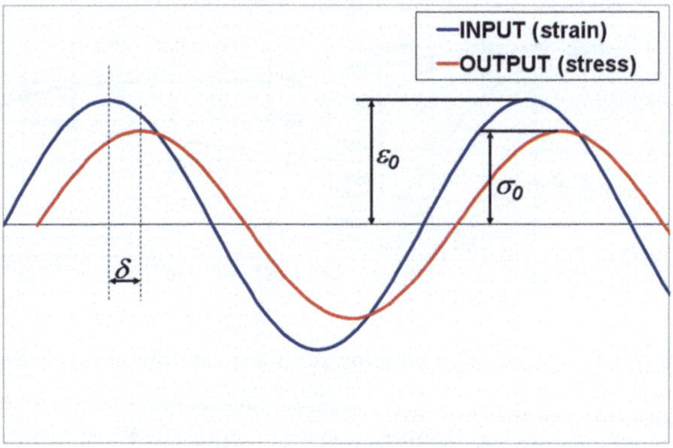

Fig. 53 Input–output offset in viscoelastic materials

Storage modulus is related with the stored energy that can be recovered in strain terms and corresponds to elastic response. On the other hand, loss modulus is related to dissipated energy per cycle in heating terms when specimen is deformed, corresponding to viscous response. The relationship between both of them is:

$$\tan\delta = \frac{E'}{E''} \tag{188}$$

This relationship represents the offset between elastic and loss components, being called damping as well and representing the material energy loss in its molecular reordering and in internal frictions. When the variation of modulus and offset in function of temperature is analysed, maintaining fixed work frequency, relaxations with implicit changes in molecular mobility can be identified in the obtained curves.

The whole frequency range behaviour of a viscoelastic material is unfeasible to be covered. To solve this problem, master curves are used. These curves allow estimating the mechanical properties of the elastomer in wide ranges of time or frequency from mechanical tests of short time at different temperatures.

The theoretical basis used to build up a master curve is the time–temperature superposition principle, that is: the material response at different temperatures is similar as that one at other perturbation frequencies. The building of these master curves implies carrying out studies at different temperatures in a previously defined frequency range. These curves are displaced to superimpose with the next one resulting the master curve (see Fig. 54). The displacement factor in abscise direction is not fixed, but usually is constant for each elastomer and is known as a_T. The relationship between these fittings follows the WLF (Williams-Landel-Ferry) equation [210] (184).

$$\log(a_T) = \frac{-C_1\left(T - T_{ref}\right)}{C_2 + (T - Tref)} \tag{189}$$

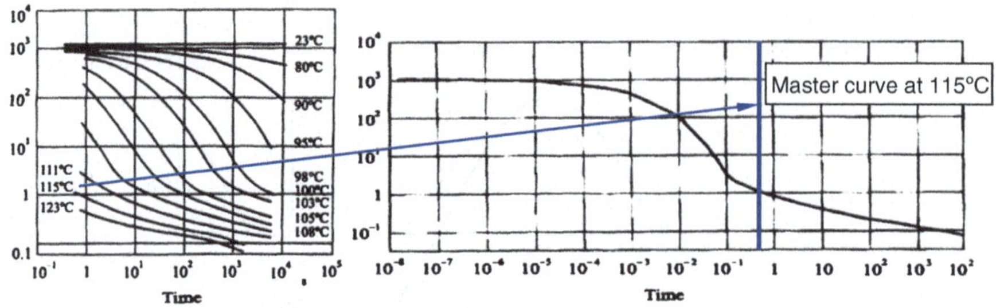

Fig. 54 Master curve building from tests at different temperatures

where T_{ref} is the reference temperature (in K) at which the master curve is built up, a_T is the displacement factor and C_1 and C_2 are constants with values, respectively, of 17.4 and 51.6 for many polymers near glass transition temperature.

Williams-Landel-Ferry equation is used to set up time–temperature relationship for polymers near glass transition temperature. This equation is based on the hypothesis that free volume increases linearly with temperature over glass transition temperature. The model also assumes the abruptly decrease of the polymer viscosity when the material free volume increases.

Other model used to set up a relationship between displacement factor and temperature is Arrhenius relationship (190):

$$\log(a_T) = \frac{E}{R(T - Tref)} \tag{190}$$

where E corresponds to the activation energy associated with transition of relaxation, being R gas constant (8.314 J/mol °C), T_{ref} the reference temperature at which the master curve is built up and a_T is the displacement factor. Arrhenius equation is used to set up the glass transition temperature of crystalline polymers.

In the master curve building up, different material and experimental, coming from instruments, factor must be taken into consideration. A master curve is built up from accelerated tests, depending on the mechanical properties aimed to obtain. These tests can be multifrequency dynamic tests or creep/relaxation tests.

The definition of a DMTA test involves the selection of the deformation mode of the test, the specimen geometry, the test conditions in terms of frequency ranges, temperature, strain/stress amplitude and other parameters to be defined in the test.

Before the test definition, several important matters must be taken into account:

- Test temperature should range from lower temperature than glass transition temperature to 100 °C higher than glass transition temperature. Therefore, a wide material stiffness range should be covered.
- The different curves obtained from tests (E', E'' or $tan\delta$ vs. frequency) should present curvature in order to clearly define the displacement and the superposition of the curves.

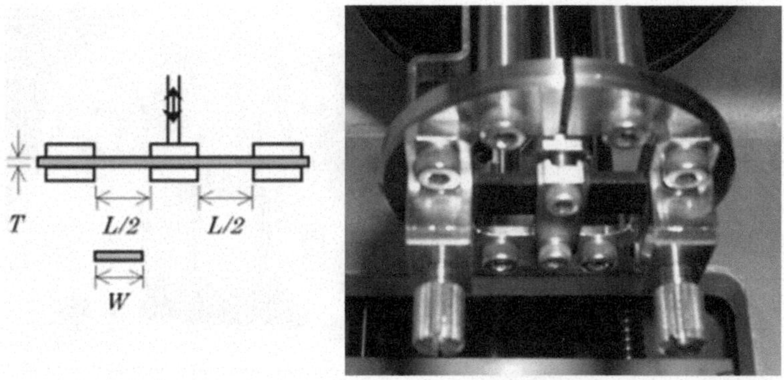

Fig. 55 Double bending deformation mode in DMTA tests

- The temperature increment between tests near the glass transition temperature should be lower because the changes of the mechanical properties with the temperature are much higher at this zone of the curve.
- The tests should lie within the measurement range in temperature and force at all the times, especially at lowest and highest temperatures.
- Before testing, and if it is possible, it is convenient to obtain as much material information as possible, especially stiffness data in glass and amorphous states and an estimation of the glass transition temperature in order to choose the deformation mode, the specimen geometry and other test conditions. Besides, the industrial application of the material is also relevant in order to choose the test and the deformation mode.
- Another important task in the tests definition is the liquid nitrogen required in the whole testing plan. Tests conditions must be defined in order to fit the liquid nitrogen quantity to be used.

In the test definition, the deformation mode must be chosen firstly. The deformation mode is determined by the industrial application of the component and by the availability of material for testing. However, both matters depend on the fact that the specimen stiffness can be measured in glass and amorphous states in order to obtain a master curve in a wide range of frequencies. To do that, there are recommendations in DMTA devices to use different deformation modes depending on the kind of tested material. It is worth pointing out that shear test is the most convenient for most of elastomeric materials. Two deformation modes usually used in DMTA devices correspond to double bending and shear. Those are shown in Figs. 55 and 56.

Another important task is the choice of the specimen and its dimensions. There are several factors to take into account: available material volume, if sheets to extract specimens are available or not, if the dimension of the specimen is enough to be put into the DMTA device, the nitrogen consumption and the precision in load and temperature measurements.

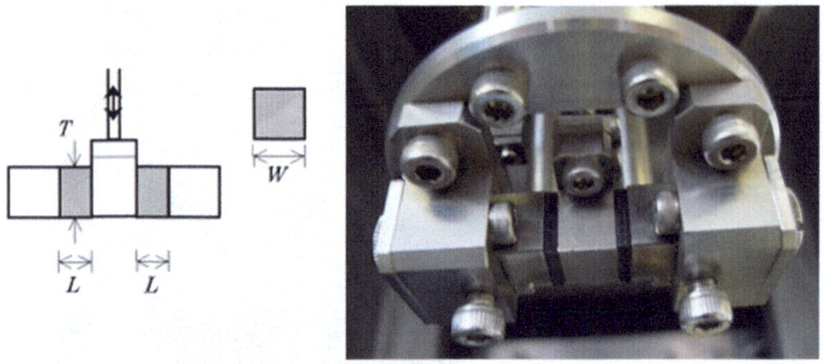

Fig. 56 Shear deformation mode in DMTA tests

In case of elastomeric materials, it is necessary to carry out a mechanical preconditioning to the specimen in order to eliminate Mullins effect to the material. To do that, a static test is carried out to the specimen with a load or deformation between 30 and 50 % higher than that one used in the dynamic tests and eliminating after that the load or deformation. This test is recommended to be carried out three times in order to assure the loss of the initial material stiffness, that is, to remove the Mullins effect.

The number of DMTA tests on a same material will depend on the repeatability of master curves. It is recommended to obtain two master curves with repeatability. The test control is another parameter to choose before testing. The choice of a particular one: by deformation, by load or by displacement, will depend on the type of material and on the previous experience. Many times, it is recommended to change from one to another is the resulting master curve is not well defined, mainly for high frequencies, corresponding with temperature lower than glass transition temperature. The cyclic amplitude is another parameter to control, being interesting to choose a value at which the test is stable for the highest temperature range as possible.

Permanent Set, Mullins Effect and Material Hysteresis

The mechanical behaviour of elastomeric materials is highly non-linear, depending on the strain rate, exhibiting hysteresis, permanent strains and stress softening under cyclic load conditions. From curves obtained in uniaxial tests, those effects can be characterised. When the specimen is loaded under uniaxial load, the stress in the second load is lower than that one required in the first load for higher strains than in the initial load. This effect is known as Mullins effect and it takes place in the first load cycles. On the other hand, permanent set is set up as the remaining strain in the last cycle of the uniaxial test carried out in universal test machines. Hysteresis corresponds to the different behaviour between in the same cycle of application for loading and unloading conditions. Figure 3 included in Sect. 1 shows these effects in curves obtained from a uniaxial test.

For calibrating the constitutive models able to model these inelastic effects, the same type of tests above described can be used. The key of this task is in the procedures and methodology used for managing the experimental data from the material for the calibration of these models. Some methodology for calibration of these models is described in Olsson and Austrell [58] and Gracia [199, 211] for the overlay method, in Bergström and Boyce [12] for the Bergström-Boyce model and in Hibbit et al. [97] and Gracia [199, 212] for Ogden and Roxburgh model and Simo model.

3.2.3 Fatigue Tests

The experimental test required to characterise fatigue behaviour in elastomeric materials are grouped in crack growth tests to characterize the rubber material crack growth properties and in uniaxial fatigue test to fit the initial crack size and to check the validity of the material properties fitting carried out from the data obtained in the previous set of experiments.

Crack Growth Characterisation

Different specimens can be used to analyse the crack growth behaviour on elastomers [64], the typical configurations are pure shear and simple tension specimens. The pure shear specimens are particularly well suited since the equation that characterizes its behaviour is independent from the crack length, because the energy release rate T is calculated taking into account the specimen height and the strain energy density only, according to the next equation:

$$T = W \cdot h \tag{191}$$

where W is the strain energy density and h the height of the specimen.

Figure 57 shows, as example, pure shear specimens to perform the crack growth characterization of the elastomeric material.

The test methodology consists of applying a set of cyclic displacements to the elastomeric material specimen with increasing amplitude and registering the crack length evolution. Different amplitudes can be tested to cover the power law crack growth regime as well as the linear regime. Crack growth can be automatically recorded at varying intervals via a high speed camera and image analysis system.

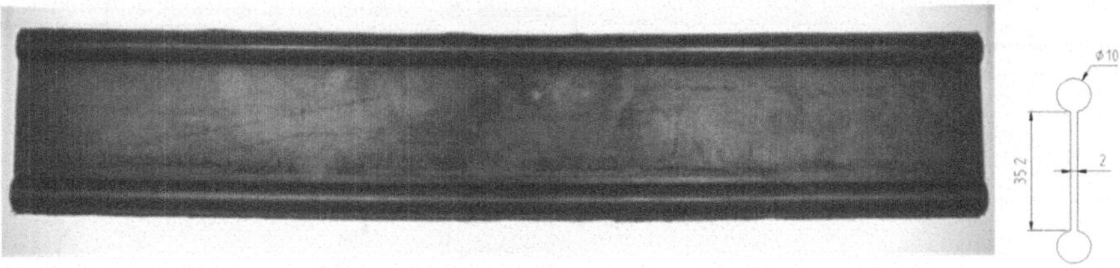

Fig. 57 Pure shear specimens for crack growth tests

Fig. 58 Crack growth tests

The tearing energy T corresponding to each strain level is calculated according to Eq. (191), evaluating the strain energy density W in the pure shear specimen as the area under the stress–strain curve for the previous cycle where the crack length is measured. An example of the test is shown in Fig. 58.

Uniaxial Tensile Fatigue Tests

The initial flaw size is adjusted by means of uniaxial fatigue tests carried out in simple tension specimens. The test samples to be tested are dumbbell shape specimens obtained from pure shear specimens.

The number of cycles for each strain level and specimen until the failure is obtained as a result of the experiment. An example of the geometry of the uniaxial fatigue test specimens is shown in Fig. 59, while Fig. 60 shows the mounting of the specimens in the test rig.

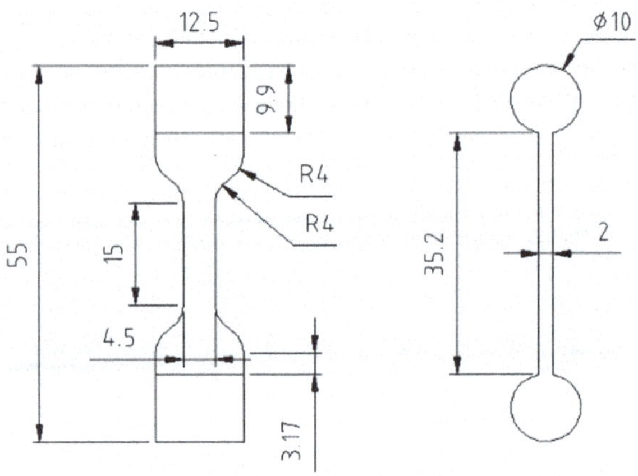

Fig. 59 Simple tension specimens used in the uniaxial fatigue tests

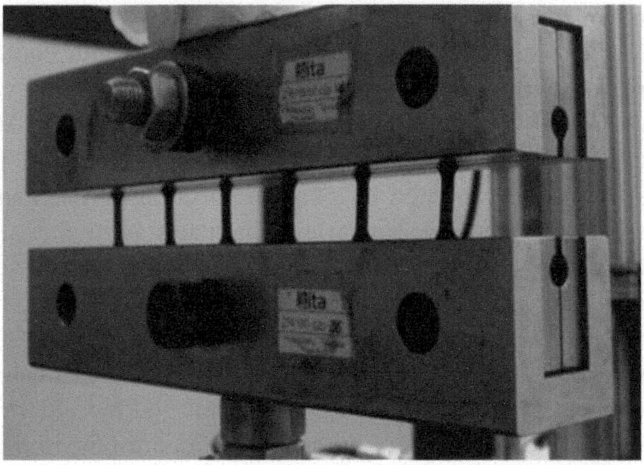

Fig. 60 Mounting of the specimens in the uniaxial fatigue test rig

3.2.4 Tests on Tribometer

Tribometer constitutes a versatile test instrument for the evaluation of friction and wear properties of elastomers in dry and lubricated conditions under reciprocating or rolling movement. The type of test will depend on the relative movement between the elastomer and the counter material in the real application. If it is possible, specimens used in the test are extracted by machining from the real application in order to maintain the same contact pair as in component. Typical test configurations used in tribometers are detailed in Fig. 61:

Depending on the tribometer and on the configuration, parameters such as frequency, vertical load, temperature, testing time and stroke can be specified in the test. Along the test, other parameters such as friction force and contact temperature can be measured, as well as the real values of vertical load and frequency in order to compare the possible deviations with the theoretical values specified to the test.

Usually, tribometers are also equipped with high speed data acquisition module which enables to acquire friction force values at a high frequencies rate in continuum signal. These additional options enable the storing and displaying of complete friction cycles during very low speed Stribeck Curve (see Sect. 1.3 Fig. 12) and stick-slip tests, apart from providing typical RMS (acronym for root

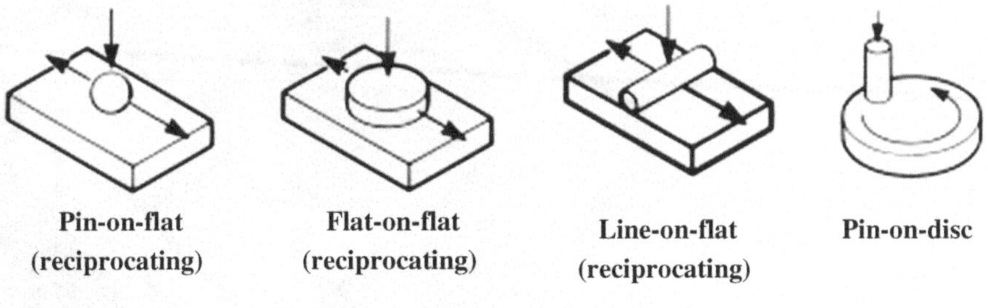

Fig. 61 Typical test configurations in tribometer

mean square, it gives the mean value of a signal) friction curves along the whole test. In addition, tribometers are equipped with cooler system, which enables to perform tribological tests at very low temperatures.

In reciprocating configuration, according to ASTM G133-05 standard, the test consists on a horizontal arm with reciprocating motion. The moving elastomer specimen is mounted in a carrier, oscillating mechanically against a fixed lower specimen of a harder material, which acts as counter surface. The mechanical drive comprises a motor driven cam and scotch yoke assembly, providing pure sinusoidal motion. The drive mechanism can run inside a bath in order to perform tests under lubrication conditions. The fixed specimen is located in a stainless steel reservoir. The reservoir is clamped to a block that is heated by electrical resistance elements and the temperature is monitored by a thermocouple pressed against the side of the specimen or holder. Movement in the horizontal direction is resisted by a piezo-electric force transducer, which measures the friction forces in the oscillating contact and the output range is set to match expected friction levels in the contact. Figure 62 shows an image of this configuration in a commercial tribometer.

Regarding pin-on-disc configuration, according to ASTM G99-05 standard, an adapter replaces the standard reciprocating head on the machine and allows the performance of rolling tests, using the machine drive motor and automatic loading system. In this case, the elastomer specimen is vertically loaded and fixed in the test, while the harder counter material moves rotating in the test. Figure 63 shows an image of the pin-on-disc configuration in a commercial tribometer.

To follow a procedure of testing in tribometer, elastomer samples and counter material must be cleaned before testing by agents like ethanol, petroether or acetone. Additionally, in wear tests may be necessary to include air blowing over the count surface in order to eliminate debris, which can acts as a third body inducing a different wear type. In general, both in friction and wear tests, preconditioning steps are also recommended to set up test conditions, in load and

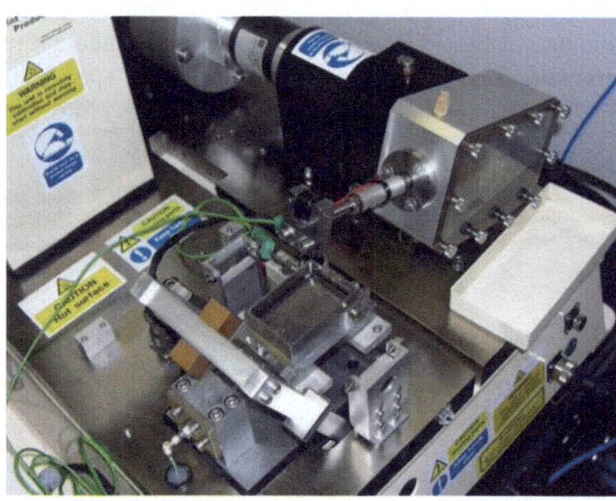

Fig. 62 Tribometer under reciprocating conditions

Fig. 63 Tribometer under pin-on-disc conditions

frequency, in a smooth way, looking for repeatability results under the same conditions. Remarkable investigations on tribometer friction tests are carried out by Blau and de Vore [213] and by Plint [214], while investigations on wear tests, with specimens extracted from a real component, are carried out by Song et al. [215], by Franklin [216] and by Burris and Sawyer [156].

The results obtained from friction tests are the friction coefficient, obtained from the ratio between the friction force and the applied load, versus time as a RMS signal, and the continuum evolution of the friction coefficient. From wear tests, the weight loss of the specimen is obtained as weight difference before and after testing.

3.3 Material Characterisation from Experimental Tests

Data obtained from experimental tests detailed in Sect. 3.2 are used to characterize the behavior of the elastomer. The results of the tests on universal test machines are used to obtain the elastic and inelastic behavior of the material, those of the DMTA tests to obtain its viscoelastic behavior, and those of tribometer tests to obtain the friction and wear behavior of the contact pair between the elastomer and the corresponding counter material.

3.3.1 Hyperelastic Characterisation of the Material

The elastic behaviour of the elastomer is modelled by means of an energy density function (EDF). Usual available functions in FE codes are: Arruda-Boyce, Mooney-Rivlin, Neo-Hookean, Polynomial Ogden, Reduced Polynomial, Van der Waals and Yeoh. The values for the parameters fitting of these models are obtained from experimental data, allowing the use, at the same time, of data from uniaxial, biaxial, volumetric or plane tests. Material can be considered as totally incompressible or quasi-incompressible, characterising the model from data obtained in a volumetric compression test or including Poisson's ratio in the material definition.

In this characterisation, the material is considered as isotropic because the material molecules are distributed in a random way. Effects like permanent deformation, viscoelasticity or hysteresis are not included in the material model.

Regarding the experimental data to fit the parameters of the selected EDF, depending on the predominant deformation mode which is present in the component to simulate, it is recommended to use experimental data obtained under the same deformation mode. In case of there is not a predominant deformation mode, it is necessary to use a predictive material model (Reduced Polynomials, Arruda-Boyce and Van der Waals), characterised with data from tensile test, compressive test or a combination of both of them, checking the obtained prediction with experimental data of each one of the tests. For polynomial material models, the polynomial grade is based on the available experimental data. For experimental curves with not too much data (points separated at 10 or 20 %), low grade polynomials are recommended (grades from 1 to 3), while experimental data have higher acquisition data, better results are obtained with higher grade polynomials. In any case, the range of the curve stability must be checked.

To select an EDF, the behaviour of the selected EDFs must be evaluated under simple deformation modes, so that the selected function shows a stable behaviour in the strain range of interest (strain attained in the simulation), as well as a minimum deviation with regard to the experimental data is obtained. If the material is specially confined in a particular application, the material compressibility will be also important and must be characterised previously.

As summary, to define the hyperelastic and inelastic behaviour of an elastomeric material, the next steps should be followed:

1. Definition of the hyperelastic behaviour: providing parameters model directly or fitting them from the next data sets:
   - Elastic behaviour (and its dependence on temperature, in required case).
   - Volumetric behaviour (in required case).
   - Thermal expansion coefficient (in required case).
2. Selection of an EDF from a set of simple deformation mode characterisations (for instance, see Figs. 41, 43 or 44).

3.3.2 Inelastic Characterisation of the Material

Viscoelastic Properties

The viscoelastic characterisation of the material is carried out from master curves of the storage modulus (G'), loss modulus (G'') and damping factor ($tan\ \delta$) obtained in DMTA tests. Once the set of tests are carried out in the range of frequencies under study and multiple test temperatures, the subsequent results must be treated to construct master curves of G', G'' and tan δ at a reference temperature which is selected to be the glass transition temperature.

To carry out this task, commercial software can be used to deal with the data in an automatic way. With this software, master curves at the glass transition temperature as reference temperature from DMTA test data can be constructed. The program calculates the corresponding shift factors from the temperature of each cyclic test to the reference temperature (Tg) of the master curve. The program also provides the value of the glass transition temperature by evaluating the DMTA results.

To implement the viscoelastic material characterisation in a FE code, a user subroutine must be programmed to enable the shifting of the master curve at the reference temperature to any other desired temperature of analysis directly inside the main FE simulation. By means of this user subroutine, parameters corresponding to the expression for the Williams Landel Ferry calculation and the Arrhenius correction term for temperatures below glass temperature (Tg) are specified.

Once master curve is built up, the complete fitting of the FE material model must be carried out from both hyperelastic and viscoelastic characterisation tests, carried out with universal test machines and DMTA tests respectively. In this model, the quasi-static and time-dependent mechanical responses or the rubber material are included. Both parts are available in FE codes by selecting the most adequate models for each one and providing the adequate constants for their parameters.

For the hyperelastic part, a simple material model, like for instance a Neo-Hookean, can be selected. It is known from literature that its corresponding parameter C_{10} equals to the shear modulus of the rubber material divided by a factor of two. The hyperelastic model offers the possibility to fit the parameter values to the instantaneous response of the rubber material (modulus at very high frequencies) or to the relaxed response (very low frequencies). Based on the assumption that the relaxed response of the rubber obtained in a uniaxial test should take several hours or even days, and therefore to fit the model parameter from the uniaxial quasi-static tests probes not valid under a strictly theoretical point of view (although provides good approximation for quasi-static FE analysis where time has no physical sense in simulation), the instantaneous value of the shear modulus provided by the DMTA tests can be used to fit the C_{10} parameter of the Neo-Hookean model.

For the viscoelastic part, a Prony series can be adjusted by means of the software. The Prony series terms, corresponding to a generalized Maxwell model of 30 terms, can be obtained by fitting to the shear storage modulus master curve (G').

Once both hyperelastic and viscoelastic models are developed, a checking with one-element FE simulation for simple tests is carried out, in order to analyse that the dynamic response (quasi-static and time-dependent mechanical response, including temperature dependent behaviour) of the FE material model predicts adequately the experimental results obtained in universal test machine. The FE simulations must be therefore executed at the same strain rates and temperatures than the uniaxial tests.

In addition, correlation between experiment results obtained from universal test machine and FE simulations of one element will be used to re-adjust the value of the C_{10} Neo-Hookean parameter. This is necessary since the corresponding cyclic frequencies equivalent to the universal test machine tests, although certainly low, cannot be fixed with enough accuracy. Therefore, several FE simulation loops are carried out until the readjusted Neo-Hookean parameter value enables good correlation with the universal test machine tests. The importance of the viscoelastic characterisation can be observed, for instance, in the Sect. 4, examples of industrial applications, in the simulation of a rubber block absorber in automotive industry.

3.3.3 Characterisation of Fatigue

According to the definition of the crack growth characteristic proposed by Lake and Lindley [71] and expressed in (162), a fitting of parameters is required to feed the elastomeric material fatigue model in the different crack growth regimes. Additionally, it is necessary to define the initial and final crack growth lengths in order to define the integration limits of the equation properly.

3.3.4 Characterisation of Friction and Wear of Contact Pair

Once friction and wear tests on tribometer are carried out, friction and wear laws must be characterised from test results to be implemented in a FE code.

Friction Characterisation

Classic Coulomb's and Amonton's friction laws, which mainly establish that the friction coefficient is independent of the contact area, are proven to be invalid in the case of rubber-like materials. Friction coefficient can be expressed as function of several variables such as contact pressure (by means of vertical load F), sliding speed (v), temperature (T), lubrication regime (by means of lubricant viscosity υ) or roughness of counter material (R_a) (192).

$$\mu = \mu(F, T, v, \upsilon, R_a, \ldots) \tag{192}$$

The dependence with the contact pressure is associated to the varying ratio of real contact area, at microscopic level, to apparent contact area, at macroscopic level, when vertical load rises. The problem increases in complexity when neither the contact pressure distribution nor the ratio of real to apparent contact area are uniform along the apparent contact area, as in case of a contact geometry different from flat to flat configuration.

Rubber-like materials in general and rubbers in particular, have high friction characteristics, a consequence of their low elastic modulus and their viscoelasticity. Thus, they deform in a large extent, resulting in high values of the real contact area. Hence, classical models for metals are no longer valid for the case or rubber friction.

As it was previously referred, Amonton and Coulomb established that friction force is proportional to the vertical load and independent of the contact geometry. Coulomb defined the friction coefficient μ as the ratio between friction and vertical load. For materials obeying this law, μ is independent of the vertical load and thus of the normal stress. Rubber does not obey Amonton's and Coulomb's laws since the friction coefficients falls markedly when increasing of normal stress. For this particular behaviour, an analytical law, which later would be of wide use, is due to Thirion [85] (193).

$$\frac{1}{\mu} = a + b\left(\frac{P}{E}\right) \tag{193}$$

where μ is the friction coefficient, P is the normal stress, E is the elastic modulus of the rubber and a and b are empirical constants. Schallamach [133] showed later how behaviour described in (16) may be explained on the assumption that the friction force is proportional to the true contact area, resulting in (194)

$$\mu = const \cdot \left(\frac{P}{E}\right)^{-1/n} \tag{194}$$

where the value of n is derived from a model, which considers the deformation of the rubber on the asperities of the metallic counterpart and depends on the geometry and distribution considered for peaks and valleys. In general, n depends on the nominal normal stress, but for restricted ranges is considered to be constant. At sufficiently high normal stresses, the real contact area becomes equal to the apparent contact area, so that the frictional force becomes constant and μ is inversely proportional to P, as described in (194). This particular condition is referred to as "saturation".

A methodology to implement friction coefficient can be obtaining frictional force from tribological test results, and the contact pressure distribution and the contact area for each load level from FE simulations. The relationship between the frictional force and the contact pressure can be set up according to the following expression (195).

$$F_f = \int_A \mu(P) \cdot P \cdot dA \approx \sum\nolimits_i \mu(P_i) \cdot P_i \cdot A_i \tag{195}$$

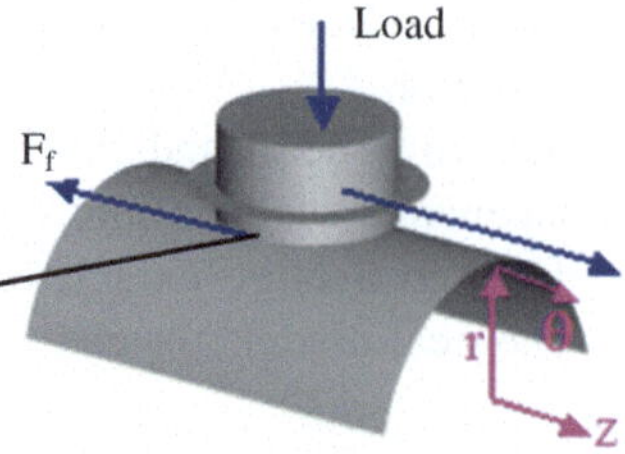

Fig. 64 Sketch of friction characterisation in a "flat-on-cylinder" configuration

where F_f is the frictional force measured by the tribometer, A is the contact area, P is the contact pressure and $\mu(P)$ is the friction coefficient dependent of the contact pressure.

The integral can be transformed into a sum due to the FE discretization. Next step is to approximate the friction coefficient by an analytical function. Choosing a polynomial form allows the friction coefficient curve to take the form desired and gives some advantages when manipulating the sum above (195). This way, approximating the friction coefficient by an order n polynomial form (196) and substituting in (195), a system of $n + 1$ unknowns is obtained. Figure 64 shows the identification of the variables involved in the next equations.

$$\mu(P) = \sum_{j=0}^{n} k_j \cdot P^j \tag{196}$$

$$F_f = \sum_{i} \left(\sum_{j=0}^{n} k_j \cdot P_i^j \right) \cdot P_i \cdot A_i \tag{197}$$

Manipulating adequately (197):

$$F_f = \sum_{j=0}^{n} \left(\sum_{i} A_i \cdot P_i^{j+1} \right) \cdot k_j \tag{198}$$

Taking into account the number of different load levels used in the tribological test, and the order of the polynomial chosen, a different number of equations systems can be obtained. In general, N being the order of the chosen polynomial, to solve the system it is necessary to consider $N + 1$ tests, all of them at different values of normal load.

Wear Characterisation

The widespread characterisation about wear modelling in different materials was carried out by Meng and Ludema [95], which classified more than three hundred wear model equations along last century. These authors consider in the classification three main approximations about wear modelling: models based on empirical relationships, models based on contact mechanics and models based on material failure mechanisms.

According to the results obtained in the tribometer tests, the most appropriate models to set up a starting point from which a wear model can be developed are those based on contact mechanics. Regarding the rest of models, those based on empirical relationships, like those developed by Barwell [217] and Rhee [218] are specific to the particular test for which they are fitted and it is difficult to be adapted to new test conditions; on the other hand, models based on material failure mechanisms, like those described by Cantizano et al. [219] and by Torrance [220], include material parameters such as fatigue properties, shear failure or surface parameters obtained from surface characterisation techniques, so that available tribometer test data are not enough to characterise this kind of models. Therefore, models based on contact mechanics are the most suitable to characterise an initial wear model because most of them relate worn volume with a material property like Young's modulus or hardness, properties that can be easily obtained from test data.

Within models based on contact mechanics, that one proposed by Archard [221] is one of the most broadly used in literature, being expressed in (199):

$$W = \frac{k}{H} \cdot F \cdot \gamma \qquad (199)$$

where W (mm^3) represents material worn volume, F (N) is the applied normal load, γ (m) is the travelled distance, k [−] is a wear constant particular to the contact pair and H (N/mm^2) is the material hardness. When experimental results are interpreted, hardness of the contact layer may not be exactly known and therefore, a more appropriate parameter to express wear behaviour in Archard's model is the ratio k/H (mm^3 N^{-1} m^{-1}), known as K and so-called dimensional wear coefficient of specific wear rate. Therefore, (199) can be expressed as (200):

$$W = K \cdot F \cdot \gamma \qquad (200)$$

Archard model referred in (200) is applied to the stationary stage of the wear curve shown in Fig. 65, where the wear rate on the surface is attained in a constant and uniform way. According to Archard model, and with the rest of variables of

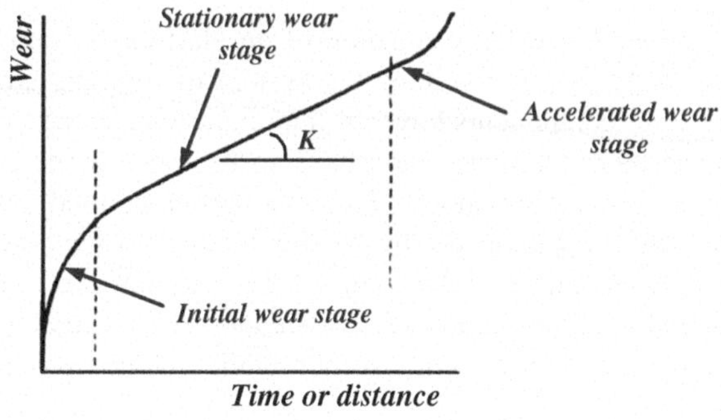

Fig. 65 Wear curve applied to Archard model

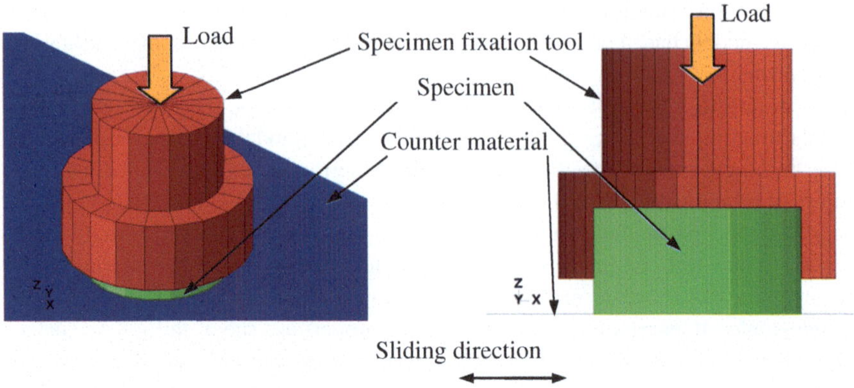

Fig. 66 Sketch of wear characterisation in reciprocating configuration

(200) known, K can be considered as the characteristic wear coefficient of the wear process. Figure 66 shows an example of application of the Archard's law stated in equations (199) and (200).

Other authors have proposed modifications and improvements of the Archard's model: Greenwood and Williamson [222] included particles deformation, Sarkar [223] included the friction coefficient dependency, Liu and Li [224] proposed an Archard's law modification in order to take in consideration material properties for high elastic materials by means of its strain energy and Molinari et al. [225] included the dependency of material hardness with temperature using an elasto-plastic material model for contact bodies.

With the tribometer test results, Archard's model can be taken as starting point to firstly study the dependencies, applied force and travelled distance, obtained in these results. Depending on the obtained dependency between variables, relationship stated by Archard can be modified if the linearity is not fulfilled, stating the dependency that better characterises tribometer test results.

3.4 FE Model and Related Aspects

In general, the FE simulation of an elastomeric component requires to know as precisely as possible how the system to simulate works. Generally, as the elastomeric components have an important interference in the housings in which are assembled, auxiliary parts have to be used in the mounting of the elastomeric component in the model. This fact will be pointed out in the Sect. 4, examples of industrial applications, in the simulation of a window seal of a railway coach.

The geometry of the component to simulate has to be simplified if it is possible, including fillet radius in sharp edges so that the mechanical behaviour of the component is not significantly modified. A compromise between the accuracy of the results obtained with the model and the calculation time must be set up, in order to build up a suitable model fine enough at the zones of interest but within a reasonable computation time.

Elastomeric material is modelled from results obtained in elastic or ineslastic characterisation tests. If only elastic behaviour is available, the model that better reproduces the experimental characterisation will be used in the FE simulation. If inelastic behaviour is available, it can be implemented with the data detailed in Sect. 1.2.2 by means of material user subroutines available in FE codes.

Models to be used in FE simulations with elastomeric components are usually three dimensional models. Hyperelastic elements are considered as deformable bodies, while in the case of other parts made of a material with much higher stiffness than elastomeric material, such as steel of rigid plastic elements, and if a stress analysis is not focused on them, can be modelled as rigid parts and modelled as rigid surfaces. When the model shows symmetry in geometry and loads, axi-symmetric models or models with half or a quarter of symmetry must be used to reduce the computational cost of the simulations, especially important task in non-linear calculations like those in which elastomeric components are included (non-linearities are present in material formulation and in contact algorithms).

Deformable elements are modelled so that the mesh reproduces accurate enough the zones of interaction between surfaces, with a suitable precision in the calculation results. In general, linear hexahedron elements with full integration and hybrid formulation are used in three dimensional problems, while axisymmetric elements with four nodes and hybrid formulation are used in problems with axisymmetry.

The simulation of an elastomeric component, in general, can be divided into two parts. Firstly, the component is assembled in its housing in order to obtain the assembly or nominal position. If the friction between the elastomeric component and the housing is significant, the assembly must be carried out accurate enough in order to obtain in the simulation the same position as in real conditions. Secondly, the working conditions of the system are applied to the simulation in subsequent steps.

Loads and boundary conditions are applied by means of contact pairs between the elastomeric component and the housing. The housing is usually built up by means of analytical rigid surfaces because their stiffness is higher than that one of the elastomeric component. Each analytical rigid surface is defined with a reference node, used to define the boundary conditions and loads over the component.

In the definition of the contact pairs, the elastomeric component modelled as deformable body and that one of the rigid surfaces of the housing, sharp edges and zones with an incorrect definition of the normal vector must be avoided in order to obtain enough convergence in the simulation. The definition of the analytical rigid surfaces must be done with its normal opposite to that one of the surface of the deformable part. The friction model will be characterised from, if it is possible, tribometer test results or, if not, estimated from previous results or from literature under similar conditions. In case of characterisation from tribometer test results is used, the dependencies obtained between the friction coefficient and the rest of parameters with the model detailed in Sect. 1.3.2 can be implemented by means of a friction user subroutine available in FE codes.

In case of wear effects in elastomeric components are to be taken into consideration and a wear model from tribometer tests is available, it can be implemented in a FE code via user subroutine in order to simulate the wear by means of a FE simulation.

3.5 Results Interpretation

The results of the FE simulations that include elastomeric components are usually analysed by means of the same results that can be available in experimental tests. It can included, for instance: curves of reaction force–displacement in the rigid parts, variable that can be usually validated with experimental results, with the same conditions than those stated in the FE simulation. Other variables that can be analysed are true strain distribution maps in the elastomeric component modelled as deformable part, real strains, real or Cauchy stresses, contact pressures and shear stresses at the contact surfaces with the rigid parts and variables which relate both stress and strain variables such as strain energy density.

Results from tribometer friction tests are usually used to characterise the friction model that feed the FE simulation. Anyway, the results attained in these tests can be also validated by comparing the results obtained from tests, like friction force or friction coefficient. Similarly, results from tribometer wear tests are also used to characterise the wear model, but these results can also be post-processed in the wear simulation of an elastomeric component by means of wear distribution maps.

4 Examples of Industrial Applications

In the present section, some examples related to the application of FE techniques for the analysis and design of components for automotive and railway applications will be presented, as demonstration of the advantages which these techniques provide compared to the experimental design procedures used until recent past in the industry.

As will be shown, the main advantages of using virtual prototyping (FE based) techniques constitute mainly the significant reduction in cost and time of the design cycles, compared to the large and expensive experimental testing series performed in industry on component prototypes. Other advantages are related to the increase of knowledge in the mechanical functional behaviour of the component and also of its constituting material, which allow obtaining designs with optimised mechanical functionalities and with extended functionalities.

The following examples are collected as demonstrators of FE analysis as virtual prototyping tool applied to design of industrial components: rubber seal of a railway coach, silent-block in automotive industry, rubber block on wheel of railway coach and rubber shock absorber in automotive industry. Each example describes the scope in the complexity of the applied rubber material model and the FE simulation assumptions, according to the boundary conditions and the needed accuracy of the results. The FE code used in each simulation depends on its suitability to reproduce the desired effects. These FE codes are ABAQUS Standard [97] and MSC.MARC [61].

All the examples that are included in the present chapter have been extracted from real design and development projects, which have been carried out by the Instituto Tecnológico de Aragón (Spain) in collaboration with industrial clients.

4.1 Example 1: Window Seal of Railway Coach

The first example consists of the analysis of the functional behaviour of a rubber seal located in the window of a railway coach. The analysis aims to study the seal in terms of strain distributions and to identify critical areas under wind pressure conditions inside and outside the coach (static condition), studying also the mounting position on the window frame.

4.1.1 Finite Element Model Description

The elements that are included in the FE model are the main glass window, the aluminium frame and the rubber seals between glass and aluminium. Both rubber components are characterised by a *Shore A* 65 hardness. Figure 67 represents the mentioned elements included in the model development.

Since the requirements for the loading conditions are considered as static, as well as the FE analysis itself, the mechanical stress–strain response of the rubber will be modelled by means of a hyperelastic model, assuming therefore isotropic and incompressible behaviour with no dissipative effects included in the model (viscoelasticity and hysteresis). In this case, the range of strains that is expected to be obtained in FE simulation suggests using a reduced order polynomial with three constants (N = 3; *Yeoh* model).

The *Yeoh* hyperelastic model constants can be fitted in two ways, by means of fitting to experimental quasi-static tests performed on samples of the same material, or using analytical (experimental based) laws which relate *Shore A* hardness values to estimations of the shear modulus. If the latter case, the material model fitting is considered only as estimation which in most of the cases probes valid enough. In the present case, the Yeoh model constants are estimated as $C_{10} = 0.86\ \text{N/mm}^2$, $C_{20} = 0.07\ \text{N/mm}^2$ and $C_{30} = -0.0015\ \text{N/mm}^2$. Figure 68 shows the stress–strain response assumed at room temperature:

Fig. 67 Sketch of the rubber seals

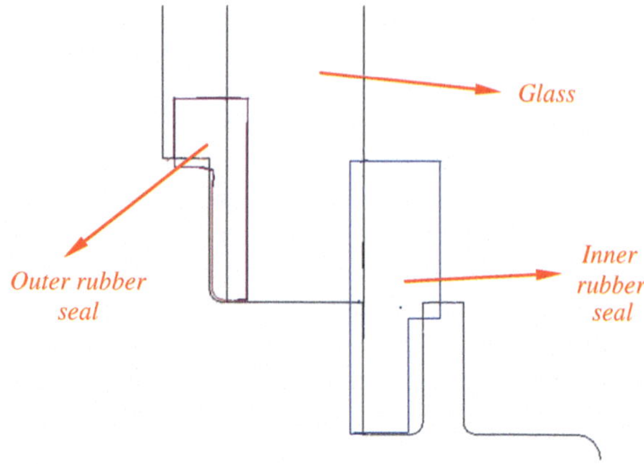

According to the analysed geometry of the window structure, a 2D model which considers a plane strain deformation state is constructed to spare computational cost. For this reason, the glass window and the aluminium frame are also considered as rigid bodies since their stiffness is assumed to be several orders of magnitude higher than that of the rubber material. The model is constructed using plane strain elements with hybrid formulation, resulting in a model size of 34000 nodes and 33000 elements. The analysis is performed by means of the commercial FE code ABAQUS [97].

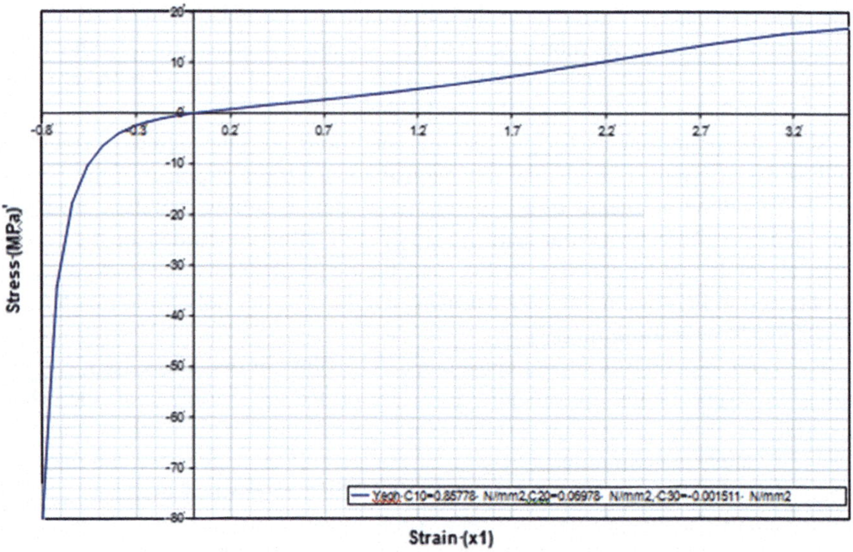

Fig. 68 Mechanical stress–strain response modelled by Yeoh hyperelastic model

The simulation cases to be considered correspond to the static pressure conditions on the outer and inner sides of the window mentioned previously, corresponding to component test specifications: Starting from the mounted position, the first case considers therefore the static pressure applied on the outer side of the window and, in a similar way, the second case considers the static pressure applied on the inner side.

4.1.2 Results of the Static Finite Element Simulations Cases

The present section shows a summary of the most relevant quantitative and qualitative results that can be obtained from any of the two performed simulation cases. In this case, the results shown correspond to the first condition of outer pressure, being the type of analysis the same as in the second simulation case.

Static Pressure on the Outer Window Side

Figure 69 shows the zones at which results are analysed and compared.

The values of the most important variables of analysis (principal strains and stresses and also contact pressures) can be analyzed for a specific FE simulation instant, as well as during their evolution along FE simulation progress. In Fig. 69, the critical zones of the seals sections have been labeled. In these zones, the most relevant variables of analysis can be studied in detail, comparing the initial assembled position with the final one after pressure application. For example, in the critical "*zones 1 and 2*" of the outer seal, the maximum values of principal

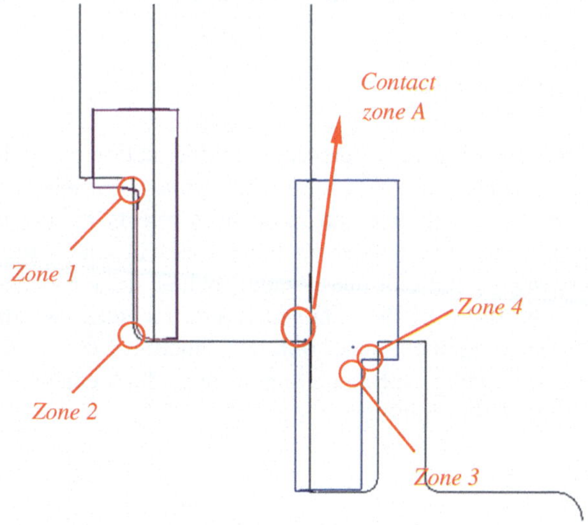

Fig. 69 Analysis zones

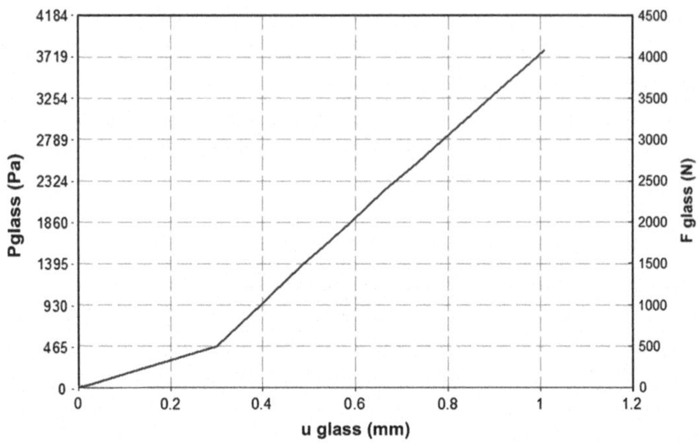

Fig. 70 Curve of equivalent force and pressure on glass for the case of 3800 Pa outer side pressure

strains and stresses drop 10 % in average, whereas in the inner seal, there is an increase of them around 50 % at the labeled critical zones. Values of contact pressures between the rubber and rigid bodies can be also studied in FE simulation. As example, the labeled "*contact zone A*" suffers from a significant increase in the contact pressure due to the displacement of the glass towards inside.

Finally, Fig. 70 shows the evolution of equivalent force and pressure on glass against its main displacement due to the uniformly applied outer pressure.

Among the conclusions that can be obtained from the previous curve, it can be drawn for example that the glass displacement value at which the change in the slope of the curve occurs correspond to the point of the simulation at which the gasket and the inner seal contact each other (contact zone A).

4.1.3 Conclusions

From the FE analysis presented in this section, it can be concluded that modelling the mechanical stress–strain behaviour of rubber, assuming isotropy, incompressibility and with no dissipative effects by means of a hyperelastic model probes accurate enough in those cases in which the load specifications can be considered static or quasi-static, such as the presented one.

The scope of the obtained results includes the stress and strain states of the rubber component under different functional conditions and their interaction with the surrounding metallic components, which enable to check the accomplishment of the design limits.

4.2 Example 2: Silent-Block in Automotive Industry

The second example consists of the analysis of the static, including inelastic effects in the analysis, and dynamic mechanical behaviour of a vibration isolation mount made out of filled rubber. Filled rubber is a material composed of a matrix of vulcanised gum where different filler particles are gathered. These fillers, such as carbon black, silica, etc. provide to the material an increase on stiffness and at the same time, an increase of inelasticity.

Due to its special properties, filled rubber is commonly used in the manufacture of vibration isolation mounts. Silent-blocks usually experience small oscillatory loads superimposed on large static deformation and their static and dynamic stiffness are required in order to meet the performance of load bearing and isolation. These characteristics of a filled rubber mount are often very complex in nature, due to the fact that the filled rubber material response is dependent on several variables, such as frequency, amplitude [57, 56], pre-strain and temperature.

In the present section, two FE analysis approaches are presented: first one considers a static analysis which includes also the typical inelastic effects which characterise the mechanical rubber behaviour to predict the axial stiffness of the rubber component. The second approach focuses on the prediction of its dynamic stiffness under cyclic loading.

4.2.1 Static Analysis Including Inelastic Effects

Figure 71 shows the silent-block analyzed in this section. The component consists of an inner steel ring over which the rubber is moulded. The resulting assembly is then mounted on an outer ring as shown:

The selected experimental conditions to be later simulated by FE method correspond to the axial stiffness test. This test consists of applying an axial displacement on the inner ring while outer ring is maintained fixed. The history of

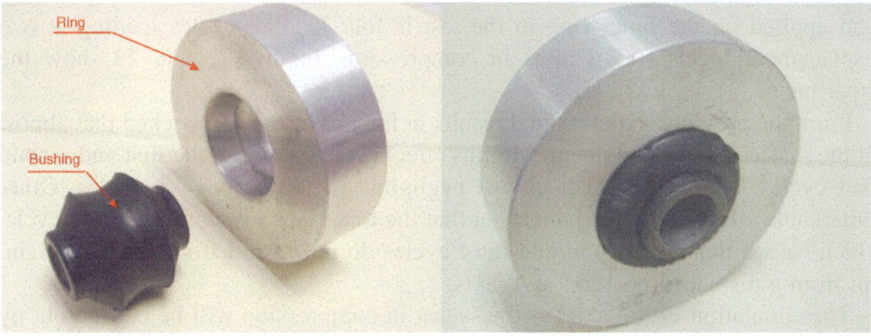

Fig. 71 Silent-block studied in the present FE analysis

Fig. 72 Axial stiffness test

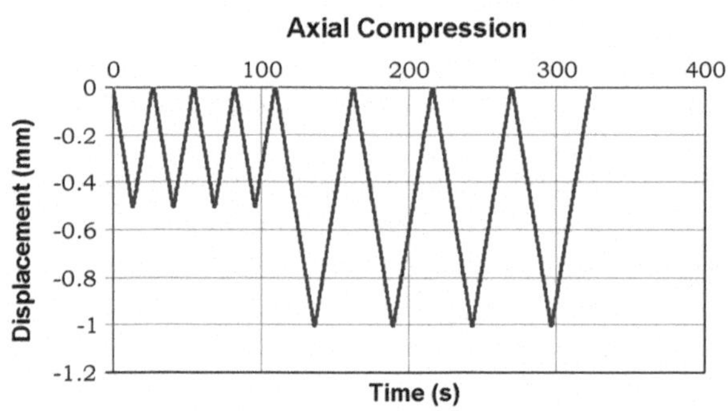

Fig. 73 Load history in axial stiffness test

load applied on the inner ring in the test is four loading cycles loading at two displacement levels (0.5–1 mm) in compression. Figures 72 and 73 show the experiment.

From the previous experimental results in Fig. 74, it can be checked that almost all the material softening due to Mullins effect occurs between the first and second load cycles, being this effect almost negligible in the successive cycles. Other remarkable effect is the permanent set that the material shows after the first cycle, which causes that the second and third cycles do not restart from the XY origin, but from a displaced position at Y axis.

The simulation of the axial stiffness test in compression will be carried out by using the following material models: the overlay method for modelling the

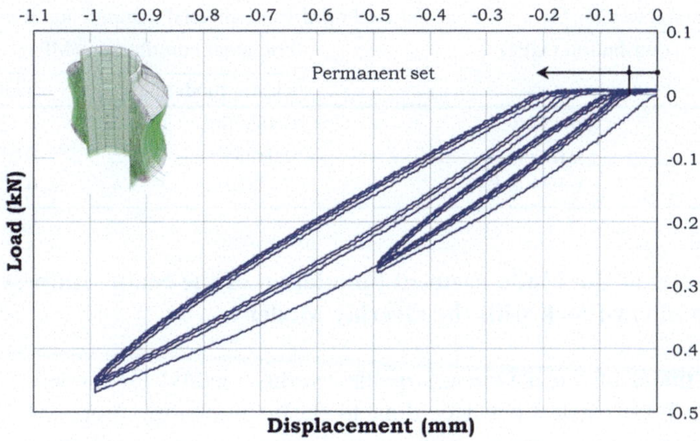

Fig. 74 Force–displacement curves obtained for axial stiffness test

Fig. 75 FE mode of the silent-block in the FE code ABAQUS

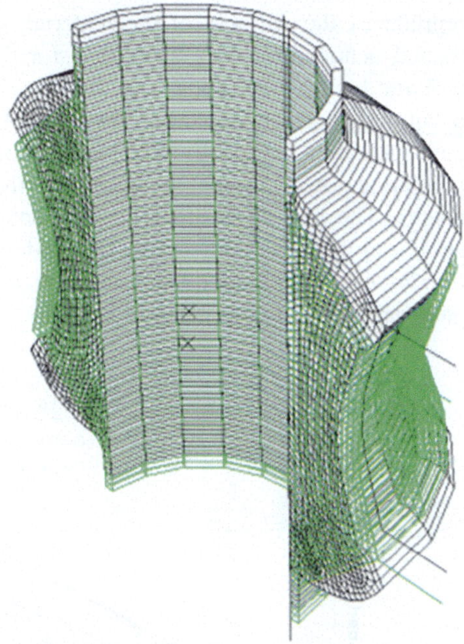

hysteresis of the material [211] and the Ogden-Roxburg or Simo models for modelling of Mullins effect [212]. Figure 75 shows the FE model that, with this purpose, is developed with the FE code ABAQUS [97]: the rubber component has been considered as the deformable body and both the inner and the outer rings as rigid bodies.

Table 2 Parameters for the *overlay model* calibrated from uniaxial compression tests [201]

Hyperelastic contribution (MPa)		Frictional contribution (MPa)	
		Elastic modulus	Yield strength
C_{10}	0.45789	0.485760	0.0244235
C_{20}	0.01687	0.085889	0.0103116
C_{30}	−6.97406 E-4	0.148014	0.0355299

4.2.2 Results of the Finite Element Simulation of the Static Stiffness of a Silent-Block with the Overlay Model

The adjustment of the parameters of the overlay method for modeling the hysteresis has been carried out according to the methodology proposed by Gracia [212], by means of calibration with uniaxial compression tests. Table 2 shows the values for the overlay model parameters.

The results for this model compared with the experimental results are displayed in Fig. 76. Results with the *overlay model* show that this model adequately reproduces the stiffness of the material, with differences, regarding the experimental behaviour, lower than 10 % for 0.5 mm of compression and lower than 5 % for 1 mm, what points out that the characterization of the hyperelastic contribution is correct. Regarding the amount of dissipated energy in the cycle, variable governed by the frictional contribution of the material, the *overlay model* predicts a lower amount of hysteresis than that exhibited by the experimental results. The hysteresis of the experimental cycles from second to forth is more similar to that one predicted by the model. However, it can be concluded that the *overlay model* calibrated from uniaxial compression test data is able to predict in an acceptable way the material behaviour in the test.

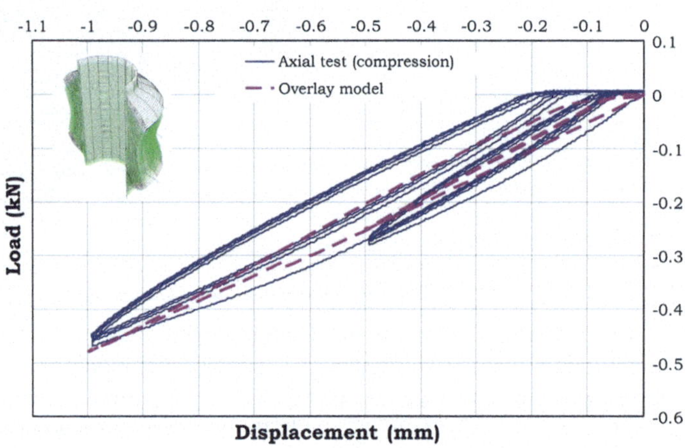

Fig. 76 Force–displacement curves obtained for axial stiffness test, compared to FE simulation with the *overlay model* [201]

Table 3 Parameters for the *Ogden–Roxburgh model* calibrated from uniaxial compression tests [202]

Hyperelastic contribution (MPa)		Mullins effect (O-R)	
C_{10}	0.40	β	2.3524
C_{20}	0.0158745	m	0.0
C_{30}	−6.97406 E-4	r	1.001

Logically, the *overlay model* does not reproduce the permanent deformation of the material observed in the experimental response because this is not a feature incorporated in the model.

4.2.3 Results of the Finite Element Simulation of the Static Stiffness of a Silent-Block with the Ogden–Roxburgh Model

The adjustment of the parameter of the *Ogden–Roxburgh model* is carried out according to the methodology proposed by Gracia [212], by means of calibration with uniaxial compression tests. Table 3 shows the *B-model* parameter values.

Figure 77 shows the results in terms of force–displacement obtained for the *Ogden–Roxburgh model* compared with the experimental results of the axial stiffness test on the silentblock.

The overall stiffness predicted by the *Ogden–Roxburgh model* is appropriate with differences of 10 % for −0.5 mm displacement and only 1 % for −1 mm, what again indicates that the characterization of the hyperelastic contribution is suitable. In this model the dissipated energy, that is, the area enclosed by the cycle, is only determined by the damage phenomenon (136), because of this model does not include any other dissipation mechanism (neither hysteretic, nor viscous). This model predicts acceptable energy dissipation within the cycle, what also indicates that the characterization of the parameters of the *Ogden–Roxburgh model* from uniaxial compression tests is also suitable for this application.

Neither *Ogden–Roxburgh model* reproduces permanent deformation of the material, because it does not include this feature.

4.2.4 Dynamic Experimental Characterisation

The analysis aims to reproduce the effect of these dependencies, which can be absolutely critical in capturing the dynamic stiffness of the component. The presented analytical method for predicting these response characteristics makes possible to design elastomeric components without performing costly design iterations which involve design, testing and redesign.

To predict the dynamic stiffness of filled rubber mounts, a modification of the viscoelastic mechanical behaviour model, typically implemented in FE codes, is

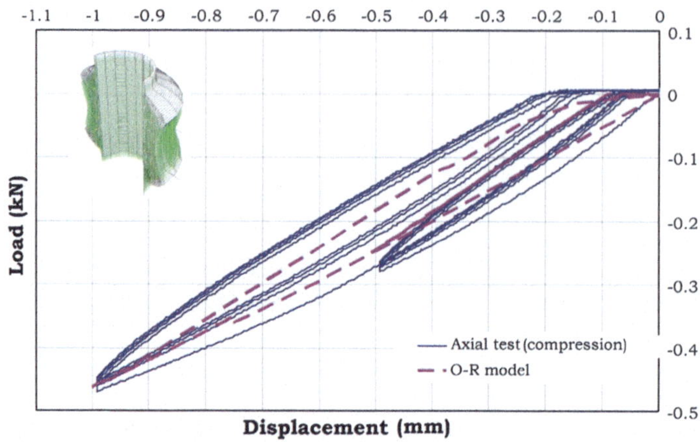

Fig. 77 Force–displacement curves obtained for axial stiffness test, compared to FE simulation with the *Ogden–Roxburgh-model* [202]

proposed, considering not only the static pre-strain but also the dynamic strain-dependent properties. This is accomplished by changing the viscoelastic material parameters element by element, depending on the strain level, due to pre-strain and amplitude of each element.

As requirement to feed the FE material model, experimental characterisation of these dependencies is carried out and these data are applied to the prediction of the dynamic stiffness of the filled rubber under different vibration conditions (pre-strain, amplitude, frequency). Dynamic tests on components are also performed in order to validate the material model. To characterize the viscoelastic behaviour of the material and its cyclic response, tests are conducted to evaluate the dependencies of excitation parameters: pre-strain, amplitude and frequency.

The material used is a *Shore A 67.5* rubber hardness natural rubber compound (NR) used in rubber-metal parts for automotive industry. Data from harmonic displacement controlled loading of compression specimen, cylindrical block compressed by two parallel plates as shown in Fig. 78, are obtained for a wide range of frequency, pre-strains and strain amplitudes.

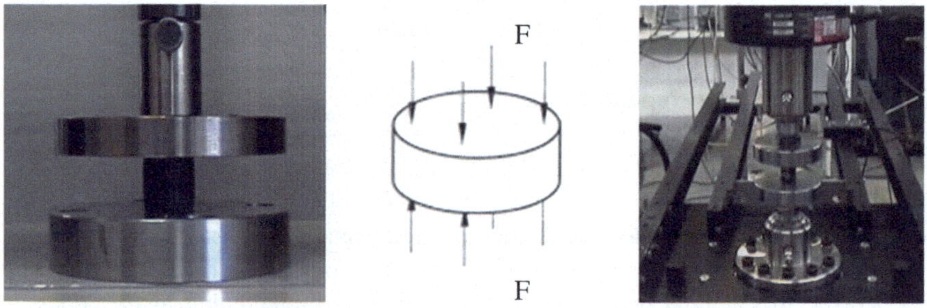

Fig. 78 Compression test specimen used in material characterization tests

Table 4 Parameters for dynamic material characterization tests

Parameter	Levels
Pre-strain[a] (%)	10, 15, 20, 25, 30, 35
Amplitude[a] (%)	0.125, 0.25, 0.5, 1, 2.5, 5, 10, 15, 25
Frequency (Hz)	2, 5, 15, 20, 50, 100, 150, 200, 250, 300

[a] Engineering strains

The dynamic modulus and damping are obtained for each condition according to (6) and (7). The complex shear modulus is obtained as a function of the complex uniaxial modulus with (201):

$$G^* = \frac{3K^*E^*}{9K^* - E^*} \tag{201}$$

Considering that the bulk modulus is very large compared to the shear modulus, the material can be considered to be incompressible and the expressions simplify further to (202):

$$G^* \approx \frac{E^*}{3} \tag{202}$$

By means of an Instron servo hydraulic universal testing machine, dynamic tests are carried out with all combinations of the parameters listed in Table 4. Experiments are performed with harmonic excitation to determine the dynamic modulus of the material.

First, the minimum static pre-strain is applied and the dynamic excitation starts with the smallest strain amplitude. The frequency varies from the smallest to the highest value. After the frequency variation the static pre-strain remains constant and the amplitude is increased to the next higher value. The frequency sweep is carried out again and this procedure is applied until the highest amplitude is reached. The static pre-strain is then set to the next level and the test continues in a similar way until the highest pre-strain level in compression has been reached. Representation of the different dependencies obtained experimentally is shown in Figs. 79, 80 and 81

4.2.5 Dynamic Material Model Development and Validation

The material model is implemented in MSC.MARC [226], which allows performing mechanical analysis of incompressible solids in which a small amplitude time harmonic oscillation is superposed on a static finite deformation field. The amplitude of the superposed vibrations is considered to be sufficiently small that the relevant equations can be linearized and the response to the static preload is computed on the basis of purely elastic behavior in the elastomeric parts of the model. This FE code treats the finite elasticity material behavior using the finite linear viscoelasticity constitutive equations proposed by Morman [60] for predicting the response of

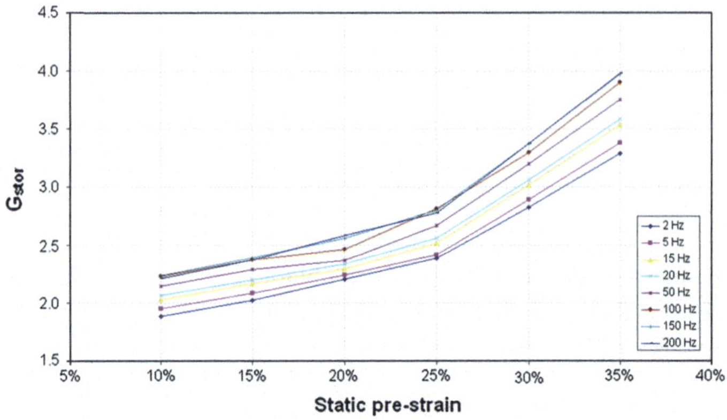

Fig. 79 Measured dynamic stiffness for the compression test (NR material). Pre-strain dependence

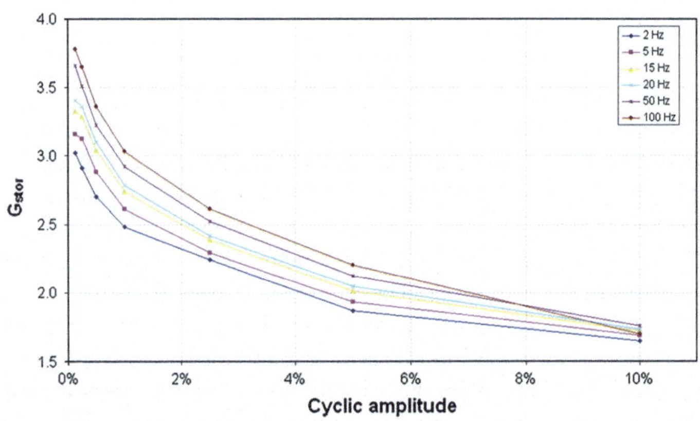

Fig. 80 Measured dynamic stiffness for the compression test (NR material). Amplitude dependence

statically deformed components subjected to small amplitude vibrations in the frequency domain.

According to the Morman's model, the frequency domain viscoelastic material model implemented in MSC.MARC describes frequency-dependent material behavior in small steady-state harmonic oscillations for those materials in which dissipative losses caused by "viscous" (internal damping) effects must be modeled in the frequency domain and assumes that the shear (deviatoric) and volumetric behaviors are independent in multiaxial stress states. The FE code linearizes the problem around the equilibrium state and considers all effects of the nonlinear deformation on the dynamic solution. These effects include the initial stress,

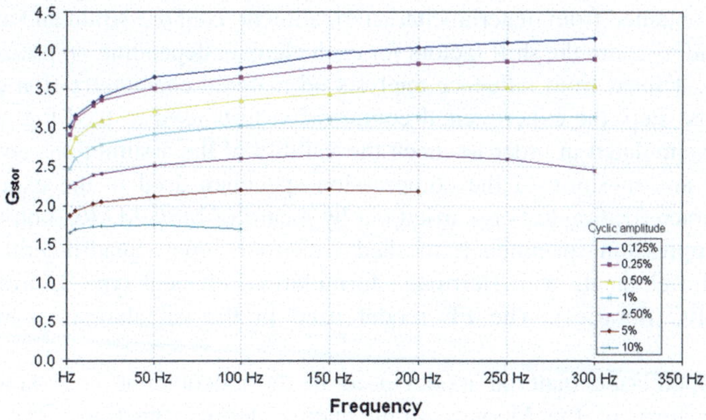

Fig. 81 Measured dynamic stiffness for the compression test (NR material). Frequency dependence

change of geometry and influence on constitutive law. The vibration problem can be solved as a linear problem using complex arithmetic.

As mentioned, the frequency-domain viscoelastic model implemented in MSC.MARC is only able to consider the frequency dependence and the dynamic behavior is characterized with only one dynamic simple shear test, through the complex shear modulus (storage (G') and loss (G'') moduli) versus frequency. These variables are defined in MSC.MARC by means of the UPHI subroutine. The model presented here proposes to modify this subroutine for considering the pre-strain and amplitude effects, including them in the material parameters for each element, depending on the strain level, due to pre-strain and amplitude.

As shown in the Morman's model definition, strain and frequency effects are separated, and the frequency-dependent part of the material's response, defined above, is not affected by the magnitude of the pre-strain. This separability assumption is not suitable for filled rubber. To consider it, first it is needed to modify the long-term shear modulus, which varies with the amount of static pre-strain. This is obtained from the static (long term) characterization of the material.

$$G^{\infty} = G^{\infty}(\varepsilon_0) \tag{203}$$

To consider pre-strain and amplitude effects on the complex shear modulus G^*, shift factors are introduced to the constitutive equation in order to describe both effects, assuming that these effects are separable. These shift factors, fitted with experimental data and self-computing inside the subroutine, allow correcting the data introduced to feed the viscoelastic model at fixed pre-strain and amplitude conditions, to consider the real strain levels of individual elements.

$$G^*(\omega, \varepsilon_0, Amp) = G^*(\omega) \cdot \alpha_{\varepsilon_0} \cdot \alpha_{Amp} \tag{204}$$

where $G^*(\omega, \varepsilon_0, Amp)$ is the material viscoelastic parameter defined for each element depending on strain levels, $G^*(\omega)$ is the material viscoelastic parameter of input file, obtained from material characterization at fixed pre-strain and amplitude and α_{Po} and α_{Amp} are the shift factors for each element depending on pre-strain and amplitude, defined from influence analysis on material characterization curves

As a first step, the experimental compression tests used to feed the numerical model are simulated in order to check the validity of the assumptions considered. Geometry and meshing of the compression specimen used in the experimental cyclic characterization tests is carried out by means of MSC.MARC non linear FE code. Compression specimen is meshed with four nodes, quadrilateral axisymmetric solid elements with Herrmann formulation (element type 82 available in MSC.MARC libraries). The FE model used in the calculations is shown in Fig. 82.

The hyperelastic material model used in the calculations is a Yeoh model (particular form of the Mooney strain energy density function). The material viscoelastic properties (and frequencial dependence) are considered by means of the G_{store} and G_{loss} experimental values for fixed static pre-strain and cyclic amplitude through the UPHI user subroutine fitted. The material is assumed as isotropic and incompressible. Different experimental test are simulated in order to validate the fitting of the viscoelastic model for different static preloads, cyclic amplitudes and frequencies. Table 5 resumes the different simulations performed in terms of static pre-strain and cyclic strain.

The simulations results obtained are enclosed from Figs. 83, 84, 85 and 86 in terms of dynamic stiffness calculated as the ratio reaction force/harmonic amplitude imposed. These simulations have been run with the same input data, taken from experimental curves at fixed pre-strain and amplitude, changing the preload of the static case and the amplitude of the harmonic one. The experimental results corresponding to the same conditions simulated are also included in order to compare the predicted results with the experimental ones.

The FE simulation results obtained for a simple geometry under uniaxial compressive load show an excellent agreement with the experimental values; therefore the next step is the application to the FE analysis of the dynamic stiffness of the real industrial component: rubber silent-block.

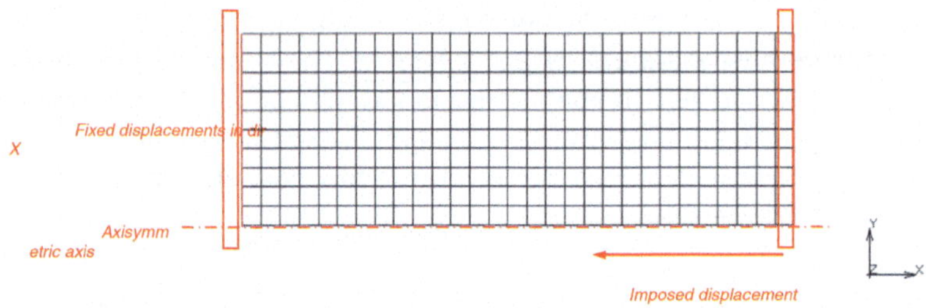

Fig. 82 Axisymmetric FE model used in the uniaxial compression test simulations

Table 5 Experimental cyclic test conditions simulated

	Static pre-strain (%)	Cyclic amplitude (%)
Simulation 1	10	5
Simulation 2	20	10
Simulation 3	30	0.125
Simulation 4	30	15

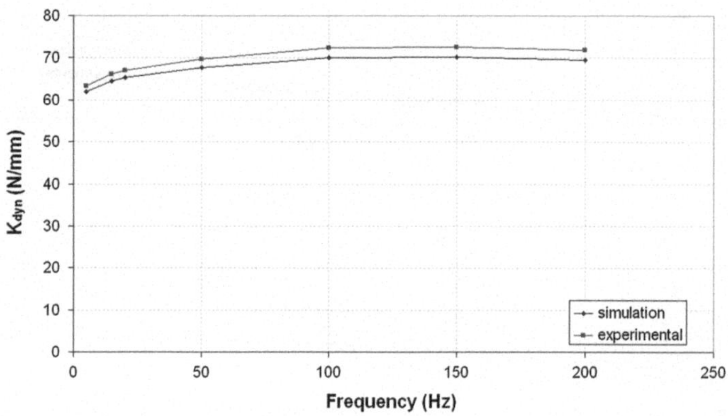

Fig. 83 10 % of static preload and 5 % of cyclic amplitude

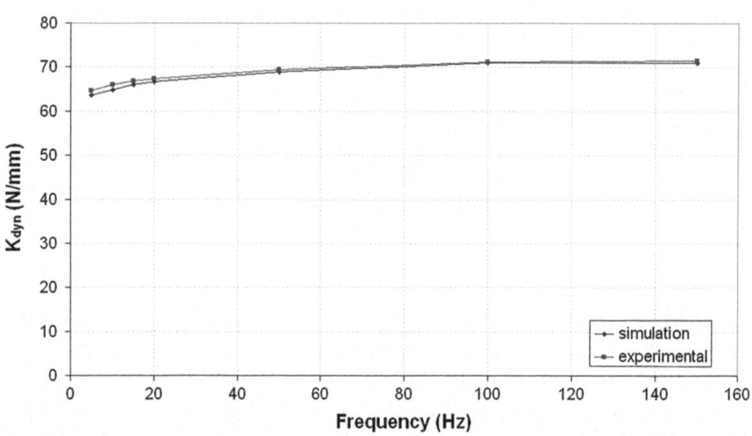

Fig. 84 20 % of static preload and 10 % cyclic amplitude

4.2.6 Results of the Finite Element Simulation of the Dynamic Stiffness of a Silent-Block

The rubber silent-block used in this study (see Fig. 71) consists of a part of the same NR compound that was characterized and is subjected to axial, radial, torsional or conical loads, giving an inhomogeneous multiaxial state of deformation.

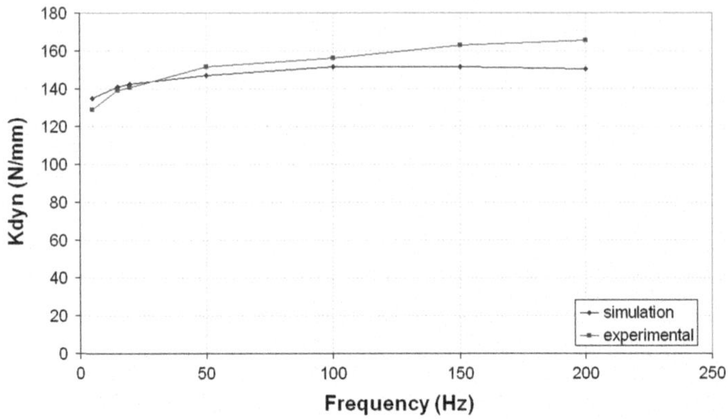

Fig. 85 30 % of static preload and 0.125 % cyclic amplitude

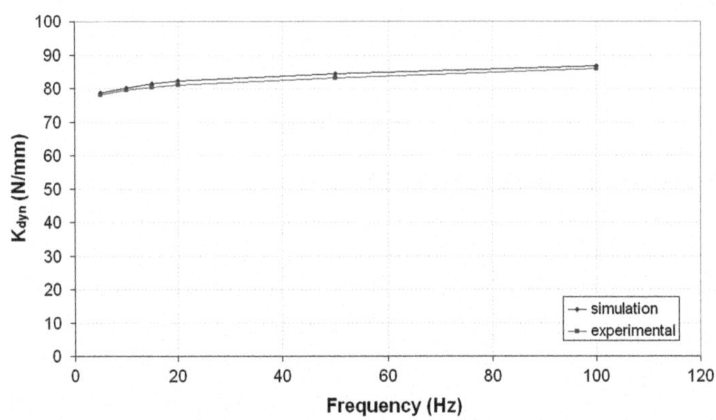

Fig. 86 30 % of static preload and 15 % cyclic amplitude

The FE model is created in MSC.MARC [226]. To simulate the assembly of the bushing with the test ring, an axisymmetric model with half section is used, considering a static radial assembly as shown in Fig. 87.

Once the bushing is assembled with the ring, the model is revolved 180° with regard its axisymmetry axis (due to symmetry only one-fourth of the bushing is modeled) and the initial conditions extended (axi to 3D). On this 3D configuration (Fig. 88) harmonic load case is calculated.

Material properties are the same than those defined in previous section. A modification in the user subroutine has to be incorporated to relate the shear strain amplitudes in the material characterization with the element load level in the multi-axial case in each FE.

After the assembly, that means a static pre-strain on the rubber, the bushing is loaded harmonically in the radial direction at frequencies ranging from 5 to

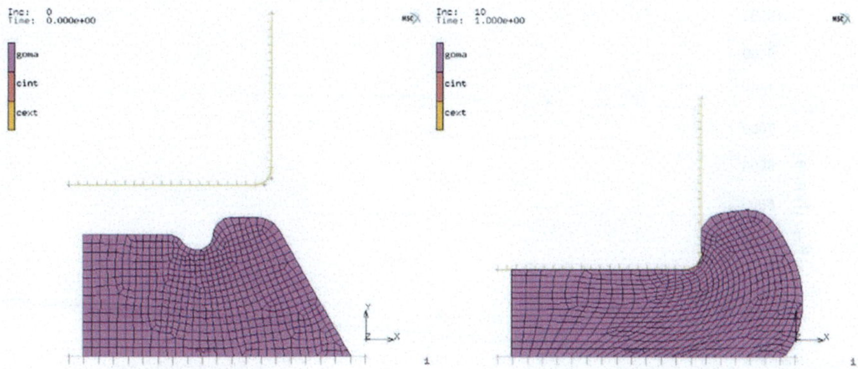

Fig. 87 Axisymmetric FE model of the bushing before and after assembly

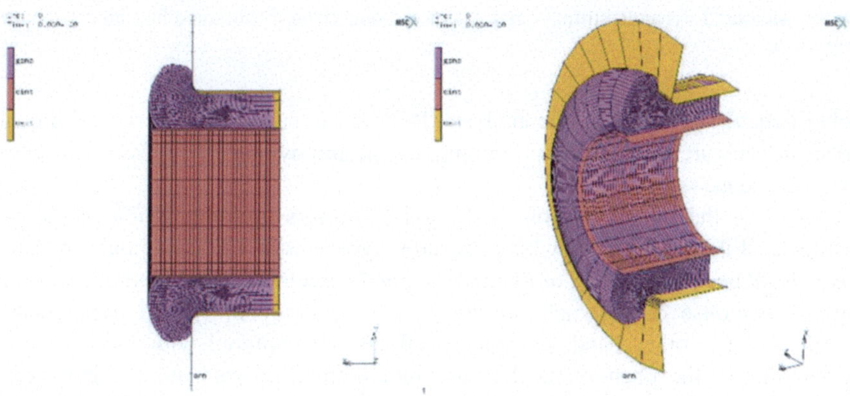

Fig. 88 3D FE model of the bushing for the harmonic analysis

105 Hz and considering an amplitude of 0.2 mm. Results, in terms of dynamic stiffness (N/mm), are compared with experimental data obtained in the same component (Fig. 89).

It is important to note that if the pre-strains or amplitudes or frequencies of the FE model get out of the measured range, the data has to be extrapolated, that means a certain deviation from the results obtained on tests depending on the method used. To avoid that, it is recommended to extend the material characterization ranges to those obtained on the real part.

4.2.7 Conclusions

In the present section, two FE analysis approaches are presented: static analysis, which includes also the typical inelastic effects which characterise the mechanical

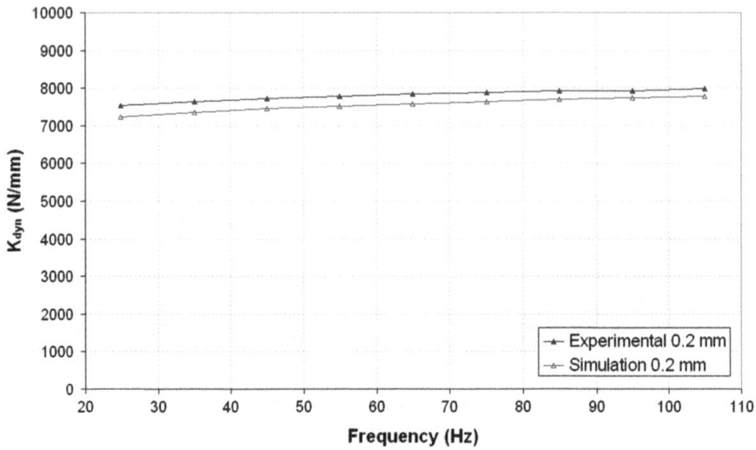

Fig. 89 Measured dynamic stiffness and calculated one given by the modified model for the silent-block

rubber behaviour, and dynamic analysis. Both are intended to predict mechanical rubber behaviour: axial stiffness in compression and dynamic stiffness of the rubber component.

Regarding the FE simulation of the axial stiffness in compression of the silentblock, different models which include hyperelastic response and inelastic effects have been used. Most of them show good agreement in the prediction of the stiffness response of the material due to the accuracy of typical hyperelastic models. On the other hand, dissipative effects are captured with more or less success due to the phenomena that are implemented on each model. However, none of them is able to reproduce the permanent set in compression. The FE code used in this phase is ABAQUS due to its flexibility to develop FE models with high complexity degree, as is the case of the overlay methodology.

Regarding the FE simulation of the silentblock dynamic stiffness, the FE modelling procedure presented in this section shows good agreement with experimental data for the prediction of dynamic stiffness of filled rubber parts, as shown in the current industrial application: a silent-block of the automotive industry.

For this, it has been necessary to apply a modification of the frequency domain viscoelastic model implemented in a commercial FE code (MSC.MARC) through user subroutine. This model considers the vibration as a harmonic perturbation of small dynamic strain amplitude around the elastically predeformed state. The used FE code is in this case suitable due to the possibility of programming the dynamic behaviour of the rubber material through user subroutine.

This model provides an accurate description of the dynamic behavior of filled rubber material, by means of the modification of input data through shift factors to consider the individual static pre-strain and amplitude strain of each element. Because the model is implemented in frequency domain, that means solve a linear system of equations (with complex coefficients), is computationally very efficient, yielding very fast solutions compared to time stepping procedures.

The material model developed is applied to predict the dynamic stiffness of a silent-block, showing good agreement with experimental data obtained from dynamic tests on the same component.

4.3 *Example 3: Analysis of Rubber Block on Wheel of Railway Coach*

In the present section, the analysis by means of FE simulation of the static mechanical behaviour under complex compression conditions of a rubber block is presented. The rubber block is part of the damping system of a classical *elastic wheel* concept of a railway coach, which is shown schematically in Fig. 90.

The herewith presented analysis procedure is similar to the two previous examples collected in the present section: first, a material model for predicting the rubber mechanical behavior is developed, according to the design functional requirements and boundary conditions of the real component, from experimental characterization tests on samples of the same rubber used in the block. Once validated, the rubber material model is applied to the FE simulation of the real component mechanical behavior under service conditions and design specifications to be later applied to virtual prototyping design cycles.

4.3.1 Finite Element Model Description

The FE model developed for the analysis consists of a 3D model of one single block (the elastic wheel consists of several blocks placed in circumferential direction), mounted on its placement between the wheel inner and outer tires (Fig. 91).

Two assumptions have been applied for the development of the rubber material model for the block, depending on the validity of incompressibility in the mechanical behaviour of the rubber, therefore leading to two different FE material models which will be evaluated in parallel. First, the used hyperelastic material

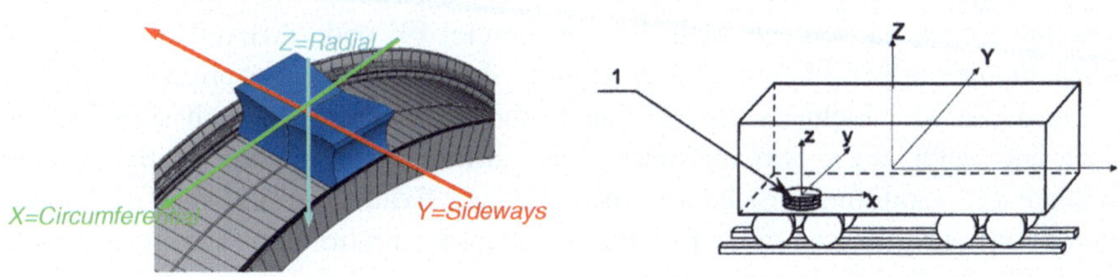

Fig. 90 Sketch of one rubber block mounted on the wheel inner tire

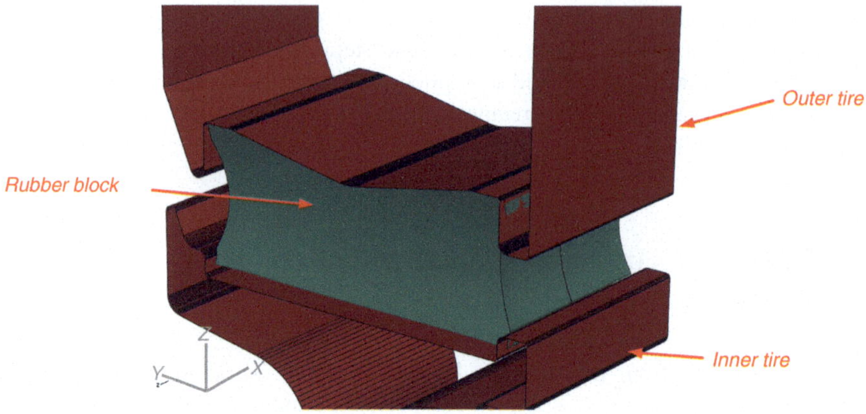

Fig. 91 Sketch of the rubber block assembly

model is characterised by means of the Marlow expression for the strain energy density function in both materials. Second, in one case, the rubber material is considered incompressible, whereas in the other, compressibility is assumed by means of a volumetric behaviour model. This model is therefore fed with experimental data from a volumetric compression test.

In both cases, the mechanical behaviour is considered isotropic with no dissipative effects (viscoelastcity and hysteresis). Figures 92 and 93 show the stress–strain curves obtained from experimental uniaxial compression tests in two different rubber materials: data from one of them will feed the incompressible material model, whereas the other data will feed the compressible one, for which volumetric compression tests are also available.

In the present case, the hyperelastic Marlow model is fed with compression data, for which the raw data from the stress–strain curve are given as input instead of performing a fitting of constants.

The 3D FE model which has been constructed considers the rubber block as a deformable body, meshed by means of linear 8-node hexahedrons with hybrid formulation and constant pressure, resulting in a model size of 30000 nodes and 27000 elements, including half symmetry. In addition, both inner and outer tires have included in the FE model as rigid bodies since their stiffness is assumed to be several orders of magnitude higher than that of the rubber material. The simulations have been carried out with the commercial FE code ABAQUS. Figure 94 shows the developed FE model according to the previous description.

The FE model includes, in addition to the rigid surfaces which simulate the inner and outer tires, two auxiliary rigid surfaces which numerically ease to simulate the mounting procedure (they are deactivated for the rest of the FE simulation). This procedure is usually considered a preliminary step to the main FE simulation, being of great importance the final mounting position correlation with the real one. Figure 95 shows the rigid bodies included in the model.

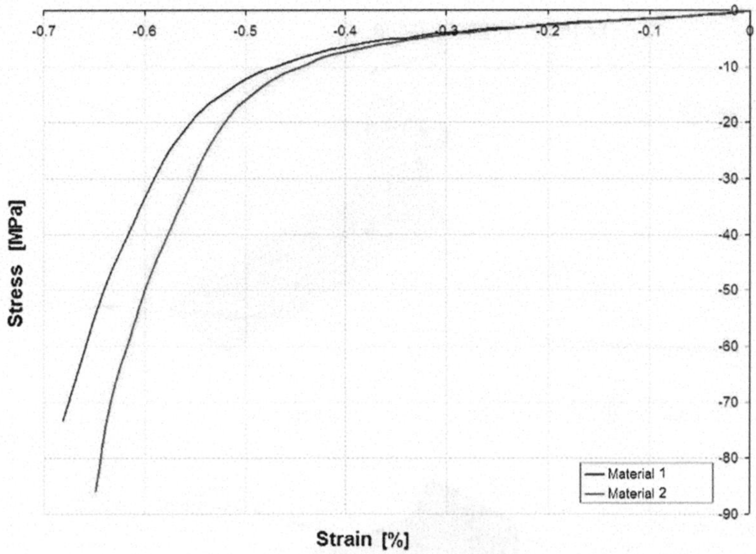

Fig. 92 Stress–strain curves obtained from experimental uniaxial compression tests on both materials

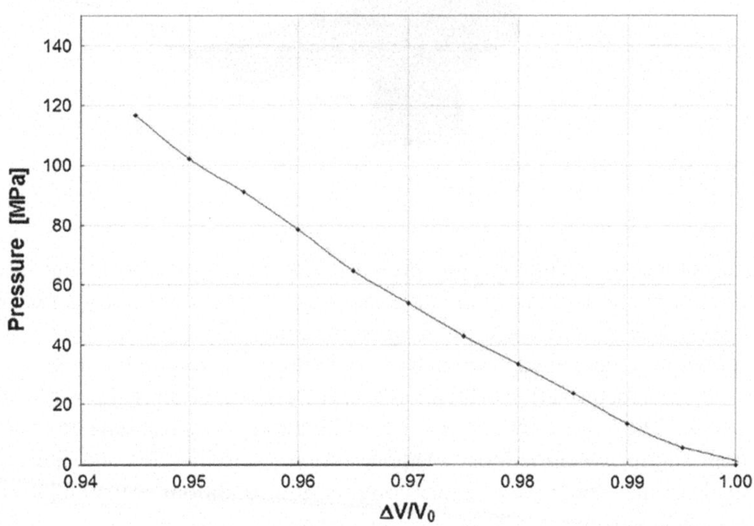

Fig. 93 Stress–strain curve obtained from experimental volumetric compression test

The simulation cases to be considered correspond to two different typologies depending on the complexity of the compression conditions: Simulation of type 1: compression test from an initial position in which the block is downloaded to a radial compression of 15.5 mm from its nominal mount position at which it is

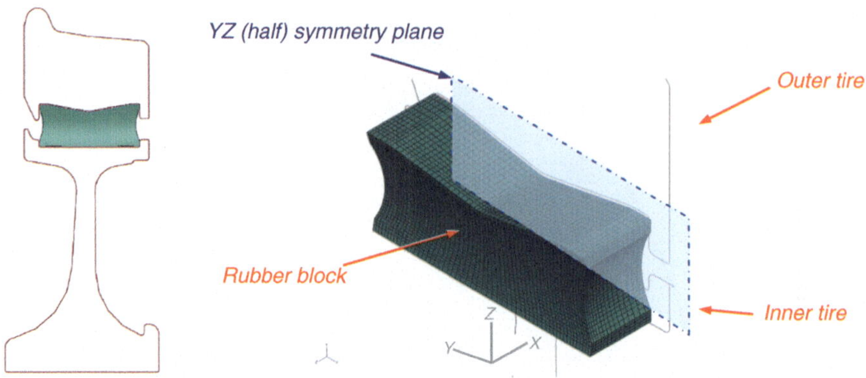

Fig. 94 FE model of the rubber block + inner and outer tires

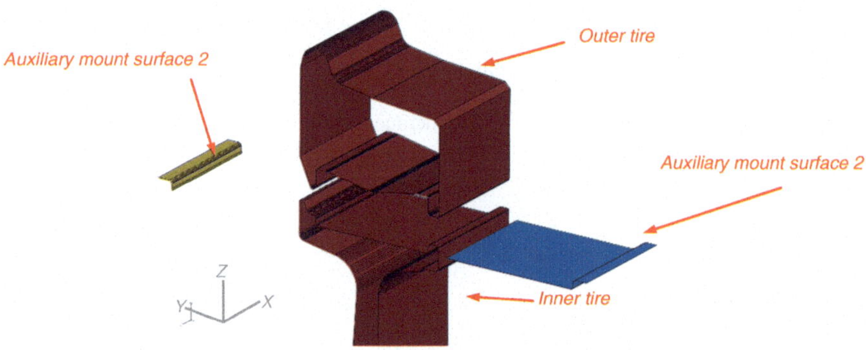

Fig. 95 Rigid bodies included in FE simulation

already subjected to a compression preload. Simulation of type 2: the block is subjected to a cyclic loading in positive and negative directions around an initial position (0.4, 1 or 2 mm) corresponding to the nominal one.

The previously described compression conditions correspond to two different tests of static characterization which have been carried out on real samples of the rubber block. The objective of these FE simulations is to validate the material model used and to select the most adequate one for applying this procedure to the design cycle of the rubber block under compression conditions. All simulations are of static type.

4.3.2 Results of the Static Finite Element Simulations Cases

The results shown in this section are presented in terms of force–displacement curves in radial direction (Z-axis on Fig. 91) and deformed configurations compared to real experiments performed on real samples of the rubber block.

Simulation of Type 1: Compression Test

Figure 96 shows the results obtained from the static FE simulation of the compression case type 1, compared to the experimental results from the rubber block sample of the first material (assumed incompressible in FE simulation).

In Fig. 97 the deformed configuration obtained for the nominal position of the rubber block in the FE simulation of the test is shown in the figure, compared with the deformed configuration obtained from the experimental one.

Comparing the deformed configuration obtained by FE simulation with the one observed experimentally, there are discrepancies in the amount of rubber that is extruded between outer and inner tires (see zones A and B in Fig. 97), being considerably higher quantity of extruded from the simulation to test material.

Figure 98 shows the results obtained from the static FE simulation assuming compressible rubber material, compared to the experimental results from the rubber block sample of the second material.

In Fig. 99, the deformed configuration obtained for the nominal position of the rubber block in the FE simulation of the test is shown, compared with the deformed configuration obtained from the experiment.

In the case of considering compressible material, it is observed that the deformed shape obtained through simulation presents a level of extrusion of material in areas A and B (see Fig. 99) very similar.

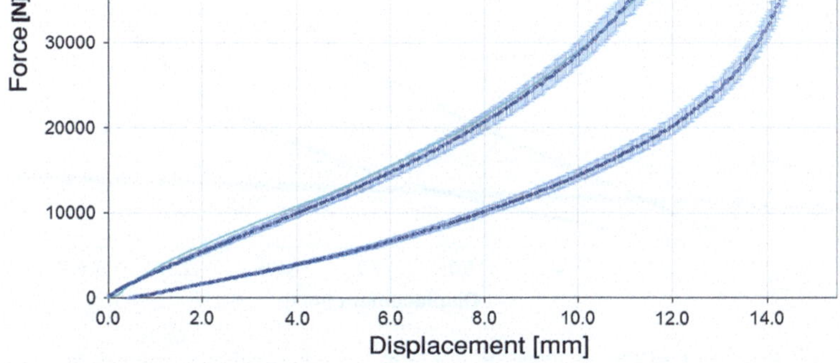

Fig. 96 FE simulation of the compression test type 1. Incompressibility assumption

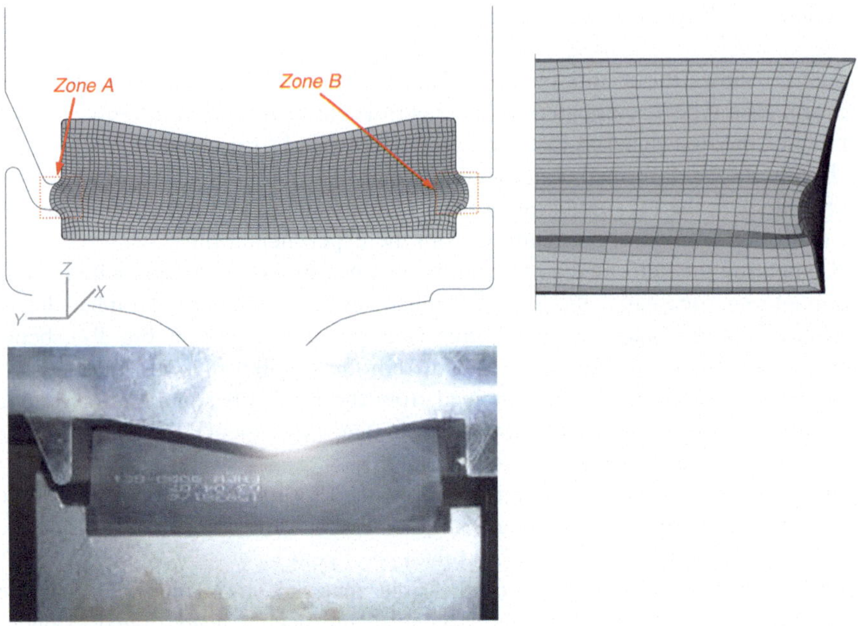

Fig. 97 Deformed configuration from the FE simulation of the compression test type 1. Incompressibility assumption

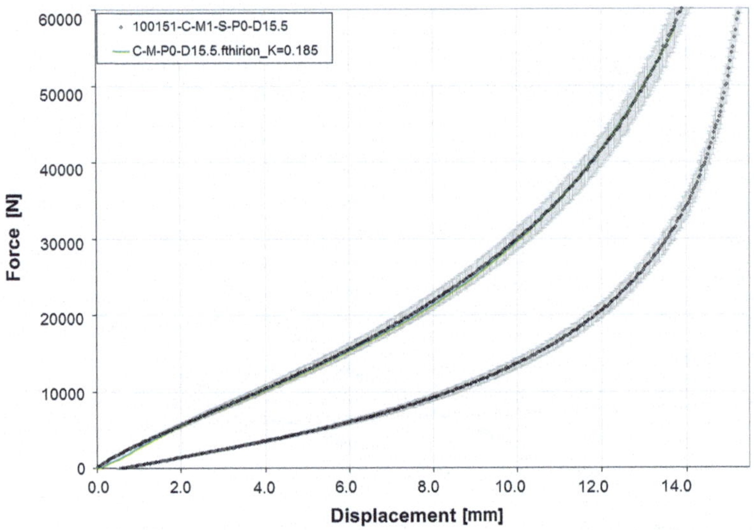

Fig. 98 FE simulation of the compression test type 1. Compressibility assumption

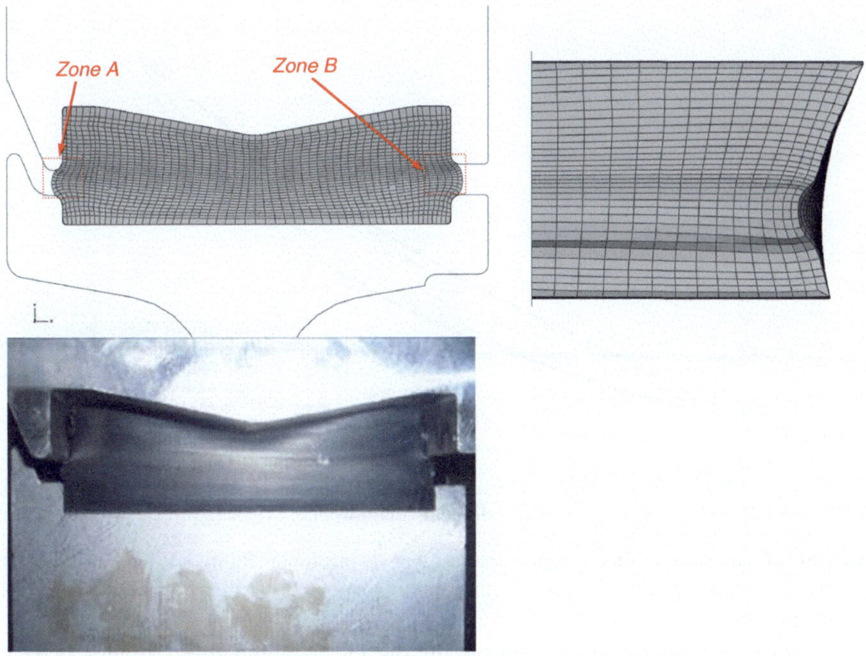

Fig. 99 Deformed configuration from the FE simulation of the compression test type 1. Compressibility assumption

Simulation of Type 2: Cyclic Compression Test

Figure 100 shows the results obtained from the static FE simulation of the case 2 of cyclic compression around the nominal position of the rubber block for amplitudes of 0.4, 1 and 2 mm, compared to the experimental results from block samples of the first material (assumed incompressible in FE simulation).

In Fig. 100 it can be seen a good correlation between numerical and experimental data for the three analyzed cyclical amplitudes.

It is necessary to clarify that, although numerical results presented in Fig. 100 show a hysteresis cycle similar to that observed experimentally, it is due only to the effect of the friction between contact surfaces present in the FE model (so as sliding). While in the experiment, apart from the contribution of friction and slip (if any), the contribution of the viscoelasticity and the hysteresis of the material are present.

Results in terms of rigidity, calculated as the maximum force least minimum divided by the total displacement of the simulations, can be analyzed by this procedure: The deviations in terms of stiffness from FE simulation to experimental results are less than 3 % in the case of the amplitudes of 1–2 mm. In the case of the amplitude of 0.4 mm, obtained deviation is 18 %. Experimental results show static rigidity component versus cyclic amplitude dependence, showing behaviour

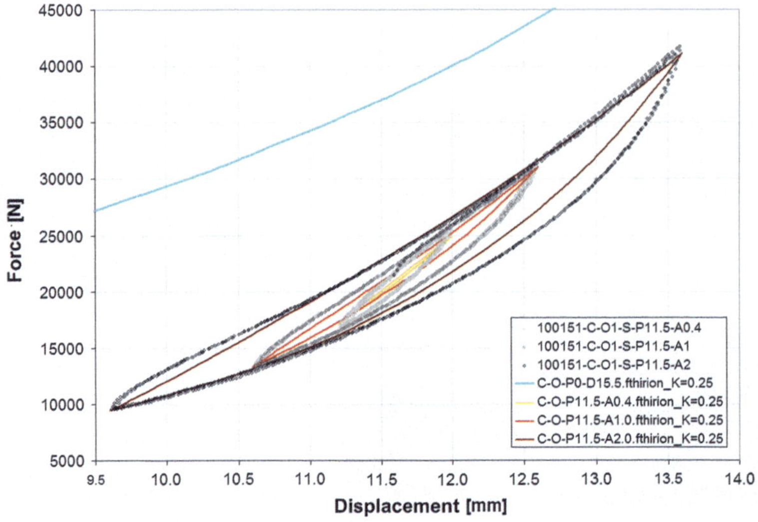

Fig. 100 FE simulation of the compression test type 2. Incompressibility assumption

similar to that seen in other materials in the case of dynamic stiffness (a less dynamic stiffness, increased amplitude).

Figure 101 shows the results obtained from the static FE simulation of the case 2 of cyclic compression around the nominal position of the rubber block for amplitudes of 0.4, 1 and 2 mm, compared to the experimental results from block samples of the second material (assumed compressible in FE simulation).

Also in this case, the results in terms of rigidity (calculated as the maximum force least minimum divided by the total displacement of the simulations) are analyzed: yielding to conclusions which can be similar to the commentary for the first assumed incompressible material. The FE simulation results show a dependency of static stiffness versus the cyclic amplitude imposed, increasing stiffness value as amplitude decreases. However, deviations between experimental and numerical results show are significantly higher for the assumed compressible rubber, being the minimum deviation around 18 %.

4.3.3 Conclusions

From the results presented in the current example, it can be concluded that the incompressibility assumption probes accurate enough for the simulation of the two radial compression cases. However, it should be remarked that in the simulation type 2 there are some inelastic effects (such as relaxation, viscoelasticity and hysteresis) that are not included in FE simulation and therefore treated just as fitting parameters, thus, the application of FE analysis should be restricted to static simulation and only for the specified boundary conditions.

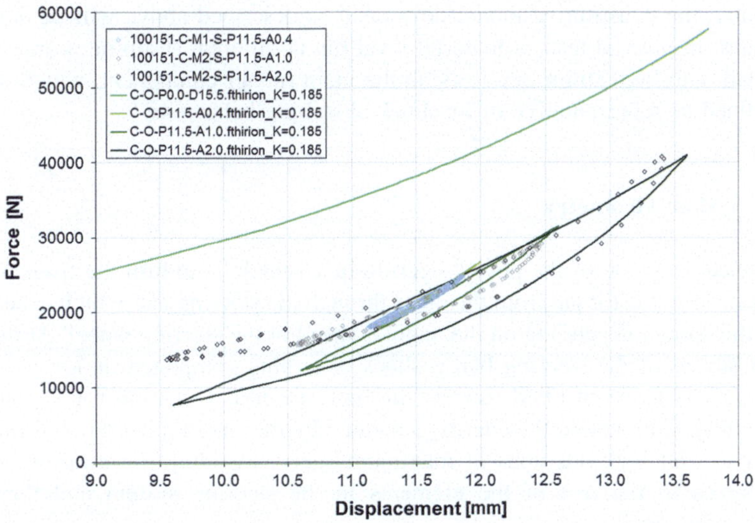

Fig. 101 FE simulation of the compression test type 2. Compressibility assumption

4.4 Example 4: Rubber Shock Absorber in Automotive Industry

In the present example, the FE analysis of a rubber automotive component working on shock absorbing conditions is described. The shock absorber is a filled rubber component which is used in the power steering system of the vehicle to absorb the impact generated by the steering bar when it travels to its end lock positions at maximum speed, avoiding the transmission of peak forces to the rest of the elements in the steering system and therefore protecting them. The rubber component, which is then assumed to work in high strain rate conditions due to the impact of the steering bar, will be modelled by means of the commercial FE code ABAQUS.

The basic elastic behaviour of the rubber material will be modelled by means of one of the hyperelasticity models which are available in the referred FE code. With regard to the dissipative effects caused by the high strain rate conditions of the impact, it will be assumed a rate dependent mechanical response which will be modelled by means of a linear viscoelastic behaviour model. This model is based on a generalized Maxwell model which is formulated by a Prony series. The parameters of the Prony series terms will be evaluated by fitting to mastercurves of storage and loss moduli obtained from dynamic tests run on dynamic mechanical thermal analysis equipment (D.M.T.A.). Since the complex modulus characterises the stiffness and the viscous behaviour of a rubber material, dynamic test results should, in principle, yield characteristic data which would describe the viscoelastic behaviour and therefore allow calculations to predict these strain rate effects.

Finally, the constitutive model developed as described above will be adjusted by means of uniaxial tests at material level run at different strain rates and finally validated with high strain rate tests on the main rubber component, which in fact are defined as representative of its shock absorbing behaviour.

4.4.1 System Geometry

The typical function of the power steering in a vehicle is to turn the front wheels by transferring the torque from the drive through the steering bar, which executes a linear movement depending on the angular speed of the steering wheel. At the end lock positions of the steering bar, a shock absorbing component is located. This component is made of filled rubber (the base material is natural rubber) and its arrangement is to absorb the energy generated by the steering bar when it impacts against its end lock positions at maximum speed, avoiding the transmission of peak forces to the rest of the elements in the steering system and therefore avoiding damage.

Figure 102 shows a sketch of the shock absorbing system where the rubber component is placed. This systems exhibits circumferential symmetry and is composed of a main housing made out of steel where the rubber damper is inserted between the housing walls. An external actuator clamped to the steering bar impacts directly on the rubber part at maximum linear speed.

The FE model developed in this study has been reduced to the rubber component and its nearest metallic housing, which is composed by actuator and housing (see Fig. 102). The system is simulated with the shock absorber as the deformable body and with the rest of the parts (actuator and housing) considered as rigid surfaces. The model of the rubber shock absorber is meshed with 2D-axisymmetric solid continuum hybrid elements with four nodes and reduced integration with hourglass control. The surfaces defined in the simulation are considered as rigid bodies and are modelled as axisymmetric rigid surfaces. The whole model geometry and FE discretization are shown are shown schematically in Fig. 103.

4.4.2 Development of Material Model

As commented previously, the developed numerical rubber model consists of a hyperelastic part to account for the non linear elastic stress–strain response of the rubber, and of a viscoelastic part to account for the strain rate dependent effects. The material model has been developed from experimental tests at material level: The mechanical behaviour has been modelled by means of uniaxial tests (compression), in order to obtain the hyperelastic properties, as well as DMTA tests, in order to analyse the dynamic behaviour and sensitivity to strain rate of the rubber material.

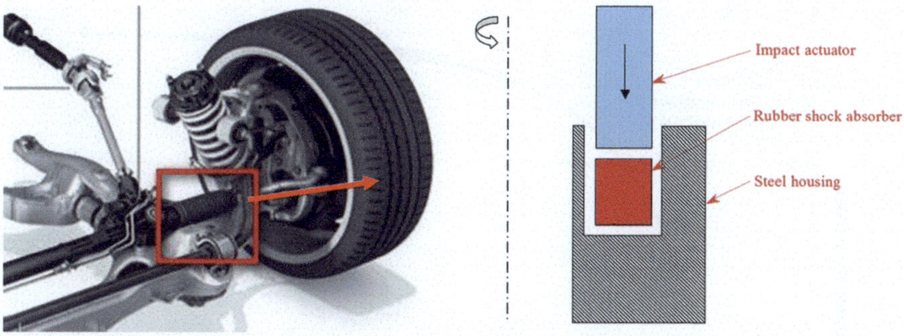

Fig. 102 Sketch of a section of the shock absorbing system of the steering system (circumferential symmetry)

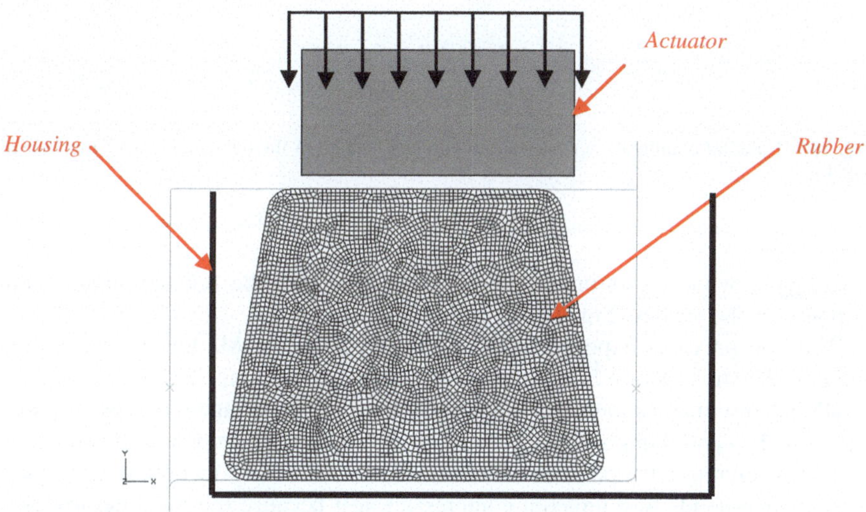

Fig. 103 Axisymmetric FE model of the rubber shock absorber (circumferential symmetry)

The hyperelastic part, which is assumed to be independent of the strain rate, is modelled by means of the Marlow formulation for the strain energy density function. This formulation has been chosen since, from one side, the available experimental tests comprise only uniaxial tests. From the other side, the main deformation mode of the component is compression; therefore it is preferred to have a prediction as accurate as possible until high strain levels in compression, although loosing quality in the prediction of the behaviour in other deformation modes.

The fitting of the parameters of the Marlow model is performed internally by the ABAQUS code, for which uniaxial test data are provided. The test plan for the quasi-static uniaxial compression characterization of the rubber material in universal test machine has been arranged for a selected test speed of 10 mm/min

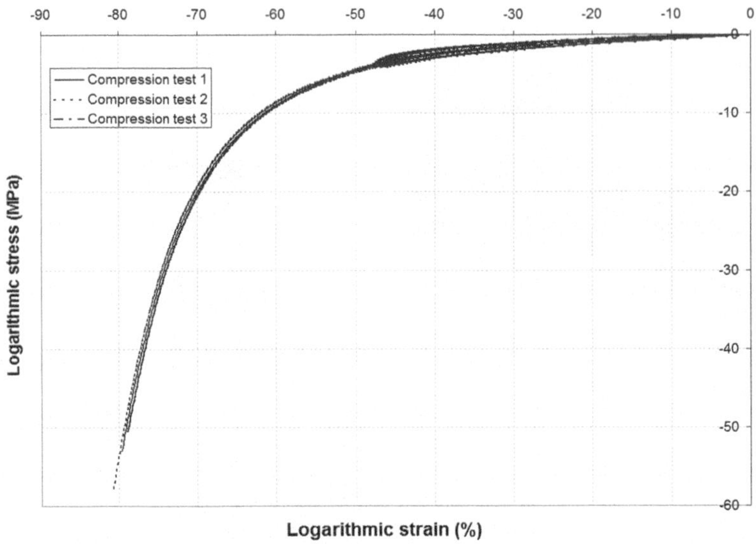

Fig. 104 Uniaxial compression stress–strain curves obtained in the universal testing machine at $0.013\ s^{-1}$

($0.013\ s^{-1}$), which is assumed slow enough to obtain the relaxed stress–strain response of the rubber (Fig. 104).

With the previous experimental data, the hyperelastic Marlow model is then defining the stress–strain behaviour of the rubber as the long term response.

DMTA test enables to obtain the master curve of the storage modulus (E'), loss modulus (E'') and damping factor ($\tan \delta$) over a range of frequencies from cyclic tests at a reference temperature (RT). The cyclic test used to obtain the master curves corresponds to a uniaxial compression test because the main deformation mode in the component is compression. In order to define the viscoelastic FE model parameters, a Prony series for each material has been adjusted. The Prony series terms, corresponding to a generalized Maxwell model of 15 terms, are obtained by fitting to the storage modulus master curve (E').

Figure 105 shows the master curve E' and the checked prediction of E'' with the Prony series.

Using previous hyperelastic and viscoelastic FE rubber material models, a one-element simulation (patch test) of simple compression test has been carried out in order to check the response of the FE material model. The FE simulation, which has been run considering that time has physical sense (dynamic, implicit integration procedure), has been executed at the same strain rates than the experimental uniaxial compression tests performed specifically for validation at 10 mm/min ($0.013\ s^{-1}$) and 1000 mm/min ($1.3\ s^{-1}$). Comparative curves are shown in Fig. 106.

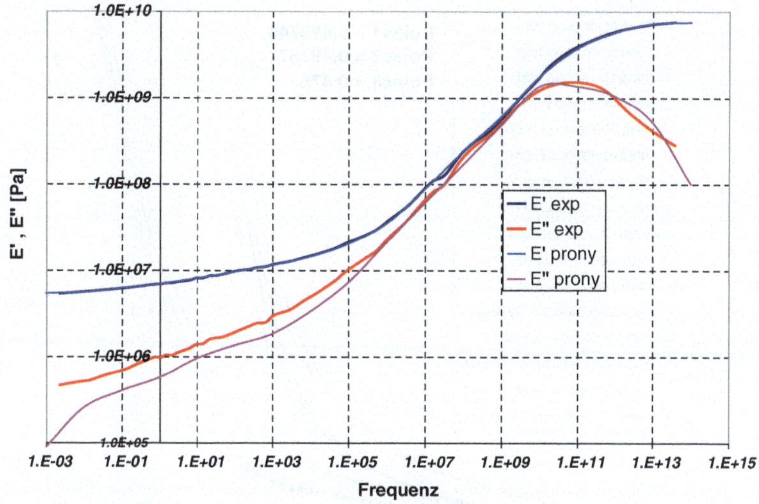

Fig. 105 Master curves

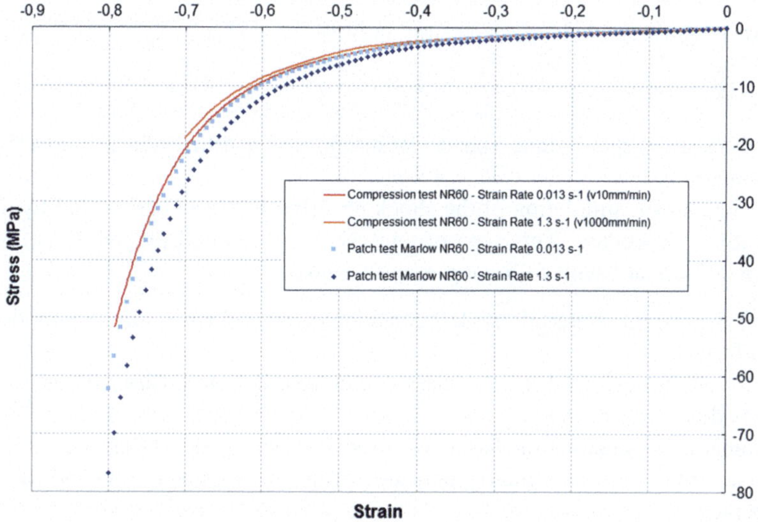

Fig. 106 Comparison one-element simulation with uniaxial compression tests

4.4.3 Results of the Dynamic Finite Element Simulation of the Impact

The experimental characterization results at component level that will be used to validate the whole FE simulation has been obtained in terms of axial force—axial displacement response. Experimental test consists of: three pre-cycles (10 mm/s),

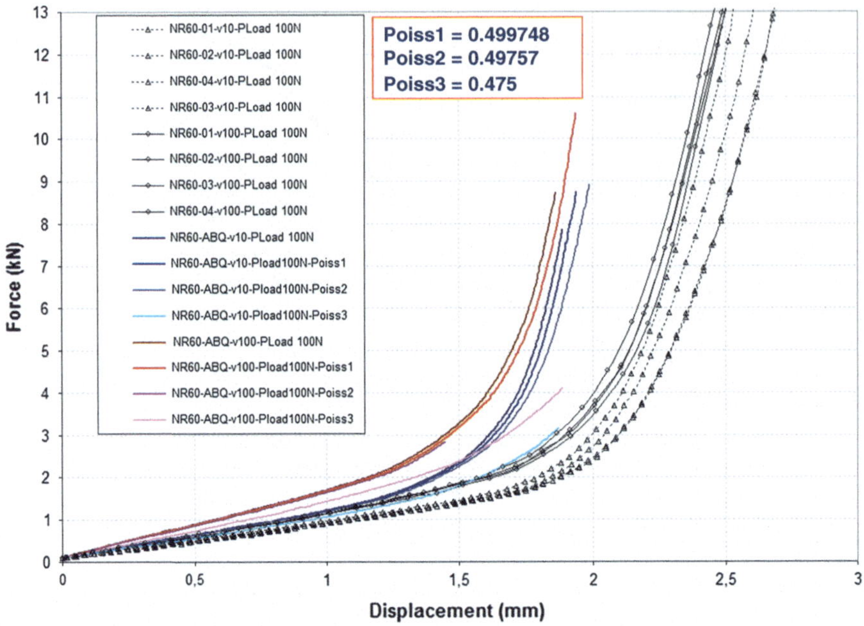

Fig. 107 Force versus displacement comparison curves on tests on component at impact rates

a fourth cycle at quasi-static speed (10 mm/s) and a fifth cycle at impact speed (100 mm/s).

The impact FE simulation of the shock absorber has been carried out by means of a VISCO procedure. This type of analysis is an implicit procedure defined in ABAQUS/Standard with following assumptions:

- Time-dependent material model considered, so velocities in the model get significance.
- Inertia effect neglected, so the mass of actuator is not considered and accelerations don't get meaning.
- Automatic incrementation has been used by means of a tolerance parameter which limits the maximum inelastic strain rate change allowed over an increment.

FE simulations of the shock absorber have been done according to defined characteristics of the FE model and of the corresponding FE material model. Results has been analysed in terms of force–displacement response and compared to experimental result in order to obtain a validated FE simulation procedure. In order to adjust the adequate compressibility level of rubber, three Poisson coefficients are considered: 0.499748, 0.49757 and 0.475 to feed the simulation. Comparative force–displacement responses for the component at two velocities and different compressibilities are shown in Fig. 107.

4.4.4 Conclusions

A tool for virtual prototyping to be used to optimise the shock absorbing behaviour of the analysed rubber component using FE method has been developed, including an advanced material model and a methodology of impact FE simulation using dynamic implicit integration algorithm.

The full strain-rate dependency on the rubber material under impact conditions has been successfully predicted in FE simulation by means of an hyperelastic model for the non linear stress–strain response and a viscoelastic model for the time-dependent response, both developed from tests at material model, and also from a compressibility model, implemented by means of the Poisson coefficient which has been treated as a fitting parameter to real component tests of the impact.

The fully incompressible assumption is not adequate for the rubber material due to the deviations from FE simulated stiffness response and the experimental component test at the same strain rates. The compressibility has been fitted from experimental tests at component level by means of adjusting the Poisson coefficient.

Acknowledgments The authors gratefully acknowledge the financial support from the Spanish Ministry of Science and Technology through Research Projects DPI2001-2406, DPI2004-06747, and DPI2008-02335 as well as the cortesy of the companies Industrias E. Díaz, S. A., Caucho Metal Productos, Construcciones y Auxiliar de Ferrocarril, S. A. and TRW Automotive for allowing to publish their industrial examples.

References

1. Hamed, G.R., Hatfield, S.: On the role of bound rubber in carbon black reinforcement. Rubber Chem. Technol. **62**, 143–156 (1989)
2. Meinecke, E.A., Taftaf, M.I.: Effect of carbon black on the mechanical properties of elastomers. Rubber Chem. Technol. **61**, 534–547 (1988)
3. So, H., Chen, U.D.: A nonlinear mechanical model for solid filled polymers. Polym. Eng. Sci. **6**, 410–416 (1991)
4. Zienkiewicz, O.C., Taylor, R.L.: The FE Method, vol. 1: Basic Formulation and Linear Problems. McGraw-Hill, London (1989)
5. Zienkiewicz, O.C., Taylor, R.L.: The FE Method, vol. 2: Solid and Fluid Mechanics, Dynamics and Non-Linearity. McGraw-Hill, London (1991)
6. Govindjee, S., Simo, J.: Transition from micro-mechanics to computationally efficient phenomenology: carbon black filled rubbers incorporating Mullins effect. J. Mech. Phys. Solids **40**, 213–233 (1992)
7. Govindjee, S., Simo, J.: Mullins effect and the strain amplitude dependence of the storage modulus. Int. J. Solids Struct. **29**, 1737–1751 (1992)
8. Lion, A.: A constitutive model for carbon black filled rubber: experimental investigations and mathematical representation. Continuum Mech. Thermodyn. **8**, 153–169 (1996)
9. Lion, A.: Thixotropic behaviour of rubber Ander dynamic loading histories: Experiments and theory. J. Mech. Phys. Solids **46**(5), 895–930 (1998)
10. Kim, S.J., Kim, K.S., Cho, J.Y.: Viscoelastic model of finitely deforming rubber and its FE anlysis. J. Appl. Mech. **64**, 835–841 (1997)

11. Bergström, J.S., Boyce, M.C.: Constitutive modelling of the large strain time-dependent behaviour of elastomers. J. Mech. Phys. Solids **46**, 931–954 (1998)
12. Bergström, J.S., Boyce, M.C.: Large strain time-dependent behavior of filled elastomers. Mech. Mater. **32**, 627–644 (2000)
13. Miehe, C., Keck, J.: Superimposed finite elastic-viscoelastic-plastoelastic stress response with damage in filled rubbery polymers. Experiments, modelling and algorithmic Implementation. J. Mech. Phys. Solids **48**, 323–365 (2000)
14. Mars, W.V., Fatemi, A.: Observations of the constitutive response and characterization of filled natural rubber under monotonic and cyclic multiaxial stress states. J. Eng. Mater. Technol. **126**(1), 19–28 (2004)
15. Paige, R.E., Mars, W.V.: Implications of the Mullins effect on the Stiffness of a Pre-loaded Rubber Component. In: Proceedings of the 17th Abaqus User's Conference, Boston, Massachusetts, USA (2004)
16. Rivlin, R.S., Saunders, D.W.: Large elastic deformations of isotropic materials. VII. Experiments on the deformation of rubber. Philos. Trans. R. Soc. Lond., Ser. A **243**, 251–288 (1951)
17. Treolar, L.R.G.: The Physics of Rubber Elasticity. Oxford University Press, Oxford (1975)
18. James, A.G., Green, A., Simpson, G.M.: Strain energy functions of rubber. I. Characterization of gum vulcanizates. J. Appl. Polym. Sci. **19**, 2033–2058 (1975)
19. Yeoh, O.H.: Characterisation of elastic properties of carbon-black filled rubber vulcanizates. Rubber Chem. Technol. **63**, 792–805 (1990)
20. Mooney, M.: A theory of large elastic deformation. J. Appl. Phys. **11**, 582–592 (1940)
21. James, H.M., Guth, E.: Theory of the elastic properties of rubber. J. Chem. Phys. **11**(10), 455–481 (1943)
22. Wall, F.T., Flory, P.J.: Statistical thermodynamics of rubber elasticity. J. Chem. Phys. **19**(12), 1435–1439 (1951)
23. Flory, P.J.: Theory of elasticity of polymer networks. The effect of local constraints on junctions. J. Chem. Phys. **66**(12), 5720–5729 (1977)
24. Ogden, R.W.: Large deformation isotropic elasticity I: on the correlation of theory and experiment for incompressible rubber-like materials. Proc. R. Soc. Lond. Ser. A **326**, 565–584 (1972)
25. Blatz, P.J., Ko, W.L.: Application of finite elastic theory to the deformation of rubbery materials. Trans. Soc. Rheol. **6**, 223–251 (1962)
26. Ogden, R.W.: Non-linear Elastic Deformations. Dover Publications, Ellis Harwood Ltd, New York (1984)
27. Yeoh, O.H.: Some forms of the strain energy function for rubber. Rubber Chem. Technol. **66**, 754–771 (1993)
28. Arruda, E., Boyce, M.C.: A three-dimensional constitutive model for the large stretch behaviour of rubber elastic materials. J. Mech. Phys. Solids **41**(2), 389–412 (1993)
29. Gent, A.N., Thomas, A.G.: Forms of the stored (strain) energy function for vulcanized rubber. J. Polym. Sci. **28**, 625–637 (1958)
30. Valanis, K.C., Landel, R.F.: The strain-energy function of a hyperelastic material in terms of the extension ratios. J. Appl. Phys. **38**, 2997–3002 (1967)
31. James, A.G., Green, A.: Strain energy functions of rubber. II. The characterization of filled vulcanizates. J. Appl. Polym. Sci. **19**, 2319–2330 (1975)
32. Ferry, J.D.: Viscoelastic Properties of Polymers. John Wiley & Sons, Inc., New York (1980)
33. Hausler, K., Sayir, M.B.: Nonlinear viscoelastic response of carbon black reinforced rubber derived from moderately large deformations in torsion. J. Mech. Phys. Solids **43**(2), 295–318 (1995)
34. Johnson, A.R., Quigley, C.J., Freese, C.E.: A viscohyperelastic FE model for rubber. Comput. Methods Appl. Mech. Eng. **127**, 163–180 (1995)
35. Simo, J.: On a fully thee-dimensional finite-strain viscoelastic damage model: formulation and computational aspects. Comput. Methods Appl. Mech. Eng. **60**, 153–173 (1987)

36. Kaliske, M.: Zur Theorie und Numerik von Polymerstrukturen unter statischen un dynamischen Einwirkungen. Mitteilung Nr. 41-95 des Instituts für Statik der Universität Hannover (1995)
37. Holzapfel, G.A., Stadler, M., Ogden, R.W.: Aspects of stress softening in filled rubbers incorporating residual strains. In: Dorfman, A., Muhr, A. (eds.) Constitutive Models for Rubber I, pp. 189–193. ©1999 Balkema, Rotterdam, ISBN 90 5809 113 9 (1999)
38. Besdo, D., Ihlemann, J.: Zur Modellierung des Stoffverhaltens von Elastomeren. Kautschuck un Gummi Kunstoffe **49**, 495–503 (1996)
39. Dannenberg, E.M.: The effect of surface chemical interactions on the properties of filler reinforced rubbers. Rubber Chem. Technol. **44**, 440–478 (1975)
40. Vidal, A., Donnet, J.B.: Carbon black: surface properties and interactions with elastomers. Adv. Polym. Sci. **76**, 104–106 (1996)
41. Mullins, L.: Effect of stretching on the properties of rubber. Rubber Chem. Technol. **21**, 281–300 (1948)
42. Mullins, L., Tobin, N.R.: Stress softening in rubber vulcanizates. Part I. J. Appl. Polym. Sci. **9**, 2993 (1965)
43. Harwood, J.A.C., Mullins, L., Payne, A.R.: Stress softening in natural rubber vulcanizantes. Part II. Stress softening in pure gum and filler loaded rubbers. J. Appl. Polym. Sci. **9**, 3011–3021 (1965)
44. Harwood, J.A.C., Payne, A.R.: Stress softening in natural rubber vulcanizates. Part II. Carbon black-filled vulcanizates. J. Appl. Polym. Sci. **10**, 315–324 (1966)
45. Harwood, J.A.C., Payne, A.R.: Stress softening in natural rubber vulcanizates. Part IV. Unfilled vulcanizates. J. Appl. Polym. Sci. **10**, 1203–1211 (1966)
46. Bueche, F.: Molecular basis for the Mullins effect. J. Appl. Polym. Sci. **4**(10), 107–114 (1960)
47. Bueche, F.: Mullins effect and rubber-filler interaction. J. Appl. Polym. Sci. **5**(15), 271–281 (1961)
48. Marigo, J.J.: Modelling of brittle and fatigue damage for elastic material by growth of microvoids. Eng. Fract. Mech. **21**(4), 861–874 (1985)
49. Kachanov, L.M.: Introduction to Continuum Damage Mechanics. Martinus Nijhoff Publishers, Dordrecht (1986)
50. Lemaitre, J.: A Course on Damage Mechanics. Springer, Berlin (1992)
51. Ogden, R.W., Roxburgh, D.G.: A pseudo-elastic model for the Mullins effect in filled rubber. Proc. R. Soc. Lond. Ser. A **455**, 2861–2877 (1999)
52. Ogden, R.W., Roxburgh, D.G.: An energy based model of the Mullins effect. In: Dorfmann, A., Muhr, A. (eds.) Constitutive Models for Rubber I, pp. 23–28. ©1999 Balkema, Rotterdam, ISBN 90 5809 113 9
53. Austrell, P.E.: Modelling of elasticity and damping for filled elastomers. Ph.D. Dissertation, Report TVSN-1009, Lund University, Division of Structural Mechanics, Sweden (1997)
54. Austrell, P.E., Olsson, A.K., Jönsson, M.: A method to analyse the non-linear dynamic behaviour of carbon-black-filled rubber components using standard FE-codes. In: Besdo, D., Shuster, R.H., Ihlemann, J. (eds.) Constitutive Models for Rubber II, pp. 231–235. ©2001 Swets & Zeitlinger, ISBN 90 2651 847 1
55. Besseling, J.F.: A theory of elastic, plastic and creep deformation of an initially isotropic material. J. Appl. Mech. **25**, 529–536 (1998)
56. Fletcher, W.P., Gent, A.N.: Non-linearity in the dynamic properties of vulcanised rubber compounds. I.R.I. Trans. **29**, 266–280 (1953)
57. Payne, A.R.: The dynamic properties of carbon black-loaded natural rubber vulcanizates. Part I. J. Appl. Polym. Sci. **VI**, 57–63 (1962)
58. Olsson, A.K.: Austrell, P.E.: A fitting procedure for a viscoelastic-elastoplastic material model. In: Besdo, D., Shuster, R.H., Ihlemann, J. (eds.) Constitutive Models for Rubber II, pp. 261–266. ©2001 Swets & Zeitlinger, ISBN 90 2651 847 1
59. Ahmadi, H.R, Kingston, J.G.R., Muhr, A.H., Gracia, L.A., Gómez, B.: Interpretation of the high low-strain modulus of filled rubbers as an inelastic effect. In: Busfield, J., Muhr, A.

(eds.) Constitutive Models for Rubber III, pp. 357–364. ©2003 Swets & Zeitlinger, Lisse, ISBN 90 5809 566 5
60. Morman, K.N., Nagtegaal, J.C.: FE analysis of sinusoidal small-amplitude vibrations in deformed viscoelastic solids. Part I. Theoretical development. Int. J. Num. Methods Eng. **19**, 1079–1103 (1983)
61. MSC MARC User's manual (2005)
62. Miller, K.: Testing and Analysis. Measuring the Dynamic Properties of Elastomers for Analysis. Dynamic Review. Axel Products, Ann Arbor (2000)
63. Gómez, J., Royo, J.M.: Prediction of dynamic stiffness of filled rubber mounts. III European Conference on Computational Mechanics, Lisbon, Portugal, 5–8 June 2006
64. Ellul, M.D.: Mechanical fatigue. Engineering with Rubber: How to Design Rubber Components (Chap. 6), pp. 130–167. Hanser, New York (1992)
65. Lake, G.J.: Mechanical fatigue of rubber. Rubber Chem. Technol. **45**, 309 (1972)
66. Lake, G.J.: Fatigue and fracture of elastomers. Rubber Chem. Technol. **68**, 435–460 (1995)
67. Cadwell, S.M., Merrill, R.A., Sloman, C.M., Yost, F.L.: Dynamic fatigue life of rubber. Ind. Eng. Chem. (Anal. Ed.) **12**, 19–23 (1940) (reprinted in Rubber Chem. Technol. **13**, 304–315)
68. Roach, J.F.: Crack growth in elastomers under biaxial stress. Ph.D. dissertation, University of Akron (1982)
69. Lake, G.J.: Aspects of fatigue and fracture of rubber. Prog. Rubber Technol. **45**, 89 (1983)
70. Lake, G.J., Yeoh, O.H.: Effect of crack tip sharpness on the strength of vulcanized rubbers. J. Polym. Sci.: Polym. Phys. Ed. **25**, 1157 (1987)
71. Lake,G.J., Lindley, P.B.: The mechanical fatigue limit for rubber. J. Appl. Polym. Sci. **9**, 1233–1251 (1965) (reprinted in Rubber Chem. Technol. **39**, 348–364 (1966))
72. Gent, A.N., Lindley, P.B., Thomas, A.G.: Cut growth and fatigue of rubbers. I. The relationship between cut growth and fatigue. J. Appl. Polym. Sci. **8**, 455–466 (1964) (reprinted in Rubber Chem. Technol. **38**, 292–300 (1965))
73. Mars, W.V.: Multiaxial fatigue on rubber. Ph.D. thesis, University of Toledo (2001)
74. Flamm, M., Steinweger, T., Weltin, U.: Schadeakkumulation bei Elastomeren. Kautschuk Gummi Kunststoffe **55**(12), 665–668 (2002)
75. Mars, W.V., Fatemi, A.: Multiaxial fatigue of rubber—part I: Equivalence criteria and theoretical aspects. Fatigue Fract. Eng. Mater. Struct. **28**, 515–522 (2005)
76. Mars, W.V., Fatemi, A.: Multiaxial fatigue of rubber: Part II: Experimental observations and life predictions. Fatigue Fract. Eng. Mater. Struct. **28**, 523–538 (2005)
77. Mars, W.V., Kingston, J.G.R., Muhr, A.: Fatigue analysis of an exhaust mount. In: Austrell, P.E., Kari, L. (eds.) Constitutive Models for Rubber IV, pp. 23–29. ©2005 Taylor & Francis Group, London, ISBN 0 415 38346 3
78. Thomas, A.G.: Rupture of rubber. VI. Further experiments on the tear criterion. J. Polym. Sci. **31**, 467 (1958)
79. Lindley, P.B.: Relation between hysteresis and the dynamic crack growth resistance of natural rubber. Int. J. Fract. **9–4**, 449–462 (1973)
80. Lindley, P.B.: Non-relaxing crack growth and fatigue in a non-crystallizing rubber. Rubber Chem. Technol. **47**, 1253–1264 (1974)
81. Rabinowicz, E.: Friction and Wear of Materials, 2nd edn. Wiley-Interscience, New York (1995)
82. Stachowiak, G.W., Batchelor A.W.: Engineering Tribology, 3rd edn. Elsevier (2011)
83. Zhang, S.W.: State-of-the-art of polymer tribology. Tribol. Int. **31**, 49–60 (1998)
84. Persson, B.N.J.: Sliding Friction: Physical Principles and Applications, 2nd edn. Springer, Heidelberg (2000)
85. Thirion, P.: Les coefficients d'adhérence du caoutchouc. Rubber Chem. Technol. **21**, 505–515 (1948)
86. Persson, B.N.J.: Theory of rubber friction and contact mechanics. J. Chem. Phys. **115–118**, 3840–3861 (2001)

87. Mofidi, M., Prakash, B., Persson, B.N.J.: Albohr O. Rubber friction on (apparently) smooth lubricated surfaces. J. Phys.: Condens. Matter **20**(8), 085223 (2008)
88. Kragelskii, I.V.: Friction and Wear, p. 458. Pergamon Press, Elmsford (1982)
89. Blau, P.J.: Friction and wear transitions of materials. Noyes Publication, New York (1989)
90. Zhang, S.W.: Tribology of elastomers. In: Briscoe, B.J. (ed.) Tribology and Interface Engineering, Series no. 47, pp. 37–177. Elsevier, Amsterdam (2004)
91. Myshkin, N.K., Petrokovets, M.I., Kovalev, A.V.: Tribology of polymers: adhesion, friction, wear and mass-transfer. Tribol. Int. **38**, 910–921 (2005)
92. Je, J.H., Gyarmati, E., Naoumidis, A.: Scratch adhesion test of reactively sputtered TiN coatings on a soft substrate. Thin Solid Films **136**, 57–67 (1986)
93. Viswanath, V., Bellow, D.G.: Development of an equation for the wear of polymers. Wear **181–183**(1), 42–49 (1995)
94. Nah, C.: Ph.D. dissertation, Wear mechanisms of rubber compounds, The University of Akron (1995)
95. Meng, H.C., Ludema, K.C.: Wear models and predictive equations: their form and content. Wear **181–183**, 443–457 (1995)
96. Muhr, A.H.: Dynamic properties of rubber. In: Proceedings, ACEM, NR in Engineering Workshop. Kulalumpur, Sept. 1991
97. Hibbitt, Karlsson and Sorensen. Abaqus Standard Theory and User's Manual v6.5 (2005)
98. Freakly, P.K., Payne, A.R.: Theory and Practise of Engineering with Rubber. Applied Science Publishers, London (1978)
99. Layouni, K., Laiaridrasana, L., Piques, R.: Compressibility induced by damage in carbon black reinforced natural rubber. In: Busfield, A., Muhr, A. (eds.) Constitutive Models for Rubber III, pp. 273–281. ©2003 Swets & Zeitlinger, Lisse, ISBN 90 5809 566 5
100. Muhr, A.H.: Properties of rubber compounds for engineering applications. J. Nat. Rubber Res. **7**(1), 14–37 (1992)
101. Ahmadi, H.R., Muhr, A.H.: Modelling dynamic properties of filled rubber. Plast. Rubber Compos. Process. Appl. **26**, 451–461 (1997)
102. Harris, J.A.: Dynamic testing under non-sinusoidal conditions and the consequences of nonlinearities for service performance. In: Proceedings of the Rubber Division Meeting. American Chemical Society, Montreal, Quebec, Canada, 26–29 May 1987
103. Harris, J., Stevenson, A.: On the role of non-linearity in the dynamic behaviour of rubber components. Rubber Chem. Technol. **59**, 741–764 (1986)
104. Mullins, L.: Softening of rubber by deformation. Rubber Chem. Technol. **42**, 339–362 (1969)
105. Mullins, L., Tobin, N.R.: Theoretical model for the elastic behaviour of filler-reinforced vulcanized rubber. Rubber Chem. Technol. **30**, 555–571 (1957)
106. Spencer, A.J.M.: Constitutive theory for strongly anisotropic solids. In: Continuum Theory of the Mechanics of Fiber-Reinforced Composites, pp. 1–32. Springer, Wien (1984)
107. Holzapfel, G.A., Eberlein, R., Wriggers, P., Weizsäcker, H.W.: A new axisymmetrical membrane element for anisotropic, finite strain analysis of arteries. Commun. Numer. Methods Eng. **12**, 507–517 (1996)
108. Weiss, J.A., Maker, B.N., Govindjee, S.: FE implementation of incompressible, transversely isotropic hyperelasticity. Comput. Methods Appl. Mech. Eng. **135**, 107–128 (1996)
109. Alastrué, V., Calvo, B., Peña, E., Doblaré, M.: Biomechanical modelling of refractive corneal surgery. J. Biomech. Eng. T ASME **128**(1), 150–160 (2006)
110. Roland, C.M.: Dynamic mechanical behaviour of filled rubber at small strains. J. Rheol. **34**, 25 (1990)
111. Medalia, A.I.: Effects of carbon-black on dynamic properties of rubber. Rubber Chem. Technol. **51**, 437 (1978)
112. Brown, M.W., Miller, K.J.: A theory for fatigue under multi-axial stress-strain condition. Proc. Inst. Mech. Eng. **187**, 745–755 (1973)
113. Fatemi, A., Socie, D.F.: A critical plane approach to multiaxial fatigue damage including out-of-plane loading. Fatigue Fract. Eng. Mater. Struct. **14**, 149–166 (1988)

114. Smith, R.N., Watson, P., Topper, T.H.: A stress-strain parameter for the fatigue of metals. J. Mater. **5**, 767–778 (1970)
115. Wang, C.H., Brown, M.W.: A path-independent parameter for fatigue under proportional and non-proportional loading. Fatigue Fract. Eng. Mater. Struct. **16**, 1285–1298 (1993)
116. Wang, C.H., Brown, M.W.: Life prediction techniques for variable amplitude multiaxial fatigue, Part 1: Theories. J. Eng. Mater. Technol. **118**, 367–370 (1996)
117. Wang, C.H., Brown, M.W.: Life prediction techniques for variable amplitude multiaxial fatigue, Part 2, comparison with experimental results. J. Eng. Mater. Technol. **118**, 371–374 (1996)
118. Chen, X., Xu, S.-Y., Huang, D.-X.: Critical plane-strain energy density criterion of multiaxial low-cycle fatigue life. Fatigue Fract. Eng. Mater. Struct. **22**, 679–686 (1999)
119. Saintier, N., Cailletaud, G., Piques, R.: Crack initiation and propagation under multiaxial fatigue. Int. J. Fatigue **28**, 61–72 (2006)
120. Saintier, N., Cailletaud, G., Piques, R.: Multiaxial fatigue life prediction for a natural rubber. Int. J. Fatigue **28**, 530–539 (2006)
121. Findley, W.M., Mathur, P.N., Szczepanski, E., et al.: Energy versus stress theories for combined stress—a fatigue experiment using a rotating disk. ASME Trans. J. Basic Eng. **83**, 10–14 (1961)
122. Rivlin, R.S., Thomas, A.G.: Rupture of rubber. I. Characteristic energy for tearing. J. Polym. Sci. **10**, 291–318 (1953)
123. Ro, H.S.: Modeling and interpretation of fatigue failure initiation in rubber related to pneumatic tires. Ph.D. dissertation, Purdue University, USA (1989)
124. Yamashita, S.: Selecting damping materials (service environment, strain and endurance). Int. Polym. Sci. Technol. **19**(4), T/41–T/56 (1992)
125. Andre, N., Cailletaud, G., Piques, R.: Haigh diagram for fatigue crack initiation prediction of natural rubber components. Kautschuk Und Gummi Kunstoffe **52**, 120–123 (1999)
126. Lindley, P.B., Stevenson, A.: Fatigue resístanse of natural rubber in compression. Rubber Chem. Technol. **55**, 337–351 (1982)
127. Gent, A.N., Wang, C.: Strain energy release rate for crack growth in an elastic cylinder subjected to axial shear. Rubber Chem. Technol. **66**, 712 (1993)
128. Fielding-Russell, G.S., Rongone, R.L.: Fatiguing of rubber–rubber interfaces. Rubber Chem. Technol. **56**, 838–844 (1983)
129. Verron, E., Le Cam, J.B., Gournet, L.: A multiaxial criterion for crack nucleation in rubber. Mech. Re. Commun. **33**, 493–498 (2006)
130. Verron E., Andriyana, A.: Definition of a new predictor for multiaxial fatigue crack nucleation in rubber. J. Mech. Phys. Solids **56**(2), 417–443 (2008)
131. Andriyana, A., Verron, E.: Prediction of fatigue life improvement in natural rubber using configurational stress. Int. J. Solids Struct. **44**, 2079–2092 (2007)
132. Wang, B., Lu, H., Kim, G.: A damage model for the fatigue life of elastomeric materials. Mech. Mater. **34**, 475–483 (2002)
133. Schallamach, A.: Friction and abrasion of rubber. Wear **1**, 384–417 (1958)
134. Rymuza, Z.: Wear in polymer micro-pairs. Proceedings of the 3rd International Conference on Wear of Materials, pp. 125–132 (1981)
135. Buckley, D.H.: Surface effects in adhesion, friction, wear and lubrication. Elsevier, Amsterdam (1981)
136. Makinson, K.R., Tabor, D.: The friction and transfer of polytetrafluoroethylene. Proc. R. Soc. Lond. Ser. A **281**, 49–61 (1964)
137. Tanaka, K., Uchiyama, Y., Toyooka, S.: The mechanism of wear of PTFE. Wear **23**, 153–172 (1973)
138. Thorpe, J.M.: Tribological properties of selected polymer matrix composites against steel surfaces. In: Friedrich, K. (ed.) Friction and Wear of Polymer Composites, Vol. 1, Composite Materials Science, pp. 137–174. Elsevier, Amsterdam (1986)
139. Jain, V.K., Bahadur, S.: Material transfer in polymer–polymer sliding. Wear **46**, 177–198 (1978)

140. Birkett, A., Lancaster, J.K.: Counterface effects on the wear of a composite dry-bearing liner. In: Proceedings of the JSLE International Tribology Conference, Tokyo, pp. 465–470. Elservier, Amsterdam (1985)
141. Dowson, D., Challen, J.M., Holmes, K., Atkinson, J.R.: The influence of counterface roughness on the wear rate of polyethylene. In: Proceedings of the 3rd Leeds-Lyon Symposium on Tribology, Wear of Non-Metallic Materials, Sept. 1976, pp. 99–102. University of Leeds, London (1978)
142. Barrett, T.S., Stachowiak, G.W., Batchelor, A.W.: Effect of roughness and sliding speed on the wear and friction of ultra-high molecular weight polyethylene. Wear **153**, 331–350 (1992)
143. Play, D.F.: Counterface roughness effect on the dry steady state wear of self-lubricating polyimide composites. Trans. ASME, J. Lubr. Technol. **106**, 177–184 (1984)
144. Blanchett, T.A., Kennedy, F.E.: The development of transfer films in ultra-high molecular weight polyethiylene/stainless steel oscillatory sliding. Tribol. Trans. **32**, 371–379 (1982)
145. Tanaka, K., Uchiyama, Y.: Friction, wear and surface melting of crystalline polymers. In: Lee, L.H. (ed.) Advances in Polymer Friction and Wear, Vol. 5B, pp. 499–531. Plenum Press, New York (1974)
146. Kar, M.K., Bahadur, S.: Micromechanism of wear at polymer-metal sliding interface. Wear **46**, 189–202 (1978)
147. Ettles, Mc C., C.M., : Polymer and elastomer friction in the thermal control regime. ASLE Trans. **30**, 149–159 (1987)
148. Mizutani, Y., Kato, K., Shimura, Y.: Friction and wear of phenolic resin up to 200 °C. In: Proceedings of the JSLE International Tribology Conference, Tokyo, pp. 489–494. Elsevier (1985)
149. Watanabe, M., Yamaguchi, H.: The friction and wear properties of nylon. Proceedings of the JSLE International Tribology Conference, Tokyo, pp. 483-488. Elsevier (1985)
150. Southern, E., Thomas, A.G.: Studies of rubber abrasion. Plast. Rubber Mater. Appl. **3**, 133–138 (1978)
151. Cohen, S.C., Tabor, D.: The friction and lubrication of polymers. Proc. Roy. Soc. Lond. Series A **291**, 186–207 (1966)
152. Evans, D.C. (1978) Polymer-fluid interactions in relation to wear. In: Proceedings of the 3rd Leeds-Lyon Symposium on Tribology, Wear of Non-Metallic Materials, Sept. 1976. University of Leeds, London, pp. 47–71 (1978)
153. Batchelor, A.W., Tan, B.P.: Effect of an oxidizing agent on the friction and wear of nylon 6 against a steel counterface, vol. I, pp. 175–180. In: Proceedings of the 4th International Tribology Conference, AUSTRIB'94. Uniprint UWA (1994)
154. Scott, N.W., Stachowiak, G.W.: Long-term behaviour of UHMWPE in hydrogen peroxide solutions, vol. I, pp. 169–174. In: Proceedings of the 4th International Tribology Conference, AUSTRIB́94. Uniprint UWA (1994)
155. Bartenevev, G.M., Lavrentev, V.V.: Friction and wear of polymers. Tribology, Tribology Series, no. 6, p. 10-260. Elsevier Scientific Publishing Company, Amsterdam (1981)
156. Burris, D.L., Sawyer, W.G.: A low friction and ultra low wear rate PEEK/PTFE composite. Wear **261**, 410–418 (2006)
157. Zhang, S.W., Deguo, W., Yin, W.: Investigation of abrasive erosion of polymers. J. Mater. Sci. **30**, 4561–4566 (1995)
158. Zhang, S.W., Deguo, W., He, R., Fan, Q.: Abrasive erosion of polyurethane. J. Mater. Sci. **36**, 5037–5043 (2001)
159. Arnold, J.C., Hutchings, I.M.: The mechanisms of erosion of unfilled elastomers by solid particle impact. Wear **138**, 33–46 (1990)
160. Bely, V.A., Sviridenok, A.I., Petrokovets, M.I., Savkin, V.G.: Friction and wear in polymer-based materials, p. 416. Pergamon Press, Oxford (1982)
161. Andrew, W.: Fatigue and Tribological Properties of Plastics and Elastomers, vol. VI. Plastics Design Library, New York (1995)

162. Swain, M.V.: Microscopic observation of abrasive wear of polycrystalline alumina. Wear **35**, 185–189 (1975)
163. Bhowmick, A.K., Basu, S., De, S.K.: Scanning electron microscopy studies of abraded rubber surfaces. J. Mater. Sci. **16**(6), 1654 (1981)
164. Kayaba, T.: The latest investigations of wear by the microscopic observations. JSLE Trans. **29**, 9–14 (1984)
165. Kuriakose, B., De, S.K.: Scanning electron microscopy studies on tensile, tear and abrasion of thermoplastic elastomers. J. Mater. Sci. **20–5**, 1864–1872 (1985)
166. Thomas, S.: Scanning electron microscopy studies on wear properties of blends of plasticized poly(vinyl chloride) and thermoplastic copolyester elastomer. Wear **116**, 201–209 (1987)
167. Singer, I.L., Wahl, K.J.: Role of third bodies in friction and wear. J. Vac. Sci. Tech. A. Vac. Surf. Films **21**(5), 232–240 (2003)
168. Zum Gahr K.H.: Wear by hard particles. Tribol. Int. **31**(10), 587–596 (1998)
169. Harsha, A.P., Tewari, U.S.: Two-body and three-body abrasive wear behaviour of polyaryletherketone composites. Polym. Tests **22**, 403 (2003)
170. Johnson, R.W.: The use of the scanning electron microscope to study the deterioration of abrasive papers. Wear **12**, 213–216 (1968)
171. Misra, A., Finnie, I.: A classification of three-body abrasive wear and design of a new tester. ASTM International Conference On Wear of Materials, 1979
172. Jain, V.K., Bahadur, S.: Surface topography changes in polymer-metal sliding. In: Proceedings of International Conference on Wear of Materials, Dearborn, p. 581 (1979)
173. Neale, M.J., Gee, M.: Guide to Wear Problems and Testing for Industry. William Andrew Publishing, New York (2001)
174. Jia, X., Ling, R.: Two-body free-abrasive wear of polyethylene, nylon 1010, epoxy and polyurethane coatings. Tribol. Int. **40**(8), 1276–1283 (2007)
175. Liu J.J., Zhou P.A., Sun, X.T., Liao, Q.C.: Adhesive Wear and Fatigue Wear of Materials, pp. 234–323. Machinery Industry Press, Beijing (1989)
176. Marchenko, E.A.: Essentials of Friction Breakage for Metals Surface, pp. 1–8. National Defence Industry Press, Beijing (1990)
177. Johnson, K.L.: Contact mechanics and the wear of metals. Wear **190**, 162–170 (1995)
178. Suh, N.P., Mosleh, M., Arinez, J.: Tribology of polyethylene homocomposites. Wear **214**, 231–236 (1998)
179. Da Silva, R.C.L., Da Silva, C.H., Medeiros, J.T.N.: Is there delamination wear in polyurethane? Wear **263**(7–12), 974–983 (2007)
180. Holzapfel, G.A.: Nonlinear Solid Mechanics. A Continuum Approach for Engineering. John Wiley & Sons, Chichester (2000)
181. Marlow, R.S.: A general first-invariant hyperelastic constitutive model. In: Busfield, A., Muhr, A. (eds.) Constitutive Models for Rubber III, pp. 157–160. ©2003 Swets & Zeitlinger, Lisse, ISBN 90 5809 566 5
182. Chaboche, J.L.: Continuum damage mechanics: present state and future trends. Nucl. Eng. Des. **105**, 19–33 (1987)
183. Kaliske, M., Rothert, H.: Viscoelastic and elastoplastic damage formulations. In: Dorfmann, A., Muhr, A. (eds.) Constitutive Models for Rubber I, pp. 159–167. ©1999 Balkema, Rotterdam, ISBN 90 5809 113 9
184. Miehe, C.: Discontinuous and continuous damage evolution in Ogden-type large strain elastic materials. Eur. J. Mech. A/Solids **14**, 697–724 (1995)
185. Desmorat, R., Cantournet, S.: Thermodynamics modelling of internal friction and hysteresis of elastomers. In: Besdo, D., Shuster, R.H., Ihlemann, J. (eds.) Constitutive Models for Rubber II, pp. 37–43. ©2001 Swets & Zeitlinger, ISBN 90 2651 847 1
186. De Souza Neto, E.A., et al.: A phenomenological three-dimensional rate-independent continuum damage model for highly filled polymers: formulation and computational aspects. J. Mech. Phys. Solids **42**, 1533–1550 (1994)

187. Aubard, X., et al.: Modelling and simulation of damage in elastomer structures at high strains. Comput. Struct. **80**, 2289–2298 (2002)
188. Reese, S., Wriggers, P.: Modelling of the thermo-mechanical material behaviour of rubber-like polymers—micromechanical motivation and numerical simulation. In: Dorfmann, A., Muhr, A. (eds.) Constitutive Models for Rubber I, pp. 13–21. ©1999 Balkema, Rotterdam, ISBN 90 5809 113 9
189. Klüppel, M., Schramm, J.: Advanced micro-mechanical model of hyperelasticity and stress softening of reinforced rubbers. In: Dorfmann, A., Muhr, A. (eds.) Constitutive Models for Rubber I, pp. 211–218. ©1999 Balkema, Rotterdam, ISBN 90 5809 113 9
190. Heinrich, G.: Statistical-mechanical basis of constitutive models for heterogeous rubber materials. In: Besdo, D., Shuster, R.H., Ihlemann, J. (eds.) Constitutive Models for Rubber II, pp. 3–10. ©2001 Swets & Zeitlinger, ISBN 90 2651 847 1
191. Achenbach, M.: A model to describe filler effects in rubber. In: Besdo, D., Shuster, R.H., Ihlemann, J. (eds.) Constitutive Models for Rubber II, pp. 21–26. ©2001 Swets & Zeitlinger, ISBN 90 2651 847 1
192. Govindjee, S., Simo, J.: A Micro-mechanically based continuum damage model for carbon black-filled rubber incorporating Mullins' effect. J. Mech. Phys. Solids **39**, 87–112 (1991)
193. Lubliner, J.: A model of rubber viscoelasticity. Mech. Res. Comm. **12**, 93–99 (1985)
194. Lianis, G.: Small deformations superposed on large deformation in viscoelastic bodies. In: Proceedings of the Fourth International Congress on Rheology, pt. 2, pp. 104–119. Interscience, New York (1965)
195. Coleman, B.D., Noll, W.: An approximation theorem for functional with applications to continuum mechanics. Arch. Ratl. Mech. Anal. **6**, 355–370 (1960)
196. Besdo, D., Ihlemann, J.: A phenomenological constitutive model for rubber like materials and its numerical applications. Int. J. Plast. **19**, 1019–1036 (2003)
197. Qi, H., Boyce, M.C.: Constitutive model for stretch-induced softening of the stress-strain behavior of elastomeric materials. J. Mech. Phys. Solids **52**, 2187–2205 (2004)
198. Marckmann, G., Verron, E., Gornet, L., Chagnon, G., Charrier, P., Fort, P.: A theory of network alteration for the Mullins effect. J. Mech. Phys. Solids **50**, 2011–2028 (2002)
199. Gracia, L.A.: Simulación por Elementos Finitos de efectos inelásticos en materiales elastómeros. Ph.D. Thesis, Universidad de Zaragoza (2006)
200. Fielding, J.H.: Flex life and crystallization of synthetic rubber. Ind. Eng. Chem. **35**, 1259–1261 (1943)
201. Standard test method for rubber property—extension cycling fatigue, ASTM D 4482-85 (1994)
202. Klenke, D., Beste, A.: Ensurance of the fatigue-life of metal–rubber components. Kautschuk und Gummi Kunstoffe **40**, 1067–1071 (1987)
203. Grosch, K.: Rolling resistance and fatigue life of tires. Rubber Chem. Technol. **61**, 42–63 (1988)
204. DeEskinazi, J., Ishihara, K., Volk, H., Warholic, T.C.: Towards predicting relative belt edge endurance with the FE method. Tire Sci. Technol. **18**, 216–235 (1990)
205. Oh, H.L.: A fatigue-life model of a rubber bushing. Rubber Chem. Technol. **53**, 1226–1238 (1980)
206. Griffith, A.: The phenomenon of rupture and flow in solids. Phil. Trans. Real Soc. Lond. Ser. A **221**, 163–198 (1920)
207. Lindley, P.B.: Ozone attack at a rubber–metal bond. J. Inst. Rubber Ind. **5**, 243–248 (1971)
208. Mars, W.V., Fatemi, A.: A phenomenological model for the effect of R ratio on fatigue of strain crystallizing rubbers. J. Rubber Chem. Technol. **76**(5), 1241–1258 (2003)
209. Pereña, J.M., Benavente, R., Cerrada, M.L.: Ciencia y Tecnología de materiales polímeros. Ed. CSIC **I**, 233–248 (2004)
210. Williams, M.L., Landel, R.F., Ferry, J.D.: The temperature dependant of relaxation mechanisms in amorphous polymers and other glass-forming liquids. J. Am. Chem. Soc. **77**(14), 3701–3707 (1955)

211. Gracia, L.A., Liarte, E., Pelegay, J.L., Calvo, B.: FE simulation of the hysteretic behavior of an industrial rubber. Application to design of rubber components. Finite Elem. Anal. Des. **46**, 357–368 (2010)
212. Gracia, L.A., Peña, E., Royo, J.M., Pelegay, J.L., Calvo, B.: A comparison between pseudo-elastic and damage models for modelling the Mullins effect in industrial rubber components. Mech. Res. Commun. **36**, 769–776 (2009)
213. Blau, P.J., De Vore, C.E.: Sliding friction and wear behaviour of several nickel aluminide alloys under dry and lubricated conditions. Tribol. Int. **23**(4), 226–234 (1990)
214. Plint, A.G., Plint, M.A.: A new technique or the investigation of stick-slip. Tribol. Int. **18**(4), 247–249 (1985)
215. Song, J., Liu, P., Cremens, M., Bonutti, P.: Effects of machining on tribological behavior of ultra high molecular weight polyethylene (UHMWPE) under dry reciprocating sliding. Wear **225–229**, 716–723 (1999)
216. Franklin, S.E.: Wear experiments with selected engineering polymers and polymer composites under dry reciprocating sliding conditions. Wear **251**, 1591–1598 (2001)
217. Barwell, F.T.: Wear of metals. Wear **1**, 317–332, 1957–1958 (1958)
218. Rhee, S.K.: Wear equation for polymers sliding against metal surfaces. Wear **16**, 431–445 (1970)
219. Cantizano, A., Carnicero, A., Zavarise, G.: Numerical simulation of wear-mechanism maps. Comput. Mater. Sci. **25**, 54–60 (2002)
220. Torrance, A.A.: A method for calculating boundary friction and wear. Wear **258**, 924–934 (2005)
221. Archard, J.F.: Contact and rubbing of flat surfaces. J. Appl. Phys. **24**, 981–988 (1953)
222. Greenwood, J.A., Williamson J.B.P.: Contact of nominally flat surfaces. Proc. R. Soc. Lond. Ser. A **295**, 300–319 (1966)
223. Sarkar, A.D.: Friction and Wear. Academic Press, London (1980)
224. Liu, R., Li, D.Y.: Modification of Archard's equation by taking account of elastic/pseudoelastic properties of materials. Wear **251**, 956–964 (2001)
225. Molinari, J.F., Ortiz, M., Radovitzky, R., Repetto, E.A.: FE modeling of dry sliding wear in metals. Eng. Comput. **18**, 592–609 (2001)
226. MSC MARC.: Theory and User Information, vol A. MSC Software Corporation, Santa Ana, CA, USA (2001)

Physical Phenomena Related to Free Volumes in Rubber and Blends

A. J. Marzocca, W. Salgueiro and A. Somoza

Abstract In the present chapter, different aspects related to free volumes and the physical phenomena involving free volumes in rubbers and blends are discussed. Experimental results were obtained using conventional experimental techniques (e.g. dynamic mechanical tests, differential scanning calorymetry and swelling) and principally a non-conventional one (positron annihilation lifetime spectroscopy—PALS). PALS has demonstrated a high capability to give direct information on free volumes. Due to its significant role in the study of nanoscopic effects in molecular systems (among them polymers), the physical grounds of the technique are explained. It is also illustrated how PALS detects free nanohole volumes and gives information on their changes as a consequence of different reactions induced in polymers. Based on the latest experience of the authors, some examples of PALS studies on NR and SBR rubbers and NR/SBR blends are presented. The results obtained are discussed using a modern scientific approach to the study of physicals processes in these elastomers; i.e. the analysis of the experimental information is given into the frame of well recognized theoretical models.

A. J. Marzocca (✉)
LP&MC, Facultad de Ciencias Exactas y Naturales, Departamento de Física, Universidad de Buenos Aires, Ciudad Universitaria, Pabellón I C1428EGA Buenos Aires, Argentina
e-mail: marzo@df.uba.ar

W. Salgueiro · A. Somoza
IFIMAT, Universidad Nacional del Centro de la Provincia de Buenos Aires, Pinto 399 B7000GHG Tandil, Argentina

W. Salgueiro · A. Somoza
Comisión de Investigaciones Científicas de la Provincia de Buenos Aires, Buenos Aires, Argentina

P. M. Visakh et al. (eds.), *Advances in Elastomers II*,
Advanced Structured Materials 12, DOI: 10.1007/978-3-642-20928-4_10,

1 Introduction

The present chapter addresses the potential of Positron Annihilation Spectroscopy (PAS) in the field of elastomers, with special attention to the physical phenomena related to free volumes in rubbers and blends. Generally speaking, in this kind of materials crosslinking has a strong influence on the physical properties of polymers and in particular on the free volume.

The free volume theory makes possible to describe qualitatively and quantitatively the polymeric behavior under different processes occurring in rubbers and blends such as physical aging, sorption and transport, plasticization and miscibility of blends.

The concept of free volume initially introduced to account for the dependence of the viscosity on the temperature in simple liquids [1] and, subsequently extended to polymers [2] turned out to be useful to explain mechanical and viscoelastic properties of macromolecules. However, the free volume in macromolecules is not unambiguously defined due to there are diverse theories considering a different behavior of the occupied volume as a function of the temperature. An interesting experimental investigation of the occupied volume in polymers was reported in Ref. [3].

Various experimental techniques have been used to get information on the free volume: photochromic labels [4], ^{129}Xe NMR [5], Electron Spin Resonance [6] and Small Angle X-ray Scattering [7]; among them, a variant of PAS, Positron Annihilation Lifetime Spectroscopy (PALS), has gained increasing popularity as a non-destructive tool [8].

PAS is a well-established high-sensitivity technique for detecting open volume sites in solids [9]. It has been applied to the study of the defect structure in solids since almost 30 years, and is presently used in many fields of materials science, from building and structural materials (for instance, light alloys [10] cement [11] and polymers [8, 12–14] to semiconductor-based systems for the information technology [15]).

In particular PALS, like many other sophisticated characterization techniques, requires specialists not only for the measurement itself, but also for analyzing raw experimental data. On the other hand, understanding PALS results and their meaning in terms of material properties does not require any specific background besides a few basic concepts, which are presented in Sect. 2 of this chapter. Section 3 addresses the free volume theory and its relation to the direct measurement of free nanohole volume by means of PALS. It is worth mentioning that this positron technique enables the study of free volume in polymers on a time scale for molecular motions slower than one nanosecond and a hole dimension bigger than about 2 Å. Section 4 is intended to motivate the attention of polymer researchers to PALS studies on rubber and blends by presenting recent results taken from the direct experience of the authors. The final Section gives a short summary of the main results presented with comments on the advantage of using PALS, complemented with specific techniques, to the study of free nanohole volumes and their related physical phenomena.

2 Positron and Positronium: Dynamics and Annihilation

Positron annihilation spectroscopy is the spectroscopy of the photons coming out from annihilation of positrons and electrons. Positrons are anti-matter. They enter in the material world if created by β^+ decay or by materialization of radiation (e^+-e^- pair production). Positrons are stable particles, but in matter they react with their ubiquitous anti-particles, the electrons. The result of the electron–positron reaction is annihilation, i.e. both particles disappear and their mass $2m_0$ is converted in electromagnetic energy $E = 2m_oc^2$.

The physical principles of PAS, and the dynamics and annihilation of positrons and positronium (Ps) in solids have exhaustively discussed elsewhere [9, 16].

In particular, in the case of molecular solids or insulators (among them polymers) the Ps formation process and the way to obtain information about the physical and chemical properties of these solids were deeply discussed by Eldrup in Ref. [9]. However, to allow the reader a better understanding on the interaction of Ps with insulators like polymers (i.e., Ps formation and its subsequent annihilation), a brief summary is given here. This information is also important to understand how PALS works for the analysis of elastomers (see also Refs. [17–19]).

A fraction of the positrons injected into an insulting system before annihilation forms with an electron a H-like bound system, called positronium (Ps), which occurs in two ground states. These bound states are instable and "annihilates". The singlet $S = 0$ state (parapositronium, p-Ps) in which electron and positron have antiparallel spins decays through the emission of two γ-rays with a characteristic lifetime of 125 ps, while the triplet $S = 1$ state (orthopositronium, o-Ps) with parallel particle spins decays via 3γ-annihilation with a characteristic lifetime $\tau_{o\text{-}Ps}$ of 142 ns (the lifetime values refer to the in-vacuum case). These two states are formed according to the ratio 1:3 between p-Ps and o-Ps. The probability of Ps formation depends on the physical and chemical properties of the materials.

The characteristic of Ps in a condensed material changes in comparison with it behavior in vacuum. In insulators this positron–electron bound system is repelled from the ionic cores of the atoms and molecules due to exchange interactions, therefore Ps tends to be pushed into low electron density sites of the host matrix, such as nanoholes forming the free volume. So, from the annihilation process of the Ps it is possible to obtain information regarding the free volume distribution (more details are given in Sect. 3). Particularly, the lifetime of p-Ps is too short to be influenced by volume factors. As the characteristic lifetime of o-Ps is long enough in vacuum, it is possible that the positron in the o-Ps state can annihilate with an electron other than its bound partner and of opposite spin following a process commonly called pick-off annihilation. In such a case, two photons are emitted with a characteristic lifetime strongly reduced (about 2 ns, depending on the molecular system) in comparison with that of the vacuum.

3 On PALS Technique and its Capability to Probe Free Nanohole Volumes

Annihilation is a random process, thus the positron lifetime, i.e. the time interval between the injection of a positron into a sample and the instant of annihilation, has a statistical distribution. This is called a lifetime spectrum. In defect-free homogeneous metals, the lifetime spectrum is a simple decaying exponential. In the presence of trapping at defects or when positron exist in different states as free positrons or positronium, the lifetime spectrum becomes complex one consisting in a superposition of decaying exponentials.

PALS consists of injecting positrons into a sample and measuring their lifetime spectrum. In most applications, the positron source is the β^+ radioactive isotope ^{22}Na. The advantage of this isotope is that its most frequent decay channel is the emission of a photon of 1.274 MeV in coincidence with the positron. This is a very useful property, because the 1.274 MeV photon gives the signal that indicates the birth of the positron. The signal of annihilation is either one of the two annihilation photons (about 0.511 MeV), coming from the conversion in electromagnetic energy of the mass of the annihilated particles. Thus, in order to measure the positron lifetime, one essentially needs two scintillation detectors coupled to a time analyzer (see experimental details in Ref. [10]).

Figure 1 gives an example of lifetime spectra, obtained with a resolution of 252 ps FWHM (full width at half maximum of the resolution function), after subtraction of the background of accidental coincidences and of a spurious component due to annihilations in the source support. The spectra have the form of a superposition of decaying exponentials convoluted with the response function of the spectrometer. The presence of points at the left of the maximum and at negative lifetimes is an effect of the limited resolution of the instrument. The data shown in Fig. 1 refer to uncured styrene butadiene rubber and a metallic alloy (a well-annealed sample of β Cu–Zn–Al single crystal shape memory alloy). These examples were chosen to demonstrate the sensitivity of the technique to the chemistry of the samples and to the positron and positronium annihilation processes occurring in different kind of materials (in metallic alloys, the main contribution comes from positron annihilation while in polymers is due to o-Ps annihilation via pick-off).

Several computer programs exist for deconvolving the time spectrum from the resolution function of the time spectrometer and for isolating the different exponential components.

In polymers PALS spectra, after a suitable correction for the positrons annihilated in the source support, are usually deconvoluted into three discrete components. In Fig. 1, a typical lifetime spectra obtained measuring a sample of uncured SBR is presented. From the decomposition of the SBR spectrum these three lifetime components can be clearly recognized. The solid line represents the longest exponential decaying component associated with the o-Ps annihilation. This line was obtained using an usual program allowing to analyze PALS spectra.

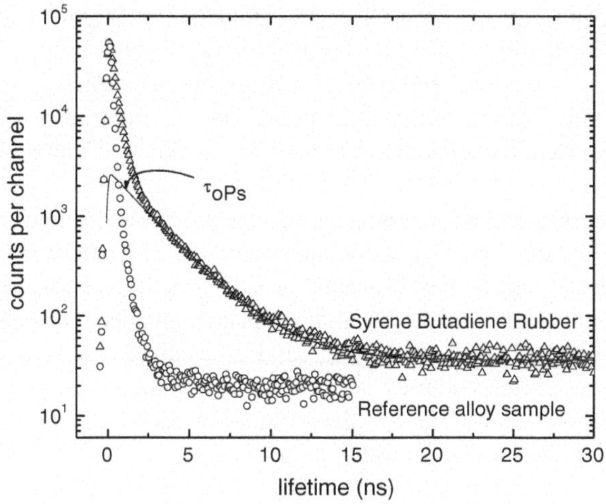

Fig. 1 Comparison of experimental lifetime spectra for positron and positronium annihilated in uncured styrene butadiene rubber (*up triangles*) and well-annealed β Cu–Zn-Al shape memory alloy (*circles*). The lifetime spectrum for SBR was decomposed into three lifetime components. *Solid line* represents the long-lived exponential decaying component of the ortho-Ps. The spectra were taken with a resolution of 252 ps (FWHM); source component was subtracted

As already mentioned, the reference spectrum was obtained measuring in our laboratory a metallic alloy characterized by only one positron lifetime.

According to the common interpretation for PALS measurements in this kind of materials, the longest component is ascribed to o-Ps decay in the nanoholes forming the free volume. The intermediate component (0.35–0.50 ns) is attributed to positrons annihilated in low electron density regions of the structure. The shortest component (0.15–0.23 ns) is due to positrons annihilated in the bulk and to p-Ps annihilations; this last contribution cannot be resolved as a distinct component.

Lifetime data can be transformed into average sizes of the free volume holes by using the Tao-Eldrup semiempirical equation [20, 21] the cavity hosting Ps is assumed to be a spherical nanohole with effective radius R. Such a Ps trap has a potential well with finite depth; however, for convenience of calculations one usually assumes the depth as infinite, but the radius increased to $R + \Delta R$, ΔR (0.166 nm) [22] being an empirical parameter which takes into account the annihilation of o-Ps with the electrons belonging to the walls of the hole ('pickoff' process). Thus, the relationship between the longest lifetime lifetime $\tau_{o\text{-}Ps}$ and the size of the nanohole is given by:

$$\tau_{o-Ps} = 0.5\left[\frac{\Delta R}{R + \Delta R} + \frac{1}{2\pi}\sin\left(2\pi\frac{R}{R + \Delta R}\right)\right]^{-1} \quad (1)$$

of course, the values of the radii obtained from Eq. (1) should be interpreted only as rough estimates, since real holes are irregularly shaped.[1]

From Eq. (1), values of the average volume of nanoholes $v_h = \frac{4}{3}\pi R^3$ can be obtained; they are a key quantity to evaluate the free volume.

Various approaches were used to get the fractional free volume from positron data.

In the simplest approach, the number of free nanohole volumes in the material is represented by the intensity associated with the o-Ps lifetime $I_{o\text{-}Ps}$ which is associated with the longest lived lifetime in a discrete decomposition of the PALS spectra in polymers (as mentioned above, spectra are decomposed in three lifetimes components). Therefore, in this kind of analysis $\tau_3 \equiv \tau_{o\text{-}Ps}$. Under this frame, the fractional free volume f in polymers is assumed to be proportional to the density of nanoholes and to the average volume of each hole. Then, the following semi-empirical equation can be used:

$$f = A v_h I_{o-Ps} \tag{2}$$

where A is a constant.

Since last decade the most used definition of the fractional free volume is:

$$f = \frac{V_f}{V} = \frac{N^f\, v_h}{N^f v_h + V_{occ}}. \tag{4}$$

[1] Actually, from the analysis of PALS spectra a continuous distribution of the o-Ps lifetimes, and therefore of hole radii, is obtained. So, according to Ref. [23] from an estimation of the distribution of the inverse lifetime $\xi(1/\tau_{o\text{-}Ps})$ the hole radius probability density fraction $f(R)$ can be obtained as follows

$$f(R) = 2\delta R\left(\cos\frac{2\pi R}{R+\delta R} - 1\right)\frac{\xi(1/\tau_{o-Ps})}{(R+\delta R)^2} \tag{3}$$

Then, it is easy to obtain the hole volume density distribution as $g(v_h) = f(R)/4\pi R^2$ which represents the volume fraction of holes as determined by o-Ps annihilation with volumes between v_h and $v_h + dv_h$ is $g(v_h)\, dv_h$ [17, 24]. Then, from $g(v_h)$ can be calculated the number fraction of holes.

In Ref. [25] we studied the evolution of the free volume during vulcanization in SBR. In this work, for each sample instead of analyzing the o-Ps lifetimes as discrete ones we assumed the corresponding $\tau_{\text{o-Ps}}$ as continuum distributions and then, using Eq. (3) we obtained the different $g(v_h)$. Then, we found that all $g(v_h)$ consisted of one skewed peak and extended from about 20 to 450 Å^3 approximately. Results are discussed in Sect. 4. Another example can be found in Ref. [19] in which the authors correlated the size, numerical concentration, and size distribution of free volumes with the nanostructure of different degree of crosslinkage cured polysiloxanes induced by thermal treatments. On the other hand, Dlubek et al. [17] estimated the free nanohole volume distribution in polycarbonate and polystyrene a room temperature.

However, it is worth mentioning that for the analysis of the o-Ps lifetime component is normally accepted as a good approximation the use of the most probable free volume size as a representative parameter of $g(v_h)$.

where N^f is the hole density, V the specific volume, V_{occ} is the occupied volume and the free volume V_f is given by

$$V_f = N^f v_h \tag{5}$$

To get information on N^f the method used in Ref. [26] can be followed. The specific volume V can be obtained using the Simha-Somcynsky equation of state for polymers [27] at atmospheric pressure, which can be approximated by the following universal scaling relationship [28] for $T > T_g$:

$$\ln \frac{V}{V^*} = a + b\left(\frac{T}{T^*}\right)^{3/2} \tag{6}$$

where $a = -0.1033$ and $b = 23.835$ are universal constants; V^* and T^* are scaling parameters, dependent on the specific structure of polymer. Concerning V^*, it results, with a very good approximation, $V^* = 1.45\ V_W$, where V_W is the Van der Waals volume [29]. In the following Section, when describing different studies on rubbers and blends the specific values of V_W and V^* are given.

Summarizing, changes of the average volume of nanoholes in polymers induced by variations of external parameters, such as temperature or mechanical stresses, can be directly monitored by studying the o-Ps lifetime in a positron annihilation lifetime spectrum.

4 Experimental Studies of Free Volumes in Rubbers and Blends

In this Section, selected studies on the influence of the free volume on the physical phenomena occurring in rubbers and blends as a consequence of the change of different external parameters are presented. The examples presented below were taken from the direct experience of the authors.

4.1 Analysis of Uncured and Cured Elastomers

The free volume in polymers systems is an important entity which has strong consequences on the mechanical, thermal and electrical behavior of this type of materials.

Since the experimental determination of the free volume is not simple, to analyze and discuss the experimental results several theoretical frameworks have been usually used to get an indirect estimation of the free volume. In the case of dynamic mechanical measurements, the William-Landel-Ferry (WLF) relationship [30] is one of the most popular methods based on the time temperature principle to evaluate the free volume fraction. It can be estimated by means of the shift factor a_T evaluated from analysis of the dependence of the storage modulus on frequency and temperature by applying the WLF relationship. The shift factor is expressed by

$$\log a_T = -c_1^0(T - T_0)/(c_2^0 + T - T_0) \tag{7}$$

where c_1^0 and c_2^0 are constants evaluated at the reference temperature T_0 and T is the temperature in K degrees.

The shift factor a_T is related to the fractional free volume as [30]

$$\log a_T = \frac{B}{2.303}\left(\frac{1}{f} - \frac{1}{f_0}\right) \tag{8}$$

being f_0 the fractional free volume evaluated at the reference temperature T_0 and B is an empirical constant near unity.

Some years ago, as it was reported in Ref. [25], we found that using the simplest expression to obtain the fractional free volume from the PALS data given in Eq. (2), it was possible to connect this parameter with the WLF parameters c_1^o and c_2^o through the following relationships [30]:

$$c_1^o = B/2.303 f_o \tag{9}$$

$$c_2^o = f_o/\alpha_f \tag{10}$$

$$\alpha_f = B/2.303 c_1^o c_2^o \tag{11}$$

where α_f is the thermal expansion of the free volume relative to the total volume.

In literature, there are not many researches on the evaluation of the free volume in elastomers by means of PALS. The annihilation characteristics of the o-Ps in *cis*-1,4-poly(butadiene) was measured by Bartoš et al. [31] in the temperature range from 15 to 313 K. These authors interpreted the data in the framework of the free volume concept. The results they reported are given in Fig. 2a: a pronounced break on the τ_3 versus T plot at the glass transition temperature T_g. As a result, two bends were observed. The first, in glassy state, is relatively slight. The other one, in an elastic state, was correlated with the melting temperature T_m of the crystalline phase of the *cis*-1,4-poly(butadiene). At temperatures up to 50 K, the authors found that $\tau_3 \approx 0.05$ ns which is a very low o-Ps lifetime value in comparison with those observed in other polymers at cryogenic temperatures. The corresponding mean free volume was $v_h = 22$ Å^3, that is about one-third of the van der Waals volume of a monomer unit, $V^W_{\text{mon}} = 62$ Å^3. This small size of the free nanohole volumes indicated a very effective packing of the chains in the glassy state of this elastomer. This result was interpreted as a consequence of the chemical structure as characterized by a relatively high internal flexibility which could be quantified by the characteristic ratio $C_\infty = 5.1$ [32]. At the glass transition temperature, $T_g = 168$ K, the value obtained for $v_h(T_g) = 42$ Å^3 was about two/thirds of that the van der Waals volume of the basic structural unit. In the

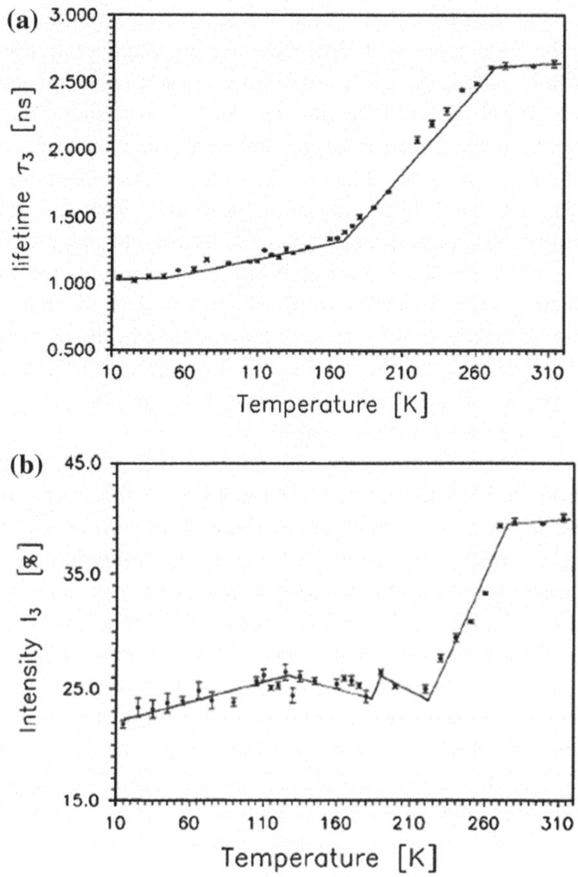

Fig. 2 o-Ps lifetime (**a**) and relative intensity (**b**) as function of temperature in cis 1,4-polybutadiene rubber. From Bartoš et al. [30]

elastic state, from 170 to 265 K, an increase of τ_3 with temperature was observed, which was attributed to an enhancement of the free volume up to 150 Å^3. Finally, the fourth high temperature region is characterized by slight temperature change.

In Fig. 2b, the decrease of the intensity associated with the o-Ps lifetime with temperature is showed for *cis*-1,4-poly(butadiene). For temperatures above 120 K Bartoš et al. [30] interpreted the data considering that about 120 K the observed change correlates with the onset temperature of the so-called fast motion from neutron scattering measurements as well as with the so-called Vogel temperature for the primary segmental dynamics. Above T_g the shape of the I_3-curve was correlated with the cold crystallization obtained from DSC measurements. In the melt region above 265 K, the slightly dependence of τ_3 and I_3 quantities on T was attributed to a formation of the bubble states of the o-Ps in the soft matrix. Furthermore, Bandžuch et al. [33] reported PALS studies of several uncured elastomers: cis-1,4-poly(butadiene), cis 1,4 poly(isoprene), poly(isobutylene), cis–trans 1,4 poly(butadiene) in the temperature range between 15 and 470 K.

On the other hand, some studies of free volume using PALS were performed on filled elastomers [34–36]. Specifically, Wang et al. [34] studied the influence of the filler content on the free volume of uncured styrene butadiene rubber (SBR) with carbon black and rectorite. The authors concluded that the dispersion of nanoscale rectorite clay in SBR largely enhances the gas barrier property in contrast to results obtained in carbon black/SBR system. These results show that gas permeability in rectorite/SBR is mainly influenced by the fractional free volume and tortuous diffusional path effects attributed to the clay platelet-like morphology.

The influence of carbon black in natural rubber, emulsion and solution SBR was also analyzed [35]. The results reported show that carbon black fillers have no effect on the o-Ps lifetime but a decrease in the associated intensity to this lifetime component was observed and it was reported that such a behavior depends on the type of carbon black. Besides, it was reported that both parameters decrease as a function of sulfur concentration.

Authors of this chapter analyzed the influence of the cure level on the free volume in SBR cured at 433 K [37]. In this work, the cure level γ, obtained by DSC tests, was changed in the rubber curing the samples for different times. Experimental measurements at room temperature using Dynamic Mechanical Spectroscopy (DMA) and PALS were performed. In Fig. 3 the evolution of the free nanohole volume distribution $g(v_h)$ as a function of v_h at different cure levels is shown. Besides, in the inset of this figure the behavior of the most probable free volume size against γ is also shown. As can be seen, at the beginning of the cure v_h significantly decreases. This behavior implies a lower freedom of movement of the macromolecular segments in the cured state compared to the uncured state. From the results described in the present paragraph, it can be concluded that the presence of crosslinks decreases the volume of holes in the sample. Once a minimum value is obtained, the free nanohole volume increases until reaching the maximum cure level. It is worth mentioning that in these samples, the cure system is composed of sulfur, an accelerator (TBBS) and an activator (zinc oxide). On the other hand, it is

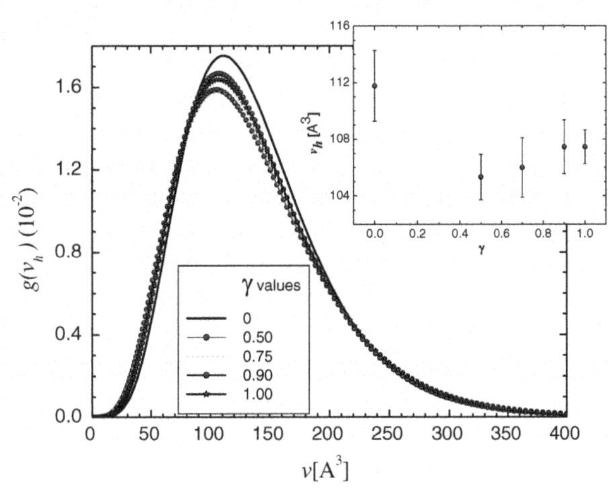

Fig. 3 Free volume distribution and average of free nanohole volumes of uncured and cured SBR at different cure levels. In the *inset*, the free volume distribution as a function of the cure level γ is shown. Data from Marzocca et al. [25])

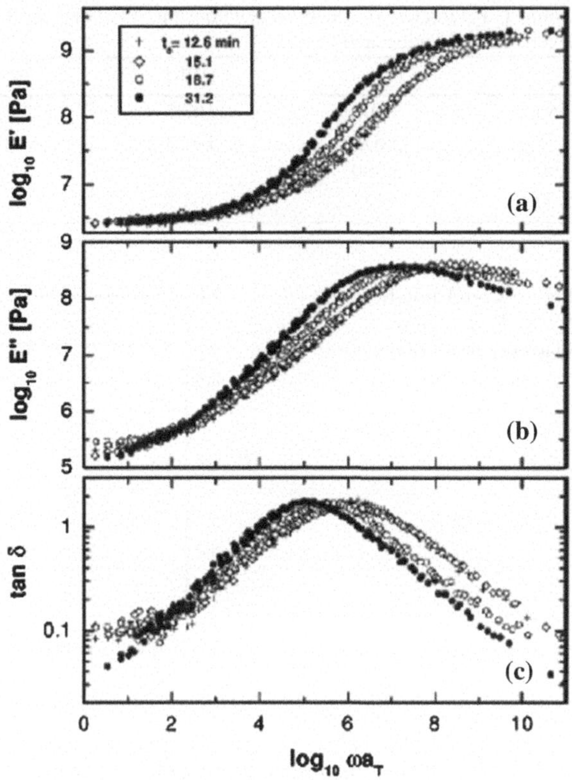

Fig. 4 Master curves of SBR vulcanizated at 298 K. From Marzocca et al. [25]

well known that in sulfur vulcanization only a part of the combined sulfur goes to form crosslinks between chains [38] and different chemical reactions take place during the curing process that lead to a variety of structures (polysulfides, disulfides, monosulfides crosslinks, pendent chains, etc.) [39]. The broadening of the free volume distribution when comparing the uncured sample with those of the cured ones can be attributed to a broadened distribution of the free volume in the polymer [40].

In Fig. 4, the master curves of the storage modulus E', the loss modulus E'', and the loss tangent *tan* δ as a function of the angular frequency ω, for the cured samples of SBR at a reference temperature of 298 K, are shown. The parameters of the WLF relationship, represented by Eq. (7), used to build the master curves, are given in Table 1.

From Eq. (9) and considering $B = 1$, the fraction of free volume f at the reference temperature T_o was calculated. The corresponding values are also reported in Table 1. The fraction of free volume is nearly constant at low cure levels and increases when the cure reaction advances. This fact could be associated not only with the creation of new crosslinks during the curing process but also to side reactions produced during sulfur vulcanization.

Table 1 Parameters of the WLF relationship, f(To) and α_f/B at $To = 298$ K for different cure levels. From Marzocca et al. [25]

γ	c_1^o	c_2^o [K]	$\alpha_f/B\,[10^{-4}\mathrm{K}^{-1}]$	f
0.50	3.600	107.7	11.2	0.121
0.75	3.636	109.3	10.9	0.119
0.90	3.380	105.3	12.2	0.128
1.00	3.123	104.7	13.4	0.139

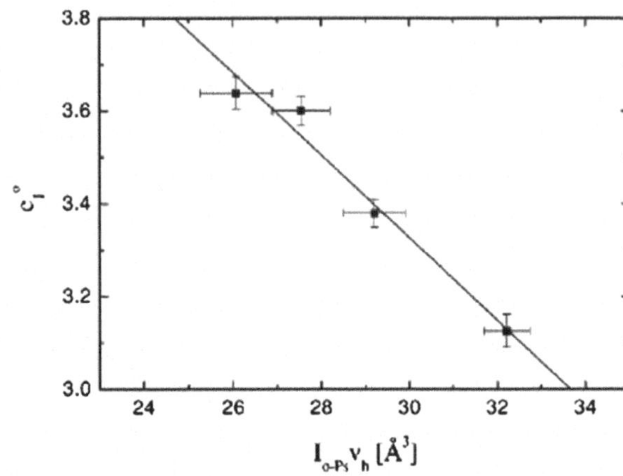

Fig. 5 Correlation between c_1^o and $I_{o\text{-}Ps}v_h$. From Marzocca et al. [25]

In order to use the information obtained using PALS measurements together with those of the dynamic mechanical tests, from Eqs. (2) and (9) it is easy to obtain values for the c_1^0 parameter as:

$$c_1^o = \frac{B}{2.303AI_{o-Ps}v_h} \tag{12}$$

As can be seen in Fig. 5, c_1^o values obtained using the WLF relationship to analyze the data obtained by dynamic mechanical tests, show a good correlation with the relationship $I_{o-Ps}\,v_h$ (in Ref. [25], this product was written as $I_3\,v_h$) obtained using PALS data.

On the basis of this comparison, we concluded that both techniques apparently give the same physical information [25].

The amount and types of crosslinks produced by the vulcanization process is not only function of the cure temperature and time but mainly of the cure system used in the compound formulation. As known, the creation of crosslinks between polymer chains can be performed by means of different types of vulcanizing agents [41] or by irradiation. Varying the amount of sulfur and accelerator, the concentration of crosslinks due to sulfides linkages between carbon atoms of the polymer chains can be adjusted [39].

In Fig. 6 we show the rheometer curves, at 433 K, of a SBR compound containing in its formulation 1.2 phr of TBBS as accelerator and different sulfur

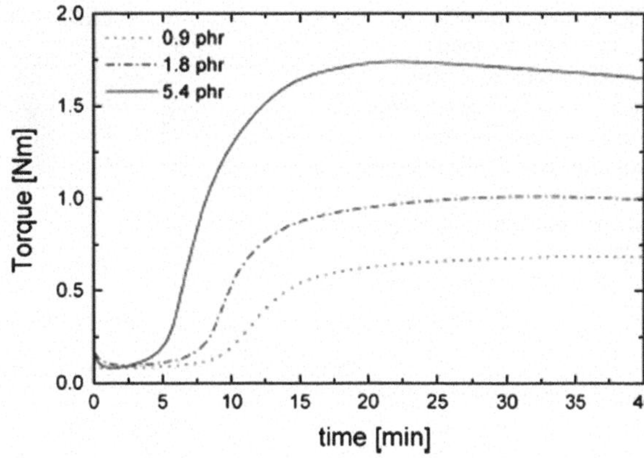

Fig. 6 Rheometer curves at 433 K of SBR with different formulations (after Salgueiro et al. [42])

concentrations. We reported these results in Ref. [42]. As it was expected, the maximum torque corresponds to the higher level of sulfur in the compound. This behavior was attributed to the fact that increasing the level of sulfur, the concentration of active crosslink increases.

In this study, the crosslink density μ_c in a 4-functional network was estimated as [43]:

$$\mu_c = \frac{\rho}{2}\left(\frac{1}{M_{cs}} - \frac{1}{M_n}\right) \tag{13}$$

where ρ is density of the polymer and M_n is average molecular weight. The molecular weight of the network chain between chemical crosslinks M_{cs} was obtained by means of swelling tests in toluene using the Flory-Rehner relationship [44, 45]

$$M_{cs} = -\frac{\rho(1 - 2/\phi)V_1\, v_{2m}^{1/3}}{\ln(1 - v_{2m}) + \chi v_{2m}^2 + v_{2m}} \tag{14}$$

being ϕ the functionality of the crosslinks, v_{2m} the volume fraction of polymer at the equilibrium (maximum) degree of swelling and V_I the molar volume of solvent. Besides, χ is an interaction parameter between the polymer and the swelling agent [46].

In Fig. 7, the variation of the total crosslink density with the sulfur content for each compound was plotted.

The polysulfide, disulfide and monosulfide crosslink densities in each sample were estimated following the procedure described in Ref. [42]. From these data, it can be concluded that the total crosslink density increases with the sulfur content. Besides, the proportion of polysulfides in the sample with 5.4 phr of sulfur is lower than those corresponding to the other two samples and there is also an increase in monosulfides crosslink density for the sample with higher sulfur content.

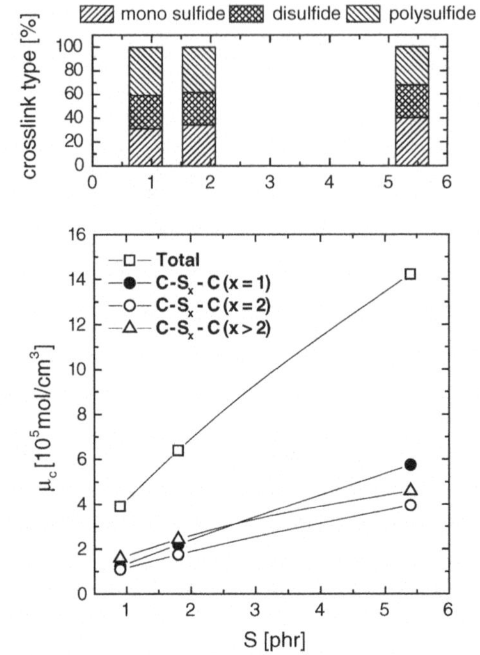

Fig. 7 *Lower panel*: Variation of the total crosslink density versus sulfur content. *Upper panel*: Percentage of the three crosslink types (di, mono or polysulfide) in the studied samples (after Salgueiro et al. [42])

The same samples were investigated by mechanical spectroscopy and differential scanning calorimetry. Figure 8 shows the dynamic moduli E', E'' and *tan* δ obtained at a frequency of 1 Hz, as a function of the temperature. As usual, the storage modulus E' increases with the degree of crosslinking both in the rubbery and in the glassy regions of the curves. The temperature scan of the loss modulus E'' and the loss factor *tan* δ show a maximum which, in both cases, can be attributed to the glass transition temperature.

In Fig. 9, for all samples the evolution of the free nanohole volume, obtained by PALS measurements, as a function of the temperature is given. A similar trend is observed for all the SBR curves but with a systematic shift to lower v_h values when the sulfur content increases. In the figure, three temperature regions can be distinguished: $T \leq 220$, $220 \leq T \leq 300$ and $T \geq 300$ K, characterized by different values of dv_h/dT which starts to increase at the glass transition temperature.

Using the techniques DMA, DSC and PALS values of the glass transition temperature as a function of the crosslink density for each SBR compound were estimated. These results are presented in Fig. 10. It can be observed that T_g increases with the crosslink density. From the analysis of the different v_h versus T curves we observed that the glass transition temperature values obtained from PALS measurements are systematically lower than those obtained by DSC or DMA.

In fact, T_g is not a thermodynamic property but a kinetic one and its value is influenced by the heating rate as DSC as well as DMA tests. Times required for carrying out PALS measurements are much longer than those necessary for thermal tests and this fact can explain the systematic lower T_g values obtained

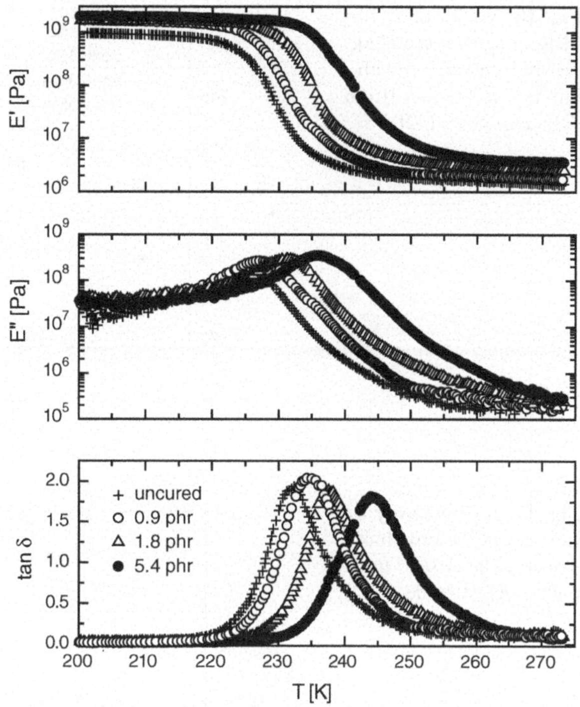

Fig. 8 Storage modulus (E'), loss modulus (E'') and loss factor (*tan δ*) versus temperature. All the dynamic moduli were obtained at a frequency of 1 Hz (from Salgueiro et al. [42])

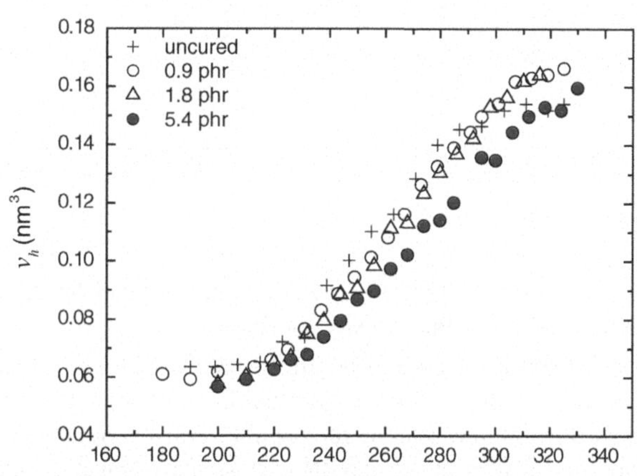

Fig. 9 Average nanohole volumes versus temperature uncured and cured SBR at 433 K for different formulations (after Salgueiro et al. [42])

using the positron technique. Furthermore, in PALS measurements the glass transition is related to a change in the amplitude of the movements of very short polymer segments which are involved in the thermal processes of the nanohole formation. These processes begin at temperatures lower than those involved in the T_g evaluated by DSC, which are related to a change of the heat capacity of the polymer [42].

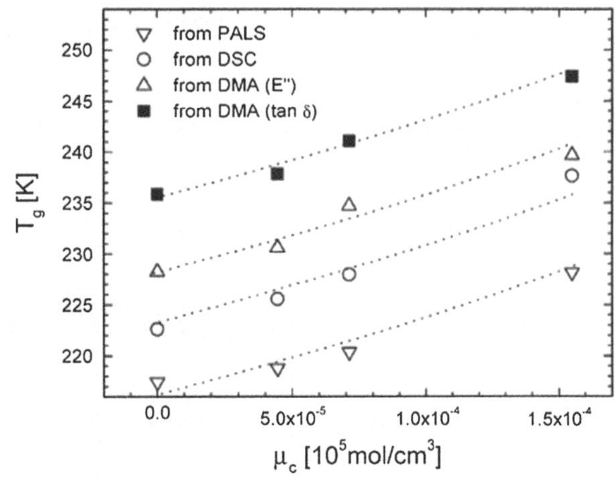

Fig. 10 Values of T_g for SBR, at several crosslink densities, measured with different techniques (from Salgueiro et al. [42])

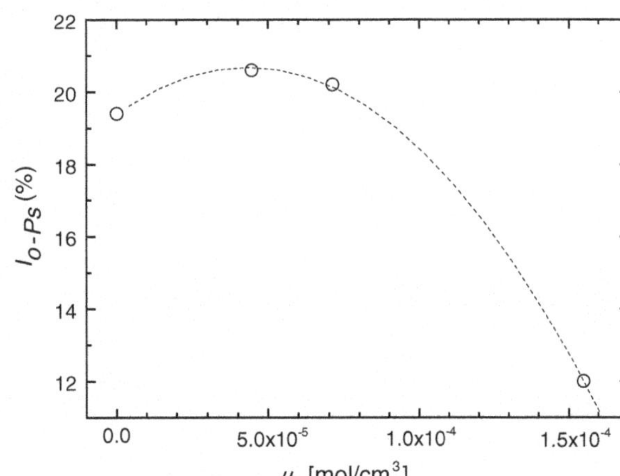

Fig. 11 o-Ps intensity as a function of the crosslink density. The *dashed line* is only a visual guide

As shown in Fig. 11, $I_{o\text{-}Ps}$ does not depend on the temperature but it decreases when the crosslink density increases.

According to Eq. (4), the specific volume is defined as $V = N^f v_h + V_{occ}$. Dealing with positron experimental data and using Eqs. (4)–(6), values of the specific volume V as a function v_h can be obtained; the different evolutions for uncured SBR and cured SBR with several crosslink densities are plotted in Fig. 12. As can be seen, the specific volume shows a linear dependence with v_h for all the samples. It follows that in Eq. (5), N^f is constant which is in total agreement with reported results [26]. From the intersection of the straight lines plotted in Fig. 12, values of V_{occ} of each sample were obtained. Finally, averaging all V_{occ} we obtained the following value: $V_{occ} = 0.931 \pm 0.10$ cm^3/g.

In Fig. 13, the evolution of the fractional free volume f as function of the temperature for uncured and cured SBR with different total crosslink densities is

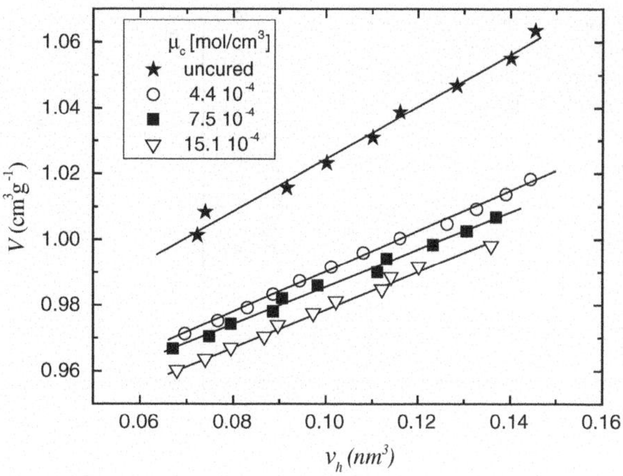

Fig. 12 Relation between specific volume and hole volume for uncured SBR and cured SBR with several crosslink densities (from Salgueiro et al. [42])

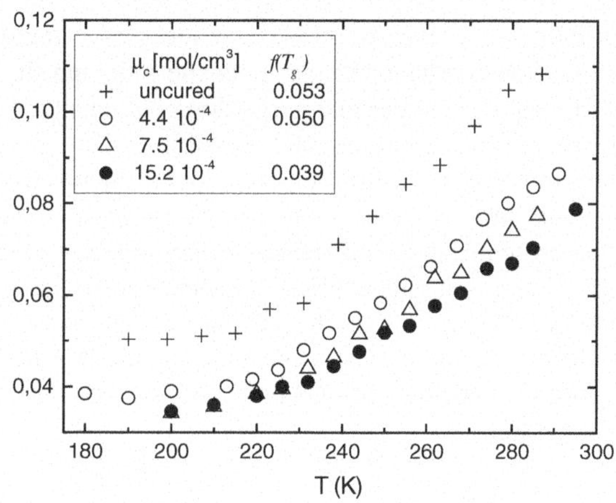

Fig. 13 Evolution of the fractional free volume as a function of the temperature for uncured SBR and cured SBR at 433 K and with different crosslink densities (from Salgueiro et al. [42]). In the *inset*, values of the crosslinks density and the free volume fraction, evaluated at T_g, for each sample measured are reported

shown. The corresponding values of f evaluated at the glass transition temperature, i.e. $f(T_g)$, are different depending on the crosslink density. This result is in correspondence with the contention that T_g is not an iso-free volume state [34]. By increasing the crosslink density of the samples, f decreases indicating an increased packing density in the rubbery phase. This result seems reasonable since the higher constraints on the polymer due to the network hinder the thermal mobility of the chains. This effect is not so pronounced in the glassy region, since the thermal mobility is almost frozen and reduced effects are expected on the free volume.

The thermal expansion coefficient of the free volume was evaluated as $\alpha = df/dT$ and it is presented in Fig. 14; it shows a constant decrease by increasing the crosslink density in the rubbery phase (α_r). This means that in the rubbery phase the thermal expansion of the polymer is slower, the higher is the crosslink density, which can be expected by considering that the presence of the network limits the

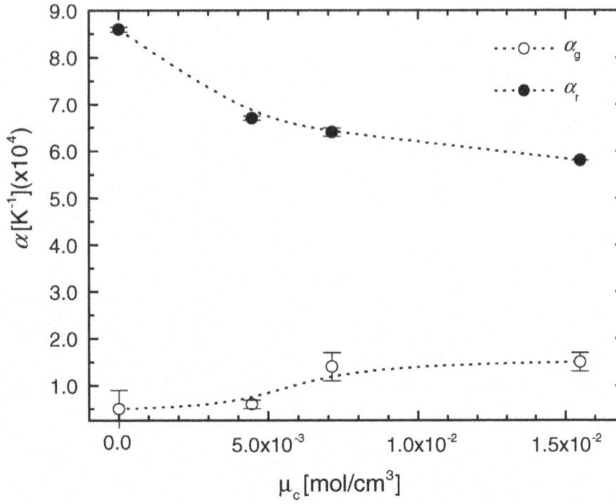

Fig. 14 Thermal expansion coefficients of the free volume as a function of the crosslink density

mobility of polymeric chains. In the glassy region the behavior of the thermal expansion coefficient (α_g) is opposite although the change is not so significant.

In Fig. 15 the thermal expansion coefficients of the free volume in the rubbery phase normalized to the total crosslink density, α_r/μ_c; versus the ratio between the polysulphide crosslink density and the monosulphide crosslink density are presented. In fact, as already shown in Fig. 7, the contribution (percentage) of the disulphide crosslink density is the same for all the cured samples. It is clear that the presence of polysulphide crosslink density in the samples increases the value of the ratio α_r/μ_c. Although a definite conclusion cannot be drawn due to the small number of investigated samples, it seems that the thermal expansion of the free volume is affected by the ratio between the different kinds of crosslinks in the compound.

4.2 Analysis of Elastomer Blends

As it is well known, miscibility of polymer blends is rare, compared to blends of low molecular weight, because of the relatively small entropy of mixing. Frequently, the formation of miscible polymer blends depends on an exothermic heat of mixing, which usually arises from strong intermolecular forces, for example, dipolar, hydrogen bonding, or charge transfer interactions [47]. The mixing properties of two polymers may generally be analyzed considering the change in the Gibbs free energy. When a system with two components A and B is contained into a volume $V = V_A + V_B$, where V_A and V_B are the volumes containing the polymers A and B respectively, the Gibbs free energy of mixing ΔG_{mix} is given by

$$\Delta G_{mix} = G_{AB} - (G_A + G_B) \tag{15}$$

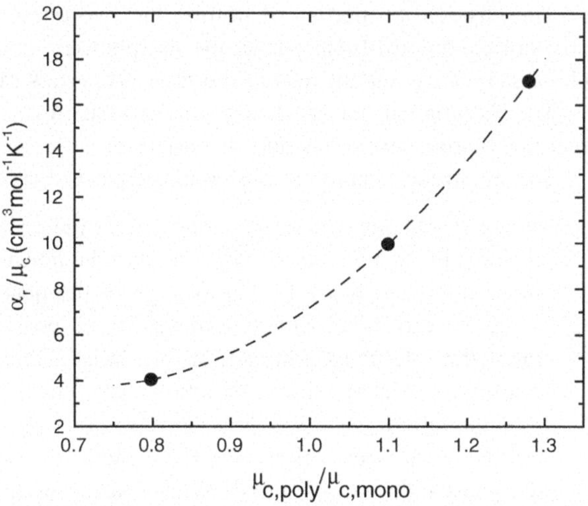

Fig. 15 Thermal expansion coefficient of the free volume in the rubbery region, normalized to the total crosslink density, versus the ratio between mono and polysulfide crosslink densities (after Salgueiro et al. [42])

where G_A, G_B and G_{AB} are the Gibbs free energies of both polymers in separate states and the mixed state, respectively.

In the frame of the Flory–Huggins theory, ΔG_{mix} contains two contributions:

$$\Delta G_{mix} = -T\Delta S_t + \Delta G_{loc} \tag{16}$$

being ΔS_t the increase in the translational entropy. The entropy associated with the motion of the centers of mass of all the polymer molecules increases in the first steps of the mixing process; then, it may change the local interactions and motions of the monomers ΔG_{loc}. The term $-T\Delta S_t$ always favors mixing. In the other side, ΔG_{loc} may contribute in a favor or unfavorable way depending on the interactions between the monomers of the species involved.

Mixing can lead to a distorted total volume $V \neq V_A + V_B$, which means that a shrinkage or an expansion takes place in the total volume associated with a change in the local mobility of the monomers. Then a change of entropy must be included in ΔG_{loc}.

Any change in the free volume due to the mixing process will affect the mechanical and transport properties of the blend. Consequently, any study on the microscopic free volume certainly will give valuable information that can be correlated with the mechanical and physical properties of the blends.

Most blends of elastomers are immiscible because mixing is endothermic and the entropic contribution is small because of the high molecular weights. Fortunately, miscibility is not a requirement for most rubber applications.

In some cases, technological difficulties are associated with mutual incompatibility between dissimilarities of the elastomers constituting a blend. The incompatibility can be due to thermodynamic aspects, viscosity mismatch and cure rate mismatch [48].

The physical properties of immiscible two-phase blends depend on the properties of each constituent phase; the morphology, dispersion and stiffness of each phase play a significant role in the final properties of the blend.

The rheological behavior of a polymer blend can be analyzed considering its thermodynamic properties and its structure.

The mentioned behavior can be classified in two categories [49]:

(i) When the blend is homogeneous due to the specific polymer–polymer interactions; in such a case, its viscosity and the normal stresses will be higher than those predicted from the log-average of the neat polymers.
(ii) In the case of heterogeneous blends, three possible cases can be considered: (a) if the interphase interactions are largely due to compatibilization, shear-grafting, partial specific interactions, etc.; then, the system will show a positive deviation from the log-additivity mixing rule for both viscosity and normal shear stress difference; (b) if there is a very small interaction between the phases, the viscosity will show a negative deviation and the normal shear stress difference will show a positive deviation from the log-additivity mixing rule; (c) if there is a concentration-dependent change of the blend structure with a phase inversion, both positive and negative deviations should be expected [49].

Although immiscible blends show distinct glass transition temperatures which could be shifted with respect to the T_g values of each polymer forming the blend [50], a detailed analysis of this behavior is complicated by the presence of two glass transition temperatures. A similar analysis is also applicable for miscible blends [51, 52].

In the case of elastomers, the advantages for blending are numerous due to an alternative of synthesizing a new elastomer. In such a case, older and better characterized rubbers can be employed, hopefully yielding an ideal complement of properties from the blended materials.

The analysis of the free volume in elastomer blends appears as an interesting way for the physical and structural characterization of these materials and a potential tool for the analysis of the interphase region between the elastomers constituting the blend.

Using DSC and PALS, Peng et al. [53] studied the free volume dependence with the temperature in high-vinyl polybutadiene/cis-polyisoprene blends. In Fig. 16 the evolution of $\tau_{o\text{-}Ps}$ and $I_{o\text{-}Ps}$ with the temperature of the samples from blends with different compositions of the constituent elastomers is shown. From the reported results, the authors concluded that at high and low temperatures the free nanohole volume size and the fractional free volume exhibit a negative deviation from the additivity tie line. From the DSC scans for all samples, only one glass transition temperature was observed. Calorimetry data were then analyzed following an idea proposed by Kovacs in Ref. [54] and using the free volume parameters obtained from the PALS data.

The same authors studied the average free volume sizes and the fractional free volume in 50–50 high-vinyl polybutadiene/cis-polyisoprene blends vulcanized at

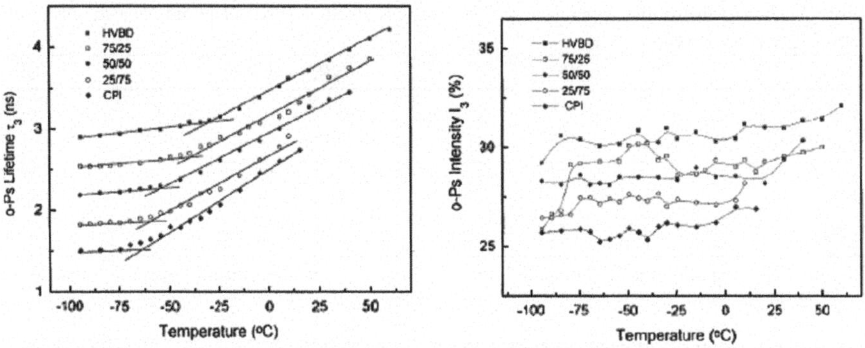

Fig. 16 PALS parameters, representative of the o-Ps lifetime component, as a function of the temperature. The data were obtained for different samples of the blend high-vinyl polybutadiene/cis-polyisoprene blends with different compositions of each elastomer (from Peng et al. [53])

443 K [26]. In order to achieve different crosslink densities in the compounds, the authors used different amounts of the curing agent dicumylperoxide. In that work [26], the authors analyzed the influence of the crosslink density on the PALS parameters observing that $\tau_{o\text{-}Ps}$ and the fractional free volume decreased as the crosslink density increased. This result could be interpreted as the packing density is increased in the rubbery state as a consequence of the loss of mobility of the chains due to the presence of crosslinks.

Uncured medium-vinyl polybutadiene/cis-polyisoprene blends were studied by Akiyama et al. [55] whom attributed the change in the apparent free volume fraction with the blend composition to the variation of the thermal expansion coefficients.

As a consequence of their use in tire applications, blends of natural rubber NR and styrene butadiene rubber SBR are of particular interest. Recently, small angle X-ray scattering technique was used to obtain valuable information about the interphase developed in the vulcanized NR-SBR blends [56]. Authors of the present chapter have used calorimetric tests and diffusivity measurements to discuss the thermal properties of the same kind of blends on the basis of the crosslink structure formed during the vulcanization process [57].

In a recent work, it was studied the behavior of the o-Ps lifetime as a function of the temperature for a NR-SBR blends; these blend were vulcanized at 433 K with sulfur and TBBS as accelerator (see Ref. [58]). In such a work, all the samples studied were cured at the optimum cure time t_{100}, i.e. the time at where the maximum torque of the rheometer curves is obtained. In all the compounds, the molecular weight between crosslinks M_{cs} was measured using swelling test in toluene [56].

In the present chapter and with the aim to show general tendencies, we only discuss the results obtained for the compounds NR, SBR and the blend 50NR-50SBR. Values of t_{100} and M_{cs} for the mentioned compounds are given in Table 2.

In Fig. 17 the $\tau_{o\text{-}Ps}$ curves versus temperature of each compound is shown. The three o-Ps lifetime evolutions show a sigmoidal shape with well-known features usually reported in PALS studies of polymers; i.e., a slow increase of $\tau_{o\text{-}Ps}$ in the

Table 2 Glass transition temperature of NR, SBR and NR50/SBR50 blend measure by DSC scans [4] and by PALS

	NR	NR50-SBR50	SBR
t_{100} [min] [57]	10.7	24.7	42.3
M_{cs} [g/mol]	5451	8080	11958
T_g^{DSC}[K] [57]	213.1	214.5/226.4	229.6
T_g^{PALS}[K]	191	196	220
T_b [K]	239	250	–

The bending temperatures T_b obtained from Fig. 17 in the glass transition region are also given

glassy state followed by a sudden increase of this parameter when the glass transition temperature T_g^{PALS} is overcome.

As can be seen in the figure, at higher temperatures, with respect to T_g^{PALS}, there is a tendency to a quasi-saturation of $\tau_{o\text{-}Ps}$ and, specifically in the case of the NR-SBR compound a saturation temperature T_S around 297 K was observed.

This behaviour of the o-Ps lifetime was attributed to the formation of a Ps bubble in the liquid phase, digging the holes by Ps itself [59] or to the fact that the relaxation time of the molecular chains is comparable to the o-Ps lifetime [60]. In this last case, Ps is no longer able to correctly probe the sizes of the nanoholes trapping Ps. However, there is a general agreement on the contention that in this temperature region o-Ps does not give reliable information on the free nanohole volume.

From the $\tau_{o\text{-}Ps}$ curves the glass transition temperature of each compound was obtained. Their values are also given in Table 2. In the case of the SBR compound, the lifetime data could be satisfactory fitted by a straight line for a temperature range between $T_g^{PALS}$and T_S.

A different situation is observed in the natural rubber compound, where a value of T_g^{PALS} (NR) ≈ 191 K was obtained. A special feature is given in the glass

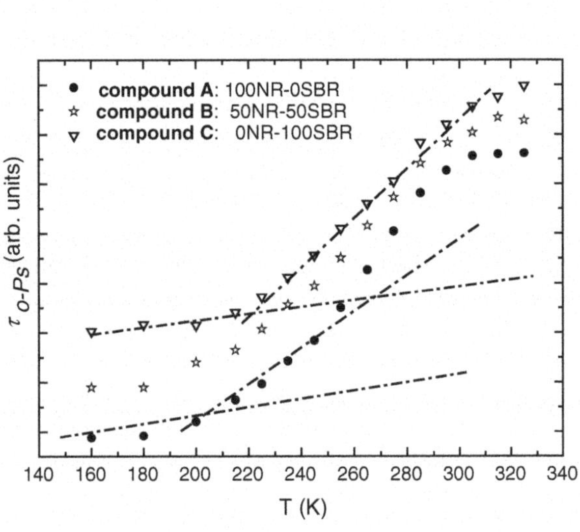

Fig. 17 o-Ps lifetime as a function of temperature for NR, SBR and NR50-SBR50 blend samples. (after Salgueiro et al. [58])

transition zone of the curve $\left(T > T_g^{PALS}\right)$: there is a change of the slope at a temperature $T_b \approx 239$ K. This situation has been observed in other polymers [26, 53, 61] and, in some cases, it was not discussed in detail or simply omitted. The origin of this change is still under debate [62].

For the 50NR-50SBR compound a bending temperature T_b was also observed and it value is given in Table 2 together with the corresponding T_g^{PALS}. As can be seen, T_b increases when it is compared with that of the NR compound. T_b is higher than T_g^{PALS} (SBR) and this fact implies that at T_b the blend has both phases (SBR and NR) in the rubbery state or at least in the glass transition region.

The change in the slope of $\tau_{o\text{-}Ps}$ curves at T_b can be ascribed to a subtle transition in the liquid phase from liquid-like to quasi-solid-like character on cooling the supercooled liquid due to the onset of certain constraints on an expansion of the free volume [61]. It is known that, at least two relaxation processes take place in glass-forming polymers: the main relaxation, or α relaxation, which is related to the segmental relaxations of the main chain; and the β relaxation (also called the Johari-Goldstein relaxation) which has been assigned to the local relaxation of the flexible parts of the chain [63]. The β relaxation would be an universal feature and an important characteristic of the dynamics of supercooled glass-forming liquids [64].

In the case of uncured and cured SBR, the β relaxation was detected by means of mechanical dynamic spectroscopy. It was reported that the position of the peak in the *tan* δ as a function of the temperature plot, at several frequencies, does not depend on the crosslink density [65]. By contrast, the β relaxation was not detected in natural rubber.

Figure 18 shows a SEM image of the cryofracture surface obtained for the NR50-SBR50 blend. As we report in Table 2, this sample was cured during $t_{100} = 24.7$ min and, if we take into account the cure times for the pure compounds, also reported in Table 2, the SBR and NR phases constituting the blend are undercured and overcured, respectively. However, this assumption would be adequate if there would exist an even distribution of curatives into the phases.

According to previous researches on similar blends [66], the diffusion of curatives from the SBR phase to the NR phase could be taking place enriching, therefore, in curatives the interphase between both elastomers and, consequently reducing the amount of curatives into the SBR phase. Then, it could be considered that a diffusion process would produce a reduction in the amount of curatives into the core of the SBR phase. Then, at the time t_{100} some portion of the SBR phase in the blend has not yet begun to crosslinking.

It is worth mentioning that the method employed for the preparation of the samples to obtain SEM images, dissolves the uncrosslinked regions [66]. On the other hand, the voids observed in the micrograph presented in Fig. 18 can be attributed to the uncrosslinked portion of the SBR phase.

In Fig. 19 the evolution of the fractional free volume f as a function of the temperature for the three cured compounds above mentioned is plotted.

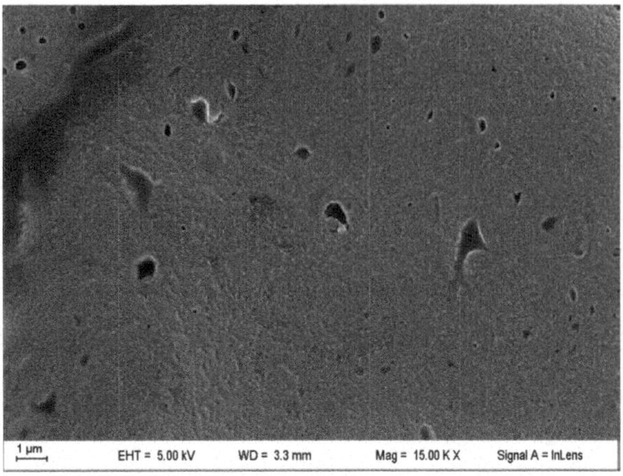

Fig. 18 SEM micrograph of cryofracture surface of NR50/SBR50 blend vulcanized at 433 K and t_{100}. (after Salgueiro et al. [58])

The approach expressed by Eqs. (4)–(6) can be used to get the fractional free volume from positron data. In the case of the uncured SBR and NR, V_W is 0.672 $cm^3\ g^{-1}$ [42] and 0.700 cm^3/g [29], respectively. Then, at 293 K, V^* (SBR) = 0.974 $cm^3\ g^{-1}$ and V^* (NR) = 1.015 cm^3/g. For the NR50/SBR50 blend, V^* was calculated according to the weighted mean of the pure components, obtaining the values 0.995 cm^3/g.

Values of T^*, N^f and the occupied volume V_{occ} are reported in Table 3 [67]. In the case of the cured SBR, these parameters were also previously evaluated but for samples with different TBBS/sulfur ratio in his composition [42]. The observed differences when comparing these values with those reported in Ref. [42], can be assigned to the different network structures obtained after curing the pure SBR samples containing different amounts of TBBS and sulfur in their formulations [39].

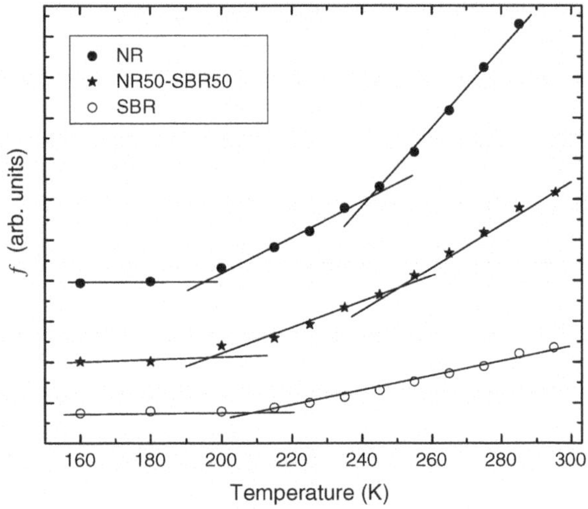

Fig. 19 Fractional hole free volume *f* as a function of the temperature for cured NR, SBR and NR50-SBR50 blend. The vertical scale is shifted in order to evidence changes in the slope of the *f* versus *T* curves

Table 3 Scaling temperature T*, hole density N^f, occupied volume V_{occ} and thermal expansion coefficients obtained for NR, SBR and NR50-SBR50 from PALS data. Data were taken from Salgueiro et al. [67]

	NR	NR50-SBR50	SBR
T^*[K]	9179	8827	8547
$N^f \times 10^{-21}[g^{-1}]$	0.51	0.66	0.75
V_{occ} [cm^3 g^{-1}]	0.976	0.947	0.925
$\alpha_{r1} \times 10^{-4}[K^{-1}]$	6.4	6.3	6.1
$\alpha_{r2} \times 10^{-4}[K^{-1}]$	2.8	5.2	–
$\alpha_g \times 10^{-4}[K^{-1}]$	0.6	1.3	0.7

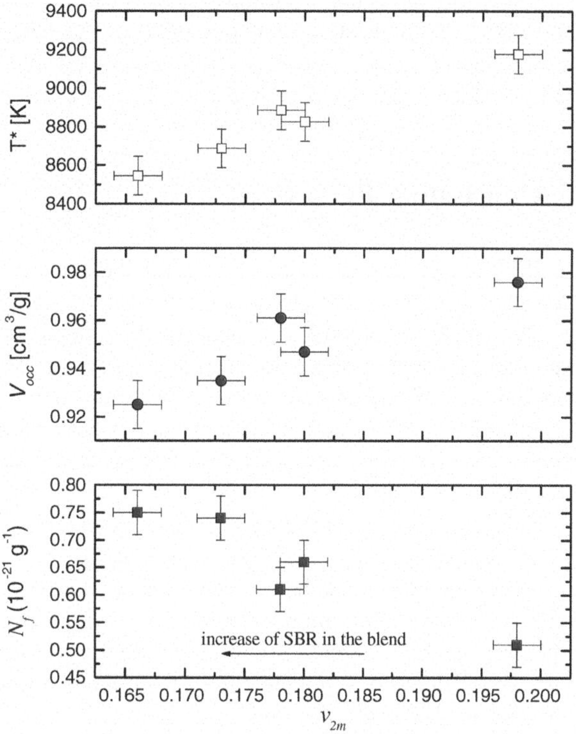

Fig. 20 Scaling temperature T*, hole density N^f, occupied volume V_{occ} as function of the polymer volume fraction at the equilibrium (maximum) degree of swelling, v_{2m} in cured NR, SBR and NR50-SBR50 blend (values of the parameters presented in the figure were taken from Ref. [67])

On the other hand, to understand some aspects of the network structure of the compounds, authors of the present work carried out SAXS and swelling measurements on the same compounds [56]. It was reported that the polymer volume fraction at the equilibrium (maximum) degree of swelling, v_{2m}, decreases when the blend is richer in SBR. This result is in good agreement with the results presented in this chapter, in which with the increasing SBR content in the compounds N^f increases and the V_{occ} decreases (see Fig. 20). In Fig. 20, we have also included values of the parameters for the cured blends NR75-SB25 and NR25-SBR75 recently published [67].

The thermal expansion coefficients of the free volumes, $\alpha = df/dT$ in the rubbery states (α_{r1}, α_{r2}) as well as in the glassy phases (α_g) are also reported in Table 3.

Looking at the α_{r1} and α_{r2} values obtained for the different compounds it can be concluded that α_{r1}, corresponding to the first rubbery state, is almost constant, for the three samples. For the pure SBR compound, there is not a thermal expansion coefficient of the free volume in the second rubbery state α_{r2}.

It can be seen that the addition of SBR in the sample NR50-SBR50 produces an increase of the α_{r2} value. This behaviour could not be only linked to the different content of every elastomer in each phase, but also to the crosslink density of each phase and to the development of an interphase that must be considered as a third phase into the analysis (see details in Ref. [67]).

In the glassy region, α_g values are very similar for the pure compounds of NR and SBR. The highest value is obtained for the NR50-SBR50 blend stressing our interpretation on the influence of the developed interphases into this temperature region. However more work must be done to elucidate this point.

5 Concluding Remarks

In the present chapter, some issues on the physical phenomena related to free volumes in rubbers and blends were presented and discussed.

Particularly, this work was addressed to the study of the influence of the free nanohole volume on the physical processes occurring in SBR and NR compounds and SBR-NR blends as a consequence of the change of different external parameters.

In order to illustrate the readers and based on the direct experience of the authors, we have chosen specific studies on: (i) uncured and cured elastomers and (ii) elastomeric blends.

Taking advantage of the well recognized capability of the Positron Annihilation Lifetime Spectroscopy nuclear technique to give direct information on free nanohole volumes, we have mainly used this experimental technique for our investigations. Furthermore, depending on the elastomer studied, we have used dynamic mechanical tests, differential scanning calorimetry and swelling tests. The use of additional or complementary techniques allowed us going deeper into the analysis of the results obtained.

The results presented were discussed taking into account other papers reported in the literature. In the discussion of the different results presented in this chapter the role of crosslinking, which has a strong influence on the physical properties of elastomers and on the free volume, was expressly taken into account. For each case studied, the analysis of the experimental data was given into the frame of well recognized theoretical models.

Acknowledgments This work was supported by Agencia Nacional de Promoción Científica y Tecnológica (PICT 2011-1088), Comisión de Investigaciones Científicas de la Provincia de Buenos Aires, SECAT (UNCentro) and University of Buenos Aires (Project UBACYT 2010–2012), Argentina.

References

1. Doolittle, A.K.: J. Appl. Phys. **22**, 1031 (1951)
2. Ferry, J.D.: Viscoelastic Properties of Polymers. Wiley, New York (1980)
3. Consolati, G.: Mater. Sci. Forum **363–365**, 244 (2001)
4. Victor, J.G., Torkelson, J.M.: Macromolecules **20**, 2241 (1987)
5. Suzuki, T., Yoshimizu, H., Tsujita, Y.: Polymer **44**, 2975 (2003)
6. Bruno, G.V., Freed, J.H.: J. Phys. Chem. **78**, 935 (1974)
7. Curro, J.J., Roe, R.R.: Polymer **25**, 1424 (1984)
8. Jean, Y.C., Mallon, P.E., Schrader, D.M. (eds.): Principles and Applications of Positron and Positronium Chemistry. World-Scientific, London (2003)
9. Brandt, W., Dupasquier, A. (eds.): Positron Solid-State Physics., North- Holland, Amsterdam (1983)
10. Dupasquier, A., Kögel, G., Somoza, A.: Acta Mater. **52**, 4707 (2004)
11. Salgueiro, W., Somoza, A., Cabrera, O., Consolati, G.: Cem. Concr. Res. **34**, 91 (2004)
12. Jean, Y.C.: Microchem. J. **42**, 72 (1990)
13. Jean, Y.C.: Mater. Sci. Forum **59**, 175 (1995)
14. Dlubek, G., Fretwell, H.M., Alam, M.A.: Macromolecules **33**, 87 (2000)
15. Krause-Rehberg, R., Leipner, H.S.: Positron Annihilation in Semiconductors. Springer, Berlin (1999)
16. Dupasquier A., Mills, A.P. Jr (eds.): Positron Spectroscopy of Solids. IOP Press, Amsterdam (1995)
17. Dlubek, G., Clarke, A.P., Fretwell, H.M., Dugdale, S.B., Alam, M.A.: Phys. Stat. Sol. (a) **157**, 351 (1996)
18. Hristov, H.A., Bolan, B., Lee, A.F., Xie, L., Gidley, D.G.: Macromolecules **29**, 8507 (1996)
19. Li, H.-L., Ujihira, Y., Yoshino, T., Yoshii, K., Yamashita, T., Horie, K.: Polymer **39**, 4075 (1998)
20. Tao, S.J.: J. Chem. Phys. **56**(1972), 5499 (1972)
21. Eldrup, M., Lightbody, D., Sherwood, N.J.: Chem. Phys. **63**, 51 (1981)
22. Nakanishi H., Wang Y.Y., Jean Y.C., Sharma S.C. (eds.): Positron Annihilation Studies of Fluids, p. 292. World Scientific, Singapore (1988)
23. Kanaya, T., Tsukushi, T., Kaji, K., Bartos, J., Kristiak, J.: Phys. Rev. E **60**, 1906 (1999)
24. Gregory, R.B.: J.Appl.Phys. **70**, 4665 (1991)
25. Marzocca, A.J., Cerveny, S., Salgueiro, W., Somoza, A., Gonzalez, L.: Phys. Rev. E **65**, 021801 (1-5) (2002)
26. Srithawatpong, R., Peng, Z.L., Olson, B.G., Jamieson, A.M., Simha, R., McGerwey, J.D., Maier, T.M., Halasa, A.F., Ishida, H.: J. Polym. Sci., Part B: Polym. Phys. **37**, 2754 (1999)
27. Simha, R., Somcynsky, T.: Macromolecules **2**, 342 (1969)
28. Simha, R., Wilson, P.S., Olabisi, O.: Kolloid-Z. Z. Polym. **251**, 402 (1973)
29. Bondi, A.A.: Physical Properties of Molecular Crystals, Liquids and Glasses. Wiley, New York (1968) (ch. 14)
30. Ferry, J.D.: Viscoelastic Properties of Polymers, p. 264. Wiley, New York (1980) (Cap.11)
31. Bartoš, J., Bandžuch, P., Šauša, O., Krištiaková, K., Krištiak, J., Kanaya, T., Jenninger, W.: Macromolecules **30**(22), 6906–6912 (1997)
32. Bartoš, J., Colloid, J.: Polym. Sci. **274**, 14 (1996)
33. Bandžuch, P., Krištiak, J., Šauša, O., Zrubcová, J.: Phys Rev B **61**, 8784 (2000)
34. Wanga, Z.F., Wang, B., Qi, N., Zhang, H.F., Zhang, L.Q.: Polymer **46**, 719 (2005)
35. Wang, J., Vicent, J., Quarles, C.A.: Nucl Instrum Methods Phys Res B **241**, 271–275 (2005)
36. Mohsen, M., Abd-El Salam, M.H., Ashry, A., Ismail, A., Ismail, H.: Polym. Degrad. Stab. **87**, 381–388 (2005)
37. Jobando, V.O., Quarles, C.A.: Phys. Stat. Sol. (c) **4**, 3767 (2007)
38. Akiba, M., Hashim, S.: Prog. Polym. Sci. **22**, 475 (1997)
39. Marzocca, A.J., Mansilla, M.A.: J. Appl. Polym. Sci. **103**, 1105 (2007)

40. Mason, P.: Polymer **5**, 625 (1964)
41. Coran, A.Y.: In: Mark, J.E., Erman, B., Eirich, F.R., (eds.) Science and Technology of Rubber, p. 339. Academic Press, San Diego (1978)
42. Salgueiro, W., Marzocca, A.J., Somoza, A., Consolati, G., Cerveny, S., Quasso, F., Goyanes, S.: Polymer **45**, 6037 (2004)
43. Gronsky, W., Hoffman, U., Simon, G., Wutzler, A., Straube, E.: Rubber Chem. Technol. **65**, 63 (1992)
44. Flory, P.J., Rehner, J.: J Chem Phys **11**, 512 (1943)
45. Flory, P.J., Rehner, J.: J. Chem. Phys. **11**, 521 (1943)
46. Mark, J.E., Erman, B.: Rubberlike Elasticity: A Molecular Primer, p. 51. Wiley, New York (1988)
47. Strobl, G.: The Physics of Polymers, 2nd edn. Springer, Berlin (1997)
48. Coran, A.Y.: Rubber Chem. Technol. **61**, 281 (1988)
49. Utracki, L.A.: Polym. Eng. Sci. **23**, 602 (1983)
50. Turi, E.A.: Thermal Characterization of Polymeric Materials. Academic Press, New York (1997)
51. Sakaguchi, T., Taniguchi, N., Urakawa, O., Adachi, K.: Macromolecules **38**, 422 (2005)
52. Zhao, J., Ediger, M.D., Sun, Y., Yu, L.: Macromolecules **42**, 6777 (2009)
53. Peng, Z.L., Olson, B.G., Srithawatpong, R., McGervey, J.D., Jamieson, A.M., Ishida, H., Meier, T.M., Halasa, A.F.: J. Polym. Sci., Part B: Polym. Phys. **36**, 861 (1998)
54. Kovacs, A.J.: Adv. Polymer Sci. **3**, 394 (1963)
55. Akiyama, S., Kawahara, S., Akiba, I., Iio, S., Li, H.-L., Ujihira, Y.: Polym. Bull. **45**, 275 (2000)
56. Salgueiro, W., Somoza, A., Marzocca, A.J., Torriani, I., Mansilla, M.A.: J. Polym. Sci., Part B: Polym. Phys. **47**, 2320 (2009)
57. Goyanes, S., Lopez, C.C., Rubiolo, G.H., Quasso, F., Marzocca, A.J.: Eur. Polym. J. **44**, 1525 (2008)
58. Salgueiro, W., Somoza, A., Consolati, G., Quasso, F., Marzocca, A.J.: Phys. Stat. Sol. (c) **10**, 3771 (2007)
59. Ito, Y., Mohamed, H.F.M., Tanaka, K., Okamoto, K., Lee, K.: J. Radioanal. Nucl. Chem. **211**, 211 (1996)
60. Bartoš, J., Šauša, O., Krištiak, J., Blochowicz, T., Rössler, E.: J. Phys.: Condens. Matter **13**, 11473 (2001)
61. Bartoš, J., Šauša, O., Bandzuch, P., Zrubcová, J., Krištiak, J.: J. Non-Crystal. Solids **307–310**, 417 (2002)
62. Winberg, P., Eldrup, M., Maurer, F.H.J.: Polymer **45**, 8253 (2004)
63. McCrum, N.G., Read, B.E., Williams, G.: Anelastic and Dielectric Effects in Polymer Solids. Wiley, London (1967)
64. Ngai, K.L.: J. Phys.: Condens. Matter **15**, 1107 (2003)
65. Ghilarducci, A., Salva, H., Marzocca, A.J.: J. Appl. Polym. Sci. **113**, 2361 (2009)
66. Mallon, P.E., McGill, W.J.: J. Appl. Polym. Sci. **74**, 1250 (1999)
67. Salgueiro, W., Somoza, A., Silva, L., Consolatti, G., Quasso, F., Mansilla M.A., Marzocca, A.J.: Phys. Rev. E. **85**, 51805 (2011)

Editors Biography

Visakh P. M. (MSc, MPhil) is a Research Fellow at the School of Chemical Science Mahatma Gandhi University, Kottayam, Kerala, India. He edited 5 books with Sabu Thomas from Wiley and Springer and more than 8 books in press (from Wiley, Springer, American Chemical Society and Royal Society of Chemistry and Elsevier). He has been invited as a visiting student in Czech Republic (2012, 2013), Italy (2009, 2012), Argentina (2010), Sweden (2010, 2011, 2012), Switzerland (2010), Spain (2011, 2012), Slovenia (2011), France (2011), Belgium (2012) and Austria (2012) for his research work and he published more than 5 publications and more than 10 book chapters. He has attended and presented more than 25 conferences. Now he is working in Charles University, Czech Republic as pre-post doc. His research interests include: polymer nanocomposites, bio-nanocomposites, rubber based nanocomposites, fire retardant polymers and liquid crystalline polymers and silicon sensors.

Sabu Thomas (Ph.D) is a Professor of Polymer Science and Engineering at the School of Chemical Sciences, as well as the Director of Centre for Nanoscience and Nanotechnology, Mahatma Gandhi University, India. He received his Ph.D. in 1987 in Polymer Engineering from the Indian Institute of Technology (IIT), Kharagpur, India. He is a fellow of the Royal Society of Chemistry, London and a member of the American Chemical Society. He has been ranked no 5 in India with regard to the number of publications (most productive scientists). He also received the coveted Sukumar Maithy Award for the best polymer researcher in the country for the year 2008. The research group of Prof. Thomas has received numerous awards and honors for excellent work in polymer science and engineering.

P. M. Visakh et al. (eds.), *Advances in Elastomers II*,
Advanced Structured Materials 12, DOI: 10.1007/978-3-642-20928-4,

Arup K. Chandra (M. Tech, Ph.D). Rubber Technology from IIT (Indian Institute of Technology), Kharagpur is presently Working as Global Head (RM and Compound Development) in M/s. Apollo Tyres Ltd., He worked in different capacities in Technical Department and R&D Centre of M/s. J. K. Industries Ltd. from 1983 to 1997. He is one of the founder members of Indian Rubber Institute (IRI)—Rajasthan Branch and Postgraduate Diploma in Polymer Science and Technology course in Mohanlal Sukhadia University, Udaipur.

He remains the Honorary Secretary of IRI, Rajasthan Branch till he left J. K. Industries. He is member and fellow members of different professional bodies both in India and Abroad. He is a regular visiting Professor, question setter and examiner of polymer Science and Technology Course of different Universities and PGDIRI, Conducted by IIT—Kharagpur. He has contributed more than 120 Papers (including conference papers) in different Journal of National/International Repute. He received extensive training/exposure at different leading Research Centre/Institute/Universities and travelled across the Globe (USA, Canada, The Netherlands, Belgium, Germany, Spain, Italy, Korea, Kenya, Malaysia, Singapore, Thailand, South Africa, Hong Kong and China to mention a few).

Aji. P. Mathew (Ph.D) is an Associate Professor in the field of Bionanocomposites at Luleå University of Technology, Luleå, Sweden. She recieved her Ph.D in Polymer Chemistry, in 2001 from Mahathma Gandhi University, Kottayam, Kerala, India. She worked as Postdoc. at CNRS, Grenoble, France and Norwegian University of Science and Technology (NTNU), before taking up the faculty position at Luleå University of Technology. Prof. Mathew is active in the field of biobased nanoreinforcments, nanocomposites, membranes and medical implants and products and published about 50 papers in these research areas.

The manufacturer's authorised representative in the EU is Springer Nature Customer Service Centre GmbH, Europaplatz 3, 69115 Heidelberg, Germany. If you have any concerns regarding our products, please contact ProductSafety@springernature.com

Printed and bound by CPI Group (UK) Ltd, Croydon, CR0 4YY
25/09/2026
02222884-0006